CHEMICAL SIGNALS IN VERTEBRATES 9

CHEMICAL SIGNALS IN VERTEBRATES 9

Edited by

Anna Marchlewska-Koj
Jagiellonian University
Kraków, Poland

John J. Lepri
University of North Carolina at Greensboro
Greensboro, North Carolina

and

Dietland Müller-Schwarze
State University of New York
Syracuse, New York

Kluwer Academic / Plenum Publishers
New York, Boston, Dordrecht, London, Moscow

Library of Congress Cataloging-in-Publication Data

Chemical signals in vertebrates 9/edited by Anna Marchlewska-Koj, John J. Lepri, Dietland Müller-Schwarze.
   p.   cm.
Based on papers from the 9th international symposium.
Includes bibliographical references and index.
ISBN 0-306-46682-1
   1. Chemical senses—Congresses. 2. Vertebrates—Physiology—Congresses. I. Title: Chemical signals in vertebrates nine. II. Marchlewska-Koj, Anna. III. Lepri, John J. IV. Müller-Schwarze, Dietland.

QP455 .C476 2001
573.8'7716—dc21

2001038734

Proceedings of the Ninth International Symposium on Chemical Signals in Vertebrates, held July 25–29, 2000, in Kraków, Poland

ISBN 0-306-46682-1

©2001 Kluwer Academic/Plenum Publishers, New York
233 Spring Street, New York, New York 10013

http://www.wkap.nl/

10 9 8 7 6 5 4 3 2 1

A C.I.P. record for this book is available from the Library of Congress

All rights reserved

No part of this book may be reproduced, stored in a retrieval system, or transmitted in any form or by any means, electronic, mechanical, photocopying, microfilming, recording, or otherwise, without written permission from the Publisher

Printed in the United States of America

PREFACE

It is generally accepted that the recent progress in molecular and cellular biology would not have been possible without an understanding of the mechanisms and signaling pathways of communication inside the cell and between various cells of the animal organism. In fact a similar progress occurred in the field of chemical communication between individual organisms of vertebrate species, and this volume is aimed at presenting the current state of the art on this subject. The reader can find here both original results obtained in the laboratory or field studies and comprehensive reviews summarizing many years of research. The presentations of over 60 scientists have been grouped according to their approach into nine parts covering such fields as ecological and evolutionary aspects of chemical communication, structure and neuronal mechanisms of chemosensory systems, chemical structure of pheromones and binding proteins, kin, individual and sexual recognition, predator-prey relationships, purpose and consequences of marking behavior, scent signals and reproductive processes. Expanding on former volumes of this series, entirely new chapters have been added on prenatal chemical communication describing specific effects of the intrauterine environment. In many cases a truly multidisciplinary approach was required, such as with the population analysis of polymorphic variants of the mouse's major urinary proteins that function in carrying pheromones.

All the papers presented here were discussed during the Ninth International Symposium on Chemical Signals in Vertebrates, held at the Jagiellonian University, Kraków, Poland, from July 25 to 29, 2000, and organized by Anna Marchlewska-Koj. The conference was attended by over 100 scientists coming from 16 different countries all over the world who were interested in various aspects of chemical communication in higher animals. The discussions continued for 4 days (and nights) of the conference and hopefully led to better understanding between scientists applying different techniques in the analysis of chemical signals between animals.

However, this volume does not represent typical conference proceedings since in comparison to the original presentations many chapters have been improved by the authors, who decided to take into account discussions in Kraków. We sincerely hope that the book will be useful to a wide range of scientists working in the field of chemical communication and using methods related to chemistry, molecular biology, neurophysiology, behavioral science and evolutionary biology.

The editors wish to acknowledge the help of several persons engaged in the organization of the conference and preparation of the publication, especially Hajnalka Szentgyörgyi, Elżbieta Pochroń and Joanna Kapusta.

Anna Marchlewska-Koj Kraków, March 2001
John Lepri
Dietland Müller-Schwarze

CONTENTS

ECOLOGICAL AND EVOLUTIONARY ASPECTS OF CHEMICAL COMMUNICATION

1. From individuals to populations: Field studies as proving grounds for the role of chemical signals 1
   D. Müller-Schwarze

2. The economic consequences of advertising scent mark location on territories 11
   S. C. Roberts and L. M. Gosling

3. Do chemical alarm signals enhance survival of aquatic vertebrates? An analysis of the current research paradigm 19
   R. S. Mirza and D. P. Chivers

4. Mechanisms of olfactory foraging by Antarctic procelliiform seabirds 27
   G. A. Nevitt

5. Ecological aspects of house mouse urinary chemosignals 35
   L. C. Drickamer

6. Information in scent signals of competitive social status: The interface between behaviour and chemistry 43
   J. L. Hurst, R. J. Beynon, R. E. Humphries, N. Malone, C. M. Nevison, C. E. Payne, D. H. L. Robertson, and C. Veggerby

7. Molecular approaches in chemical communication of mammals 53
   E. P. Zinkevich and V. S. Vasilieva

STRUCTURE AND NEURONAL MECHANISMS OF CHEMOSENSORY SYSTEMS

8. Neural mechanisms of communication: From pheromones to mosaic signals 61
   R. E. Johnston

9. A unique subfamily of olfactory receptors 69
   J. Strotmann

10. Odours are represented in glomerular activity patterns: Optical imaging studies in the insect antennal lobe 77
    C. G. Galizia, S. Sachse, and H. Mustaparta

11. Spatial representations of odorant chemistry in the main olfactory bulb of the rat 85
    B. A. Johnson and M. Leon

12. Prenatal growth and adult size of the vomeronasal organ in mouse lemurs and humans 93
    T. D. Smith, M. P. Mooney, A. M. Burrows, K. P. Bhatnagar, and M. I. Siegel

13. Size of the vomeronasal organ in wild *Microtus* 101
    L. M. Maico, D. L. Roslinski, A. M. Burrows, M. P. Mooney, M. I. Siegel, K. P. Bhatnagar, and T. D. Smith

14. Oxytocin, norepinephrine and olfactory bulb mediated recognition 107
    D. E. Dluzen, Y. Shang, and R. Landgraf

15. A possible humoral pathway for the priming action of the male pheromone androstenol on female pigs 117
    T. Krzymowski, S. Stefańczyk-Krzymowska, W. Grzegorzewski, J. Skipor, and B. Wąsowska

CHEMICAL STRUCTURE OF PHEROMONES AND BINDING PROTEINS

16. Demonstration of volatile C_{19}-steroids in the urine of female Asian elephants, *Elephas maximus* 125
    M. Dehnhard, M. Heistermann, F. Göritz, R. Hermes, T. Hildebrandt, G. Strauss, C. Weisgerber, and H. Haber

17. The pheromone of the male goat: Function, sources, androgen dependency and partial chemical characterization 133
    R. Claus, M. Dehnhard, U. Götz, and M. Lacorn

18. Analysis of volatile compounds in scent-marks of spotted hyenas *(Crocuta crocuta)* and their possible function in olfactory communication 141
    H. Hofer, M. East, I. Sämmang, and M. Dehnhard

19. Mice, MUPs and myths: Structure-function relationships of the major urinary proteins 149
    R. J. Beynon, J. L. Hurst, S. J. Gaskell, S. J. Hubbard, R. E. Humphries, N. Malone, A. D. Marie, L. Martinsen, C. M. Nevison, C. E. Payne, D. H. L. Robertson, and C. Veggerby

20. Polymorphic variants of mouse major urinary proteins 157
    C. Veggerby, C. E. Payne, S. J. Gaskell, D. H. L. Robertson,
        J. L. Hurst, and R. J. Beynon

21. Differential responses elicited in male mice by MUP-borne odorants 165
    A. Cavaggioni, C. Mucignat-Caretta, and A. Caretta

22. Characteristics of ligand binding and release by major urinary proteins 169
    D. H. L. Robertson, A. D. Marie, C. Veggerby, J. L. Hurst,
        and R. J. Beynon

23. Chemical communication in the pig 177
    D. Loebel, A. Scaloni, S. Paolini, S. Marchese, C. Fini, L. Ferrara,
        H. Breer, and P. Pelosi

PRENATAL CHEMICAL COMMUNICATION

24. Inter-fetal communication and adult phenotype in mice 183
    J. G. Vandenbergh and A. K. Hotchkiss

25. The role of the main and accessory olfactory systems in prenatal olfaction 189
    D. M. Coppola

26. Fetal olfactory cognition preadapts neonatal behavior in mammals 197
    B. Schaal, G. Coureaud, L. Marlier, and R. Soussignan

27. Olfaction in premature human newborns: Detection and discrimination abilities
        two months before gestational term 205
    L. Marlier, B. Schaal, C. Gaugler, and J. Messer

28. Intrauterine position effects on rodent urinary chemosignals 211
    L. C. Drickamer

KIN, INDIVIDUAL AND SEXUAL RECOGNITION

29. Social status, odour communication, and mate choice in wild house mice 217
    N. Malone, C. E. Payne, R. J. Beynon, and J. L. Hurst

30. Effects of inbreeding and social status on individual recognition in mice 225
    C. M. Nevison, C. J. Barnard, R. J. Beynon, and J. L. Hurst

31. Heterogeneity of major urinary proteins in house mice: Population and sex
        differences 233
    C. E. Payne, N. Malone, R. Humphries, C. Bradbrook, C. Veggerby,
        R. J. Beynon, and J. L. Hurst

32. Genetic differences in odor discrimination by newborn mice as expressed by ultrasonic calls 241
    J. Kapusta and H. Szentgyörgyi

33. Enhanced immune function decreases odor attraction of male laboratory mice .. 249
    K. Litvinova, I. Kolosova, V. Mak, and M. Moshkin

34. The olfactory sexual preferences of golden hamster (*Mesocricetus auratus*): The effects of early social and sexual experience 255
    A. V. Surov, A. V. Solovieva, and A. N. Minaev

35. Mate recognition via waterborne chemical cues in the viviparous caecilian *Typhlonectes natans* (Amphibia: Gymnophiona) 263
    A. Warbeck and J. Parzefall

36. Maternal anogenital licking in rats: Exploring the dam's differential sexual treatment of pups 269
    I. Brouette-Lahlou, E. Vernet-Maury, F. Chastrette, and J. Chanel

PREDATOR - PREY RELATIONSHIPS

37. Predator diet cues and the assessment of predation risk by aquatic vertebrates: A review and prospectus 277
    D. P. Chivers and R. S. Mirza

38. Field observations confirm laboratory reports of defense responses by prey snakes to the odors of predatory snakes 285
    W. H. N. Gutzke

39. Spatial responses of field (*Microtus agrestis*) and bank (*Clethrionomys glareolus*) voles to weasel (*Mustela nivalis*) odour in natural habitat 289
    Z. Borowski and E. Owadowska

40. Do newts avoid conspecific alarm substances: The predation hypothesis revisited 295
    J. R. Rohr and D. M. Madison

41. Responses to nitrogen-oxides by Characiforme fishes suggest evolutionary conservation in Ostariophysan alarm pheromones 305
    G. E. Brown, J. C. Adrian, Jr., I. H. Kaufman, J. L. Erickson, and D. Gershaneck

MARKING BEHAVIOR: PURPOSE AND CONSEQUENCES

42. Anal scent gland secretion of the European otter (*Lutra lutra*) 313
    A. Bradshaw, M. Beckmann, R. Stevens, and F. Slater

43. Scent-marking behaviour of the European badger *(Meles meles)*: Resource defence or individual advertisement? 321
   C. D. Buesching and D. W. Macdonald

44. Scent marking behaviors of the striped mongoose, *Mungos mungo* 329
   I. Ianovschi

45. Increased social dominance in male rabbits, *Oryctolagus cuniculus,* is associated with increased secretion of 2-phenoxy ethanol from the chin gland 335
   R. A. Hayes, B. J. Richardson, and S. G. Wyllie

46. The response of individuals to over-marks of conspecifics differs between two species of Microtine rodents 343
   M. H. Ferkin

47. The secretion of the supplementary sacculi of the dwarf hamster *Phodopus campbelli* 347
   R. Apfelbach, U. Schmidt, and N. Y. Vasilieva

48. The role of urinary proteins and volatiles in competitive scent marking among male house mice 353
   R. E. Humphries, D. H. L. Robertson, C. M. Nevison, R. J. Beynon, and J. L. Hurst

SCENT SIGNALS AND REPRODUCTIVE PROCESSES

49. Invading pest species and the threat to biodiversity: Pheromonal control of Guam brown tree snakes, *Boiga irregularis* 361
   R. T. Mason and M. J. Greene

50. Annual and seasonal variation in the female sexual attractiveness pheromone of the red-sided garter snake, *Thamnophis sirtalis parietalis* 369
   M. P. LeMaster and R. T. Mason

51. Hypothalamic and ovarian response to pheromone application in seasonal anoestrous German mutton merino sheep 377
   K.-H. Kaulfuß, R. Süß, P. Schenk, E. Berger, and E. von Borell

52. Males' olfactory discrimination of receptive state of females in rat-like hamsters *(Cricetulus triton)* 385
   J.-X. Zhang, Z.-B. Zhang, and Z.-W. Wang

53. Pheromonal regulation of bank vole *(Clethrionomys glareolus)* reproduction ... 391
   A. Marchlewska-Koj

54. Olfactory communication in Brandt's vole *(Microtus brandti)* 397
   L. Zhang, J. Fang, and R. Sun

55. Effects of social dominance and female odor on sperm activity of male mice.... 403
    S. Koyama and S. Kamimura

56. Exposure of juvenile male Campbelli's hamsters and house mice to cat urine
    elicits species-specific responses in reproductive development 411
    N. V. Sokolskaja, R. Apfelbach, D. von Holst, and N. Y. Vasilieva

CHEMOSIGNALS AND BEHAVIORAL RESPONSES

57. The role of olfaction in the feeding behavior of human neonates....................... 417
    R. H. Porter, H. Varendi, and J. Winberg

58. Pig responses to taste stimuli 423
    D. Glaser, M. Wanner, J. M. Tinti, and C. Nofre

59. Imprinting on native pond odour in the pool frog, *Rana lessonae* Cam. 433
    S. V. Ogurtsov and V. A. Bastakov

60. Chemical discrimination in *Liolaemus* lizards: Comparison of behavioral
    and chemical data... 439
    A. Labra, C. A. Escobar, and H. M. Niemeyer

61. Simplified tests of aggression chemosignals in male house mice suggest that
    a melanocortin-dependent product of the preputial gland reduces
    attacks 445
    H. K. Caldwell, L. Wang, and J. J. Lepri

62. Remote monitoring of badgers (*Meles meles*) for testing discrimination between
    urine samples from donors of different age and sex categories 451
    K. M. Service and S. Harris

63. The development of a simple associative test of olfactory learning and
    memory 459
    H. M. Schellinck, C. A. Forestell, V. M. LoLordo, P. Guidry, and R. E. Brown

64. Olfactory communication between man and other animals............................... 467
    B. A. Sommerville and D. M. Broom

Author Index.. 473

Subject Index 475

FROM INDIVIDUALS TO POPULATIONS: FIELD STUDIES AS PROVING GROUNDS FOR THE ROLE OF CHEMICAL SIGNALS

Dietland Müller-Schwarze

State University of New York
College of Environmental Science and Forestry
Syracuse, New York 13210

The point of this essay is not to simply report and review solid results, but to stimulate further research by pointing out promising paths of inquiry, the need for confirming laboratory findings in the "real world", and discrepancies between field and laboratory results that need explanation. While we are obligated to consider the possibility that phenomena observed in the laboratory may be artifacts, we should not prematurely dismiss them as such. Confirmation in the field may take some time. Conversely, field findings that cannot be replicated in the laboratory may nevertheless be "real"; the laboratory conditions may just not be sufficient.

In free-living populations, individual animals differ in their responses to chemosignals, and the composition of chemical signals they produce and emit. For instance, North American beavers, *Castor canadensis*, vary greatly in the composition of their castoreum, a paste from the castor sacs used in territorial marking. Certain compounds occur in only a few individuals, while others are widespread in the population (Sun, unpublished). Whether and how individuals produce and send chemosignals, and how they respond to such cues, depends on their individual developmental trajectories. Genetically anchored differences, hormonal influences, early social and sexual imprinting, individual experiences with a variety of environmental agents, social interactions, diet, age, parasites, and diseases produce unique animals that compete for reproductive opportunities. Such individual variation drives fitness contests. Not only the animals themselves vary, but they also operate in ever-changing meteorological and biotic conditions. At the time of the field observation or experiment, a multitude of external stimuli, the "context", further influence how an animal responds. As we proceed from studies of individuals to whole populations, and even comparing several populations, we may find they differ in their behavioral strategies, depending on differences in habitat and resources. An example are the scent marking strategies employed by several species of hyena (see below). Because of this variability of external conditions and internal states of animals, they employ flexible responses. We often find adaptive redundancy of communication signals, meaning that the same behavior can be controlled by more than one sensory modality, or different sets of one or more chemical compounds.

In the laboratory, by contrast, we take great pains to limit the number of such variables to the greatest extent possible. We use animals of the same genetic stock, feed them the same diet, and raise and test them under "standard conditions". From these differences between laboratory and field arise some of the advantages of field studies, but also considerable challenges.

DEFINITIONS

Field: Free-ranging, unconfined animals in natural populations. Advantages: Animals behave in their natural social and physical context. Disadvantages: many confounding variables; animals not always present; logistics often difficult. Observer effects may interfere with field studies: footprints and other odor contamination during and after observations or recording of behavior may be significant, especially in chemosensory studies.

Semi-field: Captive animals; outdoors; exposed to natural celestial and meteorological regimes; but confined to pens, enclosures, pastures, aviaries, or tanks; and typically protected from predators. For domestic animals, pastures can be identical with "natural environment". Advantages: Ease of experimentation; animals with known background. Disadvantages: Limited space; animals not in contact with a greater population of conspecifics; maintenance often difficult, especially cleaning.

Laboratory: Indoors; on restricted space; and with limited contact with other animals. Controlled conditions such as light, temperature, humidity, food supply. Advantages: Individual animals with known history available at any time; controlled conditions; ease of observation, manipulation, and cleaning. Disadvantages: Artificial setting; small space; no escape routes; frequent handling and cage-cleaning; controlled conditions may lack some factors necessary for responses, such as burrows, other cover, or opportunities for climbing.

1. WHEN ONLY FIELD STUDY WILL WORK

Context and previous experiences give meaning to a chemical stimulus. Both intraspecific scent marks and predator odors may not be intrinsically informing, but a function of prior experience. Odors may be memorized as associated with specific experiences and individuals, as in scent matching. Especially predator odors may not mean much to an animal in an artificial laboratory setting.

1.1. Territorial Marking: Descriptive Studies

Some observational and correlational projects are possible (or make sense) only in the field, especially if they explore natural spatial relationships, as in the mapping of scent marks in gazelle or wolf pack territories, or the pattern of scent marking by hyenas.

1.1.1. Antelopes. Fritz Walther (1978) pioneered real-world studies of territorial marks by laboriously mapping the scent marks in a territory of a Thomson's gazelle buck. Single male Thomson gazelles (*Gazella thomsoni*) mark their entire territory. A 8000 m^2 large territory had 18 dung piles and 110 preorbital marks. Dung piles were most frequent in border sections of frequent agonistic encounters with neighboring territory owners. Only one dungpile was in the center of the territory, near a bedding site. The preorbital marks were arranged in a "broad belt" around the territory. A central area remained unmarked. Walther concluded that the sent marks were more for the owner's own orientation than for territorial defense (Walther, 1978).

Studies of scent mark distribution in the territories of other antelope species include gerenuks (Gosling, 1981), klipspringer (Roberts and Lowen, 1997), and oribi, *Ourebia ourebi*, (Brashares and Arcese, 1999).

1.1.2. Hyenas. The mapping of scent marks, termed "pastings", by the brown hyena is a fine example of integrated field and laboratory work. The investigators mapped the pastings by a hyena clan on their approx. 500-km^2 large territory. The animals placed their marks most often in the center of this area (Mills *et al.*, 1980). Other hyena species, however, marked most often at the periphery of their territory. Comparison of four species of hyenas in five different habitats revealed that distribution of scent marks depends on resources in relation to group size. Large groups with abundant resources, such as the spotted hyena in Ngorongoro crater, need only a small territory to support themselves. A small territory is easily patrolled and marked at its periphery, because on average, each individual's share of "border duty" amounts to only a small section, here about 0.44 km per individual. A species living in poor habitat, on the other hand, needs a large territory to support itself. The group is likely to be small, and it becomes prohibitive in terms of time and energy to patrol and mark the entire periphery. For instance, striped hyenas at Serengeti live singly, and such an animal would have to cover 26 km boundary line (Gorman and Mills, 1984). In this case, scent marking is limited to the center of the territory, as opposed to the periphery.

1.2. Territorial Marking: Experimental Studies

Experiments usually are done more easily in the laboratory or under semi-field conditions. Why experiment in the field? The context for proper responses may be available only in the field.

1.2.1. Transfer of scent marks. Effects of moving scent marks from one antelope territory to another provide an example for mammals. In lek species, female odors can also attract females to males and increase the males' mating success. In both Uganda kob (*Kobus kob thomasi*) and Kafue lechwe (*K. leche kafuensis*) males attract estrous females to small breeding territories within a lek. Mating success of a male is site-specific: it is predicted by the success of the previous occupant of the same territory, and not by the male's own previous success. Males do not scent mark, but breeding territories are dotted with yellow urine patches from females. Transfer of soil from a successful territory to an unsuccessful one increased mating success in the latter more than 10 fold. The more successful the soil-donating territory had been, the greater the increase in success at the soil-receiving territory. The mean number of females on the treated territory also increased. On the territories from which soil had been taken, mating success decreased. Successful territories had about 50% more oestrone-3-sulphate in the soil than unsuccessful ones. Oestrone-3-sulphate is a metabolite of estrogen that occurs in urine of estrous female ungulates. Chance events such as predation by lions, or environmental factors such as predator-harboring cover, may funnel breeding activity to certain territories where females will concentrate and affect males' mating success without enhancing sexual selection (Deutsch and Nefdt, 1992). Even though this laborious experiment has not been replicated so far, it is an example for future field studies.

1.2.2. Experimental territorial scent marks. A fourth case study deals with the effect of artificial scenting of beaver territories on colonization. Potential beaver sites along streams with a history of previous occupation were each scented in spring with six scent mounds containing castoreum and anal gland secretion from males and females. This mimicked occupancy by a breeding pair. At the end of the summer season, newly arriving beavers had colonized significantly fewer scented sites than control sites that had received only unscented mud mounds. (Welsh and Müller-Schwarze, 1989). This shows that dispersal and colonization are possibly guided by scent signals that reflect the status of the population.

2. FIELD AND LABORATORY STUDIES OF THE SAME PHENOMENON YIELD DIFFERENT RESULTS

2.1. Kin Recognition

Studies of kin recognition in fish and tadpoles have yielded conflicting results in laboratory and field. In some species, discrimination occurs in the field, but not in the laboratory (Table 1).

Table 1. Kin recognition: Different results in field and laboratory.

Species	Field	Laboratory	Reference
Bufo americanus	yes	no	Waldman, 1982
Rana cascadae	yes	no	O'Hara and Blaustein, 1985

2.2. Sexual Behavior

A variety of species exhibit different sexual behavior in the field than in the laboratory or in captivity. Even in insects this presents complications. Males of the bark beetle *Ips paraconfusus* approach females on the basis of 2 pheromone compounds in the laboratory, but need three in the field (Silverstein, 1981). Among fish, male carp spermiate near

Table 2. Sexual behavior: Different results in field and laboratory.

Species	Behavior	Field	Laboratory	Reference
Bark beetle *Ips paraconfusus*	females attracting males	3 compounds necessary	2 compounds suffice	Silverstein, 1981
Carp *Cyprinus carpio*	♂ spermiates in presence of ♀ injected with pituitary homogenate	In earthen ponds: yes	In bare tanks: no	Billard et al., 1989
Gray-tailed vole, *Microtus canicaudus*	Bruce Effect (Pregnancy block)	No evidence of Bruce Effect	Occurs in other *Microtus* species	De la Maza et al., 1999
Domestic cattle	♂ mounts ♀ treated with estrus urine	On pasture: ♂ mounts only treated ♀	In stalls: ♂ mount ♀ indiscriminately	Sambraus and Waring, 1975

females injected with pituitary homogenate in outdoor earthen ponds, but do not do so in bare tanks in the laboratory (Billard *et al.*, 1989). Sex pheromones in cattle provide another example: In pastures, bulls will mount females treated with estrous urine, but not control cows. In stalls, however, they mount cows indiscriminately (Sambraus and Waring, 1975). For a summary, see Table 2.

2.3. Feeding Behavior

Feeding behavior can differ considerably between field and captivity. This does not surprise, given the vastly greater choice of foods available to free-ranging animals. Brown tree snakes, *Boiga irregularis*, respond to several food lures in the laboratory, but not in the field (Chiszar, 1994, pers. comm.). Since control of the snake population in the wild is the goal of this work, such discrepancies must be cleared up by field experiments. Some rodents, such as Townsend's vole, *Microtus townsendi*, avoid repellent-treated food only in the open, but not in cover, usually available in the wild (Merkens *et al.*, 1991). Mongoose avoid toxin-treated eggs by smell in the field, but not in the laboratory where they will break the shell, and eat them (Nicolaus and Nellis, 1987). Badger feces repel hedgehogs, *Erinaceus europaeus* for two days in enclosures, but only for minutes or hours in open farmland (Ward *et al.*, 1996; 1997).

What does account for such different results in field and laboratory? Each case is different, depending on animal species, phenomenon investigated, and experimental technique used, among others. Still, some general statements are possible. The field provides a much wider array of non-biological and biological (including intra- and interspecific) factors than the laboratory.

Table 3. Feeding responses: Differences between field and laboratory.

Species	Behavior	Field	Laboratory	Reference
Brown tree snake *Boiga irregularis*	Responds strongly to fish attractants and trapper's lures	no	yes	Chiszar, pers. comm., 1994
Townsend's vole	Feeding in presence of repellent (R)	Avoids R only in open, not in cover	Test without cover may be misleading	Merkens *et al.*, 1991
Mongoose	Aversive conditioning to scented, toxic eggs	Avoid scented, toxic eggs from a distance	Will not avoid at distance, break shell, eat still	Nicolaus and Nellis, 1987
Hedgehog *Erinaceus europaeus*	Avoiding area tainted with badger (predator) feces	Farmland, golf courses: Avoid treated area for minutes or hours only	Enclosure: Avoid treated area for 2 days	Ward *et al.*, 1996; 1997

Table 4. Antipredator responses: Differences between field and laboratory.

Species	Behavior	Field	Laboratory	Reference
Chinook salmon	Conditioned to recognize predator (cutthroat trout)	Better survival in creek; but only if raised in complex habitat	No survival benefit in hatchery raceway; and not if raised in simple habitat	Berejikian *et al.*, 1999
Gray-tailed voles (*Microtus canicaudus*)	Mink odor effect on reproductive rate, sexual maturation, juvenile recruitment	No effect	Reproductive behavior suppressed in *M. agrestis* and bank voles, *Clethrionomys glareolus*	Wolff and Davis-Born, 1997

Such *"context"* may be necessary for a meaningful response. Because of the greater *complexity* of the natural environment, responses depend on the choses available, particularly when feeding is involved. In the laboratory, animals can be deprived of food an then forced to choose between only two foods. Responses to conspecific alarm odors or predator odors may vary with the predation pressure at a certain locale. *Genetic heterogeneity* of field populations may result in considerable variability of behavioral responses and signal composition. By contrast, forced choices are possible in the laboratory on the one hand, and responses may be absent or habituate fast because, for example, a scent mark is not backed up by, or associated with an encounter of a real animal. Likewise, if in the laboratory alarm odors are not preceded or followed by an encounter with a predator, the animal may respond differently, and not necessarily more weakly. For instance, hedgehogs responded more strongly to feces of Eurasian badgers (*M. meles*), their predator, in enclosures than did free-ranging hedgehogs on golf courses and farmland. In the pens, the hedgehogs avoided the scented area for two days, but in the field only for hours, or even just minutes (Ward *et al.*, 1997). Furthermore, laboratory or captive animals experience *different upbringing* that influences social and feeding responses.

3. DIFFERENCES BETWEEN FIELD STUDIES

Even two coordinated field studies may yield different results. We found that male muskrat scent in traps attracted primarily adult males in a field study in the Netherlands, but mostly juveniles in the Montezuma Marsh in New York. This may reflect different population phases in the two areas. In Holland, the muskrats were immigrating, with males spearheading the invasion (Ritter *et al.*, 1982). In New York, however, territorial males in a saturated population avoided male scent while immature muskrats wandered about with impunity (van den Berk and Müller-Schwarze, 1984).

Areas that differ in their predator guilds offer rewarding opportunities for separating genetically anchored behavior from responses acquired under specific and varying predator pressures. Response to alarm odors from injured conspecifics varies with local predation pressure. This has been described for the Western toad, *Bufo boreas* (Belden *et al.*, 2000), salamanders (Madison *et al.*, 1999), and Pacific tree frogs (Puttlitz *et al.*, 1999).

4. FROM LABORATORY TO FIELD: SAME RESULTS

Many studies start in the laboratory. Obviously, any study that requires well-defined individuals with precisely known genetic and experiential background, and also homogenous and finely tuned hormonal and nutritional status, and/or to be treated at certain times in a certain manner, has to be carried out in the laboratory. The red-spotted newt (*Notophthalmus viridescens*) responds to macerated newt extract with anti-predator behavior in both field and laboratory (Rohr and Madison, this volume). Among reptiles, the responses of cottonmouth (water) moccasins (*Agkistrodon piscivorus*) to odors of their predator, the king snake, in the field echoes that in the laboratory (Gutzke, this volume). Among mammals, blind mole rats in laboratory colonies have served well in behavioral studies. To cite one such study, they shift their urination and defecation sites to the border adjacent to other, new, neighboring mole rats (Zuri *et al.*, 1997).

Increasingly more laboratory findings have been subjected to field or semi-field tests. These include (1) preferences for odors of certain age and sex classes, (2) the roles of the Major Histocompatibility Complex (MHC) in individual odors and mate choice, and (3) priming pheromones that were first discovered in the laboratory.

4.1. Discrimination of Age and Sex Classes

Odor-baited live traps, placed in 0.1 ha enclosures served in elucidating the preferences of different age and sex classes of house mice for odors from certain classes of donors. Adult male mice entered traps with odor from juvenile and estrous females, while adult anestrous and estrous females preferred the odor of adult males. The latter also avoided odors of non-estrous, pregnant and lactating females. Juvenile females, on the other hand, preferred the scent of other juvenile females, and avoided all other odors. These results paralleled those of earlier laboratory experiments (Drickamer, this volume).

For instance, intrauterine position relative to male siblings has androgenizing effects on females that are expressed as varying anogenital distances. In outdoor enclosures, males preferred traps scented with urine from females of short anogenital distances, i.e. those with no male neighbors *in utero*, while females preferred the odor from males that had developed between two males, therefore exposed to a maximum of androgens (Drickamer, this volume).

4.2. Major Histocompatibility Complex

In outdoor enclosures, about 5 x 10 meters large, MHC-based nonrandom mating resulted in a deficiency of MHC homozygotes in the population. Settlement patterns account for part of the deficiency. A high frequency of extra-pair matings, initiated by the female's traveling to another territory, suggests selection for disassortative mating with partners more MHC-disparate than the female's own territorial male. Such MHC-based mating preferences may maintain MHC heterozygosity in natural populations (Potts *et al.*, 1991).

4.3. Priming Pheromones

Primer pheromones, first observed in the laboratory, would hardly have been discovered in the field. Bronson (1979) and later Keverne (e.g. Keverne and Rosser, 1986) have conceptually organized the physiological laboratory findings in mice into functional systems that are ecologically adaptive. A current challenge is to test these models in the field. Few such experiments have been performed. After decades of laboratory experiments, the work was carried into the field. An early, ingenious example is the testing in the wild of the effect of urine of crowded females on puberty delay in young females. In the *cloverleaf experiment*, Coppola (1986) introduced additional females into a house mouse population living on a "highway island". Females of this artificially augmented high-density population produced urine that delayed puberty in other females in the same fashion as urine from crowded laboratory mice.

4.4. Bioassays in the Field

4.4.1. Indirect monitoring. Bioassays by recording entry into scented traps is the method of choice for small nocturnal mammals (Drickamer, this volume). An earlier example are the field experiments that tried to elucidate the behavioral role of "musk" from the preputial glands of muskrats in Europe (Ritter *et al.*, 1982) and North America (van den Berk and Müller-Schwarze, 1984) with the aid of scented traps. Sensitive video recording equipment permits indirect observation of nocturnal mammals such as badgers at scent marks (Buesching and Macdonald, this volume).

4.4.2. Direct observation of responses. In many cases, bioassays of whole secretions with signal value, isolated fractions, or single compounds are best conducted under standard laboratory conditions. However, the confounding complexities of the field need not preclude field bioassays. On the contrary, they may provide "context" and be preferable to the challenges of taking wild animals into captivity for testing. Experiments examining kin recognition in North American beavers, *Castor canadensis*, on the basis of two secretions, castoreum and anal gland secretion, were successful in the field (Sun and Müller-Schwarze, 1998). Likewise, the behavioral effects of castoreum from colony members, neighbors, and strangers showed that beavers can discriminate between these categories in their natural setting (Schulte, 1998). Fractions, single compounds and mixtures of compounds have been successfully bioassayed in free-ranging beavers, (Müller-Schwarze et al., 1986; Müller-Schwarze and Houlihan, 1991; Schulte et al., 1995). Fractionated temporal gland secretions of alpine marmots (*M. marmota*) have recently been bioassayed *in situ* at the burrows of a free-ranging population in the French Alps (Bel et al., 1999).

5. FROM INDIVIDUALS TO POPULATIONS

5.1. From Trophic Levels to the Information Web

Most chemical ecologists start out with testing individuals for their responses to various semiochemicals. This is still the method of choice, even though we are now able to place these responses in larger ecological contexts. In a trophic web, we distinguish traditionally six trophic levels. But, as Vet (1999) pointed out, an information web connecting these trophic levels, can contain as many as 18 information links between and among the various levels. Plant volatiles alone provide six of these information links (Vet, 1999). We are now learning to understand ever more complex chemical information webs (e.g. Baldwin, 1999). It no longer suffices to test odorants on individuals only and leave it there. From a moth in a wind tunnel or a mouse in a cage we need to progress to awareness of a larger system. We have to consider population and community interactions. Such interactions can be staged in the laboratory only for small animals, mostly invertebrates. Laboratory tests are still necessary, but are not an end in itself. Where more than one social unit such as a family or clan form the focus of the investigation, there is no substitute for the field setting. Some pioneering studies already have gone beyond the level of the individual, family or group and taken into account whole populations. The two examples cited here deal with scent marking in mammalian populations.

5.2. Wolf Territories

In the classic field study Peters and Mech (1975) mapped the distribution of urine scent marks of a wolf pack in their territory. Their survey extended into parts of neighboring wolf territories, bringing this study into the realm of population biology. To appreciate the meaning of "field study" here, we have to remind ourselves that the wolf territories in their study area, the Superior National Forest in Minnesota, USA, ranged in size from 125 to 310 km^2, the latter measuring about 17 by 18 km.

5.3. Beaver Scent Mounds and Population Density

Both the North American (*Castor canadensis*) and Eurasian (*C. fiber*) beavers build scent mounds at the edges of their ponds and streams. Numbers of scent marks in beaver family territories correlate with population density. This is true for the North American beaver (Houlihan, 1989) and the Eurasian beaver (Rosell and Nolet, 1997). The number of scent mounds at each colony correlated positively with the number of other beaver colonies within 5 km up- and downstream, and negatively with the distance to the nearest neighbor.

This we interpret as intensification of scent marking as a result of frequent encounters and/or friction between beaver families, usually neighbors, but also with dispersing two-year-olds as potential immigrants.

MAJOR CONCLUSIONS

Important laboratory results such as the role of priming pheromones in population regulation, or the function of alarm pheromones, need to be confirmed in the field. We need field tests, especially when the goal is to apply the findings in natural populations, as in the control of brown tree snakes. Discrepancies between laboratory and field studies, as listed in Tables 1 - 4, need to be explained. With increasing elucidation of responses by individuals to semiochemicals we will be able to place such behavior in larger contexts and understand the effect of chemical signaling on composition and functioning of entire populations and communities. This includes clusters of neighboring territories of the same species, but also different trophic levels of organisms. In terms of methods, the best is yet to come. In the future, hitherto unattainable field observations and experiments will be feasible by utilizing electronic tagging devices, sophisticated recording techniques and "electronic noses" that detect semiochemicals in the wild. In most research on chemical signals, laboratory and field studies need to proceed in tandem.

REFERENCES

Baldwin, I. T., 1999, Inducible nicotine production in native *Nicotiana* as an example of adaptive phenotypic plasticity, *J. Chem. Ecol.* **25**:3-30.

Bel, M. C., Coulon, J., Sreng, L., Allainé, D., Bagneres, A. G., and Clément J. L., 1999, Social signals involved in scent-marking behavior by cheek-rubbuing in Alpine marmots (Marmota marmota), *J. Chem. Ecol.* **25**: 2267-2283.

Belden, L. K., Wildy, E. L., Hatch, A. C., and Blaustein, A. R., 2000, Western toads, *Bufo boreas*, avoid chemical cues of snakes fed juvenile, but not larval, conspecifics, *Anim. Behav.* **59**:871-875.

Berejikian, B. A., Smith, R. J. F., Tezak, E. P., Schroder, S. L., and Knudsen, C. M., 1999, Chemical alarm signals and complex hatchery rearing habitats affect antipredator behavior and survival of chinook salmon (*Oncorhynchus tshawytscha*) juveniles, *Can. J. Fish Aquat. Sci.* **56**:830-838.

Billard, R., Bienarz, K., Popek, W., Eppler, P., and Saad, A., 1989, Observations on a possible pheromonal stimulation of milt production in carp (*Cyprinus carpio* L.), *Aquaculture* **77**:387-392.

Brashares, J. S., and Arcese, P., 1999, Scent marking in a territorial African antelope: II. The economics of marking with feces, *Anim. Behav.* **57**:11-17.

Bronson, F. H., 1979, The reproductive ecology of the house mouse. *Q. Rev. Biol.* **54**:265-299.

Coppola, D. M., 1986, The puberty delaying pheromone of the house mouse: field data and a new evolutionary perspective, in: *Chemical Signals in Vertebrates IV* (D. Duvall, D. Müller-Schwarze, and R. M. Silverstein, eds.), Plenum, NY., pp. 457-462.

Deutsch, J. C., and Nefdt, R. J. C., 1992, Olfactory cues influence female choice in two lek-breeding antelopes, *Nature* **356**:596-598.

Gorman, M. L., and Mills, M. G. L., 1984, Scent marking strategies in hyaenas (Mammalia), *J. Zool. Lond.*, **202**:535-547.

Gosling, L. M., 1981, Demarkation in a gerenuk territory: an economic approach. *Z. Tierpsychol,* **56**:305-322.

Houlihan, P. W., 1989, Scent mounding by beaver (*Castor canadensis*): Functional and semiochemical aspects, M.S. Thesis: SUNY College of Environmental Science and Forestry, Syracuse, New York.

Keverne, E. B., and Rosser, A. E., 1986,. The evolutionary significance of the olfactory block to pregnancy. in: *Chemical Signals in Vertebrates*, vol. 4 (D. Duvall, D. Müller-Schwarze, and R.M. Silverstein eds.), New York: Plenum, pp. 433-439.

Madison, D., Maerz, J. C., and McDarby, J. H., 1999, Optimization of predator avoidance by salamanders using chemical cues: Diet and diel effects, *Ethology* **105**:1073-1086.

De la Maza, H. M., Wolff, J. O., and Lindsey, A., 1999, Exposure to strange adults does not cause pregnancy disruption or infanticide in the gray-tailed vole, *Behav. Ecol. Sociobiol.* **45**:107-113.

Merkens, M., Harestaed, A. S., and Sullivan, T. P., 1991, Cover and the efficacy of predator-based repellents for Townsend's vole, *Microtus townsendii*, *J. Chem. Ecol.* **17**:401-412.

Mills, M. G. L., Gorman, M. L., and Mills, M. E. J., 1980, The scent marking behaviour of the brown hyaena, *Hyaena brunnea*, *S. Afr. J. Zool.* **15**:240-248.

Müller-Schwarze, D., Morehouse, L., Corradi, R., Zhao, C.-H., and Silverstein, R. M., 1986, Odor images: responses of beaver to castoreum fractions. in: *Chemical Signals in Vertebrates 4*. (D. Duvall, D. Müller-Schwarze, and R. M. Silverstein, eds.), Plenum, New York, pp. 561-570.

Müller-Schwarze, D., and Houlihan, P., 1991, Pheromonal activity of single castoreum constituents in beaver, *Castor canadensis*, *J. Chem. Ecol.* **17**:715-734.

Nicolaus, L. K., and Nellis, D. W., 1987, The first evaluation of the use of conditioned taste aversion to control predation by mongooses upon eggs, *Appl. Anim. Behav. Sci.* **17**:329-346.

O'Hara, R. K., and Blaustein, A. R., 1985, *Rana cascadae* tadpoles aggregate with siblings: an experimental field study, *Oecologia* **67**:44-51.

Peters, R. P., and Mech, L. D., 1975, Scent-marking in wolves. *Amer. Sci.* **63**:628-637.

Potts, W. K., Manning, C. J., and Wakeland, E. K., 1991, Mating patterns in seminatural populations of mice influenced by MHC genotype, *Nature* **352**:619-621.

Puttlitz, M. H., Chivers, D. P., Kiesecke, J. M., and Blaustein, A. R., 1999, Threat-sensitive predator avoidance by larval Pacific tree frogs (Amphibia, Hylidae), *Ethology* **105**:449-456.

Ritter, F. J., Brüggemann, I. E. M., Gut, J., Persoons, C. J., and Verwiel, P. E. J., 1982, Chemical stimuli of the muskrat, in: *Determination of Behaviour by Chemical Stimuli*. (J. E. Steiner and J. R. Ganchrow, eds.), IRL Press, London, pp. 77-89.

Roberts, S. C., and Lowen, C., 1997, Optimal patterns of scent marks in klipspringer (*Oreotragus oreotragus*) territories, *J. Zool.* **243**:565-578.

Rosell, F., and Nolet, B. A., 1997, Factors affecting scent-marking in Eurasian beaver (*Castor fiber*), *J. Chem. Ecol.* **23**:673-689.

Sambraus, H. H., and Waring, G. H., 1975, Der Einfluss des Harns brünstiger Kühe auf die Geschlechtslust von Stieren, *Z. Säugetierk.* **40**:49-54.

Schulte, B. A., 1998, Scent marking and responses to male castor fluid by beavers, *J. Mammal.* **79**:191-203.

Schulte, B. A, Müller-Schwarze, D., Tang, R., and Webster, F. X., 1995, Bioactivity of beaver castoreum constituents using principal components analysis, *J. Chem. Ecol.* **21**:941-957.

Silverstein, R. M., 1981, Pheromones: Background and potential for use in insect pest control, *Science* **213**:1326-1332.

Sun, L., and Müller-Schwarze, D., 1998, Anal gland secretion codes for relatedness in the beaver, *Castor canadensis*, *Ethology* **104**:917-927.

Van den Berk, J., and Müller-Schwarze, D., 1984, Responses of wild muskrats (*Ondatra zibethica* L.) to scented traps, *J. Chem. Ecol.* **10**:1411-1415.

Vet, L. E. M., 1999, From chemical to population ecology: Infochemical use in an evolutionary context, *J. Chem. Ecol.* **25**: 31-49.

Waldman, B., 1982, Sibling association among schooling toad tadpoles: Field evidence and implications, *Anim. Behav.* **30**:700-713.

Walther, F. R., 1978, Mapping the structure of the marking system of a territory of the Thompson's gazelle, *E. Afr. Wildl. J.* **16**:167-176.

Ward, J. F., Macdonald, D. W., Doncaster, C. W., and Mauget, C., 1996, Physiological response of the European hedgehog to predator and nonpredator odour, *Physiol. Behav.* **60**:1469-1472.

Ward, J. F., Macdonald, D. W., and Doncaster, C. P., 1997, Responses of foraging hedgehogs to badger odour, *Anim. Behav.* **53**:709-720.

Welsh, R. G., and Müller-Schwarze, D., 1989, Experimental habitat scenting inhibits colonization by beaver, *Castor canadensis*, *J. Chem. Ecol.* **15**:887-893.

Wolff, J. O., and Davis-Born, R., 1997, Response of gray-tailed voles to odours of a mustelid predator: a field test, *Oikos*, **79**:543-548.

Ylönen, H., and Wolff, J.O., 1999, Experiments in behavioural ecology and the real world, *TREE* **14**: 82.

Zuri, I., Gazit, I., and Terkel, J., 1997. Effect of scent-marking in delaying territorial invasion in the blind mole-rat *Spalax ehrenbergi*, *Behaviour* **134**:867-880.

THE ECONOMIC CONSEQUENCES OF ADVERTISING SCENT MARK LOCATION ON TERRITORIES

S. Craig Roberts and L. Morris Gosling

Evolution and Behaviour Research Group
Dept of Psychology
University of Newcastle
Newcastle-upon-Tyne, NE1 7RU, UK

INTRODUCTION

The role of scent-marking in the maintenance of mammalian territories is well-documented (eg. Ralls, 1971; Brown and Macdonald, 1985; Gosling, 1982; Gosling and Roberts, in press). Scent marks are a form of status signal, advertising territory ownership, and recent evidence suggests they may also be condition-dependent signals of quality and competitive ability (Gosling *et al.*, in press). In common with animal signals in other sensory modalities, scent marks thus provide a means of assessment which informs signal receivers about the signaller's quality. Information about the location, density, freshness and chemical properties of scent marks are all likely to contribute to the appraisal of the signaller by the receiver before the participants meet. In a territorial context, the product of this appraisal may be a decision to avoid the risk of meeting the signaller by withdrawing from the territory, for example, if the signaller appears to be of far higher competitive ability. Alternatively, a receiver may defer a decision until it has more information, or until an opportunity arises to check this information in a face-to-face situation. Since scent mark detection often occurs in the signaller's absence, receivers may need to confirm the identity of any opponent as the resource-holder before deciding on their next move. This can be achieved by comparing the odour of the marks with that of their opponent, a process known as scent-matching (Gosling, 1982, 1990; Gosling and McKay, 1990). A positive match, which unambiguously confirms the owner's status, may be sufficient to settle the contest conventionally and scent-marking thus helps to reduce the costs of territory defence (Gosling, 1986; Stenstrom 1998).

While the benefits to owners of scent-marking a territory are clear, a major constraint is the limited spatial range over which marks are typically detectable. This is at least partly due to the need for marks to persist for extended periods, especially in species where territories are large and intervals between successive visits to marking sites are long (Alberts, 1992). Mark detection is therefore frequently probabilistic and signallers benefit if they deploy their marks in such a way that maximises the likelihood of receivers finding them, while at the same time minimising the distance that intruders travel before detection (Gosling, 1981; Roberts and Lowen, 1997). Field studies which have mapped patterns of scent marks show that marks are often clustered around or within territorial boundaries, sometimes forming characteristic rings or "bowls" (eg Peters and Mech, 1975; Walther, 1978; Gosling, 1981; Roberts and Lowen, 1997; see also Brashares and Arcese 1999). In very large territories, however, the chance of intruders missing widely interspersed boundary marks selects for centrally clustered marking patterns (Gorman, 1990).

If intruders also benefit by detecting scent marks, for example by avoiding costly encounters with opponents of higher competitive ability, it will be in their interest to actively seek out marks when they enter a territory. Detection of marks will not then simply depend on the gradient of airborne volatiles emanating from marks and the probabilistic (at least with respect to scent mark location) movements of intruders, but also on the receiver's motivation and psychology (Guilford and Dawkins, 1991). Intruders would be expected to seek marks provided that the costs of searching are outweighed by the benefits associated with additional information obtained. Observational evidence confirms that receivers seek out marks. For example, Müller-Schwarze (1974) observed that black-tailed deer search for marks after entering a new area, while a study of ring-tailed lemurs, *Lemur catta*, found that 62% of scent marks were investigated within 10 min, with a median latency of only 30s (Kappeler, 1998). This aspect of communication through scent marks is perhaps undervalued, but if intruders do search for marks, it would have profound implications for the economics of scent-marking. In particular, owners should not only deposit scent marks to intercept intruders, but should also advertise the presence of their marks in order to facilitate their detection and maximise the resulting benefits.

Here, we review briefly ways in which signallers might advertise the presence of their scent marks. We then use a simple spatial model to explore the effects of variable mark detectability as a result of receiver searching on the economics of scent-marking and resource-defence territoriality. Finally, we discuss this signalling system in terms of the debate about the evolution of honest signalling.

MARK ADVERTISEMENT AND DETECTABILITY

Two main kinds of scent-marking behaviour are consistent with the expectation that signallers advertise the location of their marks. The first is that signallers frequently deposit marks in sites that are locally conspicuous or which have characteristic topographical features. For example, klipspringer antelopes prefer to scent-mark on dead trees or branches (mainly of preferred food species), in an area slightly elevated above its surroundings and immediately above a significant break in slope (Roberts, 1997). Marks are often placed within a narrow vertical distribution, in spite of the fact that they are physically able to mark above or below the preferred height and that alternative sites may be available (Gosling, 1981; Roberts, 1997). Where suitable sites are not locally available, signallers may be able to manufacture them (Gosling, 1972). As a final example, signallers may occasionally place their scent marks at signalling sites used by other species (Gosling, 1980; Paquet, 1991), thus gaining in detectability at no additional cost to themselves.

Secondly, signalers actively create visual anomalies near their scent marks. For example, some ungulates disturb nearby vegetation by antler thrashing (Kile and Marchinton, 1977; Johansson and Liberg, 1996). In some felids and ungulates, signallers paw or scrape the ground with claws or hooves (Gilbert, 1973; Feldman, 1994; Johansson and Liberg, 1996). A more striking example is where signallers damage or tear off strips of bark before marking, creating visible wounds to trees at scent-marking sites (Graf, 1956;

Barette, 1977; Bowyer *et al.*, 1994). These wounds are generally separate from the secretion and thus appear to be unrelated to the olfactory function of the marks. There remains no direct evidence that receivers are attracted to such visual features but it would be difficult to explain their widespread existence if receivers did not respond to them.

Both kinds of marking-associated behaviour could increase detectability. In the first, marks are placed in conventional sites where receivers are more likely to find them; receivers may thus be able to form a visual search image of likely scent-marked sites (Gosling, 1981; Roberts, 1997). In the second, this visual element is actively reinforced. If receivers use these cues, scent marks become a multimodal signal (Rowe, 1999), having a visual alerting component which draws the attention of receivers to the presence of the olfactory, semantic component. Such behaviours will have the effect of enlarging the distance over which marks are detectable, beyond that attributable solely to the mark's chemical properties. In the following section, we explore the implications of this for scent-marking economics.

DETECTABILITY AND MARKING ECONOMICS

Our model incorporates movements of a single owner and intruder within a territory and variations in the amount of information about the opponent's identity. It plots the probabilities of (a) detection by intruders of owner's scent marks within the territory and (b) the frequency with which owners and intruders will meet while intruders cross the territory. The model distinguishes between territorial encounters occurring in the presence of absence of previous mark detection by the intruder. In the former case, the intruder is able to correctly identify its opponent as the owner, while in the latter, it has relatively little information about its opponent or its competitive ability.

Following Roberts and Lowen (1997), the model assumes the owner defends a circular territory of radius R_0, within which it positions and maintains n scent marks of equal volume and efficacy. These are non-overlapping and evenly distributed along the circumference of a circle, which is concentric with the territory boundary and with a radius R, such that $R \leq R_0$. The area within the ring of scent marks is termed the defended area. Mark detection is taken to be a function of the distance between the intruder and the mark, such that detection always occurs at a distance of less than or equal to a_0 and never occurs at distances greater than a_0. Thus, the probability of detecting a scent mark as an intruder crosses into the defended area is $(2na_0/2\pi R)$, and the probability of entering the defended area without detecting a mark is $[1-(2na_0/2\pi R)]$.

The probability of encounters between owner and intruder (P_E) are calculated using a modification of Waser's (1977) gas model (see also Barrett and Lowen, 1998), in which

$$P_E = (4.p.\mathrm{k}v/\pi)d \quad (1),$$

where v is travel speed, d is the distance at which intruders are detected by the owner and k is a constant. Since we are only concerned with pairwise encounters between individuals, p, the population density, is given by 2/territory area (m^{-2}) and we omit s, the term for group dispersion in the original model. Term (1) actually calculates the expected frequency of encounters given random movement. However, as expected frequencies are usually less than 1 with our parameter values and as we are solely concerned with the first encounter, we can treat these as probabilities and convert to $P_E=1$ any frequencies greater than 1. For simplicity, we assume animals travel at a speed of 1ms^{-1} and we substitute v for t, the time taken to cross a defined distance. The intruder's travel time between the boundary and the defended area, following a straight-line trajectory, is then t^{Ro-R} and that for crossing the remainder of the territory is t^{Ro+R}. Note that this produces an approximation of the

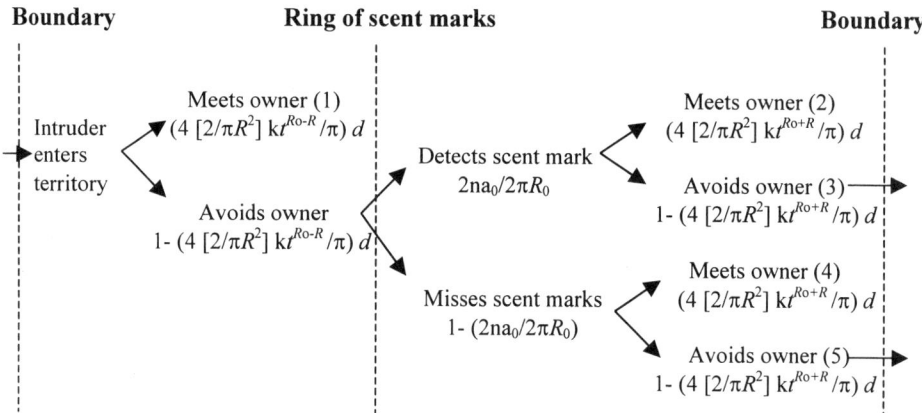

Figure 1. Summary of possible outcomes of a territorial intrusion. Numbers in parentheses relate to the categories of outcomes as outlined in the text. Terms used in functions are explained in the text.

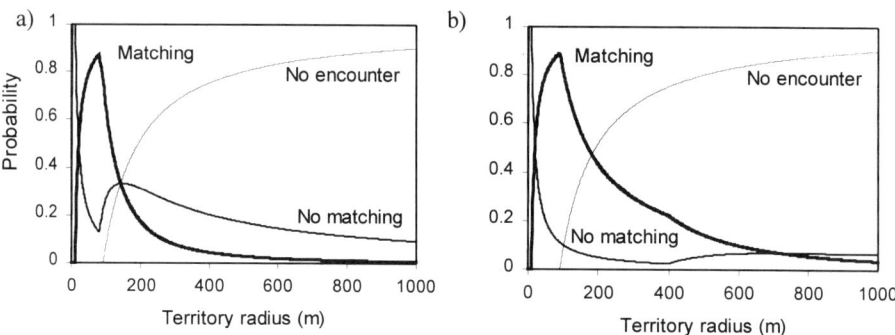

Figure 2. Probability of the three main outcomes of intrusions onto scent-marked territories in relation to mark detectability. Intruders may leave the territory without meeting the owner. Alternatively, they will encounter the owner having either detected or missed its scent marks; in the former case, they are thereby able to correctly identify the owner by scent-matching. Effective radius of marks, a_0 is (a) 2m and (b) 10m. Here, $R=0.8Ro$, $d=50$ and $n=100$.

duration of an intruder's presence within the territory before and after crossing into the defended area; because intruders do not necessarily move in a straight line, the constant k (set here at 1.25) estimates additional travel time resulting from random deviations. Probabilities of intrusion outcomes can now be calculated. After entering the territory, it takes intruders kt^{Ro-R} seconds to reach the demarcated ring, during which they may meet the owner (outcome 1, see Figure 1). If not, intruders may detect a mark while entering the defended area, and may subsequently meet the owner during the kt^{Ro+R} seconds it takes to cross the territory. We can then find the probability of encounters without opportunity for scent-matching (outcomes 1+4), where matching can occur (2) or in which intruders are not detected (3+5).

The model demonstrates the relationship between the opportunity for scent-matching and territory size (Figure 2). Since mark detection decreases exponentially with increasing territory size (assuming constant n), there is, on small territories at least, a high probability of encounters in which intruders have already encountered owner's marks. Indeed, this is usually the most probable outcome. The range of territory sizes in which this is true clearly depends on the model's parameters, but given reasonable values (those in Figure 2 are within the ranges found in our field studies on African antelopes: Gosling, 1981; Roberts and Lowen, 1997), the opportunity for matching is inversely related to territory size and eventually becomes less likely than either alternative outcome. The main exception to this trend lies in a narrow range of extremely small territories, where encounters without matching are the predominant outcome. This is because, despite the fact that intruders would be certain to detect marks in the ring, owners intercept intruders before they have the opportunity to do so.

Figure 2 also demonstrates the impact of mark advertisement on the context in which encounters occur. As detectability increases, there is a linear increase in the range of territory sizes over which intruders will be able to scent-match should an encounter occur (for a given probability of matching). In addition, mark advertisement will have knock-on effects for other aspects of marking economics. For example, for any given territory size, the number of marks required to return the same probability of mark detection decreases exponentially (Figure 3). Thus, if competitors are prepared to incur the costs of seeking out marks, signalers are able to mark at lower density.

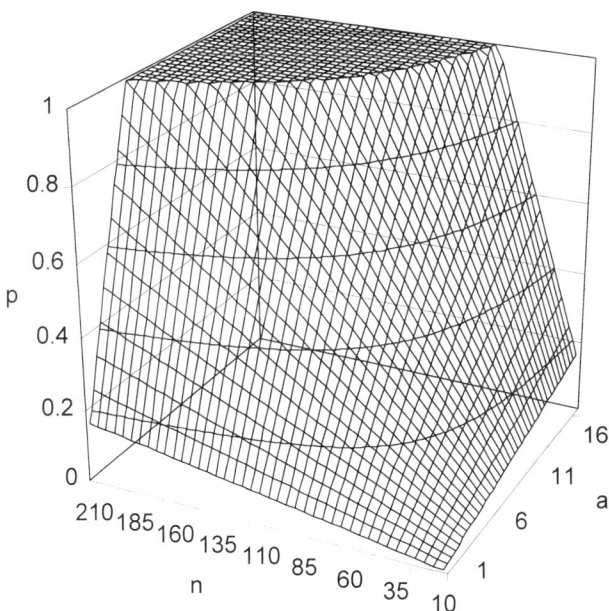

Figure 3. Surface plot of the relationship between the probability of mark detection by intruders (p), the number of marks (n) and the effective radius of each mark (a).

CONCLUSIONS

We have argued that intruders stand to gain by actively seeking out marks, because this helps to avoid costly fights. We have also shown that signallers which advertise scent mark location could benefit by their marks being more likely to be detected, particularly if intruders actively search for them. This will tend to increase the frequency of encounters in which intruders can identify owners' status, which in turn will reduce the frequency and costs of escalated resource defence. In addition, the increased detectability of scent marks conferred by advertisement and receiver searching means that owners can scent-mark at lower densities, hence offsetting to receivers some of the costs of signal transmission. It may also carry consequences for the chemistry of scent marks: selection for volatility and large effective range is reduced since the interest of receivers is gained visually. Information from marks is therefore only gained at close range, often by licking the marks to release soluble scent constituents (Alberts, 1992; Roberts, 1998), while the reduced volatility prolongs mark persistence time.

We have suggested elsewhere, in view of the apparently universal link between mammalian territoriality and scent-marking, that marking may be a prerequisite for viable territoriality (Gosling and Roberts, in press). Furthermore, our model suggests that advertisement of marks and receiver searching, through increasing the range of territory sizes over which scent-matching is possible, would appear to play important roles in the economics of territoriality. However, receivers may differ in the extent to which they are willing to seek out marks. Those of poor quality may stand to gain more by detecting marks than those of higher quality, being more likely to withdraw immediately. Conversely, good competitors may need to maximise information about the owner in preparation for a take-over attempt. Whichever is true in a particular case, such assessment can only be evolutionarily stable if the signals are reliable. Scent-marking can be seen as an honest form of signalling since owners must occupy and defend the territory at least for long enough to demarcate it (Gosling, 1982, 1990). If advertisement of marks helps to increase the likelihood of matching, then it must also reinforce the reliability of scent marks as honest signals.

Finally, advertisement of scent mark location could provide other benefits to signallers beyond that of reducing costs of escalated territory defence. While we have framed our argument in terms of an owner-intruder paradigm, improved detectability and accuracy of assessment would apply to other categories of receiver, notably potential mates. In addition, while we have considered the detection of only a single mark, advertisement of mark location is also likely to increase the chances of receivers detecting multiple marks. This may be necessary for receivers to predict owners' movements, navigate across or between territories, or to reduce the possibility of error during subsequent assessment.

ACKNOWLEDGMENTS

SCR was supported by a grant from the Leverhulme Trust while this paper was written.

REFERENCES

Alberts A. C., 1992, Constraints on the design of chemical communication systems in terrestrial vertebrates, *Am. Nat.* **139** Suppl.:S62-S89.

Barrett, L., and Lowen, C. B., 1998, Random walks and the gas model: spacing behaviour of grey-cheeked mangabeys, *Func. Ecol.* **12**, 857-865.
Barette, C., 1977, Scent-marking in captive muntjacs, *Muntiacus reevesi*, *Anim. Behav.* **25**:536-541.
Bowyer, R. T., Vanbellenberghe, V., and Rock, K. R., 1994, Scent marking by Alaskan moose - characteristics and spatial distribution of rubbed trees, *Can. J. Zool.* **72**: 2186-2192.
Brashares, J. S., and Arcese, P., 1999, Scent marking in a territorial African antelope: II. The economics of marking with faeces, *Anim. Behav.* **57**:11-17.
Brown, R. E., and Macdonald, D. W., 1985, *Social odours in mammals*, Clarendon Press, Oxford.
Feldman, H. N., 1994, Methods of scent marking in the domestic cat. *Can. J. Zool.* **72**:1093-1099.
Gilbert, B. K., 1973. Scent marking and territoriality in pronghorn (*Antilope americana*) in Yellowstone National Park, *Mammalia* **37**:25-33.
Gorman, M. L., 1990, Scent marking strategies in mammals, *Rev. suisse. Zool.* **97**:3-29.
Gosling, L. M., 1972, The construction of antorbital gland marking sites by male oribi (*Ourebia ourebia*, Zimmerman 1783), *Z. Tierpsychol.* **30**:271-276.
Gosling, L. M., 1980, Defence guilds of savannah ungulates as a context for scent communication. *Symp. Zool. Soc. Lond.* **45**:195-212.
Gosling, L. M., 1981, Demarkation in a gerenuk territory: an economic approach, *Z. Tierpsychol.* **56**:305-322.
Gosling, L. M., 1982, A reassessment of the function of scent marking in territories, *Z. Tierpsychol.* **60**:89-118.
Gosling, L. M., 1986, Economic consequences of scent marking in mammalian territoriality, in: *Chemical Signals in Vertebrates IV* (D. Duvall, D. Müller-Schwarze, and R. M. Silverstein, eds.), Plenum Press, New York, pp. 385-395.
Gosling, L. M., 1990, Scent marking by resource holders: alternative mechanisms for advertising the costs of competition, in: *Chemical Signals in Vertebrates V* (D.W. Macdonald, S. Natynczuk, and D. Müller-Schwarze, eds.), Oxford University Press, Oxford, pp. 315-328.
Gosling, L. M., and McKay, H. V., 1990, Competitor assessment by scent matching: an experimental test, *Behav. Ecol. Sociobiol.* **26**:415-420.
Gosling, L. M., and Roberts, S. C., Scent-marking by male mammals: cheat-proof signals to competitors and mates, *Adv. Stud. Behav.* (in press)
Gosling, L. M., Roberts, S. C., Thornton, E. A., and Andrew, M. J., Life history costs of olfactory status signalling in mice, *Behav. Ecol. Sociobiol.* (in press)
Graf, W., 1956, Territorialism in deer, *J. Mammal.* **37**:165-170.
Guilford, T., and Dawkins, M. S., 1991, Receiver psychology and the evolution of animal signals, *Anim. Behav.* **42**:1-14.
Johansson, A., and Liberg, O., 1996, Functional aspects of marking behaviour by male roe deer (*Capreolus capreolus*), *J. Mammal.* **77**:558-567.
Kappeler, P. M., 1998, To whom it may concern: the transmission and function of chemical signals in *Lemur catta*, *Behav. Ecol. Sociobiol.* **42**:411-421.
Kile, T. L., and Marchinton, R. L., 1977, White-tailed deer rubs and scrapes: spatial, temporal and physical characteristics and social role, *Amer. Midl. Nat.* **97**:257-266.
Müller-Schwarze, D. M., 1974, Social functions of various scent glands in certain ungulates and the problems encountered in experimental studies of scent communication, in: *The behaviour of ungulates and its relation to management* (V. Geist, and F.R. Walther, eds.), IUCN, Morges, pp. 107-11.
Paquet, P. C., 1991, Scent marking behaviour of sympatric wolves (*Canis lupus*) and coyotes (*C. latrans*) in Riding Mountain National Park, *Can. J. Zool.* **69**:1721-27.
Peters, R. P., and Mech, L. D., 1975, Scent-marking in wolves, *Amer. Sci.* **63**:628-37.
Ralls, K., 1971, Mammalian scent marking, *Science* **171**:443-449.
Roberts, S. C., 1997, Selection of scent-marking sites by klipspringers (*Oreotragus oreotragus*), *J. Zool.* **243**:555-564.
Roberts, S. C., 1998, Behavioural responses to scent marks of increasing age in klipspringer *Oreotagus oreotragus*, *Ethology* **104**: 585-592.
Roberts, S. C., and Lowen, C., 1997, Optimal patterns of scent marks in klipspringer (*Oreotragus oreotragus*) territories, *J. Zool.* **243**:565-578.
Rowe, C., 1999, Receiver psychology and the evolution of multicomponent signals, *Anim.Behav.* **58**:921-931.
Stenström, D., 1998, Mating behaviour and sexual selection in non-lekking fallow deer (*Dama dama*), PhD Thesis, University of Uppsala.
Walther, F. R., 1978, Mapping the structure and the marking system of a territory of the Thomson's gazelle, *E. Afr. Wildl. J.* **16**:167-176.
Waser, P. M., 1977, *Cercocebus albigenus*: site attachment, avoidance and inter-group spacing, *Am. Nat.* **110**:911-935.

DO CHEMICAL ALARM SIGNALS ENHANCE SURVIVAL OF AQUATIC VERTEBRATES? AN ANALYSIS OF THE CURRENT RESEARCH PARADIGM

Reehan S. Mirza and Douglas P. Chivers

Department of Biology
112 Science Place
University of Saskatchewan
Saskatoon, Saskatchewan, S7N 5E2, Canada

1. INTRODUCTION

Smith (1992, 1999) defines „alarm signalling" as situations in which individuals detect a hazard and produce some response that warns other nearby individuals of potential danger. The individual that emits the signal is the sender and the individual that detects the signal is the receiver. The receiver may respond to the sender's escape response or alternatively a chemical, auditory, mechanical or visual cue released by the sender.

It is common for a wide variety of aquatic organisms, including numerous species of fishes and amphibians, to release chemical cues that may serve as alarm signals (Chivers and Smith, 1998). Chemical cues may be particularly important when visual cues are limited, such as at night, in highly structured habitats or in areas of high turbidity (Smith, 1992). Aquatic media are well suited for chemical signals because a large number of compounds can dissolve in water, giving a large number of potential chemical signals to be detected (Kleerekoper, 1969; Hara, 1994).

Upon detecting chemical cues that are released by prey that are attacked or captured by a predator, alarm signal receivers typically respond with what seems to be appropriate antipredator behavior. Behavioral responses can include dashing, freezing, area avoidance, tighter shoaling, increased shelter use, decreased foraging activity and altered movement patterns (Lima and Dill, 1990; Mathis and Smith, 1993a, b; Chivers *et al.*, 1995). Receivers of alarm signals may also alter their life history patterns (Chivers *et al.*, 1999; Wildy *et al.*, 1999) and/or alter their morphology (Stabell and Lwin, 1997; Mirza, 1998). In this review we restrict our discussion to behavioral responses of aquatic vertebrates in response to chemical alarm cues.

Chivers and Smith (1998) divided chemical alarm signals into two categories based upon what point in the predation sequence the signal is released. Chemical signals may be emitted by the sender if the individual is disturbed or stressed prior to capture by the predator. These types of chemical signals are referred to as disturbance signals. Disturbance signals have been demonstrated in only a few aquatic vertebrates, (two fishes and one amphibian) including Iowa darters, *Etheostoma exile* (Wisenden et al., 1995), brook trout, *Salvelinus fontinalis* (Mirza and Chivers, unpublished data) and red-legged frog tadpoles, *Rana aurora* (Kiesecker et al., 1999). Disturbance signals have received little attention from researchers and the extent to which they are present in aquatic vertebrates or how these signals are used is unknown.

Chemicals released by prey animals only upon being captured by a predator are referred to as damage-released alarm signals (Chivers and Smith, 1998). These signals are thought to be widespread in a variety of aquatic vertebrates. In fishes, damage-released alarm signals are used as a taxonomic characteristic for members of the Superorder *Ostariophysi* (Schutz, 1956; Smith, 1992), which comprise 64% of all freshwater fishes (Nelson, 1994). Damage-released alarm signals are also present in other families of fishes such as salmonids, gobies, poecilids, sticklebacks, percids, cottids, and cichlids (See review: Chivers and Smith, 1998). Damage-released signals have also been documented in several families of anurans including hylids, ranids and bufonids as well as in salamanders (Woody and Mathis, 1997; Summey and Mathis, 1998).

Despite the considerable number of studies documenting that prey animals release chemical cues upon being attacked or captured by a predator, a limited number of studies have explicitly tested whether receivers of alarm signals gain a survival benefit (Chivers and Smith, 1998). Prey animals may respond to alarm cues with what seems to be appropriate antipredator behavior, but it is not clear that changes in behavior translate to greater survival in the presence of a predator. This is particularly problematic if the prey show plasticity in antipredator responses. For example, some researchers have documented that some tadpoles increase activity upon detecting cues of injured conspecifics, while others have documented tadpoles decrease activity. Would both of these responses lead to increased survival of receivers? In this paper, we review the surprisingly limited number of studies that explicitly tested whether detection of alarm cues increases the survival of the receiver. We examined 80 studies documenting responses of fishes and amphibians to chemical alarm cues. Of these studies only a handful explicitly test the survival benefit to receivers. We suggest that it is time for researchers to consider whether this current research paradigm is sufficient.

2. SURVIVAL BENEFIT FOR INDIVIDUALS 'WARNED' BY ALARM SIGNALS

There are several different ways that the benefit to receivers of alarm cues could be manifested. Prey could benefit by altering their behavior to decrease the probability of encountering a predator. Alternatively prey could alter their behavior such that they have a greater probability of escape if they were attacked by a predator (Lima and Dill, 1990). There are documented examples of each of these types of effects.

Mathis and Smith (1993a) staged encounters between fathead minnows and northern pike, *Esox lucius*, to determine whether minnows warned by alarm cues gained a survival benefit in the presence of a predator. Encounters were staged between groups of four fathead minnows and a single pike. Minnows were exposed to either minnow alarm cue or a control of distilled water, prior to interaction with the pike. Each trial ended when the first minnow was captured. Minnows warned by alarm cues survived 39.5 % longer than minnows exposed to the control treatment. There was no difference in predator capture efficiency.

Prey animals may also benefit from detecting alarm cues by evading capture by the predator. Hews (1988) demonstrated that western toad tadpoles, *Bufo boreas*, exposed to conspecific alarm cues were better able to evade predatory dragonfly naiads, *Aeshna umbrosa*. Groups of 24 tadpoles were exposed to either injured conspecific alarm cue or distilled water, and then allowed to interact with the predator. Although there were no significant differences in the latency to capture between the alarm cue and control treatments, the proportion of successful predator attacks was significantly lower when tadpoles had been warned by the alarm cue. Dragonfly larvae had to strike multiple times in order to successfully capture a tadpole.

In a similar study, Mirza and Chivers (unpublished data) found that brook trout were able to evade capture during staged encounters with chain pickerel, *Esox niger*. Groups of 3 brook trout fry were exposed to either damage-released alarm cues from conspecifics or a control of swordtail, *Xiphophorous helleri*, skin extract prior to interactions with pickerel. Swordtails were used as control fish because they are a common aquarium species that are not sympatric with brook trout. There were no differences in latency to capture between treatments, but significantly more pickerel had lower capture success when trout were warned by alarm cues than with swordtail skin extract.

Prey animals may benefit from being warned by alarm cues before they encounter the predator or during the encounter itself. The minnows in the Mathis and Smith (1993a) study decreased their activity upon detecting alarm cues before encountering the pike. This decrease in activity would most likely make it more difficult for the predators to locate their prey. Alternatively, tadpoles in the Hews (1988) study and the brook trout in the Mirza and Chivers (unpublished data) study were better able to evade capture from predators when they were warned by alarm cues. These studies provide evidence that different types of antipredator responses can lead to a survival benefit for receivers.

3. USING ALARM SIGNALS TO TRAIN NAIVE FISHES TO RECOGNIZE PREDATORS

Chemical alarm signals have been shown to be important in facilitating learned recognition of predators. For example, Göz (1941) found that blinded European minnows, *Phoxinus phoxinus*, exhibited antipredator behavior to chemical stimuli from northern pike, only after pike attacked minnows in their presence. Release of alarm cues in the presence of pike, conditioned the blind minnows to respond to the chemical stimuli of pike in later tests. Similarly, Magurran (1989) demonstrated that predator naive European minnows were conditioned to respond to chemical stimuli from pike or non-predatory tilapia, *Tilapia mariae*, by exposure to a paired stimulus of pike or tilapia odor with a conspecific alarm signal. Learned recognition occurs after a single exposure to the paired stimulus.

Learned recognition of predators through conditioning with a chemical alarm signal has also been shown in a variety of other fishes including fathead minnows (Chivers and Smith, 1994a), brook stickleback, *Culaea inconstans* (Chivers *et al.*, 1995), rainbow trout, *Oncorhynchus mykiss* (Brown and Smith, 1998), chinook salmon, *O. tshawytscha* (Berejikian *et al.*, 1999) and brook trout (Mirza and Chivers, unpublished data), in one species of amphibian, redspotted newts, *Notophthalmus viridescens* (Woody and Mathis,

1998) and various aquatic invertebrates (Chivers and Smith, 1998). This suggests that the ability to learn potential predators through conditioning with alarm signals may be a widespread phenomenon.

Researchers examining the importance of alarm signals in learned predator recognition have speculated that prey animals conditioned to recognize predators should have higher survival during subsequent encounters with the predator. Prey animals that are better able to recognize predators will be better able to assess predation risk; hence survival should be higher for these individuals. However, only two studies have examined whether prey trained to recognize predators with alarm signals gain a survival benefit during encounters with predators. The results of these studies are somewhat mixed. Additional tests are clearly needed.

Berejikian *et al.* (1999) tested survival of predator-trained fish in two experiments. First, they conditioned chinook salmon smolts to recognize cutthroat trout, *Oncorhynchus clarkii*, and then allowed interaction with live predators. Predator-trained and non-trained salmon smolts were stocked into two parallel 25 X 3.1-m hatchery raceways containing predatory cutthroat trout. Salmon were allowed to interact with the trout for 6 days. Berejikian *et al.* (1999) found no significant difference in survival rates between predator-trained and non-trained fish in this experiment. In a second experiment, Berejikian *et al.* (1999) conditioned chinook salmon to recognize cutthroat trout, and then released them along with non-trained smolts into a small creek where cutthroat trout were the primary predators. Fish were recaptured at a weir 21-km downstream over a period of 76 days. Chinook smolts conditioned to recognize cutthroat trout had higher survival than controls. However, the rearing environment of the smolts influenced the result. The survival effect was observed if the smolts were raised in a complex habitat but not a simple habitat.

Mirza and Chivers (unpublished data) conducted a series of experiments to determine whether training naive brook trout fry to recognize predatory chain pickerel enhanced their survival. Trout were exposed to either conspecific alarm cues or a control of swordtail skin extract combined with pickerel odor prior to staging encounters with predators in 500-L aquaria. Predator capture success rate was significantly lower for trout conditioned with conspecific alarm cues. In a second laboratory experiment, brook trout were tested in mixed groups of 2 predator-trained and 2 non-trained fish and encounters staged with chain pickerel. Pickerel captured significantly more non-trained fish than trained fish. In a third experiment mixed groups of predator-trained and non-trained brook trout fry were placed in enclosures with predators, within a small steam. A single pickerel was placed within each enclosure (1.5 x 1.5 x 1.5-m) along with 10 predator-trained and 10 non-trained brook trout and allowed to interact for 4 hrs. Brook trout trained to recognize pickerel had a significantly higher survival rate than non-trained fish.

When placed in a confined environment, prey may be able to recognize predators and respond, but their responses may be limited. This may severely limit the utility of laboratory studies to test whether training enhances survival. Mirza and Chivers (unpublished data) showed that pickerel had to strike significantly more often in order to capture brook trout that were trained than those that were non-trained. Although brook trout were able to better evade capture after being trained to recognize pickerel, a trout was captured in almost every trial. If the trials had been conducted outside the confines of the aquaria, trout may have been able to evade the pickerel completely and flee the area. The predator-trained chinook smolts in the Berejikian *et al.* (1999) study were also in a confined environment in the raceways. The trained fish may have been better able to recognize predators, but were not able to evade predators indefinitely. This could explain why there were no significant differences in survival rates between trained and non-trained smolts. Additional studies are needed in order to address whether prey trained to recognize predators with alarm signals have higher survival than those that are not trained. We suggest that these studies need to be conducted on a large scale such that prey have the opportunity to escape.

4. DISCUSSION

Aquatic vertebrates respond to alarm cues with a diversity of behavioral responses. The responses to the alarm cues may decrease the probability of being detected by the predator or else may alter the ability of the prey to evade the predator (Lima and Dill, 1990). Several studies have documented behavioral responses to alarm cues, but have not documented whether detection of alarm cues translates into a direct survival benefit for receivers (Chivers and Smith, 1998). Instead, general inferences are made concerning the survival benefit for receivers. In the previously mentioned studies it has been shown that prey animals do benefit from detection of alarm cues or being conditioned to recognize predators with alarm cues. Prey animals increased survival by making themselves harder to detect via decreased activity, or else more difficult to capture by evading the predator during an encounter. Thus 'live predation' trials allow researchers to verify the survival benefit for receivers.

Conducting live predation trials may clarify conflicting interpretations of experimental results. For example, Hews and Blaustein (1985) document that Cascade frog tadpoles (*Rana cascadae*) respond with an increase in activity to injured conspecific cues and conclude the response is an alarm reaction. Hokit and Blaustein (1995) also show an increase in activity in Cascade frog tadpoles exposed to injured conspecific cues, but interpret their findings as a feeding response, not an alarm reaction. An increase in activity may be consistent with appropriate antipredator behavior, but conflicting interpretations in the same species do not allow for any conclusions to be drawn. Live predation trials may allow researchers to verify the presence or absence of an alarm response in Cascade frog tadpoles or other organisms where conflicting interpretations are found.

The use of live predation trials may also clarify seemingly conflicting results in closely related animals. Studies of larval anurans provide a good example for discussion. Some researchers have documented that tadpoles increase activity upon detecting alarm cues released by injured conspecifics (e.g. Hews and Blaustein, 1985; Hokit and Blaustein, 1995) while other researchers have shown that tadpoles decrease activity upon detecting alarm cues (e.g. Wilson and Lefcort, 1993). Do both of these types of responses lead to an increase in survival? By decreasing activity tadpoles may avoid detection by the predator. By increasing activity the prey may be able to flee from the vicinity of a predation event. Both of these could be adaptive responses. We suggest that it is critical for researchers to stage encounters between predators and prey that are warned by alarm signals to determine whether different types of responses lead to reduced capture by predators. Perhaps increases in activity are feeding responses and do not represent alarm responses. We also emphasize that researchers need to incorporate several different types of predators into their studies. An appropriate behavioural response to alarm cues in the presence of one type of predator may be ineffective against another type of predator.

There is some controversy whether some species of fishes release cues that act as alarm signals. Live predation trials may be useful in determining whether prey animals are releasing and responding to alarm cues. Magurran *et al.* (1996) suggests that if no overt behavioral response in the prey animal is observed, then no detection of alarm cues by the prey occurs. However, Brown and Smith (1996) showed that fathead minnows food-deprived for 24 hrs do not show an overt behavioral response to a paired stimulus of conspecific alarm cue and pike odor. When fed to satiation and subsequently tested to pike odor alone days later, minnows that had previously shown no response, responded to pike odor with antipredator behavior. This suggests that prey animals can learn to recognize predators by detecting alarm signals even if they do not exhibit an overt behavioral response to the signals. Live predation trials would allow researchers to determine whether prey animals are detecting alarm cues by examining survival and not overt behavioral responses.

Prey animals respond to damage-released alarm cues from members of the same prey guild or species that are closely related (Schutz, 1956; Chivers and Smith, 1998). For example, fathead minnows and brook stickleback that co-occur in the same habitat both respond with antipredator behavior to damaged-released cues from the other species (Mathis and Smith, 1993b; Chivers and Smith, 1994b). The ability to detect alarm cues from heterospecifics should enhance survival of receivers. Survival of prey animals to heterospecific alarm cues from the same prey guild or closely related taxa have not been tested.

Another area that has received little attention is that of disturbance cues. Disturbance cues have been interpreted as low intensity indicators of threat and the extent to which these types of alarm cues exist in aquatic vertebrates is unknown (Kiesecker *et al.*, 1999). More-over, in those species in which disturbance cues have been documented, no tests of survival have been conducted. Live predation trials would give an indication as to the nature of disturbance cues and if these cues are used in a similar manner to damage-released cues.

5. CONCLUSION

The benefit to receivers of alarm signals has generally been inferred from the observation that responses to alarm cues are consistent with known anti-predator responses. There is some evidence to support this conclusion. However, we ask whether we should be comfortable continuing with this research paradigm. Should we make conclusions that thousands of species of fishes and amphibians release alarm cues that function to warn other conspecifics of danger when so few experiments have directly tested whether the receivers gain a survival benefit from the detection of the cues? Conducting live predation trials is obviously more difficult and perhaps there is a general reluctance to conduct live predation trials as a result of animal care concerns. Nevertheless, we argue, as did Chivers and Smith (1998), that it is critical to rigorously test the assumption that alarm signal receivers benefit from detection of the signals.

REFERENCES

Berejikian, B. A., Smith, R. J. F., Tezak, E. P., Schroder, S. L., and Knudsen, C. M., 1999, Chemical alarm signals and complex hatchery rearing habitats affect antipredator behavior and survival of chinook salmon (*Oncorhynchus tshawytscha*) juveniles, *Can. J. Fish. Aquat. Sci.* **56**:830-838.

Brown, G. E., and Smith, R. J. F., 1996, Foraging trade-offs in fathead minnows (*Pimephales promelas*, Osteichthyes, Cyprinidae): Acquired predator recognition in the absence of an alarm response, *Ethology* **102**:776-785.
Brown, G. E., and Smith, R. J. F., 1998, Acquired predator recognition in juvenile rainbow trout (*Oncorhynchus mykiss*): conditioning hatchery-reared fish to recognise chemical cues of a predator, *Can. J. Fish. Aquat. Sci.* **55**:611-617.
Chivers, D. P., and Smith, R. J. F., 1994a, The role of experience and chemical alarm signalling in predator recognition by fathead minnows, *Pimephales promelas*, *J. Fish Biol.* **44**:273-285.
Chivers, D. P., and Smith, R. J. F., 1994b, Intra- and interspecific avoidance of areas marked with skin extract from brook sticklebacks (*Culaea inconstans*) in a natural habitat, *J. Chem. Ecol.* **20**:1517-1524.
Chivers, D. P., and Smith, R. J. F., 1998, Chemical alarm signalling in aquatic predator-prey systems: a review and prospectus, *Écoscience* **5**:338-352.
Chivers, D. P., Brown, G. E., and Smith, R. J. F., 1995, Acquired recognition of chemical stimuli from pike, *Esox lucius*, by brook stickleback, *Culaea inconstans* (Osteichthyes, Gasterosteidae), *Ethology* **99**:234-242.
Chivers, D. P., Kiesecker, J. M., Marco, A., Wildy, E. L., and Blaustein, A. R., 1999, Shifts in life history as a response to predation in western toads (*Bufo boreas*), *J. Chem. Ecol.* **25**:2455-2463.
Göz, H., 1941, Über den Art-und Individualgeruch bei Fischen, *Z. Vergl. Physiol.* **29**:723-730.
Hara, T. J., 1994, The diversity of chemical stimulation in fish olfaction and gustation, *Rev. Fish Biol. Fish.* **4**:1-35.
Hews, D. K., 1988, Alarm response in larval western toads, *Bufo boreas*: release of larval chemicals by a natural predator and its effect on predator capture efficiency, *Anim. Behav.* **36**:125-133.
Hews, D. K., and Blaustein, A. R., 1985, An investigation of the alarm response in *Bufo boreas* and *Rana cascadae* tadpoles, *Behav. Neural Biol.* **43**:47-57.
Hokit, D. G., and Blaustein, A. R., 1995, Predator avoidance and alarm response behaviour in kin-discriminating tadpoles, *Ethology* **101**:280-290.
Kiesecker, J. M., Chivers, D. P., Marco, A., Quilchano, C., Anderson, M. T., and Blaustein, A. R., 1999, Identification of a disturbance signal in larval red-legged frogs, *Rana aurora*, *Anim. Behav.* **57**:1295-1300.
Kleerekoper, H. A., 1969, *Olfaction in Fishes*, Bloomington, Indiana University Press.
Lima, S. L., and Dill, L. M., 1990, Behavioral decisions made under the risk of predation: a review and prospectus, *Can. J. Zool.* **68**:619-640.
Magurran, A. E., 1989, Acquired recognition of predator odour in the European minnow (*Phoxinus phoxinus*), *Ethology* **82**:216-233.
Magurran, A. E., Irving, P. W., and Henderson, P. A., 1996, Is there a fish alarm pheromone? A wild study and critique, *Proc. Royal Soc. Lond.* **263**:1551-1556.
Mathis, A., and Smith, R. J. F., 1993a, Chemical alarm signals increase the survival time of fathead minnows (*Pimephales promelas*) during encounters with northern pike (*Esox lucius*), *Behav. Ecol.* **4**:260-265.
Mathis, A., and Smith, R. J. F., 1993b, Intraspecific and cross-superorder responses to chemical alarm signals by brook stickleback, *Ecology* **74**:2395-2404.
Mirza, R. S., 1998, *Induced morphological changes in fishes mediated by chemical stimuli associated with predation*, M.Sc. thesis, University of Saskatchewan, Saskatoon, Canada.
Nelson, J. S., 1994, *Fishes of the World*. 3rd ed., John Wiley and Sons, New York.
Schutz, F., 1956, Vergleichende Untersuchungen über die Schrekreaktion bei Fischen und deren Verbreitung, *Z. Vergl. Physiol.* **38**:84-135.
Smith, R. J. F., 1992, Alarm signals in fishes, *Rev. Fish Biol. Fish.* **2**:3-63.
Smith, R. J. F., 1999, What good is smelly stuff in the skin? Cross function and cross taxa effects in fish „alarm substances", in: *Advances in Chemical Signals on Vertebrates*, (R. E. Johnston, D. Müller-Schwarze, and P.W. Sorensen, eds.), Plenum Press, New York, pp. 475-487.
Stabell, O. B., and Lwin, M. S., 1997, Predator-induced phenotypic changes in crucian carp are caused by chemical signals from conspecifics, *Env. Biol. Fish.* **49**:145-149.
Summey, M.-R., and Mathis, A., 1998, Alarm responses to chemical stimuli from damaged conspecifics by larval anurans: Tests of three neotropical species, *Herpetologica* **54**:402-408.
Wildy, E. L., Chivers, D. P., and Blaustein, A. R., 1999, Shifts in life-history traits as a response to cannibalism in larval long-toed salamanders (*Ambystoma macrodactylum*), *J. Chem. Ecol.* **25**:2337-2346.
Wilson, D. J., and Lefcort, H., 1993, The effect of predator diet on the alarm response of red-legged frog, *Rana aurora*, tadpoles, *Anim. Behav.* **46**:1017-1019.
Wisenden, B. D., Chivers, D. P., and Smith, R. J. F., 1995, Early warning in the predation sequence: a disturbance pheromone in Iowa darters (*Etheostoma exile*), *J. Chem. Ecol.* **221**:1469-1480.

Woody, D. R., and Mathis, A., 1997, Avoidance of areas labeled with chemical stimuli from damaged conspecifics by adult newts, *Notophthalmus viridescens*, in a natural habitat, *J. Herpetol.* **31**:316-318.

Woody, D. R., and Mathis, A., 1998, Acquired recognition of chemical stimuli from an unfamiliar predator: Associative learning by adult newts, *Notophthalmus viridescens*, *Copeia* **1998**:1027-1031.

MECHANISMS OF OLFACTORY FORAGING BY ANTARCTIC PROCELLIIFORM SEABIRDS

Gabrielle A. Nevitt

Section of Neurobiology, Physiology and Behavior
Department of Biological Sciences
1 Shields Avenue
University of California
Davis, CA 95616

INTRODUCTION

Procellariiform or „tube-nose" seabirds - the petrels, albatrosses and shearwaters - have been a curiosity to sailors and biologists alike for centuries. Their typical life history is somewhat unusual among birds: most procellariiform seabirds spend nearly all of their lives in flight over the ocean and come to shore only for a few months each year to breed. Procellariiform seabirds share common adaptations for this lifestyle. Many species have extremely efficient flight styles, and an excellent sense of smell. These abilities aid them in locating patchily distributed prey. In addition, these birds tend to be long-lived, highly philopatric, and seem to have a sophisticated spatial knowledge of their foraging habitat (reviewed by Warham, 1996). With the application of satellite telemetry to studies of the foraging ecology of many members of this order, it is now well established that many species routinely forage over distances ranging from hundreds to thousands of kilometers (reviewed by Weimerskirch, 1998). Understanding how these birds are able to accomplish this task has been a primary focus of investigation in my laboratory for the last five years.

Based on a combination of experimental and empirical observation of natural odor profiles measured at sea, I have recently presented a new conceptual model for how procellariiform seabirds use olfaction to forage (Nevitt, 2000). This paper will review the basic features of this model, primarily with respect to work that has been conducted in the Southern oceans. While this report is not intended to be a comprehensive review of the literature on this subject, my hope is that it will generate new thinking in this field, and provide a framework for further investigation of these fascinating birds.

BACKGROUND NATURAL HISTORY

Most of my work has been done in the Southern oceans near South Georgia Island (54°S, 36°W). South Georgia and the surrounding habitat is populated by 28 species of seabirds making up a population of about 30 million individuals (Croxall and Prince, 1980). Other animals that inhabit this region include a variety of pinnipeds (most notably Antarctic Fur Seals, *Arctocehalus gazella*) and cetaceans. Nearly all of these species rely to some extent on Antarctic krill (*Euphausia superba*), though procellariiform species differ substantially in the amount of krill they consume as compared to other food resources (e.g., fish and squid).

While Antarctic krill is abundant in the diets of many procellariiform seabirds, little is known about how these birds are able to exploit this and other prey resources so efficiently. Detailed observational studies from ships have been conducted on mixed-species feeding aggregations that form in association with inshore patches of krill near the island (Harrison *et al.*, 1991). This work suggests that aggregations are typically dominated by several species of procellariiform seabirds (particularly Black-browed albatrosses *Diomedea melanophris*, and prions *Pachyptila sp.*) and include penguins and fur seals in some cases. A common speculation is that these large, mixed-species feeding flocks form in association with krill swarms. These krill swarms either aggregate near the surface or are forced there by seals and penguins foraging from below. Other behavioral strategies for foraging include scavenging (Croxall and Prince, 1994), or foraging at night when krill tends to be closest to the surface of the ocean (e.g., Croxall and Prince, 1980).

How the sensory abilities and complex inter-specific behavioral interactions allow procellariiforms to effectively exploit scattered prey patches are still open questions. An even more daunting challenge is presented by data collected from satellite tracking studies. These studies show that, in addition to foraging inshore, several species of albatross (most notably Black-browed *D. melanophris*, Gray-headed *D. chrysostoma*, and Wandering Albatross *D. exulans*) routinely travel thousands of kilometers to forage in productive areas of ocean during incubation shifts or while provisioning chicks on the nest (reviewed by Prince *et al.*, 1998). How these birds know where to go to forage and the mechanisms they use to navigate to these productive areas is not known.

A NEW MODEL

Most research to date has focused on establishing that procellariiform seabirds have a sense of smell (reviewed by Warham, 1996). We now have the ability to begin to integrate these data with a working knowledge of diet, foraging ecology and distribution, as well as atmospheric chemistry. The emerging picture suggests that procellariiform seabirds use odors in different contexts to forage at different spatial scales.

Large Scales: Recognizing the Productive Area

Over large scales (hundreds or thousands of kilometers) I suggest that procellariiform seabirds use odor cues as landscape indicators that alert them that they have arrived at a productive area to forage (Nevitt *et al.*, 1995; Nevitt, 1999a; Nevitt, 2000). This behavior does not imply that these birds track changes in odor gradients over these distances to locate a likely feeding target, but that potentially rich feeding grounds share an olfactory signature that the long distance forager recognizes upon arrival (Figure 1).

Small Scales: Locating the Prey Patch using Area-Restricted Search

Once the seabird has arrived at a productive area, the problem becomes one of locating exploitable prey patches. A change in the large-scale olfactory landscape thus may trigger seabirds to begin a relatively small-scale „area-restricted" search of the region using a combination of visual, olfactory and other sensory cues to pinpoint a prey patch (described in Nevitt and Veit, 1999). Some species may zigzag upwind to focus activity near the source of an odor plume (e.g., Hutchison and Wenzel, 1980) while others are likely to use visual cues to locate prey patches, either by spotting prey directly or by seeing aggregations of other foraging seabirds alighting on the water (Figure 2). Area-restricted search is likely to be used at in-shore feeding locations as well, and to be influenced by the formation of mixed-species feeding aggregations (Nevitt and Veit, 1999; Nevitt, 1999b).

This conceptual model thus provides a mechanistic framework whereby hypotheses about the sensory ecology of foraging can be addressed experimentally in the field.

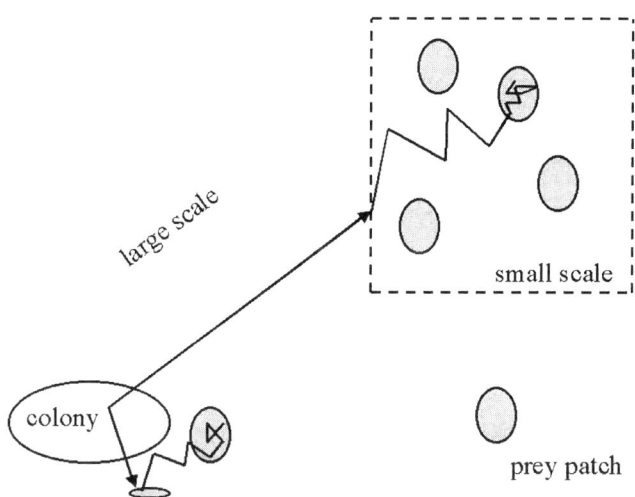

Figure 1. Behavioral predictions for foraging at large scales. Shaded areas indicated prey patches occurring inshore or distant to the colony. The dotted square indicates a productive area. Solid lines indicated the theoretical flight trajectory of the seabird. As the seabird arrives at the foraging area, the changing olfactory landscape is predicted to trigger an area-restricted search. Elements are not drawn to scale.

WHAT IS THE CURRENT EVIDENCE FOR THIS MODEL?

A key to unraveling this problem has been in identifying what specific odors different species can detect and in determining how these odor compounds are associated spatially and temporally with prey patches. Unfortunately, working at sea typically carries with it some restrictions that make detailed behavioral studies difficult do (Junger, 1997). Rather than having strict control to direct where or when research is to be conducted, researchers often have to make due with using research ships as platforms of opportunity. Months at sea may be required to produce a single, controlled experiment (e.g., Nevitt et al., 1995; Nevitt,

1999b). These limitations need to be considered when interpreting results of various studies, particularly in terms of the degree to which experimental designs are rigorously controlled.

Presenting seabirds with small (1-2 liter) scented vegetable oil slicks produces repeatable data if presentation times are limited to ~20 minutes and presentations are paired with controlled, unscented slicks. Working at South Georgia, we have found that after 20 minutes, visual cueing by birds aggregating at slicks becomes problematic in interpreting data. Birds can also be negatively conditioned to repeated presentations of the same odor (Nevitt, unpublished). Alternatively, odors can be presented using wicks or odor-impregnated sponges mounted on buoys (e.g., Grubb, 1972; Hutchison and Wenzel, 1980; Lequette *et al.*, 1989; Nevitt and Hunt, 1996), or as aerosols (Nevitt *et al.*, 1995). In the later case, step length (turning rate) has been used as an indicator of area restricted search mediated via olfaction (Nevitt *et al.*, 1995; see also Nevitt and Veit, 1999).

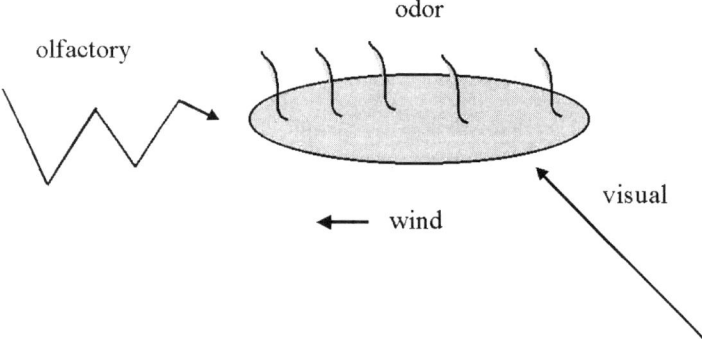

Figure 2. Behavioral predictions for area-restricted search at small scales. Seabirds using olfactory cues should turn upwind to focus activity near the source of the odor plume. Seabirds using visual cues will bee-line to the source irrespective of wind direction. Visual cues are likely to include foraging activity of other individuals.

Results from these sorts of controlled experiments performed at sea near Elephant and South Georgia Islands indicate that different species of Antarctic procellariiforms are sensitive to a variety of natural, scented compounds associated with prey. These include fishy-smelling odors (Nevitt *et al.*, 1995; see also Grubb, 1972; Hutchison and Wenzel, 1980; Lequette *et al.*, 1989), macerated krill and related aromatics (Nevitt, 1994, 1999b) as well as biogenic sulfurous compounds associated with phytoplankton (dimethyl sulfide, DMS; Nevitt *et al.*, 1995) and krill (Daly and DiTullio, 1996).

While little information is available about the natural distribution of fish and krill-related aromatics over the ocean, patterns of DMS and other biogenic sulfurous compounds have been studied because of their relevance to global climate regulation. The relevance of

DMS as a foraging cue for procellariiform seabirds has been described in detail elsewhere (Nevitt et al., 1995; Nevitt, 1999a; 2000). In brief, DMS is a byproduct of the metabolic decomposition of dimethylsulfionopropionate (DMSP) in marine phytoplankton. DMS dissolved in seawater is rapidly transferred to the atmosphere and serves as an indicator of high primary productivity (McTaggart and Burton, 1992). Large-scale DMS features can persist for several days (Berresheim, 1987), suggesting that highly productive areas smell differently than less productive areas to foraging procellariiforms and could serve as indicators for beginning an area-restricted search (Nevitt et al., 1995; Nevitt and Veit, 1999; Nevitt, 2000).

WHERE DO WE GO FROM HERE?

Large Scales: Remote Monitoring of Behaviors

While recent advances in satellite telemetry and remote behavioral monitoring have done much to advance our understanding of the foraging ecology and macro-scale distribution of several procellariiform species (particularly albatrosses), these technologies have not yet been applied to sensory questions. This is obviously an exciting area of future research and offers many avenues for further exploration. Remote instrumentation can provide instantaneous measures of global position, turning rate, feeding activity and wind velocity – parameters that will contribute to establishing behavioral algorithms for tracking odor sources over the ocean (e.g., Weimerskirch and Wilson, 1992; Weimerskirch et al., 1995; Wilson et al., 1995). While micro-sensors for specific biogenic odors are not currently available, this technology coupled to precision online measurements of behavioral tracking will be an exciting area to implement in the not too distant future (reviewed by Nevitt, 2000).

Small Scales: Defining Sensory Constraints on Area-Restricted Search Strategies

At smaller scales, sensory parameters that dictate area-restricted search strategies need to be better defined. Procellariiforms are likely to use a combination of olfactory and visual cues to locate prey patches (reviewed by Nevitt and Veit, 1999), but the relative advantages of using an odor-based search for foraging has not been rigorously studied from the point of view of odor transport models.

According to these models, odors in atmospheric flow tend to be dispersed laterally and downwind in turbulent plumes. Turbulence transports odorant-containing parcels of air in the form of eddies, which retain their coherence as they are transported long distances from the source (Tennekes and Lumley, 1982; Schetz, 1984). Because turbulent transport processes operate more rapidly than molecular diffusion can dissipate the eddies, plumes of airborne odors have an irregular, patchy, concentration distribution. In particular, the instantaneous concentration distribution meanders laterally under the influence of larger scale horizontal eddies (Kristensen et al., 1981). Thus, the edges of plumes are marked not only by low average concentrations, but also by strong concentration fluctuations. These fluctuations mean that even in a part of the plume in which an average odorant concentration is low, the seabird might still occasionally encounter detectable, high-concentration eddies (see also Weissburg, 2000).

Tracking the distribution of high concentration eddies over the ocean to a source is thus a complex problem (for review, see Vickers, 2000) and probably one that varies considerably between species depending on their olfactory abilities and flight styles (Pennycuick, 1982; 1987). For example, dynamic soarers, such as albatrosses, might encounter high-concentration eddies at different rates than storm-petrels pattering across the surface of the water. While such differences may contribute to the unique foraging strategies they employ (Nevitt, 1999b), these questions are difficult to approach experimentally in the field. As a first approximation, correlating species to odor distributions in nature has been useful in establishing that some species associate with elevated levels of particular scented compounds while foraging (Nevitt, 2000). However, in interpreting these data, it is critical to consider that empirical measurements of chemical concentrations in the field are usually made from fixed sensors positioned several meters above the surface of the ocean – not where a storm-petrel, for example, is foraging. These measurements also typically represent averaged air samples, so much of the fine-scale variability in odor concentration that the foraging seabird would normally encounter is likely to be lost. (see Bates *et al.*, 1992; Nevitt 2000).

With a clearer picture of the behavioral mechanisms different species use to track odors, it will be interesting to reconsider how different olfactory, visual or other sensory modalities mediate intra- and inter-specific interactions between foraging seabirds and other animals. For example, species that are highly vulnerable to predation (e.g., storm-petrels) are also highly attracted to DMS, whereas more aggressive species such as Cape petrels (*Daption capense*) and Black-browed albatrosses (*D. melanophris*) respond more readily to scents associated with macerated krill. These differences in behaviors may reflect different adaptations for exploiting mixed-species feeding aggregations (discussed in Nevitt, 1999b, 2000). Whether these species have different physiological sensitivities to these suites of potential odor cues will be a rich area to explore.

ACKNOWLEDGMENTS

This work was supported in part by The National Geographic Society and the National Science Foundation (OPP #9615061 and OPP #9814326 to GAN). Thanks also to the captains, officers, scientists and crews of the R/V Polar Duke, R/V Surveyor, R/V Nathaniel B. Palmer, RRS James Clark Ross, and the British Antarctic Survey for logistical support at South Georgia. The manuscript was much improved by the comments of G. Cunningham, S. Lema and J. Watters. Additional thanks to D. Grunbaum for his many discussions and helpful insights about odor transport.

REFERENCES

Bates, T. S., Calhoun, J. A., and Quinn, P. K., 1992, Variations in the methanesulfonate to sulfate molar ratio in submicrometer marine aerosol particles over the South Pacific ocean, *J. geophys. Res.* **97**:9859-9865.
Berregsheim, H., 1987, Biogenic sulfur emissions from the Subantarctic and Antarctic oceans, *J. geophys. Res.* 92: **13**:245-13,262.
Croxall, J. P., and Prince, P. A., 1980, Food, feeding ecology and ecological segregation of seabirds at South Georgia, *Biol. J. Linn. Soc.* **14**:103-131.
Croxall, J. P., and Prince, P. A., 1994, Dead or alive, night or day: How do albatrosses catch squid? *Antarct. Sci.* **6**:155-162.
Daly, K. L., and DiTullio, G. R., 1996, Particulate dimethylsoniopropionate removal and dimethyl sulfide production by zooplankton in the Southern Ocean, in: *Biological and Environmental Chemistry of DMSP and Related Sulfonuim Compounds* (R. P. Kiene, P. T. Visscher, M. D. Kellor, and G. O. Kirst, eds.), Plenum Press, New York, pp. 223-238.

Grubb, T. C., 1972, Smelling and foraging in petrels and shearwaters, *Nature* **237**:404-405.
Harrison, N. M., Whitehouse, M. J., Heinemann, D., Prince, P. A., Hunt, G. L. Jr., and Veit, R. R., 1991, Observations of multispecies seabird flocks around South Georgia, *Auk* **108**:801-810.
Hutchison, L. V., and Wenzel, B. M., 1980, Olfactory guidance in foraging by Procellariiforms, *Condor* **82**:314-319.
Junger, S., 1997, *The Perfect Storm*, W. W. Norton and Company, New York.
Kristensen, L., Jensen, N. O., and Petersen, E. L., 1981, Lateral dispersion of pollutants in a very stable atmosphere – the effect of meandering, *Atmos. Env.* **15**:837-844.
Lequette, B., Verheyden, C., and Jouventin, P., 1989, Olfaction in subantarctic seabirds: Its phylogenetic and ecological significance, *Condor* **91**:732-735.
McTaggart, A. R., and Burton, H., 1992, Dimethyl sulfide concentrations in the surface waters of the Australasian Antarctic and Subantarctic oceans during an Austral summer, *J. geophys. Res.* **97**:14407-14412.
Nevitt, G. A., 1994, Evidence that Antarctic procellariiform seabirds can smell krill, *Ant. J. U.S. 1994 Review Issue*, **29**(5):168-169.
Nevitt, G. A., 1999a, Foraging by seabirds on an olfactory landscape, *Amer. Sci.* **87**:46-53.
Nevitt, G. A., 1999b, Olfactory foraging in Antarctic seabirds: A species-specific attraction to krill odors, *Mar. Ecol. Prog. Ser.* **177**:235-241.
Nevitt, G. A., 2000, Olfactory foraging by Antarctic procellariiform seabirds: life at high Reynolds numbers, *Bio. Bull.* **196**:245-253.
Nevitt, G. A., and Hunt, G. L., 1996, Olfactory sensitivities of foraging procellariid seabirds in the Aleutian Islands *Chem. Senses,* **21**:649-650.
Nevitt, G. A., and Veit, R. R., 1999, Mechanisms of prey patch detection by foraging seabirds, in: *Proceedings of the 22nd International Ornithological Congress* (N. J. Adams, and R. H. Slotow, eds.), BirdLife-South Africa, Johannesburg, pp. 2072-2082.
Nevitt, G. A., Veit, R. R., and Kareiva, P. M., 1995, Dimethyl sulphide as a foraging cue for Antarctic Procellariiform seabirds, *Nature* **376**:680-682.
Pennycuick, C. J., 1982, The flight of petrels and albatrosses (Procellariiformes) observed in South Georgia and its vicinity, *Phil. Trans. R. Soc. Lond. B.* **300**:75-106.
Pennycuick, C. J., 1987, Flight of seabirds, in: *Seabirds: Feeding Biology and Role in Marine Ecosystems* (J. P. Croxall, ed.), Cambridge University Press, Cambridge, pp. 43-62.
Prince, P. A., Croxall, J. P., Trathan, P. N., and Wood, A. G., 1998, The pelagic distribution of South Georgia albatrosses and their relationships with fisheries, in: *Albatross Biology and Conservation* (G. Robertson, and R. Gales, eds.), Surrey Beatty and Sons, Sydney, pp. 137-167.
Schetz, J. A., 1984, *Foundations of Boundary Layer Theory for Momentum, Heat, and Mass Transfer,* Prentice-Hall, Inc. Englewood Cliffs.
Tennekes, H., and Lumley, J. L., 1982, *A First Course in Turbulence*, MIT Press, Cambridge.
Vickers, N. J., 2000, Mechanisms of animal navigation in odor plumes, *Bio. Bull.* **198**:203-212.
Warham, J., 1996, *The Behavior, Population Biology and Physiology of the Petrels,* Academic Press, London.
Weimerskirch, H., 1998, Foraging strategies of southern albatrosses and their relationship with fisheries, in: *Albatross biology and conservation* (G. Robertson, and R. Gales, eds.), Surrey Beatty and Sons, Sydney, pp. 168-179.
Weimerskirch, H., Jouventin, P., and Stahl, J. C., 1986, Comparative ecology of the six albatross species breeding on the Crozet Islands, *Ibis* **128**:195-213.
Weimerskirch, H., and Wilson, R. P., 1992, When do wandering albatrosses *Diomedea exulans* forage? *Mar. Ecol. Prog. Ser.* **86**:297-300.
Weimerskirch, H., Wilson, R. P., Guinet, J. Koudil, C., 1995, Use of seabirds to monitor sea-surface temperatures and to validate satellite remote sensing measurements in the southern ocean, *Mar. Ecol. Prog. Ser.* **126**:299-303.
Weissburg, M. J., 2000, The fluid dynamical context of chemosensory behavior, *Bio. Bull.* **198**:188-202.
Wilson, R. P., Weimerskirch, H., and Lys, P., 1995, A device for measuring seabird activity at sea, *J. Avian Biol.* **26**:172-176.

ECOLOGICAL ASPECTS OF HOUSE MOUSE URINARY CHEMOSIGNALS

Lee C. Drickamer

Department of Biological Sciences
Northern Arizona University
Flagstaff, AZ 86011-5640

INTRODUCTION

Many aspects of house mouse urinary chemosignals have been tested under laboratory conditions over the past 45 years. These phenomena include puberty acceleration, puberty delay, and signals that influence estrus cycles and pregnancy in females (Whitten, 1956; Bruce, 1959; Vandenbergh, 1969; Drickamer, 1977). Some of these effects have been tested under field conditions (Massey and Vandenbergh, 1980, 1981; Coppola and Vandenbergh 1987; Drickamer and Mikesic, 1990; and see reviews by Brown, 1985; Drickamer, 1986; Vandenbergh and Coppola, 1986). Additional findings regarding the fact that phenomena that occur under laboratory conditions do occur in natural or semi-natural settings are needed to provide a more complete picture of the manner in which house mice, and possibly other rodents, use odor cues for communication.

My review summarizes recent findings with respect to four topics concerning the use of urinary odors as significant communication cues in house mice: (1) differential responses to odors in traps as an indication that mice can detect and use odor cues from conspecifics, (2) effects of odors relating to male social dominance status on capture rates of all mice and of socially dominant and subordinate males in particular, (3) odor preferences of female mice that could be related to mate finding and mate selection, and (4) changes in odor preferences with age in young female mice that may have consequences for lifetime reproductive success. My major theme is that it is possible to translate to the field the various laboratory-based findings about odor communication in house mice.

DIFFERENTIAL CAPTURES

Laboratory studies have revealed that house mice (*Mus musculus domesticus*) exhibit differential responses to odor cues based on age, sex, genetics, and individual identity

(Yamazaki *et al.*, 1976; Bowers and Alexander, 1967; Lenington, *et al.*, 1988). The notion that responses to live traps could be influenced by odors contained in those traps remaining from previous occupants was first put forward with respect to voles (*Microtus* sp.) (Boonstra and Krebs, 1976) and deer mice (*Peromyscus* sp.) (Mazdzer *et al.*, 1976). Others have also noted that soiled traps capture conspecifics more than clean traps (Stoddart, 1982; Gurnell and Little, 1992).

I used the odor-baited trap technique to ask two types of questions about potential differential responses to odors in live traps by house mice living in outdoor enclosures (Drickamer, 1995). I was trapping wild house mice on two non-consecutive nights per week for 30 weeks each summer in a series of 0.1 ha enclosures (see Drickamer and Mikesic, 1990 for enclosure details). I assumed that a trap contained a mixture of odors from previous captures during a field season and that the most recent capture provided the most salient odor cue for a period of up to one week. A trap that had not caught a mouse for a week was considered to be not odorized.

Using the odor-baited trap technique it was possible to ask two types of questions: (1) Were there any biases in trap responses contingent upon the previous capture? and (2) Were there any significant patterns of differential responses of mice of particular age, sex, or reproductive condition to trap odors? The first is a methods question and the latter is related to the biological importance of the odors to the mice.

With respect to the methods question I found few overall differences. There was a tendency for traps odorized by juvenile females to capture more adult females than expected, but there were no consistent patterns of differential responses. These findings lead to the conclusion that from a methods perspective there actually should be little concern about the potential for biases in field trapping due to the use of soiled traps.

With respect to the biological significance of the odors, I found several distinct patterns of responses that can be interpreted with regard to house mouse social biology. Adult male mice differentially selected traps with odors from juvenile and estrous females, but avoided traps odorized by other males. Adult females that were not in estrous preferred to enter traps odorized by adult males. Adult estrous females differentially entered traps that had contained adult males and avoided traps odorized by nonestrous or pregnant/lactating females. Juvenile females chose traps containing the odor of other juvenile females and avoided all other female odors. Clearly, it is possible to use the odor-bait technique to demonstrate that under field conditions house mice will exhibit differential patterns of responses to odors in live traps that are consistent with their social biology as described by Bronson (1979), Drickamer (1986), and Hurst (1989). The specific trap responses by mice of different ages and sex can provide insight into the social and spatial relations of the mice in the enclosures.

SOCIAL DOMINANCE AND CAPTURES

Spatial relations in house mice are mediated in part by olfactory cues based on urine marking (Rowe and Redfern, 1969; Bronson, 1976; Hurst, 1990a). Laboratory studies have demonstrated that house mice can discriminate dominant male odors from subordinate male odors (Poole and Morgan, 1976; Hurst, 1990a, 1993; Hurst *et al.*, 1993). Two types of questions related to dominance and associated odor cues were tested using data from two field seasons gathered in eight 0.1 ha field enclosures using the odor-baited trap technique. One set of questions pertained to whether all mice of varying ages and sex respond differentially to odors from dominant versus subordinate males. A second set of questions pertained to whether dominant and subordinate males respond differentially to odors in traps left by mice of varying ages and sex.

The social status of males was determined by paired encounters of the males in the same enclosure. The encounters were staged opportunistically on trapping days. Only adult males were tested and the same pairing was not tested again for at least 30 days. Encounters were judged by fights and displacements; an encounter was terminated immediately if five successive attacks occurred. No male was ever visibly wounded in any of these encounters. Data for the males in each enclosure were used to create a matrix of wins and losses, and from that we considered the top ranking half of the males to be dominant and the bottom half to be subordinate.

In the first set of analyses, involving responses of all mice, regardless of age or sex, mice were more attracted to the odor of dominant males. More mice tended to enter traps that had been odorized by dominant males than to traps odorized by subordinate males. The overall effect was due primarily to differential attraction of juvenile and estrous female mice to the odors of dominant males.

The second set of analyses involved potential differences between dominant and subordinate males to traps odorized by males of different ages and dominance status, and to females of different ages and reproductive conditions. Dominant males were significantly more likely to enter traps odorized by estrous females or subordinate males than they were to enter traps odorized by juveniles of either sex, females not in estrous, or pregnant/lactating females. Subordinate males were more likely to be caught in traps odorized by dominant males than in traps with any of the other types of odors. This last result is somewhat counterintuitive and still needs further examination.

Together, these analyses demonstrate that house mice can use information about social dominance status of males to make discriminations in terms of which traps to enter. The findings can be interpreted with regard to the social biology of the mice living under seminatural conditions. Females, both young and adult, appear to be attracted to the odor of dominant males, perhaps to mate with them; these males may be judged to be potentially better mates because of their social status. Also, dominant males clearly select traps odorized by estrous females, complementing the female selection of their odor. The field data support most of the conclusions that had been drawn previously from laboratory studies.

FEMALE ODOR PREFERENCES

Female house mice can use odor cues to locate and make discriminations among males as demonstrated by their preferences for particular odor cues in choice situations in laboratory settings. These preferences may relate to selection of mating partners. The best known system of odor discrimination for mate choice involve the major histocompatibility complex (Yamazaki et al., 1976; Penn and Potts, 1998). Also, female mice, based on odor cues, choose males that have a different t-haplotype than their own (Lenington, 1983; Coopersmith and Lenington, 1990). (The t complex is part of the major histocompatibility locus.) Two other reports, based on laboratory work, indicate that odor cues play a role in female preference for particular male partners (Winn and Vestal, 1986; Barnard and Fitzsimons, 1988).

Data collected in six 0.1 ha field enclosures during one field season were used to test several questions pertaining to the possible use of odor cues by adult female mice to discriminate among adult males (Mossman and Drickamer, 1996). All of our tests contrasted the responses of estrous with nonestrous adult females. The test situation involved a discrimination between two cotton squares inside mesh screening. The two-choice test was conducted inside a 40 liter aquarium placed on the ground in the field enclosure. The test was conducted opportunistically on females captured during regular trapping procedures. Odors used for discrimination tests were from other individuals from the same enclosure population.

One question involved the ability of females to discriminate between odors from juvenile and adult males. Females, regardless of estrous condition, exhibited a clear preference for adult over juvenile males. A second question involved the ability of females to discriminate between odors from males that were trapped near (having a home range that was overlapping with or adjoining the female's home range area) versus far (other areas of the enclosure) with respect to her area. Again, regardless of estrous condition, females exhibited a clear preference, in this instance for the near males. Finally, we tested dominant and subordinate male odors. Social status for males was determined within each enclosure as described in the previous section of this review. Females in estrous demonstrated a highly significant preference for the odor from a dominant versus a subordinate male. There was no similar effect for females not in estrous.

Together, these field results suggest that female mice living in enclosures can make discriminations among males. They know whether the male is a juvenile or an adult, and they know whether the male is one with whom they have some familiarity or one that is a stranger. These data are in agreement with the earlier findings by Hurst (1987, 1990b) when she explored responses to urine marks. Lastly, females can use odor cues to discriminate social dominance status when they are in estrous. We cannot conclude anything about nonestrous females and discrimination of social dominance; they may be able to make this discrimination, but simply do not do so when they are not in estrous.

There are solid implications from these field data for possible mate finding and mate selection by female mice. They appear to do more seeking of potential mates when they are in estrous. They don't waste time with juvenile males who are not likely capable of inseminating them, and they associate more with males that they know. Data gathered under field conditions again provide answers that are similar to the results that had been obtained previously in a laboratory setting.

CHANGES IN FEMALE CAPTURES WITH AGE

Under laboratory conditions, young, prepubertal female house mice demonstrate an avoidance of the odors from adult male mice (Drickamer, 1989, 1992); this preference shifts to favor adult male odor when these females are attaining puberty and as adults. This shift in odor preference with age may be due, in part, to the fact that females whose puberty is accelerated by the presence of an adult male or adult male odor suffer higher mortality early in life and a decrement in their reproductive success to the age of six months (Drickamer, 1988) compared to females whose puberty is not accelerated. The field enclosures and odor-bait trap technique were used to determine whether this same age-related shift in odor preference could be detected for wild female house mice living in seminatural conditions (Drickamer and Brown, 1998).

Data from two different years from eight 0.1 ha field enclosures were analyzed with respect to the responses by pre-pubertal, peri-pubertal, and adult females to odors in traps from pre-pubertal females, adult females, prepubertal males, and adult males. Prepubertal females responded by avoiding traps with odors from adult males and adult females, and with a strong preference to enter traps with odors from other juvenile females. As subadults (peri-pubertal) and as adults, female house mice exhibited a strong positive bias to enter traps odorized by adult males and not to enter traps odorized by prepubertal males.

These results support the notion that under field conditions, as had occurred in the laboratory, wild female house mice exhibit a pattern of change in odor preferences with age, particularly with regard to adult males. They avoid the odor of adult males until they attain puberty; this avoidance may be related to the previously demonstrated decrement in lifetime reproduction that can occur for females who begin reproduction at too early an age. By the time these females are attaining puberty, and as adults, they shift their preference and seek out the odors of adult males. In a short-lived animal in would not make sense for the young females to avoid males for very long, but since there are some clear negative consequences of associating with males prior to puberty, it appears that young female house mice are prepared to make a behavioral switch as they approach puberty.

SUMMARY

The field experiments summarized in the four preceding sections of this brief review each resulted in the conclusion that a phenomenon related to urinary chemosignals and social communication in house mice that had been elucidated initially under laboratory conditions is also valid in a seminatural setting. Together, these studies provide an excellent example of the synergistic interchange that can occur between laboratory and field studies. In this instance, initial laboratory discoveries and investigations were predicated on a knowledge of the social biology of house mice under feral and commensal conditions. Those laboratory findings were then transferred back to a field setting to ascertain their validity in a more natural setting.

ACKNOWLEDGMENTS

I especially thank Dr. John G. Vandenbergh for his mentoring when I was a post-doctoral student in his laboratory, beginning 30 years ago. I thank Williams College, Southern Illinois University at Carbondale, and Northern Arizona University for the use of their facilities and for some financial support for these investigations. The research summarized here was supported by the Harry Frank Guggenheim Foundation, the National Institutes of Health (Grant No. HD-08585), and the National Science Foundation (Grant Nos. BNS 8796315, IBN 8616204, and IBN 9896250). I thank Ami Sessions Robinson, Catherine A. Mossman, Lisa M. Springer, and Patricia Brown for assistance in both the laboratory and field venues.

REFERENCES

Barnard, C. J., and Fitzsimons, J., 1988, Kin recognition and mate choice in mice: The effects of kinship familiarity and social interference on intersexual interactions, *Anim. Behav.* **36:**1078-1090.

Boonstra, R., and Krebs, C. J., 1976, The effect of odour on trap response in *Microtus townsendii*, *J. Zool. Lond.*, **180:**467-476.

Bowers, J. M., and Alexander, B. K., 1967, Mice: individual recognition by olfactory cues, *Science* **158:** 1208-1210.

Bronson, F. H., 1976, Urine marking in mice: causes and effects, in: *Mammalian Olfaction, Reproduction, Reproductive Success, and Behavior*, (R. L. Doty, ed.), Academic Press, New York, pp. 119-141.

Bronson, F. H., 1979, The reproductive ecology of the house mouse, *Q. Rev. Biol.* **54**:265-299.
Brown, R. E., 1985, The rodents I: effects of odours on reproductive physiology (primer effects), in: *Social Odours in Mammals,* Volume 1 (R. E. Brown, ed.), Oxford University Press, Oxford, pp. 245-344.
Bruce, H. M., 1959, An exteroceptive block to pregnancy in the mouse, *Nature* **184**:105.
Coopersmith, C. B., and Lenington, S., 1990, Preferences of female mice for males whose *t*-haplotype differs from their own, *Anim. Behav.* **40**:1179-1181.
Coppola, D. M., and Vandenbergh, J. G., 1987, Induction of a puberty-regulating chemosignal in wild mouse populations, *J. Mammal.* **68**:86-91.
Drickamer, L. C., 1977, Delay of sexual maturation in female mice by exposure to grouped females or urine from grouped females, *J. Reprod. Fert.* **51**:77-81.
Drickamer, L. C., 1986, Puberty-influencing chemosignals in mice: ecological and evolutionary considerations, in: *Chemical Signals in Vertebrates,* Volume IV (D. Duvall, D. Müller-Schwarze, and R. M. Silverstein, eds.), Plenum Press, New York, pp. 441-455.
Drickamer, L. C., 1988, Long-term effects of accelerated or delayed sexual maturation on reproductive output in wild female house mice (*Mus musculus*), *J. Reprod. Fert.* **83**:439-445.
Drickamer, L. C., 1989, Odor preference of wild stock female house mice (*Mus domesticus*) tested at three ages using urine and other cues from conspecific males and females, *J. Chem. Ecol.* **15**:1971-1987.
Drickamer, L. C., 1992, Behavioral selection of odor cues by young female mice affects age of puberty, *Dev. Psychobiol.* **25**:461-470.
Drickamer, L. C., 1995, Odors in traps: does most recent occupant influence capture rates for house mice?, *J. Chem. Ecol.* **21**:541-555.
Drickamer, L. C., and Brown, P. L., 1998, Age-related changes in odor preferences by house mice living in seminatural enclosures, *J. Chem.Ecol.* **24**:1745-1756.
Drickamer, L. C., and Mikesic, D. M., 1990, Urinary chemo-signals, reproduction, and population size for house mice (*Mus domesticus*) living in field enclosures, *J. Chem. Ecol.* **16**:2955-2968.
Gurnel, J., and Little, J. I.,1992, The influence of trap residual odor on catching woodland rodents, *Anim. Behav.* **43**:623-632.
Hurst, J. L., 1987, The Functions of urine marking in a free-living population of house mice, *Mus domesticus* Rutty, *Anim. Behav.* **35**:1433-1442.
Hurst, J. L., 1989, The complex network of olfactory communication in populations of wild house mice, *Mus musculus,* Rutty: Urine marking and investigations within family groups, *Anim. Behav.* **37**:705-725.
Hurst, J. L., 1990a, Urine marking in populations of wild house mice *Mus domesticus* Rutty. I. Communication between males, *Anim. Behav.* **40**:209-222.
Hurst, J. L., 1990b, Urine marking in populations of wild house mice *Mus domesticus* Rutty. III. Communication between the sexes, *Anim. Behav.* **40**:233-243.
Hurst, J. L., 1993, The priming effects of urine substrate marks on interactions between male house mice, *Mus musculus domesticus* Schwarz and Schwarz, *Anim. Behav.* **45**:55-81.
Hurst, J. L., Fang, J., and Barnard, C. J., 1993, The role of substrate marking odours in maintaining social tolerance between male house mice, *Mus musculus domesticus, Anim. Behav.* **45**:997-1006.
Lenington, S., 1983, Social preferences for partners carrying 'good genes' in wild house mice, *Anim. Behav.* **31**:325-333.
Lenington, S., Egid, K., and Williams, J., 1988, Analysis of a genetic recognition system in wild house mice, *Behav. Genet.* **18**:549-564.
Massey, A. M., and Vandenbergh, J. G., 1980, Puberty delay by a urinary cue from female house mice in feral populations, *Science* **209**:821-822.
Massey, A. M., and Vandenbergh, J. G., 1981, Puberty acceleration by a urinary cue from male mice in feral populations, *Biol. Reprod.* **24**:523-527.
Mazdzer, E., Capone, M. R., and Drickamer, L. C., 1976, Conspecific odors and trappability of deer mice (*Peromyscus leucopus noveboracensis*), *J. Mammal.* **57**:607-609.
Mossman, C. A., and Drickamer, L.C., 1996, Odor preferences of female house mice (*Mus domesticus*) in seminatural enclosures, *J. Comp. Psychol.* **110**:131-138.
Penn, D., and Potts, W., 1998, How do major histocompatibility complex genes influence odor and mating preferences?, *Adv. Immunol.* **69**:411-436.
Poole, T. B., and Morgan, H. D. R., 1976, Social and Territorial behavior of laboratoty mice (*Mus Musculus*) in small complex areas, *Anim. Behav.* **24**:476-480.
Rowe, F. P., and Redfern, R., 1969, Aggressive behaviour in related and unrelated wild house mice (*Mus musculus* L.), *Ann. Appl. Biol.* **64**:425-431.
Stoddart, D. M.,1982, Does trap odour influence estimation of population size of the short-tailed vole, *Microtus agrestis, J. Anim. Ecol.* **51**:375-386.

Vandenbergh, J. G., 1969, Male odor accelerates female sexual maturation in mice, *Endocrinology* **84:**658-660.
Vandenbergh, J. G., and Coppola, D. M., 1986, The physiology and ecology of puberty modulation by primer pheromones, *Adv. Stud. Behav.* **16:**71-108.
Whitten, W. K., 1956, Modification of the oestrous cycle of the mouse by external stimuli associated with the male, *J.Endocrinol.* **13:**399-404.
Winn, B. E., and Vestal, B. M., 1986, Kin recognition and choice of males by wild female House mice (Mus musculus), *J. Comp. Physiol. Psychol.* **100:**72-75.
Yamazaki, K., Boyse, E. A., Mike, V., Thaler, H. T., Mathieson, B. J., Abbott, J., Boyse, J., and Zayas, Z. A., 1976, Control of mating preferences in mice by genes in the major histocompatibility complex, *J. Exp. Med.* **144:**1324-1335.

INFORMATION IN SCENT SIGNALS OF COMPETITIVE SOCIAL STATUS: THE INTERFACE BETWEEN BEHAVIOUR AND CHEMISTRY

Jane L. Hurst,[1] Robert J. Beynon,[2] Rick E. Humphries,[1] Nick Malone,[1] Charlotte M. Nevison,[1] Caroline E. Payne,[1] Duncan H.L. Robertson,[2] and Christina Veggerby[2]

[1]Animal Behaviour Group and [2]Protein Function Group
Faculty of Veterinary Science
University of Liverpool
Leahurst, Neston CH64 7TE, UK

1. HONEST SIGNALS OF COMPETITIVE ABILITY AND SOCIAL STATUS

From an evolutionary viewpoint, signals generally should be reliable or honest (Zahavi, 1987; Johnstone, 1997). Animals can gain a number of advantages from advertising high competitive ability to potential mates and to other competitors, particularly males which often compete strongly for mating opportunities (Andersson, 1994). Animals often prefer high quality mates that will increase the fitness of their offspring, both because of genetic benefits (through good genes or Fisherian selection) and because parents of high competitive ability often provide better resources and protection. Competitors will also gain an advantage if potential challengers withdraw from, or otherwise avoid, aggressive encounters with an opponent of high fighting ability. There is thus strong selection pressure on signallers to advertise high social status and competitive ability to others. However, receivers will only gain an advantage from responding to such signals if these are reliable indicators of the signaller's competitive ability. Females that mate with low quality males that dishonestly signal high competitive ability will gain no advantage for their offspring, while males that withdraw from agonistic encounters with poorer competitors will be disadvantaged. There is thus strong selection on receivers to respond only to honest signals that are resistant to cheating, and therefore for high quality animals to provide such reliable information in signals as these will be effective in attracting mates and deterring competitors.

As well as being honest, signals will evolve to be conspicuous for ease of detection (Wiley, 1983; Guildford and Dawkins, 1991). A high degree of 'redundancy' may aid detection through the use of multiple components or by frequent repetition in space and/or time. Conspicuousness can also be increased through alerting components that encode little information but which are highly detectable and alert recipients to the rest of the display.

2. USE OF SCENT SIGNALS FOR ADVERTISING COMPETITIVE ABILITY

Many mammals use scent signals to advertise their dominant social status (reviewed by Brown and Macdonald, 1985) and ability to defend a territory (Gosling, 1982). While social systems vary between species, mammals usually compete by attempting to exclude competitors from a defended territory and by establishing social dominance over any subordinate animals that live within their territory. Dominant territory owners are thus animals of high competitive ability and advertise this by scent marking their territory at a high rate (Ralls, 1971; Johnston, 1973).

Unlike most visual or acoustic signals, scent signals can be deposited in the environment and persist in the absence of the depositor over an extended period. This characteristic provides the basis for an honest signal of competitive ability that is continuously available to challenge and to inspection by mates and competitors. Since only animals that dominate a territory can ensure that their marks predominate in the area, scent marks provide proof of territory ownership (Gosling, 1982). Further, scent marks indicate the success with which an animal dominates its territory, since only males that defend their territory well can ensure that no other males deposit competing signals that might attract mates (Hurst, 1993; Hurst and Rich, 1999). The presence of competing signals that are as fresh or fresher than those of the owner thus indicate the owner's poor success in dominating the area. Accordingly, dominant animals rapidly counter-mark if they encounter competing signals from other males in their territory and exclude any competitors that might deposit competing scent marks (Ralls, 1971; Hurst and Rich, 1999).

Both the spatial and temporal pattern of competitive scent deposition therefore play a crucial role in providing an honest signal of social status and competitive ability, providing information on any challenges for dominance and the outcome of such challenges.

There is now substantial evidence that receivers use competitive scent mark signals to assess the signaller's competitive ability when choosing between potential mates or when deciding whether to mount a competitive challenge. When presented with scent marks from two potential mates, hamsters (Johnston, 1999) and meadow voles (Ferkin, 1999) prefer the individual that deposited the top counter-mark over the owner of the bottom scent when these overlap. Rich and Hurst (1998, 1999), studying house mice, showed that this extends beyond the choice between top and bottom scent marker. When presented with a choice between two male territory owners, where one male's territory was scent marked exclusively by the owner while the other territory contained some competing scent marks from an intruder, females preferred the owner of the exclusively marked territory (Rich and Hurst, 1998). Given a choice between two males whose territories both contained intruder scent marks, females preferred the owner that had counter-marked the intruder's scent over the owner whose scent had been counter-marked by the intruder (Rich and Hurst, 1999). Competitors similarly use competitive scent marks to identify and avoid challenging males that are effectively defending the area (Gosling and McKay, 1990), and will increase challenges against the owner if the territory contains fresh counter-marks from competitors (Hurst, 1993; Hurst and Rich, 1999).

3. DESIGN OF COMPETITIVE SCENT MARKS

Competitive scent mark signals need to contain several types of information: a) species and sex of the signaler, b) individual identity of the signaler, c) social status of the signaler, d) age of the scent mark and e) presence and location of the scent mark. The components of scent marks used to signal these different types of information also need to have certain qualities. From the signaller's point of view, scent marks advertising their ability to dominate an area should be persistent to minimise the effort required to scent mark the territory whilst still being easily detectable by animals entering the area. However, labile (volatile or unstable) signals will provide much more reliable proof of current occupancy and dominant status within the territory. Receivers should therefore select labile signals that yield dynamic information on competitive challenges and their outcome. In addition, they should select persistent signals that provide information concerning the longer-term ability of animals to defend the area.

Scent marks must provide information concerning the identity of the scent owner, and identity cues must be sufficiently complex and variable to allow reliable discrimination between individual competitors. Information concerning the owner's identity must be particularly stable and persistent to ensure that, as scent marks age, they are not perceived as coming from a different individual. Volatile components are needed to alert recipients to the presence and location of the scent mark and to attract them to investigate the rest of the scent display, while species and sex specific components will allow appropriate receivers to identify signals easily.

To provide these different types of information and qualities, the design of competitive scent signals must involve interplay between the scent mark deposition pattern (mark rate, mark size and counter-marking), and both labile and non-labile chemical components in the scent mark.

4. COMPETITIVE SCENT SIGNALLING IN HOUSE MICE

Our work on competitive signalling in male house mice is starting to reveal how volatile and non-volatile chemical components and scent marking behaviour all interact to provide an honest signal of individual competitive ability. Dominant male mice scent mark their territories extensively with numerous small spots and streaks of urine, exclude other males that attempt to deposit competing scent marks, and rapidly elevate their mark rate in the vicinity of any scent marks from other males to counter-mark them (Desjardins et al., 1973; Sandnabba, 1986; Hurst, 1990; Hurst, 1993; Hurst and Rich, 1999), consistent with the use of scent marks to advertise competitive ability. Adult mouse urine contains a high concentration of major urinary proteins (MUPs) (Finlayson and Baumann, 1958; Beynon et al., this volume). In males, these proteins bind a number of ligands but principally two male-specific signalling volatiles, 2-sec-butyl-4,5-dihydrothiazole (thiazole) and 3,4-dehydro-exo-brevicomin (brevicomin) (Bacchini et al., 1992; Robertson et al., 1993; Novotny et al., 1999). In addition to MUPs, the protein fraction of mouse urine contains peptides derived from major histocompatibility complex (MHC) proteins that are expressed on the surface of all cells and are involved in cell recognition by the immune system but also contribute to individually-distinctive urinary odours (Yamaguchi et al., 1981; Brown, 1995). The dominant social status of a male is also signalled by two sesquiterpenes, E,E-alpha-farnesene and E-beta farnesene, which are secreted into the urine by the preputial glands (Harvey *et al.*, 1989; Novotny *et al.*, 1990).

4.1. Detection of Male Competitive Signals

The active chemical components of competitive signalling are held in the high molecular weight fraction of male urine (>5000Da), and it is this fraction that stimulates dominant territory owners to counter-mark intruder scent marks (Humphries *et al.*, 1999; Humphries *et al.*, this volume). Over 99% of the proteins in this fraction are MUPs and we have shown that both the volatile ligands bound to MUPs and the MUPs themselves play important, though different, roles in the competitive scent signal. The thiazole and brevicom ligands bound to MUPs are only produced by adult male mice (Schwende *et al.*, 1986) thus provide an appropriate species and sex specific signal that the scent mark derives from a male mouse. Farnesenes, which are not principally bound to MUPs (unpublished data), signal that the urine mark derives from a dominant male and can inhibit counter-marking by other males, particularly among those of low competitive ability that have previously experienced social defeat (Jemiolo *et al.*, 1992).

The ligands bound to MUPs are highly volatile and serve to attract attention to the scent mark. While most thiazole and brevicomin in fresh urine is not bound to MUPs, these free volatiles are lost from scent marks in minutes (see Robertson *et al.*, this volume). MUPs provide an extended release of these male signalling volatiles (Hurst *et al.*, 1998) so that they continue to attract attention to the male's scent marks for at least 24h after the scent mark was deposited (Humphries *et al.*, 1999). On detecting volatiles emanating from a male's scent mark, the response of both competitive males and females is to approach the scent mark to investigate it very closely.

4.2. Scent Mark Age

Scent mark age is an important component of competitive signalling in male mice since females only discriminate between scent marks and counter-marks when these differ in age (Rich and Hurst, 1999). Scent mark age appears to be signalled by the amount of ligands still retained by MUPs. Males are only stimulated to mount a counter-marking response (increase their rate of scent marking) in the presence of intruder urinary proteins that continue to release detectable levels of volatile ligands. Males thus counter-mark in the presence of intruder scent marks aged up to 24h but do not respond if the intruder scent marks are all aged by seven days, even though they detect their presence and investigate them (Humphries *et al.*, this volume). However, in the presence of a fresh intruder's scent mark, males counter-mark any stimulus containing urinary proteins from the intruder. Further, they appear to locate their own counter-marks to maximise the age difference between their own fresh scent and that of the intruder. Thus, although they increase their rate of scent marking close to fresh intruder marks, they deposit significantly more marks near to aged intruder marks that retain few if any volatile ligands (Humphries *et al.*, 1999). In doing this, they ensure that fresh scent marks of the intruder are always matched by their own fresh scents nearby. However, by continually depositing their own fresh scent near to an intruder's aged scent marks, the release of volatile ligands is likely to attract attention to a site where their own scent marks are clearly fresher than the intruder's.

The ratio of volatile ligands to urinary protein theoretically provides a very reliable signal of scent mark age. The amount of any volatile component in a scent mark depends on both the amount deposited and the time since deposition (Figure 1). Thus receivers cannot assess the age of the mark without knowing the amount deposited. In contrast, the amount of protein in a scent mark does not change and can provide a timebase to assess the loss of volatile ligands from the protein (Figure 2). Since each protein molecule can only bind one ligand molecule, and ligands are slowly released and evaporate from a scent mark, the proportion of protein molecules that contain ligands will decrease with time since deposition. The surfeit of MUP ligands in fresh male urine probably serves to ensure that MUPs will be replete with ligands when the scent is deposited and evaporates (Robertson *et al.*, this volume).

The use of ligand to protein ratio as an indicator of scent freshness is consistent with the characteristic scent mark pattern of dominant male mice. Males deposit their urine in numerous small spots, many of which are tiny (less than one microlitre of urine), and they have evolved hairs on the end of their prepuce to aid this deposition pattern (Maruniak *et al.*, 1975). While this helps to distribute their scent throughout the territory, they deposit numerous marks in the same local area. When counter-marking another male's scent, they do not attempt to deposit a bigger scent mark than that of the competitor, which would contain a greater intensity of volatile ligands (Figure 2). Neither do they attempt to deposit their scent on top of the competitor's (Humphries *et al.*, 1999). Instead, they deposit many small scent marks in the vicinity over a period of several hours (Humphries *et al.*, 1999). By dribbling out their urine in this way, they maximise the freshness of their marks and the rate of replenishment (Figure 2). While a large scent mark will contain a large quantity of volatiles, it starts to age on deposition and there will be a time delay before the animal can produce sufficient urine to deposit another large mark. Continuous deposition of scent in a series of very small marks ensures that there are always fresh marks in the territory, although each contains only a small amount of male signalling volatiles. Depositing their scent in very tiny marks also makes it difficult, if not impossible, to deposit their scent directly on top of the male's scent for competitors to over-mark (unlike hamsters, Johnston, 1999; and meadow voles, Ferkin, 1999).

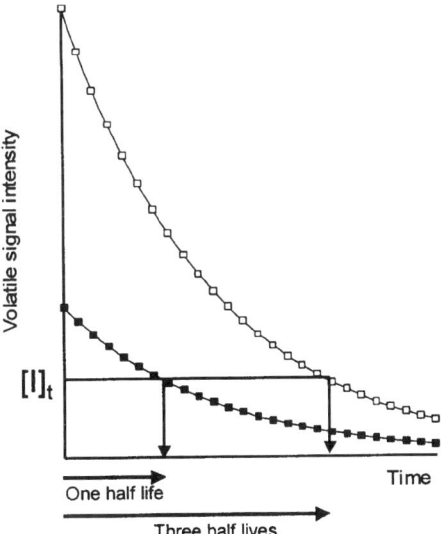

Figure 1. **The uncertainty principle in chemical signalling.** An animal sampling a volatile signal at intensity $[I]_t$ cannot discriminate between a recent, but small signal (closed symbols) and a large, but aged signal (open symbols). Thus, it is not possible to assess both the time since the signal was deposited and the amount of the deposit.

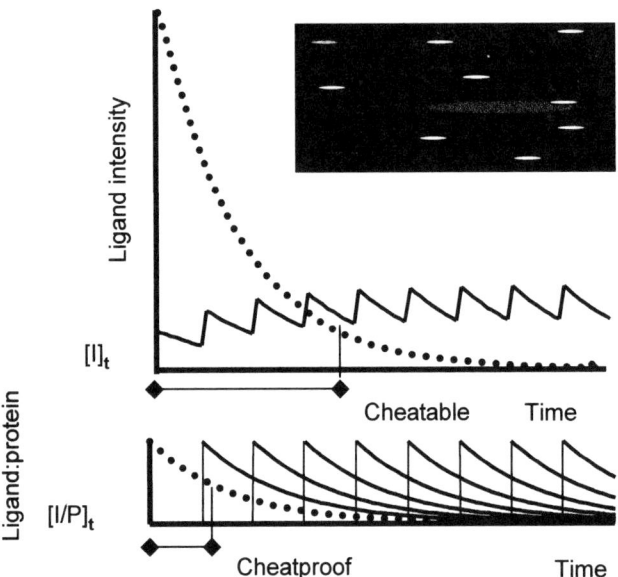

Figure 2. Cheat-proofing a scent mark system. If a volatile semiochemical must be maintained above a threshold concentration [I]$_t$ (shaded region), an animal can give the impression of sustained occupancy of a site by deposition of a single large mark (dotted line) which can be compared with repeated deposition of smaller scent marks (solid line). Thus, the signal is 'cheatable'. If however, the scent mark is assessed for volatile intensity relative to an involatile, stable component of the scent mark [I/P]$_t$, then this quantity decays at the same rate, irrespective of the size of the deposit. The signal is now 'cheatproof'.

4.3. Individual Identity

Urinary proteins provide information on the individual identity of the signaller. Genetic differences between animals can be discriminated from volatile metabolites released from protein fraction of urine (Singer et al., 1993, 1997). These protein-bound volatiles, once released, can allow animals to discriminate between individuals even when these differ at only a small number of loci in the MHC complex genes though, as yet, it is not clear whether the volatile metabolites are bound to MHC peptides in the urine, to MUPs or to both. Once animals detect volatiles emanating from the scent mark of a novel individual, they approach the odour source to investigate. However, metabolites are susceptible to disruption by environmental factors such as food type or bacterial gut flora (Brown, 1995), thus an individual's volatile identity cues appear to change when they eat different foods, for example (Schellinck et al., 1997). Metabolites thus do not provide the required stability for reliable signalling of identity but may attract attention to an interesting scent, stimulating animals to approach the scent source to investigate closely (see Nevison et al., this volume). Mice continue to recognise an intruder's scent mark for at least seven days after deposition, when most volatile ligands will have been lost (Humphries et al., 1999). This suggests that they detect the protein molecules themselves - there are specific receptors for urinary lipocalins in the vomeronasal organ (Krieger et al., 1999; Brennan et al., 1999).

Urinary proteins have all the required qualities to provide stable and persistent individual identity signals. There is considerable heterogeneity in MUPs expressed by wild mice, with each individual expressing at least four to fifteen different proteins and expressing a different combination of MUPs to other individuals in the same population (Payne *et al.*, this volume, Veggerby *et al.*, this volume). Most of the variation between MUPs is on the surface of the protein, which would be accessible to receptors (Beynon *et al.*, this volume). The MUPs are genetically determined and are very stable, with no changes detectable in the proteins expressed by an individual through time, even if the animal eats different food sources or changes social status (unpublished data). Once deposited, MUPs are very persistent and stable over many weeks or even months. MHC peptides in mouse urine are also genetically determined and exhibit similar heterogeneity, though as yet there is no evidence that these can be detected directly. However, since MUPs are present at considerably higher concentrations and their only known functions are in chemical communication, these may be the most likely candidates to provide stable identity cues (see discussion in Beynon *et al.*, this volume). If this is the case, the binding of signalling ligands to MUPs provides a direct connection between the presence, age and owner of the signal.

5. USE OF SCENT SIGNALS FOR ADVERTISING SUBORDINATE STATUS ?

While animals of high competitive ability gain an advantage from providing honest signals advertising their status, those of low competitive ability would be disadvantaged. Not only will females prefer males of higher competitive ability, resulting in reduced mating opportunities for those of poor quality (i.e. those unable to ensure their scent marks predominate over a territory), but those of low ability will also suffer a competitive disadvantage when interacting with other males. If animals can easily detect the lower competitive ability of their opponents, they are more likely to escalate aggressive encounters to defeat a weaker opponent. Under competitive pressure, we might thus expect animals of low competitive ability to withdraw from competition and from competitive advertisement. Even then, these animals might gain mating opportunities while a dominant territory owner is elsewhere and thus will still be perceived as competitors. However, animals of poor competitive ability that are not able to defend their own territory could gain a survival advantage from honestly advertising their subordinate, non-competitive status through scent marks that remain in the environment and attract investigation. While this will clearly signal their relatively poor quality as mates to females, it should increase tolerance from the owner of the territory, allowing them to reside there. They are then better placed to compete for mating opportunities in the future.

Male house mice show a dramatic and immediate change in their scent marking behaviour in response to defeat by a male of higher competitive ability (Desjardins *et al.*, 1973; Sandnabba, 1986). Instead of depositing the numerous small marks characteristic of a dominant or isolated male (see above), subordinate males greatly reduce their scent marking rate in the presence of a dominant male's odour, though they continue to deposit scent marks in larger spots and pools. These scent marks are very important for maintaining recognition and tolerance from the resident dominant male (Hurst *et al.*, 1993).

Information concerning male social status is held in labile components of their urine (reviewed by Malone *et al.*, this volume). Both males (Jones and Nowell, 1989; Nevison *et al.*, 2000) and females (Jones and Nowell, 1974) can discriminate between volatiles emanating from the urine marks of dominant and subordinate males, and a subordinate's scent stimulates much less aggression from dominant males than that of another dominant male (Jones and Nowell, 1973). In particular, subordinate males have smaller preputial glands and urinary farnesene levels than dominant males (Hucklebridge et al., 1972; Harvey et al., 1989), making their urine less attractive to females (Bronson and Caroom, 1971; Jemiolo et al., 1991) and no longer aversive to other males (Jones and Nowell, 1989; Novotny et al., 1990; Jemiolo et al., 1992). However, subordinates continue to produce MUPs at similar concentrations to dominant males (unpublished data), which slowly release thiazole and brevicomin from their scent marks. Subordinate males thus continue to actively advertise their identity and sex while signalling their subordinate status with low concentrations of farnesenes and possibly other volatiles related to dominance status (Harvey et al., 1989).

6. CONCLUSIONS

Labile and non-labile chemical components in scent marks interact with the spatial and temporal pattern of competitive scent marking behaviour to signal the ability of animals to dominate others in the locality and thus provide honest signals of individual competitive ability. Scent signals are particularly suitable for this since they persist in the environment providing a continuous record of competitive challenges and the outcome. In male house mice, urinary proteins and their volatile ligands signal the presence, owner and age of scent marks, which are all essential components of competitive scent signals. These signals allow females to select high quality and healthy mates that are able to out-compete their rivals and will provide genetic benefits (and possibly resources) for their offspring. Competitive scent marking can thus be seen as a sexual display, providing information concerning the general health and disease resistance (Hamilton and Zuk, 1982; Penn and Potts, 1998) and other genetic qualities affecting of the vigour of a male.

7. ACKNOWLEDGMENTS

This work was funded by research grants to JLH and RJB from BBSRC.

8. REFERENCES

Andersson, M., 1994, *Sexual Selection*, Princeton University Press, Princeton.
Bacchini, A., Gaetani, E., and Cavaggioni, A., 1992, Pheromone binding proteins of the mouse, Mus musculus, *Experientia* **48**:419-421.
Brennan, P. A., Schellinck, H. M., and Keverne, E. B., 1999, Patterns of expression of the immediate-early gene egr-1 in the accessory olfactory bulb of female mice exposed to pheromonal constituents of male urine, *Neuroscience* **90**:1463-1470.
Bronson, F. H., and Caroom, D., 1971, Preputial gland of the male mouse; attractant function, *J. Reprod. Fertil.* **25**:279-282.
Brown, R. E., 1995, What Is the Role of the Immune-System in Determining Individually Distinct Body Odors, *Int. J. Immunopharm.* **17**:655-661.
Brown, R. E., and Macdonald, D. W., 1985, *Social Odours in Mammals Vols 1 and 2*, Clarendon Press, Oxford.
Desjardins, C., Maruniak, J. A., and Bronson, F. H., 1973, Social rank in the house mouse: differentiation revealed by ultraviolet visualisation of urinary marking patterns, *Science* **182**:939-941.

Ferkin, M. H., 1999, Scent over-marking and adjacent-marking as competitive tactics used during chemical communication in voles, in: *Advances in Chemical Communication in Vertebrates* (R. E. Johnston, D. Muller-Schwarze, and P. Sorensen eds.), Plenum Press, New York, pp. 239-246.

Finlayson, J. S., and Bauman, C. A., 1958, Mouse proteinuria, *Amer. J. Physiol.* **192**:69.

Gosling, L. M., 1982, A Reassessment of the Function of Scent Marking in Territories, *J. Comp. Ethol.* **60**:89-118.

Gosling, L. M., and McKay, H. V., 1990, Competitor Assessment By Scent Matching - an Experimental Test, *Behav. Ecol. Sociobiol.* **26**:415-420.

Guilford, T., and Dawkins, M., 1991, Receiver psychology and the evolution of animal signals, *Anim. Behav.* **42**:1-14.

Hamilton, W. D., and Zuk, M., 1982, Heritable true fitness and bright birds: a role for parasites? *Science* **218**:382-387.

Harvey, S., Jemiolo, B., and Novotny, M., 1989, Pattern of volatile compounds in dominant and subordinate male-mouse urine, *J. Chem. Ecol.* **15**:2061-2072.

Hucklebridge, F. H., Nowell, N. W., and Wouter, A., 1972, A relationship between social experience and preputial gland function in the albino mouse, *J. Endocrinol.* **55**:449-450.

Humphries, R. E., Robertson, D. H. L., Beynon, R. J., and Hurst, J. L., 1999, Unravelling the chemical basis of competitive scent marking in house mice., *Anim. Behav.* **58**:1177-1190.

Hurst, J. L., 1990, Urine Marking in Populations of Wild House Mice Mus-Domesticus Rutty .1. Communication Between Males, *Anim. Behav.* **40**:209-222.

Hurst, J. L., 1993, The Priming Effects of Urine Substrate Marks On Interactions Between Male House Mice, Mus Musculus-Domesticus Schwarz and Schwarz, *Anim. Behav.* **45**:55-81.

Hurst, J. L., and Rich, T. J., 1999, Scent marks as competitive signals of mate quality, in: *Advances in Chemical Communication in Vertebrates* (R. E. Johnston, D. Müller-Schwarze, and P. Sorensen eds.), Plenum Press, New York, pp. 209-226.

Hurst, J. L., Fang, J. M., and Barnard, C. J., 1993, The Role of Substrate Odors in Maintaining Social Tolerance Between Male House Mice, Mus-Musculus-Domesticus, *Anim. Behav.* **45**:997-1006.

Hurst, J. L., Robertson, D. H. L., Tolladay, U., and Beynon, R. J., 1998, Proteins in urine scent marks of male house mice extend the longevity of olfactory signals, *Anim. Behav.* **55**:1289-1297.

Jemiolo, B., Xie, T. M., and Novotny, M., 1991, Socio-sexual olfactory preference in female mice: attractiveness of synthetic chemosignals, *Physiol. Behav.* **50**:1119-1122.

Jemiolo, B., Xie, T. M., and Novotny, M., 1992, Urine marking in male mice: Responses to natural and synthetic chemosignals, *Physiol. Behav.* **52**:521-526.

Johnston, R. P., 1973, Scent marking in mammals, *Anim. Behav.* **21**:521-535.

Johnston, R. E., 1999, Scent over-marking, in: *Advances in Chemical Communication in Vertebrates* (R. E. Johnston, D. Müller-Schwarze, and P. Sorensen eds.), Plenum Press, New York, pp.227-238.

Johnstone, R. A., 1997, The evolution of animal signals, in: *Behavioural Ecology An Evolutionary Approach* (J.R. Krebs and N.B. Davies, eds.), Blackwell Science, Oxford, pp. 155-178.

Jones, R. B., and Nowell, N. W., 1973, Aversive and aggression-promoting properties of urine from dominant and subordinate male mice, *Anim. Behav.* **21**:207-210

Jones, R. B., and Nowell, N. W., 1974, A comparison of the aversive and female attractant properties of urine from dominant and subordinate male mice, *Med. Weter.* **2**:141-144.

Jones, R. B., and Nowell, N. W., 1989, A potency of urine from dominant and subordinate male laboratory mice (Mus musculus): resolution of a conflict, *Aggr. Behav.* **15**: 291-296.

Krieger, J., Schmitt, A., Lobel, D., Gudermann, T., Schultz, G., Breer, H., and Boekhoff, I., 1999, Selective activation of G protein subtypes in the vomeronasal organ upon stimulation with urine-derived compounds, *J. Biol. Chem.* **274**:4655-4662.

Maruniak, J. A., Desjardins, C., and Bronson, F. H., 1975, Adpatations for urinary marking in rodents: prepuce length and morphology, *J. Reprod. Fertil.* **44**:567-570.

Nevison, C. M., Barnard, C. J., Beynon, R. J., and Hurst, J. L., 2000, The consequences of inbreeding for recognising competitors, *Proc. Royal Soc. Lond. Ser. B* **267**:687-694.

Novotny, M., Harvey, S., and Jemiolo, B., 1990, Chemistry of male dominance in the house mouse, *Mus domesticus*, *Experientia* **46**:109-113.

Novotny, M. V., Ma, W., Wiesler, D., and Zidek, L., 1999, Positive identification of the puberty-accelerating pheromone of the house mouse: the volatile ligands associating with the major urinary protein, *Proc. Royal. Soc. Lond. B Biol. Sci.* **266**:2017-2022.

Penn, D., and Potts, W. K., 1998, Chemical signals and parasite-mediated sexual selection, *Trends Ecol. Evol.* **13**:391-396.

Ralls, K., 1971, Mammalian scent marking, *Science* **171**:443-449.
Rich, T. J., and Hurst, J. L., 1998, Scent marks as reliable signals of the competitive ability of mates, *Anim. Behav.* **56**:727-735.
Rich, T. J., and Hurst, J. L., 1999, The competing countermarks hypothesis: reliable assessment of competitive ability by potential mates, *Anim. Behav.* **58**:1027-1037.
Robertson, D. H. L., Beynon, R. J., and Evershed, R. P., 1993, Extraction, characterization and binding analysis of two pheromonally active ligands associated with major urinary protein of house mouse (Mus musculus), *J. Chem. Ecol.* **19**:1405-1416.
Sandnabba, N. K., 1986, Changes in male odors and urinary marking patterns due to inhibition of aggression in male mice, *Behav. Process.* **12**:349-361.
Schellinck, H. M., Slotnick, B. M., and Brown, R. E., 1997, Odors of individuality originating from the major histocompatibility complex are masked by diet cues in the urine of rats, *Anim. Learn. Behav.* **25**:193-199.
Schwende, F. J., Wiesler, D., Jorgenson, J. W., Carmack, M., and Novotny, M., 1986, Urinary volatile constituents of the house mouse, *Mus musculus*, and their endocrine dependency, *J. Chem. Ecol.* **12**:277-296.
Singer, A. G., Tsuchiya, H., Wellington, J. L., Beauchamp, G. K., and Yamazaki, K., 1993, Chemistry of Odortypes in Mice - Fractionation and Bioassay, *J. Chem. Ecol.* **19**:569-579.
Singer, A. G., Beauchamp, G. K., and Yamazaki, K., 1997, Volatile signals of the major histocompatibility complex in male mouse urine, *Proc. Natl. Acad. Sci. USA* **94**:2210-2214.
Wiley, R.H., 1983, The evolution of communication: information and manipulation, in: *Animal Behaviour. Vol.2. Communication* (T. R. Halliday and P. J. B. Slater, eds.), Blackwell Scientific Publications, Oxford, pp. 156-189.
Yamaguchi, M., Yamazaki, K., Beauchamp, G. K., Bard, J., Thomas, L., and Boyse, E. A., 1981, Distinctive urinary odors governed by the major histocompatibility locus of the mouse, *Proc. Natl. Acad. Sci. USA* **78**:5817-5820
Zahavi, A., 1987, The theory of signal selection and some of its implications, in: *International Symposium of Biological Evolution* (V. P. Delfino, ed.), Adriatica Editrice, Bari, pp.305-327.

MOLECULAR APPROACHES IN CHEMICAL COMMUNICATION OF MAMMALS

Edward P. Zinkevich and Varvara S. Vasilieva

Institute of Ecology and Evolution Russian Academy of Sciences
Leninsky pr., 33, Moscow 117071, Russia

INTRODUCTION

Chemical communication in animals uses olfactory signals, often called pheromones, that are mainly volatile components of various excretions encoding biologically relevant information (Sokolov and Zinkevich, 1974, 1979, 1983, 1986; Sokolov et al., 1984). In general, the signals are diffused in the air and detected by specialized cells, namely, the receptors of the main olfactory system and of the accessory chemoreception system. Chemical communication is generally examined with respect to the influence of olfactory signals on the animal behavior or state, according to the classical model of "stimulus - response". An essential distinction of this communication form from the optical and acoustic communication is the fact that an olfactory signal can be characterized as a discrete marker in terms of the molecular structures, whereas optical and acoustic signals can be placed along their respective energy spectra.

Fundamental investigation of chemical communication necessitates cooperation between modern physical, chemical, biological, and semiotic approaches. This has been undertaken for the past 70 years by a multidisciplinary group of scientists in Severtsov Institute of Evolutionary Animal Morphology and Ecology (USSR). The studies were performed at various levels: organism (donor - recipient), molecular - organism (excretions - recipient and olfactory signals - recipient), molecular - cellular (olfactory signals - olfactory cells), and molecular (olfactory signals - molecular receptors).

Scents can play essential roles in the territorial, sexual, and maternal interactions of mammals by stimulating or reducing aggressive responses, retarding or accelerating sexual maturation, prolonging or shortening the estrous cycle, etc. Olfactory signals can be used by animals to discriminate species, individual identity, sex, and the physiological state. The chemical nature of mammalian olfactory signals and the mechanisms for their recognition and biological functions are unresolved problems of chemical communication. Our studies on the molecular bases of chemical communication in mammals have been aimed at understanding the mechanisms by which animal behavior and physiological state are influenced by the chemical components of the animal secretions and their artificial analogues.

Our experimental approach to chemical communication is based on the recognition that any property of a pheromone molecule is a function of its chemical structure, especially its spatial characteristics (Zinkevich et al., 1992; Aronov and Zinkevich, 1993). To date, particular parameters of the molecules of volatile compounds (including pheromones) determining their ability to stimulate olfactory cells have not been revealed.

CHEMICAL COMPOSITION OF THE MAMMALIAN EXCRETIONS

The mammalian secretions are multicomponent mixtures of almost all classes of organic compounds. Their concentration ranges within a number of orders; and the chemical structure substantially varies depending on the excretion type, the point of sample obtaining, species, sex, individuality, state, the composition of microflora, etc. (Albone, 1984; Sokolov et al., 1974, 1975, 1977, 1980a, 1987, 1994; Sokolov and Zinkevich, 1977, 1986; Zinkevich et al., 1997). The standard methods of the comparative physicochemical analysis give no way to reveal the volatile components characteristic of any animal group. The components common for different groups or states, which would be informative in predominant concentration, have not yet been revealed.

The results of the analysis of a natural mixture of volatile compounds by the mammalian olfactory system do not correspond to those obtained by standard physicochemical analysis of the same mixtures that results from the following reasons: (1) The olfactory system always analyzes only the gas phase; whereas the majority of chemical analytical studies involve investigation of the composition of the liquid excretions or their extracts. (2) Analysis by the standard physicochemical methods generally gives results that are not comparable to the results of an animal's sensory analysis, especially in terms of the sensitivity of the detectors. (3) This leads to the general discrepancy between the recognition algorithms for complex mixtures of compounds (their chemical patterns) in these two analytical systems: all components of a mixture are equivalent for the physicochemical system of recognition. For the olfactory system, the threshold concentrations of different compounds can differ by a number of orders; thus, when perceived, they are not equivalent. Resolution of chromatographic methods enables us to divide any mixture into its individual components and to characterize the mixture using an additive scheme. The olfactory system is distinguished by low resolution. Many different compounds possess the same odor. In contrast to the physicochemical systems, olfaction unpredictably integrates certain components of mixture; however, the algorithm for memorizing and recognition of mixtures by the olfactory system is not yet understood.

ODOR AND SPECIES RECOGNITION

The ability to recognize conspecifics is an indispensable condition for the appropriate delivery of sexual behavior. In many mammals, olfactory signals seem to play an essential role in recognition; however, the chemical nature of such signals is almost unknown. The chromatographic methods we have applied in studying the secretions of specific skin glands of more than 30 species of five mammalian orders have revealed that each species and certain subspecies might be characterized by a specific set of the component groups (Sokolov and Zinkevich, 1977, 1983). Chromatograms characteristic of the group of secretion are repeated in mature animals within a species, irrespective of sex, individual, physiological state, season, geographical zone, etc. This testifies that even the group composition (without separation into individual components) bears information about the mammalian species.

Lipids in the skin surface are common among mammals. Chromatomass spectrometry combined with pattern recognition, based on the principles of multidimensional analysis, has shown that, in five arbitrary mammalian species chosen irrespective of the taxonomic criteria (European bison, pig, horse, sheep, and wolf), it is possible to distinguish species-specific chemotaxonomic indicators on the basis of the ratio between certain higher fatty acids found in all studied animals (normal saturated fatty acids of the C_{14}-C_{16}, and C_{18} structure, as well as branched C_{14} and C_{15} ones; Sokolov et al., 1994b). It was also revealed that animals could recognize a conspecific individual on the basis of the complex of olfactory signals of the excretions: in house mice, this was a combination of pH-dependent and neutral substances of urine (Sokolov et al., 1979a); in Norway rats, pH-dependent compounds (certain amines) neutralized by lower fatty acids (Sokolov et al., 1978, 1979c); and in golden hamsters, volatile sulfur compounds (Sokolov et al., 1980a). This testifies that olfactory signals used by rodents to recognize conspecific animals can be complex mixtures of components that are necessary but insufficient in the absence of other compounds.

The use of dogs (trained to recognize individual humans by scent) as detectors for recognition of mammalian species revealed this possibility in all 35 species of six orders under study (Sokolov et al., 1997).

ODOR AND SEX RECOGNITION

Sex-specific pheromone components can be secretions of the organs that might be present or more highly developed in only one sex. Ordinary skin glands can also contain sex-specific chemical components. Thus, in male and female reindeer, the qualitative group composition of secretion of the interdigital gland was identical; whereas qualitative and quantitative sexual distinctions were revealed in the secretion of the tarsal gland (Sokolov et al., 1977). The secretion of the subcaudal gland of mature female and male desman differed from one another in the amount of aliphatic saturated and unsaturated propilketones, even though their qualitative composition was identical. An algorithm for sex recognition in desman by the gas chromatographic analysis of secretion of the subcaudal gland was developed on the basis of ratio between six (of 16 identified) propilketones: three of them were always in greater proportion in females, and the others were in males (Sokolov and Zinkevich, 1977, 1979).

The secretion of the anal gland of American mink is a complex mixture of many groups of compounds (proteins, carbohydrates, and lipids) containing substantial amount of volatile aliphatic acids. A set of acids containing from two to six carbon atoms is identical in both males and females. However, according to averaged data (100 animals of either sex were studied) females differed from males by relative prevalence of acetic, propionic, and n-butyric acids that enabled to reveal sex-specific characteristics (Sokolov et al., 1975).

In male Norway rats, tests of investigative responses to female olfactory signals revealed that the presence of urine acids in the sample was necessary to recognize sex and estrous state of a conspecific female (Sokolov *et al.*, 1986). Male Norway rats could distinguish sex in conspecifics and in black rats (Sokolov *et al.*, 1983c). Similarly, female Norway rats distinguished sex of relatives (Sokolov *et al.*, 1983b).

Vaginal secretions from female golden hamsters evoked copulatory responses in males even when they were applied to a mechanical model. Using the method of chemical subtraction, we revealed that, in such a behavioral test, the necessary and sufficient factors were the volatile components containing sulfur. It was revealed that hamsters and humans were highly sensitive to these substances. To determine compounds containing sulfur, we performed the gas chromatographic analysis with the use of the human olfactory system and revealed thiols and (or) mono- and polysulfides containing 4-6 carbon atoms in the composition of the vaginal secretions. Chemical subtraction of di- and other polysulfides (including the attractant, dimethyldisulfide) from the gas phase of the vaginal compounds revealed that they did not cause the sexual behavior in males. It was prepared 20-component mixture containing synthetic thiols and sulfides (in identical proportions) and applied it to a model that resulted in the same sexual behavior of males as in the case of the vaginal secretions. Using the biotest, we extracted the components of this mixture, which were necessary and sufficient to stimulate sexual behavior: tert. butanthiol, sec. isopentanthiols, and tert. pentanthiol. Only the mixture of these three components stimulated sexual behavior of males in more than 50% of cases that corresponds to the action of the vaginal secretions. Similar olfactory signal, designated by us as an artificial pheromone (Sokolov *et al.*, 1985), was a basically new biologically active composition simulating the olfactory pattern.

ODOR AND INDIVIDUAL RECOGNITION

Interrelations between animals, in particular, formation and maintenance of the social structure of communities is often based on individual recognition. Accumulation of data on the chemical composition of olfactory signals providing individual recognition by animals is at the initial stage. Individual recognition by scent is determined by the presence of the invariant characteristics of each individual, the storage system for these characteristics, and the algorithms of comparisons needed to compare a sampled odor to memorized characteristics from the other samples.

Aggressive responses of dominant male house mice and Norway rats to an individual familiar to those animals can be evoked if the familiar animals is scented with the odors of an unfamiliar male. In house mice, the neutral components of urine of strange males applied to subdominant males caused aggressive response dominant males in 100% of cases; bases (pH>7) and acids (pH<7) led to this in 71 and 33%, respectively. The mixture of dimethyl- and diethylamines (100 mkl of 0.1% water solution) prevailing among the urine bases of house mice was related to aggressive responses in more than 50% of cases. The mechanism by which aggressive responses to a subordinate animal treated with amines was not revealed, although the amines might have simply modified individual scent or otherwise impeded its recognition. In similar experiments, aggressive responses of male Norway rats were evoked to a greater extent (more than 50% of cases) by urine acids; the neutral components were half as effective; and animals were indifferent to the bases.

A universal technique of highly reliable individual identification by the odorous traces in the laboratory conditions was developed (Sulimov and Starovoitov, 1989). The experiments have revealed that trained dogs can easily recognize an individual human scent irrespective of which kind of secretions are used for memorizing (washing-off from the skin surface or blood) and which of these samples is used for recognition. Consequently, these samples contain identical components suitable for identification. The experiments performed on dogs trained and permanently used to identify humans by scent have revealed that the fraction of acids was responsible for this recognition, at least in some dogs (Zinkevich et al., 1993). A dog recipient recognized an individual human scent in the acid fraction of the secretions composed mainly of aliphatic acids and did not recognize it in the neutral and basic fractions. The higher fatty acids are perceptible by dogs in extremely low concentrations (Sulimov et al., 1995). To date, a common principle of encoding individual scents in all terrestrial vertebrates has not been proved directly; the individual scent code has not been deciphered. However, such uniformity is indirectly evidenced by similarity of individual scents in the representatives of different taxa. Thus, a dog recipient trained to recognize human scents can easily recognize animals of the other species phylogenetically remote from humans and from each other, e.g., brown frog, Norway rat, pigeon, tiger, etc. (Sokolov et al., 1990b).

To date, though high sensitivity of dog olfaction to certain compounds has been recognized, but it is not known: (1) which particular substances are used by dogs to recognize a person and (2) whether the informative characters of human odors are qualitative or quantitative or both. Moreover, there are a great number of individual distinctions in the odorous components of humans. Finally, whether or not individual dogs use the same set of informative characteristics has not been revealed.

Experiments with the dogs trained to recognize a target individual in artificially prepared mixtures of the individual scents of ten persons (controllable conditions) have corroborated that, for dogs, a particular individual scent is not masked by the scents of ten other persons. Among many mixtures of individual human scents, a dog can easily recognize the mixture containing the target scent. This may mean that the system for storage-recognition of human individual scents functions in dogs as in rats, using either qualitative distinctions in the composition of the odorous secretions, or concentration differences among the informative components of a particular individual. Theoretical models of individual chemical coding and recognition have been developed (Zinkevich and Minor, 1997).

Dogs trained to recognize individual human scents were tested for the ability to memorize and recognize artificial three-component mixtures of palmitic, stearic, and oleic acids in various concentrations (Sulimov et al., 1995). These acids compose the main component of triglycerides of mammalian fat, and present in all individuals. It was revealed that the dogs could memorize a three-component mixture of higher fatty acids and find a duplicate of the sample, which was initially memorized by them, without a special additional training on acids. Even when the concentrations of each of three acids in the analyzed samples were slightly modified (60:20:20 and 66.7:16.7:16.7%, respectively), dogs distinguished them correctly. Such ability was most similar to bouquet recognition. Extremely low concentration of determined compounds and the ease of resolving a new experimental problem by dogs are evidence for the presence of specialized olfactory receptors for the recognition of oleic, palmitic, and stearic acids. The quantitative content of these acids may be one of the informative characteristics; however, this need not be the main characteristic used in the analysis of mixed individual scent. Neither instrumental nor sensory analyses of such artificial mixtures by humans enable to detect and, hence, distinguish them because of low concentration.

THE INFLUENCE OF OLFACTORY SIGNALS ON PHYSIOLOGICAL STATE AND BEHAVIOR

Olfactory signals can quickly modify various forms of behavior by changing activity in the central nervous system. Thus, certain olfactory stimuli influence sexual maturation and ovarian cycling (Neprintseva *et al.*, 1996; Sokolov *et al.*, 1990a, 1991). When studying the onset of the first estrus juvenile female Norway rats exposed to olfactory signals from males, reliable synchronization of the first estrus occurred in response to the rhythmic influence (period of eight days) of olfactory signals.

The sex pheromone of boar, 5α-androst-16-en-3-one, initiates lordosis behavior (sexual receptivity) in estrous pigs and stimulates their reproductive functions (Zinkevich, 1988). At present, this compound is a unique olfactory signal for studies of the mechanisms of olfactory communication of pig breeding. Subsequently, boar sex pheromone was revealed not to be a species-specific signal, as it also influenced the state of human sexual cyclicity (Sokolov *et al.*, 1992), as well as that of cows (Sokolov *et al.*, 1995). The latter appeared important in practical cattle breeding (Sokolov *et al.*, 1994a).

The influence of the olfactory stimuli on an organism as a biological system depends on its physiological state. For example, the ovarian cycle results from the interaction of numerous rhythmic processes characterized by periodic changes in amplitude, frequency, degree of equilibrium, and their internal synchronization. Accumulation of knowledge concerning modifications in these parameters under the influence of the standard olfactory stimuli opens a new era in understanding the dynamic organization of the biological systems.

Thus, molecular approaches in the studies on mammalian chemical communication enable us to approach a deeper understanding of the mechanisms by which chemical communication operates at the molecular level, as well as at the level of cells, organs, organisms, and communities. As a result of these researches, the first models for communicative process have been constructed. **Qualitative chemosemantic models** are based on the conformity between the discrete chemical characteristics of an olfactory signal and meaningful units and enable to perform a new informative semantic approach to resolving such problems as discovery of the semantic structure of a signal, determination of relations between semantic markers, determination of specificity of the chemical code and its position among the other semiotic systems (Sokolov *et al.*, 1983a, 1984). **Quantitative biorhythmic models** are based on the notion of dependence of periodic modifications of the organism state on the rhythm of chemical signals and enable to optimize the parameters of influence of the olfactory signals on the reproductive state of female, as well as to control even slight quantitative modifications in the extent of synchronization of the sexual cycles (Sokolov *et al.*, 1990a, 1991).

ACKNOWLEDGMENTS

This work was supported in part by the Russian Foundation for Basic Research, projects 98-04-48928 and 99-04-48192.

REFERENCES

Albone, E. S., 1984, *Mammalian Semiochemistry*, J. Wiley, New York.
Aronov, E. V., and Zinkevich, E. P., 1993, Molecular Design of Substances with the Androstenone Odor: 2,4'-substituted 4-Cyclohexylcyclohexanones, a New Class of Androstenone-like Odorants, *Chem. Sens.* **18**:229-243.
Krutova, V. I., and Zinkevich, E. P., 1997, Individual Scent in Ontogeny of Norway Rats (Rattus norvegicus), *Sensornye Sistemy* **11**:157-167 (in Russian).
Neprintseva, E. S., Belaya, Z. A., and Zinkevich, E. P., l996, Synchronization of Entering Estrous Cycle in Female Norway Rats under the Effect of Male Chemical Signals, *Sensory Systems* **10**:67-72.
Sokolov, V. E., Albone, E. S., Flood, P. F., Heap, P. F., Kagan, M. Z., Vasilieva, V. S., Roznov, V. V., and Zinkevich, E. P., 1980a, Secretion and Secretory Tissues of the Anal Sac of the Mink, Mustela vison: Chemical and Histological Studies, *J. Chem. Ecol.* **66**:805-825.
Sokolov, V. E., Aleinikov, P. A., and Zinkevich, E. P., 1983a, Chemical Communication in Animals: The Behavior and Chemosemantic Structure of Signals. In: *Animal Behavior in Communities*, Moscow: Nauka **2**:52-55 (in Russian).
Sokolov, V. E., Aleinikov, P. A., and Zinkevich, E. P., 1984, Chemical Communication in Mammals: A Challenge to a Biologist, a Semioticist and a Chemist, *Acta Zool. Fenn,* **171**:35-37.
Sokolov, V. E., Brundin, A., Chikildin, B. S., and Zinkevich, E. P., 1974, Chemical Composition of Interdigital Gland Secretion in Rangifer tarandus, *Chem. Nat. Compounds,* 654-655 (in Russian).
Sokolov, V. E., Brundin, A., and Zinkevich, E. P., 1977, Differences in Chemical Composition of Cutaneous Gland Secretion in Reindeer (Rangifer tarandus), *Dokl. Akad. Nauk,* **237**:1529-1532 (in Russian).
Sokolov, V. E., Chernyshev, M. K., Voznesenskaya, V. V., and Zinkevich, E.P., 1990a, A Biorhythmological Approach to Estimating the Influence of Olfactory Signals on the Reproductive State of Norway Rats, Rattus norvegicus, *Izv. Akad. Nauk, Ser. Biol,* **2**:248-260 (in Russian).
Sokolov, V. E., Chernyshev, M. K., Voznesenskaya, V. V., Parfenova, V. M., Lebedev, V. S., and Zinkevich, E. P., 1991, A Method for Optimizing the Treatment with Olfactory Signals Aimed at Synchronization of Estrous Cycles in Mammals: An Example of a Group of Female Norway Rats (Rattus norvegicus), in: *Problems of Chemical Communication in Animals*, Moscow: Nauka pp. 404-415 (in Russian).
Sokolov, V. E., Chikildin, B. S., and Zinkevich, E. P., 1975, Free Volatile Aliphatic Acids in Anal Gland Secretion of American Mink (Mustela vison), *Dokl. Akad. Nauk* **220**:220-222 (in Russian).
Sokolov, V. E., Kagan, M. Z., Vasilieva, V. S., Prihodko, V. l., and Zinkevich, E. P., 1987, Musk Deer (Moschus moschiferus): Reinvestigation of Mail Lipid Components from Preputial Gland Secretion, *J. Chem. Ecol.* **13**:71-83.
Sokolov, V. E., Karavaeva, E. A., Zinkevich, E. P., Yanishevskii, L. V., Stepakin, V. M., and Barahash, R. I., 1994a, The Effect of FKRS-2, a Synthetic Sex Pheromone, on the Reproductive Function in Cattle, *Veterinariya* **8**:41-43 (in Russian).
Sokolov, V. E., Karavaeva, E. A, Belaya, Z. A., and Zinkevich, E. P., 1995, Interspecific Action of Boar Sex Pheromone: The Effect on Estrous Cycle in Cattle, *Dokl. Akad. Nauk* **341**:l40-142 (in Russian).
Sokolov, V. E., Kotenkova, E. V., and Zinkevich, E. P., 1979, A Pheromone of Conspecificity in House Mice, *Dokl. Akad. Nauk* **246**:766-768 (in Russian).
Sokolov, V. E., Krutova, V. I., Sulimov, K. T., and Zinkevich, E. P., 1997, Olfactory Identification of Individual Species Specificity in Dogs, *Dokl. Akad. Nauk* **356**:716-718 (in Russian).
Sokolov, V. E., Lyapunova, K. L., Surov, A. V., and Zinkevich, E. P., 1978, A Pheromone of Conspecificity in Norway Rats (Rattus norvegicus norvegicus): An Olfactory Image, in: *II Int. Theriol. Congress*, Brno, p. 363.
Sokolov, V. E., Lyapunova, K. L., Surov, A. V., Kuryatov, N. S., and Zinkevich, E. P., 1979b, A Pheromone of Conspecificity in Norway Rats, *Dokl. Akad. Nauk* **248**:1506-1509 (in Russian).
Sokolov, V. E., Pronkova, O. Yu., and Zinkevich, E. P., l983c, Olfactory Identification of the Sex and Physiological State of Conspecific Individuals in Female Norway Rats (Rattus norvegicus), in: *II All-Union Conf. on Chemical Communication in Animals*, Moscow, 126 (in Russian).
Sokolov, V. E., Sudeikin, V. A., Kotenkova, E. V., and Zinkevich, E. P., 1980b, The Response of House Mice to Olfactory Signals from Conspecific Animals under Natural and Simulated Natural Conditions, in: *Proc. V All-Union Conf. on Rodents*, Moscow: Nauka pp. 280-281 (in Russian).
Sokolov, V. E., Sulimov, K. T., and Krutova, V. I., 1990b, Identification by dogs of Individual Scents in Metabolic Traces Left by Vertebrates of Four Species, *Izv. Akad. Nauk, Ser. Biol.* **4**:556-564 (in Russian).

Sokolov, V. E., Surov, A. V., Vasilieva, N. Yu., and Zinkevich, E. P., 1980c, Chemical Signals in Sex and Species Recognition in Mesocricetus auratus Hamsters. in: *Proc. V All-Union Conf. on Rodents*, Moscow: Nauka, pp. 282–283 (in Russian).

Sokolov, V. E., Surov, A. V., Vasilieva, N. Yu., and Zinkevich, E. P., 1985, Stimulation of Sexual Behavior in Male Golden Hamsters (Mesocricetus auratus) by Artificial Olfactory Signals, *Dokl. Acad. Nauk* **282**:765-768 (in Russian).

Sokolov, V. E., Voznessenskaya, V. V., and Zinkevich, E. P., 1992, Olfactory Cues and Ovarian Cycles. in: *Chemical Signals in Vertebrates 6*. New York: Plenum, pp. 267-270.

Sokolov, V. E., Zaumyslova, O. Yu., and Zinkevich, E. P., 1983b, Identification of the Sex of Animals Belonging to the Same and Different Species by Olfactory Signals in Male Norway Rats, in: *The Ecology and Medical Significance of Norway rat, Rattus norvegicus*, Moscow: Nauka, pp. 50-51 (in Russian).

Sokolov, V. E., Zaumyslova, O. Yu., and Zinkevich, E. P., 1986, An Experimental Model for Studying Effects of Olfactory Signals on the Aggressive Behavior of Male Norway Rats (Rattus norvegicus), in: *Chemical Communication in Animals: Theory and Practice*, Moscow: Nauka, pp. 325-329 (in Russian).

Sokolov, V. E., and Zinkevich, E. P., 1974, Principles of Chemical Communication of Terrestrial Mammals, in: *I Int. Congress on Mammals*, Moscow, **2**:210-211.

Sokolov, V. E., and Zinkevich, E. P., 1977, Molecular Aspects of Chemical Communication in Mammals: Encoding of Information on Individual and Its Physiological State by Volatile Secretion Components, in: *IFAC Symp. on Control Mechanisms in Bio- and Ecosystems, Leipzig*, pp. 35-43.

Sokolov, V. E., and Zinkevich, E. P., 1979, Chemical Communication of Mammals, in: *Science and Humankind*, Moscow: Znanie, pp. 129–137 (in Russian).

Sokolov, V. E., and Zinkevich, E. P., 1983, Animal Communication: Scent Instead of Words, *Nauka SSSR*, **4**:84-95.

Sokolov, V. E., and Zinkevich, E. P., 1986, Main Objectives of Studies on Chemical Communication in Mammals in: *Chemical Communication in Animals: Theory and Practice*, Moscow: Nauka, pp. 213–219 (in Russian).

Sokolov, V. E., Zinkevich, E. P., Kuryatov, N. S., Vasilieva, V. S., Belaya, Z. A., Brodsky E. S., and Kluev N. A., 1994b, Chemical Taxonomic Indexes of Mammals. *Dokl. Akad. Nauk* **336**:428-430 (in Russian).

Sulimov, K. T., and Starovoitov, V. I., 1989, *Methodological Guidelines on The Use of Olfactory Evidence from the Scent of a Crime in the Investigation and Solution of Criminal Cases*, Moscow: VNII MVD SSSR (in Russian).

Sulimov, K. T., Starovoitov, V. I., Moiseeva, T. F., Poletaeva, I. I., and Zinkevich, E. P., 1995, Dogs Distinguish by Scent Quantitatively Different Mixtures of Three Higher Fatty Acids, *Sensory Systems* **9**:99-102.

Zinkevich, E. P., 1988, Effects of Olfactory Signals on the Behavior and Physiological State of Domestic Pigs, in: *Current Problems of Morphology and Ecology of Higher Vertebrates*, Moscow, **2**:327-344 (in Russian).

Zinkevich, E. P., Aronov, E. A., Kagan, M. Z., Gulevich, Yu. V., Kraevskaya, M. A., Vasilieva, V. S., Bihovskaya, M. B., and Zhorov B. S., 1992, The Dependence of Odor on Molecular Structure in a Series of 5α-Androst-16-en-3-one, *Sensornye Sistemy* **6**:115-119 (in Russian).

Zinkevich, E. P., Brodskii, E. S., Moiseeva, T. F., and Gabel, Yu. B., 1997, Volatile Components of Human Skin Surface Secretions, *Sensornye Sistemy* **11**(1):42-52 (in Russian).

Zinkevich, E. P., and Minor A. V., 1997, Human Individual Odors: Recognition by Olfaction and Artificial Sensors, *Sensory Systems* **11**:289-301.

Zinkevich, E. P., Moiseeva, T. F., Starovoitov, V. I., and Sulimov, K. T., 1993, The Individualizing Substances in Human Scent Traces, in: *Applied Forensic Science and New Research Methods*, Moscow: VNII Sudebn. Ekspert. **11**:6-13 (in Russian).

NEURAL MECHANISMS OF COMMUNICATION: FROM PHEROMONES TO MOSAIC SIGNALS

Robert E. Johnston

Department of Psychology
Uris Hall
Cornell University
Ithaca, NY 14853

INTRODUCTION

The purpose of this chapter is, first, to review concepts about the types of chemical signals that are used for communication, ranging from classic, single-compound pheromones to complex mixtures used for social recognition and, second, to summarize what we know about the neural mechanisms underlying responsiveness to these different types of signals in mammals. The specific examples come primarily from my own research and that of others on golden hamsters.

THE TYPES OF CHEMICAL SIGNALS

Scientists and other authors use the term "pheromone" in many different ways. Some use the term to refer to any chemical signal between members of a species whereas others, at the opposite extreme, use the term to refer only to a signal consisting of a single substance that has a specific effect on a receiver. In addition, the term pheromone implies to some that the chemical compound or compounds used in the signal are evolutionarily specialized for this purpose and that the sensory and central neural processes underlying responsiveness to the signal are evolved, dedicated, and genetically hard-wired. Although all of these attributes are clearly true in some cases, they are not true in all cases. Thus, considerable confusion can arise about what is meant by the term pheromone, including what is implied, if anything, about the underlying neural mechanisms, the role of genetics and environmental influences, and the nature of the causal mechanisms influencing behavior. Clarification of the terminology should increase in the precision of our thinking and may even increase the quality of our science.

I propose that the term "pheromone" should have a meaning similar to that originally proposed, i.e., a single chemical compound that elicits or facilitates a behavioral or physiological response in a member of the same species (Johnston, 2000). A "pheromone blend" is a signal that consists of a small number of chemical compounds in relatively precise proportions that has similar effects; the particular proportions of constituents is essential for a full-scale response (Linn and Roelofs, 1989). In most cases, responses to pheromones and pheromone blends will be mediated by relatively hard-wired neural mechanisms (receptors to central processing mechanisms). However, in some cases these mechanisms may also be influenced by experience, either during early development or during adulthood (Johnston and Zahorick, 1975; Johnston et al., 1978).

As so defined, the terms pheromone and pheromone blend encompass only a portion of the types of chemical signals that animals use for communication. Although I do not think that it is useful at this point in time to develop a detailed taxonomy of types of chemical signals, I do think that it is useful to characterize at least one other type, which I call "mosaic signals" (Johnston, 2000). Such signals are composed of a mixture of a large number of chemical compounds, the proportions of which may vary across individuals or classes of individuals (e.g., kin groups, colonies, castes, etc) and these mixtures provide information about individual or class identity. The significance of such signals necessarily depends on some kind of experience. Because of this, and because the perception of complex mixtures must be a more complex process than perception of one or a few ompounds, the mechanisms underlying responses to mosaic signals must differ from those underlying pheromones or pheromone blends.

Note that this way of defining the types of chemical signals is separate from various schemes to identify functional types of signals, such as primers, releasers, modulators, or information providers. These functional categories are also useful, but they should be independent of terms that characterize the types of chemical signal. If they are not kept separate there is more chance for confusion; for example, it might be assumed that one type of chemical signal (e.g., a mosaic signal), is always associated with one type of functional category (e.g., an information provider). Such assumptions are not necessarily correct; e.g., the odors of an animal's kin may release or modulate behavioral responses and will often have effects different than the odors of non-kin. Similarly, a pheromone may provide information about the sender (e.g., its sex or social status) as well as having effects on a specific behavioral or endocrine response.

In the remainder of this chapter I examine what we know about the similarities and differences in the neural mechanisms underlying responses to pheromones versus mosaic signals, using the golden hamster as a model species.

TO WHAT EXTENT IS THE VOMERONASAL ORGAN AND ACCESSORY OLFACTORY SYSTEM SPECIALIZED FOR RESPONSES TO PHEROMONES?

The Vomeronasal Organ and Responses to Pheromones

It is clear that the vomeronasal organ of several mammalian species does have a special role in mediating responses to some pheromone-like signals (Wysocki and Meredith, 1987; Wysocki, 1989; Johnston, 2000;). The vomeronasal organ seems to be particularly important for the mediation of hormonal responses to odors and for aspects of behavior related to reproduction. For example, in laboratory house mice the vomeronasal organ mediates responses to factors in male urine that modulate the female's estrous cycle, the timing of puberty, and the blocking of pregnancy by a strange male (Wysocki and Meredith, 1987; Wysocki, 1989; Keverne, 1999). In golden hamsters, males that are exposed to the vaginal secretions of females experience a surge of circulating testosterone (Macrides et al., 1974); the chemical composition of the signal is not known. This response occurs in both sexually experienced and sexually naive males, suggesting that this it is not dependent on learning about the odors of females as an adult. Removal of the vomeronasal organ eliminates this hormonal response to odors in both types of males, indicating that the vomeronasal organ and accessory olfactory system mediate these hormonal responses via a mechanism that is either entirely genetically determined or that develops early in life, perhaps dependent on interactions of male pups with their mothers' built-in mechanism (Pfeiffer and Johnston, 1994). It is interesting to note, however, that whereas sexually naive males with lesions of the VNO do not show an increase in circulating androgens after exposure to estrous females, sexually experienced males do show such a response. Since experienced males with lesions of both the VNO and the main olfactory epithelium fail to show a hormonal response to females, these experiments indicate that the main olfactory system is also capable of mediating an androgen response to odors, but only after sexual experience. In addition, they indicate that odors that are initially neutral can come to have hormone-eliciting properties. Thus, these experiments show both that the vomeronasal organ mediates a hormonal response to a putative pheromone and that the main olfactory system can mediate the same response via learned significance of odors that are not pheromones.

One of the first functions of the vomeronasal organ to be discovered was its role in the copulatory behavior of male hamsters (Powers and Winans, 1975; Winans and Powers, 1977). It was previously known that removal of the entire olfactory bulb eliminated mating by males, and it is now clear that removal of just the vomeronasal organ results in substantial deficits in copulatory performance. The effect is greater in sexually naive males than in sexually experienced males (Meredith, 1986), suggesting that the vomeronasal organ mediates relatively hard-wired increases in sexual arousal and performance in male hamsters in response to odors of females but that the VNO is not essential for mating. The main olfactory system mediates sexual responses to odors with learned significance and may also mediate some hard-wired responses to odors. The joint participation of both the main olfactory and accessory olfactory systems in mediating specific responses, such as copulation, may be a wide-spread phenomena among mammals and perhaps other vertebrates (Johnston, 1998).

The sources and chemistry of odors that influence male hamster sexual behavior have also been investigated. The vaginal secretion of females has been most thoroughly studied, and it has been shown that (a) removal of the vagina, and thereby vaginal secretions, causes a reduction in male sexual arousal and performance (Kwan and Johnston, 1980), (b) unknown substances, primarily in the volatile fraction of vaginal secretions, act as sexual attractants (Singer, et al., 1976; O'Connell and Meredith, 1984; Petrulis and Johnston, 1995), (c) substances found primarily in a high-molecular weight fraction of the vaginal secretion are effective in stimulating male copulation. Although it has been suggested that a protein in this fraction is a sex pheromone, more recent results suggest that the active signal may be either one or more ligands bound to the protein or that the effective signal may be the protein-ligand combination (Singer et al., 1987; Singer and Macrides, 1990; Jang et al., 1995; 2000). In addition, other odors influence male sexual behavior. Removal of 2 or 3 sources of odor (flank gland, ear glands, Harderian glands) from females already lacking vaginal secretions further reduces the level of sexual behavior shown toward these females. Furthermore, by thoroughly washing such females, male sexual behavior was almost completely eliminated, suggesting that odors are essential for male sexual behavior but that no one source of odor is necessary (Johnston, 1986).

In sum, these experiments indicate that there may well be one or more classic pheromones or pheromone blends from female vaginal secretions that attract males, stimulate male copulation, or increase circulating androgens in males. These odors act in part through the vomeronasal organ and accessory olfactory system. In addition, other odor sources contribute to sexual arousal, perhaps in part through learning, and some of these effects are mediated through the main olfactory system.

Vomeronasal Organ and Responses to Mosaic Signals

There has been controversy about whether the vomeronasal organ is involved in discrimination and recognition of individuals. In house mice, females that mate with one male but are then exposed to a second, unknown male (or his odors) within four days have a significantly higher frequency of failure of pregnancy than females not exposed to a strange male (the pregnancy-block effect). This effect is mediated by hormonal changes in the female upon exposure to the strange male, and it is dependent on the discrimination of a stranger (Parkes and Bruce, 1961). A number of studies have implicated the accessory olfactory bulb as a location involved in the memory for the mate and/or recognition of the stranger (Kaba *et al.,* 1989; Keverne, 1998; Keverne, 1999). In contrast, we have found in golden hamsters that the Coolidge effect (the revival of sexual appetite in sexually satiated males when they are placed with a new female) is not eliminated by removal of the vomeronasal organ but is eliminated by treatment of the olfactory mucosa with zinc sulfate, suggesting that mate recognition by male hamsters is not dependent on the VNO (Johnston and Rasmussen, 1984).

Recently we have taken a more systematic look at the role of the vomeronasal organ in discrimination of individual odors by golden hamsters (Johnston, 1998; Petrulis *et al.,* 1999a; Johnston and Peng, 2000). In this series of experiments, we investigated how removal of the vomeronasal organ affected the ability of male and female hamsters to discriminate between several different odors of male scent donors. We used a habituation paradigm in which we first presented an odor (e.g., flank gland) from male 1 on four trials and then presented the same type of odor from male 2 on a test trial. We measured the amount of time that subjects investigated these odors. Over the four habituation trials investigation decreases but then it increases on the test trial when scent from a different male is presented. This increase in investigation provides evidence for discrimination of the odors of the two scent donors. The effects of removal of the VNO were surprising. We found that these lesions did not affect the ability of females to discriminate between the odors of two males, but they did affect the ability of males to do so (Johnston and Peng, 2000; Petrulis *et al.,* 1999a). Lesioned males tested with flank gland secretions no longer discriminated between flank gland odors of other males, but they did discriminate between male urine odors and between female vaginal secretions. These results clearly show a role for the vomeronasal organ in discrimination of some odors from different individuals. They also indicate that there is a sex difference in the role of the VNO in discrimination of individual odors, at least as measured by this task.

An interesting additional twist on these findings is that the results also depended on the type of task employed. The other task we used also involved habituation to an odor of male 1 on 4 habituation trials, but on the test trial we presented the odors of both male 1 and male 2, rather than just the odor of male 2. This two-stimulus task is easier, because the familiar and novel odor can be compared simultaneously. In this two-stimulus task we found no effects of removal of the vomeronasal organ on discrimination of individual odors. These results suggest that the VNO contributes to the discrimination of some odors of individuals, and, when the task is difficult, input from the VNO is necessary for discrimination. When the task is easier, however, information from the main olfactory system is sufficient for discrimination.

These findings thus indicate a role for the vomeronasal organ in the perception and response to some mosaic signals used for discrimination of individuals in addition to its importance for responses to pheromones and pheromone-like signals. They also support the prevailing view that the main olfactory system plays a major role in the perception of mosaic signals, such as those involved in discrimination and recognition of individuals.

CENTRAL MECHANISMS OF SOCIAL DISCRIMINATION AND RECOGNITION

Although much is known about the anatomy of the projections of the main and accessory olfactory bulbs to the brain and about the secondary and tertiary structures involved in olfactory processing, little is known about the functions of these structures. For example, what particular aspects of processing occur in different areas of the brain, and to what extent are general processes such as discrimination of objects or individuals localized in particular loci? In an attempt to approach some of these questions, we have begun a program of research aimed at discovering the neural pathways that are involved in social recognition and memory.

Our initial studies were aimed at finding pathways and areas of the brain that were involved in discrimination and recognition of individuals and, as a contrast, sexual recognition (Petrulis *et al.*, 1998; Petrulis and Johnston, 1999; Petrulis *et al.*, 1999b; Petrulis *et al.*, 2000). We used habituation tasks with a two stimulus test trial to investigate the ability of female hamsters to discriminate the odors of individuals and two tasks to investigate sex recognition by odors, a Y-maze preference task and scent marking in response to odors. Because of space limitations, I will review primarily the findings on sex preferences and discrimination of individual odors in this chapter.

Lesions of the lateral olfactory tract, which interrupted projections from both the main olfactory bulb and the accessory olfactory bulb, eliminated both the ability to discriminate between odors of individuals and the preference that estrous females normally show for the whole-body odors of males over those of females (Petrulis *et al.*, 1999b). Lesions of the agranular insular cortex, which receives main olfactory projections from both the piriform cortex and from the mediodorsal nucleus of the thalamus, had no effect on either sex-odor preferences or discrimination of individual odors. This result was somewhat surprising, since the olfactory areas of the frontal cortex have been shown to be involved in some types of trained odor discriminations (Eichenbaum *et al.*, 1980; Whishaw *et al.*, 1992; Eichenbaum *et al.*, 1983). These negative results are consistent, however, with the interpretation that the frontal cortex is not involved in the basic discrimination process but rather is necessary for the use of this information in certain kinds of tasks (Eichenbaum *et al.*, 1983; Otto and Eichenbaum, 1992; Wishaw *et al.*, 1992).

Lesions of the medial amygdala did have striking effects on some of our tasks. These lesions completely eliminated the preference of estrous females for odors of males over those of females (Petrulis and Johnston, 1999). In addition, these lesions greatly reduced the level of both flank marking and vaginal marking. Interestingly, lesioned females still vaginal marked more towards the odors of males than toward the odors of females. Thus, the medial amygdala is necessary for a preference for male odors as measured in the Y-maze but is not necessary for differential marking toward the odors of males and females. This suggests that the basic discrimination between the odors of males and females does not take place in the medial amygdala but rather that this information is obtained elsewhere and distributed to a number of loci involved in particular behavioral responses. (Alternatively, sexual discrimination could be carried out in a number of different loci, each specific to certain types of responses). Finally, lesions of the medial amygdala did not influence the ability of females to discriminate between the odors of individual males.

We also investigated the effects of aspiration lesions centered on the entorhinal cortex. These lesions usually extended to some surrounding tissue as well, including the peri-rhinal cortex and temporal cortex, as well as underlying structures such as the posteromedial cortical amygdala and ventral subicular complex; we thus characterized these lesions as para-hippocampal lesions. Such lesions had the opposite pattern of effects, compared to the medial amygdala lesions, on sex preferences and discrimination of individual odors (Petrulis *et al.*, 2000). That is, para-hippocampal lesions had no effect on the preference of females for odors of males over those of females, even though they did reduce the absolute level of investigation of these odors. These lesions did, however, eliminate discrimination between the odors of two males as measured in the habituation task. Lesions of the fimbria/fornix, carried out to determine if the effects observed after para-hippocampal lesions were due to disruption of hippocampal functions, had no effects on either sex-odor preference or discrimination of individual odors.

Thus, this set of brain lesion studies suggests that an important area for processing of the mosaic signals that allow discrimination between individuals is in the para-hippocampal region. The medial amygdala, which receives direct projections from the accessory olfactory bulb and secondary projections from the main olfactory bulb via the cortical amygdala, does not seem to be involved in this process. The medial amygdala is necessary for sex-odor preferences by female hamsters, and this region is also important in mediating odor-stimulated scent marking behavior. Information about the sex of an individual is probably available in many regions of the brain that process olfactory information, including the medial amygdala.

SUMMARY

In this chapter I have summarized research that has shown some of the similarities and differences in the neural processing of pheromone-like signals and mosaic signals in a model mammalian species, the golden hamster. The vomeronasal organ appears to be the main sensory system for mediation of some pheromone-like signals, but this system also is involved in processing of some mosaic signals used for discrimination of individuals by male hamsters (but not by female hamsters). In addition, we have investigated the central brain mechanisms underlying sex and individual odor discrimination, and have discovered loci involved in both of these processes. The medial amygdala is important for at least some differences in response to male versus female odors by female hamsters, and the entorhinal cortex/parahippocampal region is involved in discrimination of individual odors by females.

REFERENCES

Eichenbaum, H., Shedlack, K. J., and Eckmann, K. W., 1980, Thalamocortical mechanisms in odor-guided behavior I. Effects of lesions of the mediodorsal thalamic nucleus and frontal cortex on olfactory discrimination in the rat, *Brain Behav. Evol.* **17**:255-275.

Eichenbaum, H., Clegg, R. A., and Feeley, A., 1983, Reexamination of functical subdivisions of the rodent prefrontal cortex, *Exp. Neurol.* **79**:434-451.

Jang, T., Singer, A. G., and Macrides, F., 1995, Induction of c-fos-gene product in the male hamster accessory olfactory bulbs by natural and bacterially cloned aphrodisin, *Chem. Senses* **20**:712-713.

Johnston, R. E., 1986, Effects of female odors on the sexual behavior of male hamsters, *Behav. Neural Biol.* **46**:168-188.

Johnston, R. E., 1998, Pheromones, the vomeronasal system, and communication, *Ann. NY Acad. Sci.* **855**:333-348.

Johnston, R. E., 2000, Chemical communication and pheromones:the types of signals and the role of the vomeronasal system, in: *The Neurobiology of Taste and Smell, 2nd edition* (T. E. Finger, W. L. Silver, and D. Restrepo, eds.), Wiley, New York, pp. 99-123.

Johnston, R. E., and Peng, M., 2000, The vomeronasal organ is involved in discrimination of individual odors by males but not by females in golden hamsters, *Physiol. Behav.*, **70**: (in press).

Johnston, R. E., and Rasmussen, K., 1984, Individual recognition of female hamsters by males: Role of chemical cues and of the olfactory and vomeronasal systems, *Physiol. Behav.* **33**:95-104.

Johnston, R. E., and Zahorick, D. M., 1975, Taste aversions to sexual attractants, *Science* **189**:893-894.

Johnston, R. E., Zahorick, D. M., Immler, K., and Zakon, H., 1978, Alterations of male sexual behavior by learned aversions to hamster vaginal secretion, *J. Comp. Physiol. Psychol.* **92**:85-93.

Kaba, H., Rosser, A., and Keverne, E. B., 1989, Neural basis of olfactory memory in the context of pregnancy block, *Neuroscience* **32**:657-662.

Keverne, E. B., 1998, Vomeronasal/accessory system and pheromonal recognition, *Chem. Senses* **23**:491-494.

Keverne, E. B., 1999, The vomeronasal organ, *Science* **286**:716-720.

Kwan, M., and Johnston, R. E., 1980, The role of vaginal secretion in hamsters sexual behavior: Males' responses to normal and vaginectomized females and their odors, *J. Comp. Physiol. Psychol.* **94**:905-913.

Linn, C. E., and Roelofs, W. L., 1989, Response specificity of male moths to multicomponent pheromones, *Chem. Senses* **14**:421-437.

Macrides, F., Bartke, A., Fernandez, F., and D'Angelo, W., 1974, Effects of exposure to vaginal odor and receptive females on plasma testosterone in the male hamster, *Neuroendocrinol.* **15**:355-364.

Meredith, M., 1986, Vomeronasal organ removal before sexual experience impairs male hamster mating behavior, *Physiol. Behav.* **36**:737-743.

O'Connell, R. J., and Meredith, M., 1984, Effects of volatile and non-volatile chemical signals on male sex behaviors mediated by the main and accessory olfactory systems, *Behav. Neurosci.* **98**:1083-1093.

Otto, T., and Eichenbaum, H., 1992, Olfactory learning and memory in the rat: A "model system" for studies of the neurobiology of memory, in: *Science of Olfaction* (M. J. Serby, and K. L. Chobor, eds.), Springer-Verlag, New York, pp. 213-244.

Parkes, A. S., and Bruce, H. M., 1961, Olfactory stimuli in mammalian reproduction, *Science* **134**:1049-1054.

Petrulis, A., and Johnston, R. E., 1995, A reevaluation of dimethyl disulfide as a sex attractant in golden hamsters, *Physiol. Behav.* **57**:779-784.

Petrulis, A., and Johnston, R. E., 1999, Lesions centered on the medial amygdala impair scent-marking and sex-odor recognition but spare discrimination of individual odors in female golden hamsters, *Behav. Neurosci.* **113**:345-357.

Petrulis, A., DeSouza, I., Schiller, M., and Johnston, R. E., 1998, Role of the frontal cortex in social odor discrimination and scent-marking in female golden hamsters (*Mesocricetus aruatus*), *Behav. Neurosci.* **112**:199-212.

Petrulis, A., Peng, M., and Johnston, R. E., 1999a, Effects of vomeronasal organ removal on individual odor discrimination, sex-odor preference, and scent marking in female hamsters, *Physiol. Behav.* **66**:73-83.

Petrulis, A., Peng, M., and Johnston, R. E., 1999b, Lateral olfactory tract transections impair discrimination of individual odors, sex odor preferences, and scent marking in female golden hamsters (*Mesocricetus auratus*), in: *Advances in Chemical Signals in Vertebrates* (R. E. Johnston, D. Müller-Schwarze, and P. W. Sorensen, eds.), Plenum Press, New York, pp. 549-561.

Petrulis, A., Peng, M., and Johnston, R. E., 2000, The role of the hippocampal system in social odor discrimination and scent marking in female golden hamsters (*Mesocricetus aruatus*), *Behav. Neurosci.* **114**:184-195.

Pfeiffer, C. A., and Johnston, R. E., 1994, Hormonal and behavioral responses of male hamsters to females and female odors: Roles of olfaction, the vomeronasal system, and sexual experience, *Physiol. Behav.* **55**:129-138.

Powers, J. B., and Winans, S. S., 1975, Vomeronasal organ: critical role in mediating sexual behavior of the male hamster, *Science* **187**:961-963.

Singer, A. G., Agosta, W. C., and Clancy, A. N., 1987, The chemistry of vomeronasally detected pheromones: Characterization of an aphrodisic protein, *Ann. N.Y. Acad. Sci.* **519**:287-298.

Singer, A. G., Agosta, W. C., O'Connell, R. J., Pfaffmann, C., Bowen, D. V., and Field, F.H., 1976, Dimethyl disulfide: An attractant pheromone in hamster vaginal secretion, *Science* **191**:948-950.

Singer, A. G., and Macrides, F., 1990, Aphrodisin: pheromone or transducer, *Chem. Senses* **15**:199-203.

Whishaw, I. Q., Tomie, J.-A., and Kolb, B., 1992, Ventrolateral prefrontal cortex lesions in rats impair the acquisition and retention of a tactile-olfactory configural task, *Behav. Neurosci.* **106**:597-603.

Winans, S. S., and Powers, J. B., 1977, Olfactory and vomeronasal deafferentation of male hamsters: histological and behavioral analyses, *Brain Res.* **126**:325-344.

Wysocki, C. J., 1989, Vomeronasal chemoreception: its role in reproductive fitness and physiology, in: *Neural Control of Reproductive Function* (J. M. Lakoski, J. R. Perez-Polo, and D. K. Rassin, eds.), A. R. Liss, New York, pp. 545-566.

Wysocki, C. J., and Meredith, M., 1987, The vomeronasal system, in: *Neurobiology of Taste and Smell* (T. E. Finger, and W. L. Silver, eds.), John Wiley and Sons, New York, pp. 125-150.

A UNIQUE SUBFAMILY OF OLFACTORY RECEPTORS

Jörg Strotmann

Institute of Physiology, University of Hohenheim
Garbenstrasse 30, D-70593 Stuttgart, Germany

INTRODUCTION

The sense of smell allows terrestrial animals to respond to an immense variety of odorant molecules. The discriminatory capacity of this sense is derived from precise information processing at different levels of the olfactory system, beginning in the chemosensory epithelium of the nose, where odorant molecules stimulate distinct sets of olfactory neurons, followed by the activation of particular glomeruli in the main olfactory bulb, and then by the activation of the piriform and the entorhinal cortices and other brain loci.

An odorant molecule in the inhaled air enters the nasal cavity and dissolves in the mucus, where it binds to a specific olfactory receptor (OR) in the ciliary membranes of an olfactory neuron. In mammals, the large repertoire of olfactory chemoreceptors appears to be encoded by about a thousand different OR genes (Buck and Axel, 1991). Each of these genes encodes a distinct seven-transmembrane domain protein, and an individual olfactory neuron is believed to express only one of the thousand OR genes (Malnick *et al.,* 1999). Analyses of the spatial patterns of OR gene expression have revealed that each receptor subtype is restricted to one of several broad but well-circumscribed zones along the anterior-posterior axis within the olfactory epithelium (Ressler *et al.,* 1993; Strotmann *et al.,* 1994; Vassar *et al.,* 1993). Recordings from the surface of the olfactory epithelium have indicated that most odorants accordingly elicit broad patterns of reactivity. With some odorants, however, a more focal reactivity was observed (Mackay-Sim and Kesteven, 1994), leading to the hypothesis that a group of receptors may be expressed in neurons segregated in highly restricted, rather than broad, areas of the olfactory epithelium.

The *OR37* Subfamily

Screening of a cDNA library from rat olfactory epithelium led to the identification of an OR gene (*OR37*) that exhibits a unique expression pattern within the nasal epithelium (Strotmann *et al.,* 1992). In contrast to other olfactory neurons, cells expressing genes from the *OR37* subfamily are concentrated in the center of the turbinate structures (Figure 1).

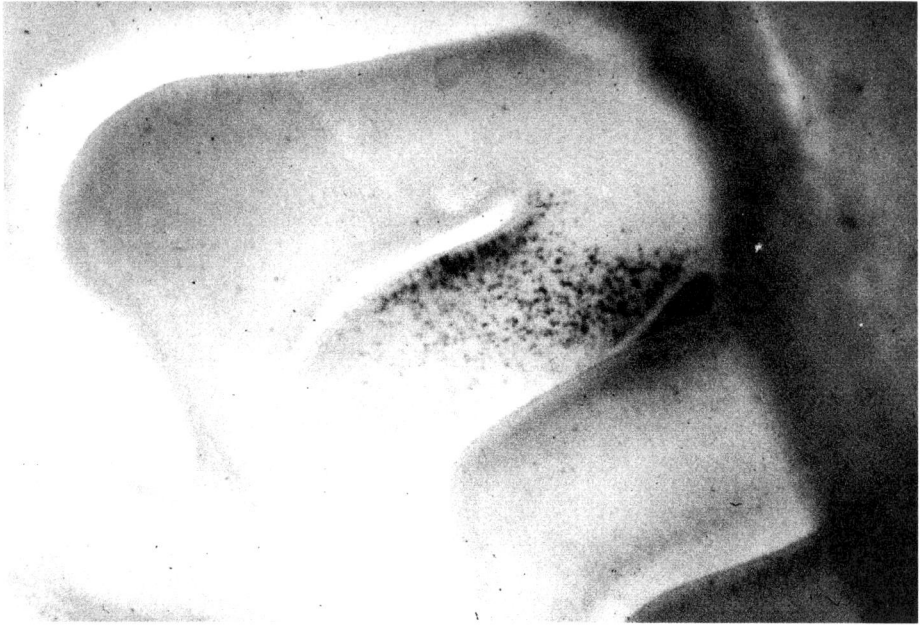

Figure 1. Whole mount *in situ* hybridization of rat turbinate structures using a digoxigenin-labeled *OR37*-specifc antisense riboprobe; reactive neurons are clustered in the center of the turbinates.

On cross sections, these cells are visible only on the tips of endoturbinate II and ectoturbinate 3, facing the nasal septum, with many reactive neurons concentrated in a small area (Strotmann *et al.*, 1994) A subsequent analysis of the rat genome for *OR37*-related genes led to the identification of four genes with sequence identities to *OR37* ranging from 68.6% to 96.9%; hence, these genes form a subfamily designated *OR37A-D*. *In situ* hybridization experiments revealed that the entire subfamily is expressed in the typical *OR37* region of the sensory epithelium. Thus, *OR37*-expressing cells in the nasal epithelium represent the first example of a 'clustered' distribution pattern.

We compared the deduced amino acid sequences of *OR37* receptors to other olfactory receptors from the rat, and found a distinct structural feature: in extracellular loop 3 (E3), the *OR37* proteins include six amino acids not found in other olfactory receptors. Thus, the E3 loop of the *OR37* proteins is significantly longer than that in the other known mammalian ORs (Kubick *et al.*, 1997) The function of the extended E3 loop is unclear. However, this domain is involved in ligand binding in other G-protein coupled receptors (Li *et al.*, 1996; Minami *et al.*, 1995), so it is possible that the ligand spectrum of the *OR37* receptors is critically determined by the extended E3 loop, probably allowing the interaction with a distinct class of odorants.

To determine whether *OR37* receptors may be specific for the rat or whether related genes also exist in other species, an RT-PCR approach was performed. *OR37* related genes were detected in house mice, gerbils, guinea pigs, cattle, and even in opossums, a phylogentically distant mammal. In contrast, these genes were not found in birds, reptiles, or amphibians, so *OR37* might be a mammalian-specific OR type. A detailed sequence analysis revealed that all of the *OR37*-related receptors share the characteristic extension of E3 by six amino acids. Furthermore, two characteristic sequence motifs within this domain (KPKS and DKXXXXDK) are conserved (Kubick *et al.*, 1997).

In situ hybridization experiments performed on the olfactory epithelium of various mammalian species using *OR37*-specific antisense riboprobes revealed that neurons expressing this unique type of receptor are always found clustered (Strotmann *et al.*, 1995). Despite the quite different morphology of turbinates, these neurons were even found in a very similar central location of the nasal cavity, suggesting that their position in the airflow of the nose may be an important parameter. These data indicate that expression of these genes is tightly controlled and its control mechanisms are phylogenetically conserved.

Genomic Organization and Expression Control of *OR37* Genes

Due to the unique and highly conserved expression pattern of *OR37* genes in the nasal epithelium, this subfamily of receptor genes might be a suitable model system to get insight into the organization of receptor genes and eventually into expression control mechanisms. Accordingly, we characterized the full repertoire of *OR37* genes in the house mouse. Screening of genomic DNA libraries led to the identification of five *OR37* subtypes in the mouse (*mOR37A-E*), comprising four functional genes and one pseudogene (Strotmann *et al.*, 1999). Genomic mapping studies revealed that all DNA clones with *mOR37* genes form a contig of DNA, indicating that the mouse *OR37* genes are clustered in the genome. Detailed analyses demonstrated that all five *mOR37* genes are indeed restricted to only 60 kilobases of DNA; within this region the genes are quite regularly spaced with an intergenic distance of about 15 kb. The transcriptional orientation of four genes is identical, only the first gene in the cluster, *mOR37A*, is positioned in opposite orientation.

In close vicinity to the *mOR37* genes, two additional olfactory receptor genes (*mOR17* and *mOR6*) are located, but these genes lack the extension of E3 characteristic of *OR37* receptors and share only about 50% sequence identity to the *mOR37* genes and to each other, so they belong to different subfamilies. However, a phylogenetic analysis revealed that these two receptors are more closely related to the *mOR37* receptors than any other mouse OR gene. *In situ* hybridization experiments employing *mOR17* and *mOR6* specific antisense riboprobes showed that neurons expressing these genes are also clustered in the typical *mOR37* region of the nose, suggesting the involvement of locus-dependent mechanisms for the spatial control of OR gene expression.

This expression control elements for this group of genes were studied because it was anticipated that common motifs are involved in this process. The gene locus was sequenced and analyzed by appropriate computer programs. In particular, the transcription start region for all *mOR37* genes was determined; we also analyzed intron/exon structure, including, for each subtype, a non-interrupted coding exon, and one or two 5' non-coding exons (Hope *et al.*, 2000). For two of the genes, *mOR37B* and *mOR37C*, alternative splice variants were found, but their functional relevance is unclear. It seems conceivable that different splice forms are involved in the subcellular targeting of RNA molecules, since olfactory receptor mRNA has been detected in different compartments of the cell (Ressler *et al.*, 1994; Vassar *et al.*, 1994); alternatively, they could be expressed during different stages of cellular differentiation.

Analyzing the putative transcription start region upstream of the initial exons of each *mOR37* gene, a short stretch of rather high sequence conservation among the *mOR37* genes was identified. Six motif blocks comprising of a pair of AT-rich motifs, GA-rich blocks and two well conserved TCCCA motifs (Figure 2) are most remarkable. Several transcription factors potentially bind to this region, suggesting that it may be involved in expression control of the *mOR37* genes. Interestingly, sequence motifs related to those in the 5' regions of the *mOR37* genes were also identified upstream of *mOR6* and *mOR17*, the two genes that share the characteristic clustered expression pattern with the *mOR37* genes. In contrast, analysis of *mOR23* (Assai *et al.*, 1996), a gene expressed zonally rather than in a clustered pattern, and not related in sequence and chromosomal position to either of these genes, revealed that no such motifs blocks are present upstream of its transcription start region.

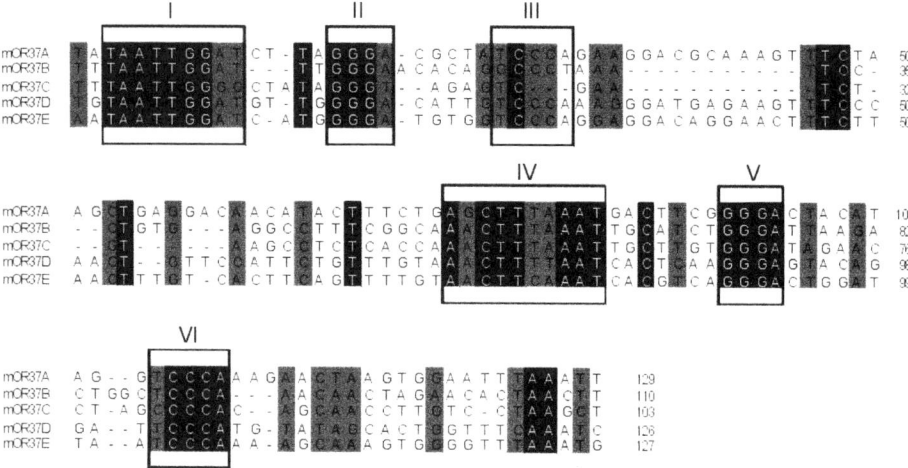

Figure 2. Conserved sequence motifs in the 5' regions of the *mOR37* genes. Identical amino acids in all sequences are shaded in black; highly conserved positions are shaded in grey. Frames (I-VI) indicate characteristic sequence motif blocks present in all five *mOR37* genes. (from Hoppe *et al.*, 2000; Copyright 2000 by Academic Press)

Monoallelic Expression of *mOR37* Genes in Different Neuron Populations

It is possible that the highly homologous *mOR37* genes are expressed in different cell populations or co-expressed within an individual neuron. To approach this question, a transgenic approach in mice was employed leading to the specific labeling, with either *tau-lacZ* or *tau-GFP*, of all neurons expressing the same *mOR37* receptor subtype. By generating double-heterozygous mice that carry two differentially tagged *mOR37* alleles, it was demonstrated on the single cell level that the highly homologous *mOR37* genes are exclusively expressed in different neuron populations (Strotmann *et al.*, 2000). Using the same approach, it was shown that *mOR37* genes are expressed monoallelically in these neurons; in compound heterozygous mice in which the same receptor gene was tagged with different markers, only neurons with one label were detectable, indicating that cells express either the maternal or the paternal allele, but never both.

Laminar Organization of *mOR37* Expressing Neurons

The analysis of *mOR37-tau-lacZ* expressing neurons within the olfactory epithelium revealed an additional level of spatial organization for these cells: their cell bodies reside within distinct epithelial layers (Figure 3) (Strotmann *et al.*, 2000). The majority of neurons expressing *mOR37A-tau-lacZ* are found in the middle layer of the epithelium, and their dendritic processes appear very similar in length (Figure 3a). In a slightly more apical position, *mOR37B-tau-lacZ* expressing neurons are located. Finally, very close to the epithelial surface, only *mOR37C-tau-lacZ* cells, which have very short dendritic processes, are situated (Figure 3b). The significance of this phenomenon is not yet understood, but it might reflect the synchronized differentiation and migration of neurons projecting to a common glomerular target. Alternatively, subpopulations of sensory cells may be trapped at a certain position in the mid-zone of the epithelium, thus resembling the migration of neurons underlying the formation of cellular layers during the development of the cerebral and cerebellar cortex (Rakic, 1972; Rakic, 1981).

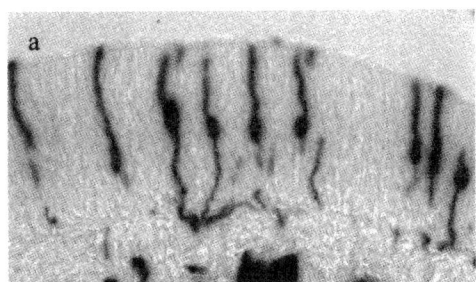
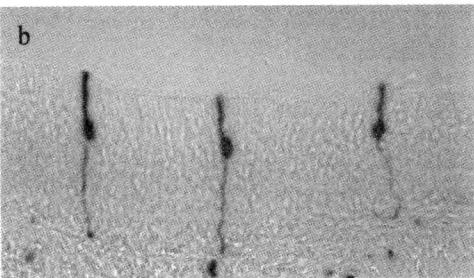

Figure 3. Laminar patterning of *mOR37A-tau-lacZ* (a) and *mOR37C-tau-lacZ* expressing neurons (b) within the nasal neuroepithelium.

Projection Pattern of *mOR37* Expressing Neurons into the Olfactory Bulb

A central question concerning the processing of information detected by olfactory sensory neurons is their connectivity with their targets in the olfactory bulb. For neuron populations that express the same olfactory receptor, and are distributed in broad zones within the olfactory epithelium, it has been demonstrated that their axons converge onto a small number of glomeruli located medially and laterally in the olfactory bulb (Mombaerts *et al.*, 1996, Wang *et al.*, 1998). Since *mOR37* expressing neurons exhibit a unique, clustered distribution within the olfactory epithelium, their axons might be expected to diverge onto a large number of glomeruli in the olfactory bulb, thereby having some kind of modulatory effect on the other neuron populations. To test this hypothesis, we examined transgenic mice, since the histological markers fused to tau, allowed us to visualize the axonal projections of these neurons. It was demonstrated that axons of neurons expressing the *mOR37A-tau-lacZ* gene converge in the olfactory bulb; surprisingly, only one glomerulus per bulb is targeted in the majority of bulbs we analyzed (Figure 4) (Strotmann *et al.*, 2000). This is different from those populations expressed in zones that usually project to at least two glomeruli per bulb. Only in a small fraction of bulbs from *mOR37A-tau-lacZ* mice were there two glomeruli that stained, in these cases they were located rather close together. It is currently unclear why the cells in most cases project to one glomerulus and in some cases to two glomeruli. It has been hypothesized that axonal convergence provides the system with the required sensitivity; if all neurons project to just one glomerulus, then an activation of even a very small percentage of them may result in an activation of its glomerulus, thereby allowing the system to detect even very weak stimuli. In this context, the position of the *mOR37A* glomerulus might be of particular interest, since its location immediately behind the cribriform plate results in strikingly short axonal connections.

The analysis of the projection pattern of two other *mOR37-tau-lacZ* mutations (*mOR37B-tau-lacZ* and *mOR37C-tau-lacZ*) revealed that in the majority of individuals, cells expressing these genes also project to single glomeruli in the olfactory bulb, similar to

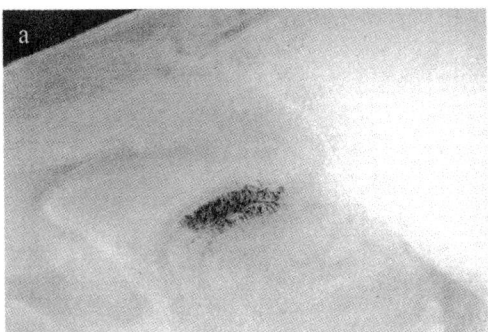 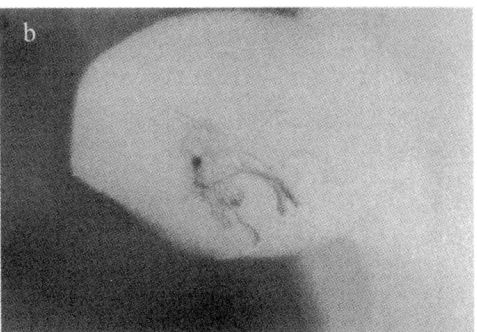

Figure 4. Projection pattern of *mOR37A-tau-lacZ* expressing neurons into the olfactory bulb. An X-gal stained whole mount preparation of the olfactory system reveals a subpopulation of stained neurons clustered in the center of the turbinates (a). Axons of *mOR37A-tau-lacZ* expressing neurons converge onto a single glomerulus on the ventral surface of the olfactory bulb (b). (b from Strotmann *et al.*, 2000; Copyright 2000 by the Society for Neuroscience).

the *mOR37A* expressing neurons (Strotmann *et al.*, 2000). Detailed study of mice double-heterozygous for the two mutations (B and C) demonstrated that each *mOR37* population projects to the same ventral region of the olfactory bulb, but each B and C neuron projects to distinct, albeit neighboring, glomeruli. This location of the different *mOR37* glomeruli in a common focal area of the bulb is probably of fundamental importance for the appropriate processing of odor information by these neuron subpopulations. Periglomerular and granule cells send dendrites into neighboring glomeruli, where they form reciprocal dendro-dendritic synaptic contact with mitral/tufted cells, and provide the basis for lateral inhibition (Mori *et al.*, 1999; Sheperd, 1972). Such inhibitory interactions between neigbouring glomeruli via local interneurons is believed to sharpen contrast between inputs that might be only slightly dissimilar. Thus, the positioning of all *mOR37* glomeruli in close proximity to each other may be of fundamental importance for a proper discrimination of very related odorants; each individual *mOR37* glomerulus may thus contribute to shape the response of the appropriate output neurons.

Differential labeling of distinct *mOR37* glomeruli allowed us to study the relative positioning of glomeruli. It was demonstrated that the position of a distinct glomerulus varies to a considerable degree among bulbs from different animals and even among the two bulbs from one individual (Strotmann *et al.*, 2000). These data are consistent with previous reports demonstrating some degree of variability in the patterning of glomeruli activated upon odorant stimulation (Rubin and Katz, 1999); however, it cannot be excluded that the absence of a strict stereotype pattern may be a special feature for *mOR37* glomeruli. Future studies comparing other related receptor families will provide further insight into the principles underlying the fine architecture of the glomerular sheet.

ACKNOWLEDGMENTS

This work was supported by the Deutsche Forschungsgemeinschaft.

REFERENCES

Asai, H., Kasai, H., Matsuda, Y., Yamazaki, N., Nagawa, F., Sakano, H., and Tsuboi, A., 1996, Genomic structure and transcription of a murine odorant receptor gene: differential initiation of transcription in the olfactory and testicular cells, *Biochem. Biophys Res. Comm.* **221**:240-247.

Buck, L., and Axel, R., 1991, A novel multigene family may encode odorant receptors: a molecular basis for odor recognition, *Cell* **65**:175-187.

Hoppe, R., Weimer, M., Beck, A., Breer, H., and Strotmann, J., 2000, Sequence analyses of the olfactory receptor gene cluster mOR37 on mouse chromosome 4, *Genomics* **66**:284-295.

Kubick, S., Strotmann, J., Andreini, I., and Breer, H., 1997, Subfamily of olfactory receptors characterized by unique structural features and expression patterns, *J.Neurochem.* **69**:465-475.

Li, X., Varga, E. V., Stropova, D., Zalewska, T., Malatynska, E., Knapp, R. J., Roeske, W. R., and Yamamura, H. I., 1996, delta-Opioid receptor: the third extracellular loop determines naltrindole selectivity, *Eur.J.Pharmacol.* **300**:R1-R2.

Mackay-Sim, A., and Kesteven, S., 1994, Topographic patterns of responsiveness to odorants in the rat olfactory epithelium, *J.Neurophysiol.* **71**:150-160

Malnic, B., Hirono, J., Sato, T., and Buck, L. B., 1999, Combinatorial receptor codes for odors, *Cell* **96**:713-723.

Minami, M., Onogi, T., Nakagawa, T., Katao, Y., Aoki, Y., Katsumata, S., and Satoh, M., 1995, DAMGO, a mu-opioid receptor selective ligand, distinguishes between mu-and kappa-opioid receptors at a different region from that for the distinction between mu- and delta-opioid receptors, *FEBS Lett.* **364**:23-27.

Mombaerts, P., Wang, F., Dulac, C., Chao, S. K., Nemes, A., Mendelsohn, M., Edmondson, J., and Axel, R., 1996, Visualizing an olfactory sensory map, *Cell* **15**:675-686.

Mori, K., Nagao, H., and Yoshihara, Y., 1999, The olfactory bulb: coding and processing of odor molecule information, *Science* **286**:711-715.

Rakic, P., 1972, Mode of cell migration to the superficial layers of fetal monkey neocortex, *J. Comp. Neurol.* **145**:61-84.

Rakic, P., 1981, Development events leading to laminar and areal organization of the neocortex, in: Organization of the Cerebral Cortex (F. O. Schmitt, G. Worden, G. Edelman, and S. G. Dennis eds.), MIT Press, Cambridge.

Ressler, K. J., Sullivan, S. L., and Buck, L. B., 1993, A zonal organization of odorant receptor gene expression in the olfactory epithelium, *Cell* **73**:597-609.

Ressler, K. J., Sullivan, S. L., and Buck, L. B., 1994, Information coding in the olfactory system: evidence for a stereotyped and highly organized epitope map in the olfactory bulb, *Cell* **79**:1245-1255.

Rubin, B. D., and Katz, L. C., 1999, Optical imaging of odorant representations in the mammalian olfactory bulb, *Neuron* **23**:499-511.

Shepherd, G. M., 1972, Synaptic organization of the mammalian olfactory bulb, *Physiol. Rev.* **52**:864-917.

Strotmann, J., Beck, A., Kubick, S., and Breer, H., 1995, Topographic patterns of odorant receptor expression in mammals: a comparative study, *J.Comp.Physiol.* **177**:659-666.

Strotmann, J., Conzelmann, S., Beck, A., Feinstein, P., Breer, H., and Mombaerts, P., 2000, Local permutations in the glomerular array of the mouse olfactory bulb, *J. Neurosci.***20**:6927-6938.

Strotmann, J., Hoppe, R., Conzelmann, S., Feinstein, P., Mombaerts, P., and Breer, H., 1999, Small subfamily of olfactory receptor genes: structural features, expression pattern and genomic organization, *Gene* **236**:281-291.

Strotmann, J., Wanner, I., Krieger, J., Raming, K., and Breer, H., 1992, Expression of odorant receptors in spatially restricted subsets of chemosensory neurones, *NeuroReport* **3**:1053-1056.

Strotmann, J., Wanner, I., Helfrich, T., Beck, A., and Breer, H., 1994, Rostro-caudal patterning of receptor-expressing olfactory neurones in the rat nasal cavity, *Cell Tissue Res* **278**:11-20.

Vassar, R., Chao, S. K., Sitcheran, R., Nunez, J. M., Vosshall, L. B., and Axel, R., 1994, Topographic organization of sensory projections to the olfactory bulb, *Cell* **79**:981-991.

Vassar, R., Ngai, J., and Axel, R., 1993, Spatial segregation of odorant receptor expression in the mammalian olfactory epithelium, *Cell* **74**:309-318.

Wang, F., Nemes, A., Mendelsohn, M., and Axel, R., 1998, Odorant receptors govern the formation of a precise topographic map, *Cell* **93**:47-60.

ODOURS ARE REPRESENTED IN GLOMERULAR ACTIVITY PATTERNS: OPTICAL IMAGING STUDIES IN THE INSECT ANTENNAL LOBE

C. Giovanni Galizia[1], Silke Sachse[1], Hanna Mustaparta[2]

[1] Institut für Biologie – Neurobiologie. Freie Universität Berlin
Königin-Luise Str. 28-30 D-14195 Berlin, Germany
[2] Department of Zoology
Norwegian University of Science and Technology
N-7194 Trondheim, Norway

INTRODUCTION

At a meeting devoted to chemical signals in vertebrates a contribution about the coding of odors in the brain of insects appears to be out of place. However, there are several reasons why a comparison of the mechanisms of olfactory coding across the animal kingdom is useful. Prominent among these reasons is the architectural similarity between the olfactory lobe (antennal lobe, AL) in insects and the olfactory bulb (OB) in mammals (reviewed in Hildebrand ans Shepherd, 1997). In both the AL and the OB afferent axons of olfactory receptor cells (ORC) extensively branch in olfactory glomeruli, with each individual neuron only innervating a single glomerulus (Hannson *et al., 1992;* Brockmann and Brückner, 1995; Mombaerts, 1996). In insects, glomeruli are interconnected by inhibitory interneurons local to the AL, which probably parallel the function of the interglomerular and the granule cells in the OB of vertebrates. From within the olfactory glomeruli of AL, projection neurons relay the processed information to higher order brain centers, such as the mushroom bodies and the lateral protocerebrum, and are comparable in function to the mitral/tufted cells in OB. A major advantage when working with insects is that individual glomeruli can be morphologically identified, and therefore their response properties can be characterized within a species by combining physiological measurements with anatomical identification of the measured glomeruli.

The best-known example of insect olfaction is undoubtedly the sexual pheromone system, which has been extensively studied in moths and in cockroaches (Boeckh and Tolbert, 1993; Hildebrand, 1996). In these species, the female attracts males by releasing a bouquet of several substances that together form the species-specific sexual pheromone blend. Males are attracted over long distances following odor plumes by flying upwind as soon as they encounter such a plume (Kaissling, 1997). Often, sympatric species use similar pheromone blends that only differ in the relative proportion of the compounds, or by the addition or lack of some compounds. The olfactory system has the task, therefore, to make an accurate perception of the precise composition of the blended signal, with high sensitivity. Within the AL of male moths, the macroglomerular complex (MGC) is devoted to processing these blends. It consists of several (often quite large) glomeruli, the number and arrangement of which are species-specific, and it is generally located at the entrance of the AL (Hansson and Christensen, 1999). Cockroaches have a single macroglomerulus (Boeckh and Tolbert, 1993), whereas drones (male honeybees) have an MGC consisting of 4 large glomerular structures(Arnold et al., 1985).

Besides the sexual pheromone system, insects have a well-developed olfactory system for general odors, which are often plant-derived substances. Again, the number of glomeruli, their shapes and sizes differ between species. It has been shown in several species that within each species individual glomeruli are identifiable (Rospars, 1988). This has allowed the construction of morphological reference atlases (Flanagan and Mercer, 1989; Rospars and Hildebrand, 1992); (Laissue et al., 1999; Galizia et al., 1999a), which then allow physiological measurements to be related to individual glomeruli, and thus be more accurately compared between individuals.

In this paper, we report about recent results obtained with optical imaging techniques in two species, the moth *Heliothis virescens* and the honeybee *Apis mellifera*. In the moth, we mapped the neural responses of glomeruli in the MGC to presentation of pheromone components, allowing us to confirm the clear functional separation between the MGC and the rest of the AL. These results also confirmed that the pheromone system and the plant-odor system do not use identical coding strategies. In the honeybee, we characterized the molecular response ranges of individual glomeruli, and thus laid the foundations for the development of a physiological atlas of odor responses in this species. Our results yielded some new speculation about the functional and spatial organization of the olfactory code in honeybees.

METHODS

Calcium imaging of the AL in live insects has been described (Jorges et al., 1997; Galizia et al., 1997; Galizia et al., 1998). Very briefly, a hole is cut into the head capsule, and the brain is flooded with saline solution containing the calcium-sensitive dye calcium-green-2-AM. This dye is taken up by cells, and increases its fluorescence with increasing intracellular calcium concentration. After one hour of incubation, excess dye is rinsed off, and the animal fixed under a microscope equipped with a 20x, NA 0.6 objective and a CCD-camera. Spatial resolution varied in different experiments between 2.4µm and 5µm in the plane. Axial resolution is determined by tissue properties, and lies in the range of 50-100µm. Temporal resolution is between 2 and 10 frames per second. The antennae are kept dry at all times. Precise amounts of odors were delivered to the antennae while changes in fluorescence were recorded in the AL. From the raw fluorescence measurements, relative changes were calculated by dividing responses to odors by the background fluorescence, in order to get values, which reflect changes in calcium concentration, indicative of neuronal activity. Only adults were measured, both male and female *Heliothis virescens*, and female forager honeybees *Apis mellifera*.

PHEROMONE CODING IN THE MOTH MGC

The macroglomerular complex of the heliothine moth *Heliothis virescens* consists of 4 glomeruli: the cumulus glomerulus, the dorsomedial compartment, and two ventral compartments. In our imaging studies, we could measure the responses in only three of these glomeruli, since one of the two ventral compartments was out of focus in our preparation. We tested four compounds: the major, Z11-16:AL (*cis*-11-hexadecenal), and minor, Z9-14:AL (*cis*-9-tetradecenal), components of the species-specific sexual pheromone, which occur naturally in a ratio of 16:1. The other two test stimuli, *cis*-11-hexadecenyl acetate (Z11-16:AC) and *cis*-11-hexadecenol (Z11-16:OH), are intraspecific antagonists, i.e., substances which are part of the pheromone blend in sympatric species, and which, if present in the blend, block the behavioral response of *Heliothis virescens* to the odor. We found that the cumulus glomerulus selectively responds to the major pheromone component, Z11-16:AL, the dorsomedial compartment selectively responds to the minor component Z9-14:AL, while the visible ventral compartment responded to the antagonist Z11-16:OH. No response was observed to the antagonist Z11-16:AC. Furthermore, no response to any of the tested pheromone substances could be observed in the remaining (non-MGC) glomeruli in the AL of male moths. Similarly, no responses to these substances could be measured in the AL of female moths. These results confirm and extend findings from single cell recordings both of receptor neurons and projection neurons (Vickers *et al.*, 1998; Berg *et al.*, 1998), which also showed a component-specificity among the MGC glomeruli. Interestingly, when testing different concentration, the effects were not equal for the glomeruli involved. While the cumulus glomerulus responded to the major component Z11-16:AL in a dose-dependent manner across the range from 0.2µg to 2µg, the response to the minor component Z9-14:AL was inversely related to stimulus-concentration. Furthermore, when repeating the stimulus after an interval of 1 min, the response to weak concentrations of the major component Z11-16:AL in the cumulus glomerlus appeared to be higher, while the response to the minor component Z9-14:AL in the dorsomedial compartment was greatly reduced at the higher concentration, indicating chemosensory adaptation in the receptor cells. This reflects the relative proportion of the two substances in the natural blend, in which the minor component Z9-14:AL is present at much lower concentrations. For this substance, the concentrations in our experiments are therefore above the range naturally encountered by animals in the field, and the high level of adaptation observed suggests that they are tuned to much lower concentrations than the receptor cells, which respond to the major component Z11-16:AL.

As was proposed long ago for this system (Hildebrand and Shepherd 1997), pheromone coding is thus realized as a labeled line system from the highly selective receptor cells to the MGC, where each glomerulus is also selective for a particular substance. The behaviorally relevant information, however, is encoded in an across-glomeruli pattern: only when there is activity in the cumulus glomerulus and in the dorsomedial compartment, but no activity in the ventral compartments, does the odor plume correspond well enough to the species-specific blend to elicit anaemotactic behavior. Therefore, a comparison of the activity in all four glomeruli of the MGC determines whether or not a behavioral response will occur.

Our finding of different ranges in concentration dependency reveals a stunning example of a hard-wired physiological filter: the minor component leads to activation of the dorsomedial compartment equal to that of the cumulus glomerulus only when presented at a lower concentration than the major component. Thus, the species-relevant blend of the two substances is encoded with equal activation of the two respective glomeruli, cumulus and dorsomedial. This may simplify the decoding of the species-specific blend, as opposed to stimuli from sympatric species with other concentration ratios of the same substances. However, this hypothesis has yet to be tested by measuring the responses to blends with varying ratios of component substances.

We did not find any responses to sexual pheromones in the female AL, which indicates that either the female is not capable of sensing the pheromone which she herself releases, or that this odor elicits activity in a region of the AL, which is out of focus in our preparations. So far, no pheromone-sensitive receptor neurons have been found in females, which again would suggest that either the female is not receptive to the odor, or only few sensilla are dedicated to its perception.

THE ACROSS-GLOMERULI OLFACTORY CODE

When stimulated with plant extracts, or with chemicals that are important components of plant odors, we found conspicuous responses in the AL of both female and male moths. In males, these responses consistently excluded the MGC-area. Responses to plant odors are combinatorial, i.e., each odor generally elicited a pattern of several activated glomeruli, so each glomerulus might play an active role in determining the pattern of neural responses to different odors. Thus environmental stimuli are coded in an across-glomeruli code, which appears as a spatial activity map in the AL. This coding strategy differs from the sexual-pheromone coding system described above. Using these techniques, we mapped the food-odor-coding patterns in the olfactory system of honeybees, *Apis mellifera*.

These spatial activity patterns in the AL of insects have two important aspects, the identity and the relative position of the glomeruli being activated. In other words, each "spatial" code can also be regarded as an "identity" code, where individual, identified glomeruli are the functional units. A similar view has been proposed for the output pattern of the glomeruli (Laurent, 1999) This does not imply that the relative position of the glomeruli is relevant for the code; rather, the spatial component must be scrutinized. What are the rules behind the spatial arrangement of the glomeruli, in terms of their response properties?

In order to tackle the first aspect (the "identity code"), we took advantage of the fact that the individual glomeruli of insects are recognizable on the basis of their shape and relative position (Rospars, 1988; Flanagan and Mercer, 1989). With the aid of a digital AL atlas (Galizia *et al.*, 1999a), we mapped the physiological responses of individual glomeruli. This allowed us to compare patterns between individuals, and we found that the responses are conserved and species-specific within the honeybee species, *Apis mellifera*. Indeed, knowing the identity of the activated glomeruli, and the magnitude of their activation, even allowed us to reliably identify the coding used as a stimulus (Galizia *et al.*, 1999b). Like a morphological atlas, which is necessary for looking up the position of a glomerulus, it is now possible to create a physiological atlas of the honeybee AL. So far, this atlas contains the response patterns to 53 odors, and includes 38 glomeruli of a total of about 160 glomeruli. For example, olfactory stimulation with 1-nonanol leads to a characteristic pattern comprising the glomeruli T1-17 and T1-33, while 1-hexanol more strongly activates T1-28 (Sachse *et al.*, 1999). For orientation to the neuroanatomical labels, the antennal nerve, when entering the AL, splits into 4 tracts: T1-T4, and the glomeruli are individually labeled with the name of the tract innervating them, and a unique number.

MOLECULAR RESPONSE PROFILES OF INDIVIDUAL GLOMERULI

Glomeruli are the site where the axons of many receptor neurons converge. With 160 glomeruli receiving about 60,000 afferent axons in the honeybee, there are on average 400 axons innervating each glomerulus (Esslen and Kaissling, 1976). It is not known whether axons innervating a particular glomerulus form a homogeneous family, i.e. whether they all express the same receptor protein, or the same complement of receptor proteins. The glomerular activity patterns elicited by individual odors confirm that glomeruli are functional units. The situation in insects might be similar to what has been found in mammals, where all axons expressing the same receptor protein converge onto a limited set of glomeruli, generally two for each olfactory bulb (Mombaerts, 1996). First results from the fruit fly *Drosophila*, whose receptor genes have been cloned, confirm this view (Gao *et al.*, 2000). Morphologically, each glomerulus is concentrically organized, with the receptor neurons preferentially innervating the outer cap of the glomeruli.

Looking at the response properties of individual glomeruli over a range of chemical substances, it is possible to measure their molecular response range. Since the calcium measurements emphasize the response properties of receptor neurons (Galizia *et al.*, 1998), these response profiles also approximate the response profiles of a receptor neuron class. We tested a range of homologous aliphatic hydrocarbons, ranging in length from five to ten carbon atoms, and varying in their functional group (primary alcohol, secondary alcohol, ketone, aldehyde, carboxylic acid, alkanes). We found that the glomeruli are broadly tuned: for example, glomerulus T1-17 preferentially responds to alcohols of intermediate chain length (C7-C9), but also responds to longer and shorter alcohol molecules, though with decreasing intensity. This response profile range with respect to carbon chain length was a common finding in all glomeruli that strongly responded to hydrocarbons (Sachse *et al.*, 1999). Individual glomeruli have preferences with respect to the functional group: glomerulus T1-17 responds more strongly to alcohols than to aldehydes, and glomerulus T1-52 preferentially responds to ketones, though it also responds to the other functional groups tested, apart from alkanes. However, also with respect to the functional group, the tuning was always broad: T1-17 responds strongest to alcohols, but does respond to aldehydes and ketones, too. Interestingly, within the sub-optimal functional group the response patterns showed the same carbon-chain-length dependency as for the preferred group, C7-C9 in the case of T1-17. Most glomeruli with a preferred functional group also showed weaker responses to at least some other functional groups. Taken together, these observations indicate that odor coding in honeybees does not work as a building-set, where a molecule is encoded by properties that are independent. If that were the case, one would expect a glomerulus to respond, for example, to an "aldehyde," irrespective of its carbon chain length, or to a "C7 carbon backbone," irrespective of functional group. Rather, for each glomerulus there appears to be a "best molecule", and changes in this molecule - be it in size or in a functional group - lead to a decrease in the glomerular response. For most glomeruli mapped so far we have most likely not yet found *the* best olfactory stimulus.

SPATIAL ARRANGEMENT OF GLOMERULAR RESPONSES

Some glomeruli with overlapping olfactory response profiles are quite distant in their position in the AL, and some are direct neighbors. A good example for the latter are the glomeruli T1-28, T1-17 and T1-33. These have their best responses to alcohols of chain lengths C6, C8 and C10, respectively. Recordings from mitral cells in rabbits have shown that these cells are spatially arranged according to their response profiles, and that the responses of neighboring mitral cells are antagonistically inhibited (reviewed in Mori *et al.*, 1999) Such a mechanism leads to a sharpening of the response profiles in the output of the olfactory neuropil. The close apposition of glomeruli T1-28, T1-17 and T1-33 in the honeybee, with their overlapping response profiles, suggests a similar mechanism in the insect AL. The inhibitory connections between glomeruli are coordinated by local interneurons. These interneurons, however, generally branch within individual glomeruli, and into the center of the AL. From the center of the AL the distance to each glomerulus is approximately equidistant. This reduces the advantage of glomeruli with similar response profiles being direct neighbors: having glomeruli farther apart does not dramatically increase the necessary „wiring" length (Galizia and Menzel, 2000). Arranging olfactory response profiles on a two-dimensional sheet, where relative distance corresponds to the degree of physiological response overlap, does not appear adequate for the multidimensional olfactory world, so there must be additional ways to sharpen the response profile.

Nevertheless, efficient olfactory coding needs glomerulus-specific connections: a scenario where only direct neighbors inhibit each other appears as unrealistic as a model where each glomerulus makes contact to all other glomeruli in the AL. Therefore, it is necessary to measure the glomerulus-specific inhibitory and/or excitatory interactions. The result of these interactions is a glomerular output pattern that does not correspond to the input pattern.

We have now started to selectively measure glomerular output patterns using calcium imaging by backfilling a large population of output cells with a calcium sensitive dye (Sachse *et al.*, 2000). These studies show that the spatial response patterns are indeed sharpened by the AL network and thereby contrast-enhanced: the similarity between the response patterns to different odors is reduced in the output of the AL as compared to the global activity patterns; this has been assessed by calculating the glomerulus-based correlation between the patterns. Furthermore, by using pharmacological tools, we could clearly show functionally active inhibitory connections between selected glomeruli. These connections were not limited to neighboring glomeruli – indeed, we found more inhibitory connections between non-neighbors than between neighbors. Even more striking, these connections were not reciprocal!

CONCLUSIONS

These studies lay the foundations for studying the functional wiring of a model olfactory system, the honeybee. This will lead to understanding the olfactory code in this species. The comparison with the pheromonal system in moths clearly shows that different species use different coding strategies, despite the apparent similarity in the architecture of their AL. Coding of plant odors appears to be similar in moths and bees, indicating that different coding strategies can be realized even within one species.

Because of the variety of solutions found across the animal kingdom, a comparative approach, where both vertebrates and invertebrates are studied, appears to be very important. It will only be through such a broad view that we will understand the principles of olfactory coding.

ACKNOWLEDGMENTS

We thank Prof. Randolf Menzel, Freie Universität Berlin, for his unbeatable enthusiasm and support.

REFERENCES

Arnold, G., Masson, C., and Budharugsa, S., 1985, Comparative study of the antennal lobes and their afferent pathway in the worker bee and the drone (*Apis mellifera*), *Cell Tissue Res.* **242**:593-605.

Berg, B.G., Almaas, T. J., Bjaalie, J. G., and Mustaparta, H., 1998, The macroglomerular complex of the antennal lobe in the tobacco budworm moth *Heliothis virescens*: specified subdivision in four compartments according to information about biologically significant compounds, *J. Comp. Physiol.* [A] **183**:669-682.

Boeckh, J., and Tolbert, L. P., 1993, Synaptic organization and development of the antennal lobe in insects, *Microsc. Res. Tech.* **24**:260-280.

Brockmann, A., and Brückner, D., 1995, Projection pattern of poreplate sensory neurones in honey bee worker, *Apis mellifera* L. (Hymenoptera: Apidae), *Int. J. Insect Morphol. Embryol.* **24**:405-411.

Esslen, J., and Kaissling, K.-E., 1976, Zahl und Verteilung antennaler Sensillen bei der Honigbiene (*Apis mellifera* L.), *Zoomorphol.* **83**:227-251.

Flanagan, D., and Mercer, A. R., 1989, An atlas and 3-D reconstruction of the antennal lobes in the worker honey bee, *Apis mellifera* L. (Hymenoptera: Apidae), *Int. J. Insect Morphol. Embryol.* **18**:145-159.

Galizia, C. G., and Menzel, R., Odor perception in honeybees: coding information in glomerular patterns, *Curr. Opin. Neurobiol.* **10**: (in press).

Galizia, C. G., Joerges, J., Küttner, A., Faber, T., and Menzel, R., 1997, A semi-in-vivo preparation for optical recording of the insect brain, *J. Neurosci. Methods* **76**:61-69.

Galizia, C. G., Nägler, K., Hölldobler, B., and Menzel, R., 1998, Odour coding is bilaterally symmetrical in the antennal lobes of honeybees (*Apis mellifera*), *Eur. J. Neurosci.* **10**:2964-2974.

Galizia, C. G., McIlwrath, S. L., and Menzel, R., 1999a, A digital three-dimensional atlas of the honeybee antennal lobe based on optical sections acquired using confocal microscopy, *Cell Tissue Res.* **295**:383-394.

Galizia, C. G., Sachse, S., Rappert, A., and Menzel, R., 1999b, The glomerular code for odor representation is species-specific in the honeybee *Apis mellifera*, *Nat. Neurosci.* **2**:473-478.

Gao, Q., Yuan, B., and Chess, A., 2000, Convergent projections of *Drosophila* olfactory neurons to specific glomeruli in the antennal lobe, *Nat. Neurosci.* **3**:780-785.

Hansson, B. S., and Christensen, T. A., 1999, Functional characteristics of the antennal lobe, in: *Insect Olfaction* (B. S. Hansson, ed.), Springer, Berlin Heidelberg New York, pp. 125-161.

Hansson, B. S., Ljungberg, H., Hallberg, E., and Löfstedt, C., 1992, Functional specialization of olfactory glomeruli in a moth, *Science* **256**:1313-1315.

Hildebrand, J. G., 1996, Olfactory control of behavior in moths: Central processing of odor information and the functional significance of olfactory glomeruli, *J. Comp. Physiol.* [A] **178**:5-19.

Hildebrand, J. G., and Shepherd, G. M., 1997, Mechanisms of olfactory discrimination: Converging evidence for common principles across phyla, *Ann. Rev. Neurosci.* **20**:595-631.

Joerges, J., Küttner, A., Galizia, C. G., and Menzel, R., 1997, Representations of odours and odour mixtures visualized in the honeybee brain, *Nature* **387**:285-288.

Kaissling, K.-E. 1997, Pheromone-controlled anemotaxis in moths, in: *Orientation and Communication in Arthropods* (M. Lehrer, ed.), Birkhauser Verlag, Basel, pp. 343-374.

Laissue, P. P., Reiter, Ch., Hiesinger, P. R., Halter, S., Fischbach, K.-F, and Stocker, R. F., 1999, Three-dimensional reconstruction of the antennal lobe in *Drosophila melanogaster*, *J. Comp. Neurol.* **405**:543-552.

Laurent, G., 1999, A systems perspective on early olfactory coding, *Science* **286**:723-728.

Mombaerts, P., 1996, Targeting olfaction. *Curr. Opin. Neurobiol.* **6**:481-486.

Mori, K., Nagao, H., and Yoshihara, Y., 1999, The olfactory bulb: coding and processing of odor molecule information, *Science* **286**:711-715.

Rospars, J. P., 1988, Structure and development of the insect antennodeutocerebral system, *Int. J. Insect Morphol. Embryol.* **17**:243-294.

Rospars, J. P., and Hildebrand, J. G., 1992, Anatomical identification of glomeruli in the antennal lobes of the male sphinx moth *Manduca sexta*, *Cell Tissue Res.* **270**:205-227.

Sachse, S., Rappert, A., and Galizia, C. G., 1999, The spatial representation of chemical structures in the antennal lobe of honeybees: steps towards the olfactory code, *Eur. J. Neurosci.* **11**:3970-3982.

Sachse, S., Galizia, C. G., and Menzel, R., 2000, Calcium imaging of glomerular properties of output neurons in the honeybee antennal lobe, *Proceeding of ISOT/ECRO 2000*, Brighton, UK, pp. 209-210.

Vickers, N. J., Christensen, T. A., and Hildebrand, J. G., 1998, Combinatorial odor discrimination in the brain: attractive and antagonist odor blends are represented in distinct combinations of uniquely identifiable glomeruli. *J. Comp. Neurol.* **400**:35-56.

SPATIAL REPRESENTATIONS OF ODORANT CHEMISTRY IN THE MAIN OLFACTORY BULB OF THE RAT

Brett A. Johnson and Michael Leon

Department of Neurobiology and Behavior
University of California, 2205 BioSci II
Irvine, CA 92697-4550

1. SYSTEMATIC STUDIES OF ODORANT-EVOKED SPATIAL ACTIVITY PATTERNS IN THE OLFACTORY BULB

Odorants of very different chemical structure that yield the perception of very different odors have long been known to evoke distinct patterns of neural activity in the rat olfactory bulb. This observation suggested that at least one step in the coding of odor information might involve spatial patterns of bulbar activity. Until recently, however, it was not understood how different spatial patterns might arise from the differences in odorant chemical structure.

Our general approach to odor coding in the rat olfactory bulb has involved looking for systematic changes in spatial activity patterns that occur with small changes in odorant chemistry. Through this approach, elements of the spatial patterns can be related to particular molecular features and/or chemical properties of the odorant stimuli. To compare the patterns of activity evoked by different odorants systematically, we have developed a procedure to map uptake of [^{14}C]2-deoxyglucose (2-DG) across the entire glomerular layer (Johnson *et al.*, 1999). By using radial grids to dictate positions of measurements in coronal sections, and by equalizing rostral-caudal positions in relation to anatomical landmarks, we generate standardized arrays of data that allow patterns to be compared quantitatively across multiple animals exposed to distinct odorants.

We review here our findings concerning oxygen-containing, aliphatic odorants that differ in carbon number, functional groups, and hydrocarbon structure. Taken together, our results demonstrate several fundamental principles underlying spatial representations of odorants in the olfactory bulb. Each of these principles will be discussed in turn below.

2. MODULAR REPRESENTATIONS OF ODORANT CHEMISTRY

Figure 1 shows glomerular activity patterns averaged across three to six rats exposed to each of 13 distinct odorants chosen to illustrate results we have obtained by using a larger number of compounds (Johnson *et al.*, 1998, 1999; Johnson and Leon, 2000a, b). The patterns shown involve straight-chained aliphatic acids differing incrementally in carbon number (top row), five-carbon aliphatic acids differing in hydrocarbon structure (middle row), and aliphatic compounds sharing a four-carbon, straight-chained portion, but differing in the functional group present at the fifth carbon (bottom row). Each row represents a distinct experiment.

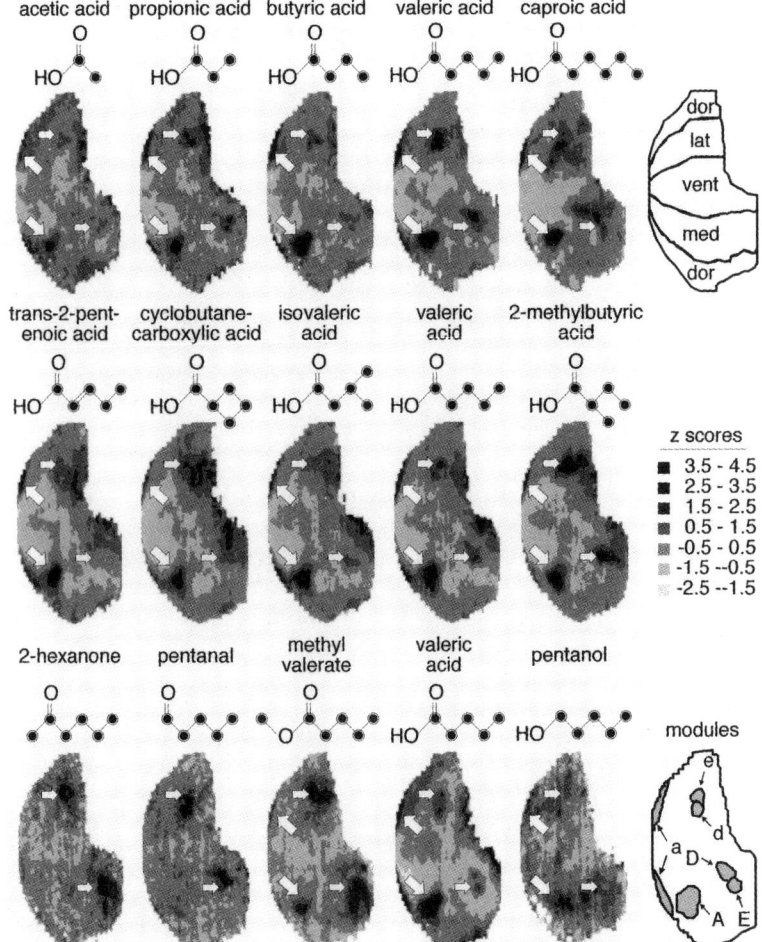

Figure 1. Average activity patterns evoked by odorants differing in carbon number (7.2 ppm, top row), hydrocarbon structure (8 ppm, middle row), and oxygen-containing functional groups (bottom row, lowest concentrations evoking a pattern as indicated in Figure 2). Structures are shown as ball-and-stick diagrams. Circles represent carbon atoms. Hydrocarbon hydrogen atoms are omitted. Rostral is to the left. All other orientation is as shown in upper right inset (dor, dorsal; lat, lateral; vent, ventral; med, medial). Bottom right inset shows a selection of glomerular modules activated by these odorants. Large arrows indicate modules a and A. Small arrows indicate modules e and E. Arrows appear in the same location on each chart.

As is readily apparent in Figure 1, even very closely related odorant chemicals evoked distinct, but overlapping patterns of glomerular activity. There are two types of differences among the odorant-evoked patterns. One type of difference involves the stimulation by one odorant of a part of the bulb (typically a cluster of adjacent glomeruli, which we will refer to as a glomerular module) that is not stimulated by another odorant. For example, at the lowest concentrations evoking patterns, valeric acid, methyl valerate, and pentanol activated rostral modules that were not activated by 2-hexanone or pentanal (Figure 1, bottom row, large arrows). Another type of difference involves small changes in the position of overlapping modules. For example, rostral modules shifted in position with increasing carbon number of straight-chained aliphatic acid odorants (Figure 1, top row, large arrows). This difference will be discussed below in the „chemotopic organization" section.

In our functional group study, higher concentrations of some of the odorants activated glomerular modules that were not activated at lower concentrations (Johnson and Leon, 2000a). The full set of modules in that study is diagrammed in Figure 2, which also shows the relative amounts of 2-DG uptake evoked in each module for each concentration of each odorant. Each module was stimulated by at least two of the five odorants in the study. Therefore, the coding of the odors of these compounds must be combinatorial. That is, a comparison of the relative levels of activity in various modules is needed to identify a given odorant as unique. Also, each odorant evoked activity in multiple modules (Figure 2). As shown in Figure 1, even simple molecules such as propionic acid stimulated uptake in at least four modules distributed across both lateral and medial aspects of the bulb (Johnson *et al.*, 1999). Two of the modules were located in the rostral part of the bulb, and at least two in the caudal part of the bulb. This parallel and spatially distributed representation of propionic acid likely explains its continued detection by rats with large experimental ablations of the olfactory bulb (Slotnick *et al.*, 1997).

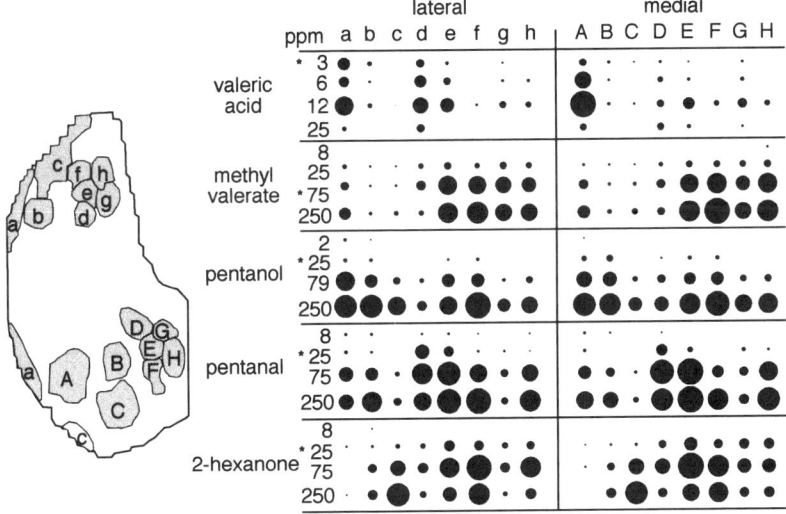

Figure 2. 2-DG uptake in glomerular modules as a function of the concentration of odorants differing in oxygen-containing functional groups. Locations of modules are shown in the left panel. Diameters of circles (right panel) indicate uptake standardized for each odorant to the largest value obtained at any concentration. Asterisks denote the lowest concentrations judged to evoke a pattern (Data from Johnson and Leon, 2000a).

The specificity of a given glomerulus probably reflects the specificity of an individual odorant receptor protein, because individual glomeruli appear to receive convergent projections from sensory neurons that express the same, single odorant receptor gene (Ressler et al., 1994). Because most receptor proteins actually detect molecular features present in multiple ligand molecules, the activation of an individual glomerulus likely indicates the presence of a particular odorant molecular feature (Johnson et al., 1998). By comparing the odorants we have investigated so far, we can begin to hypothesize the molecular features that are necessary for the stimulation of some of the modules we have identified. For example, modules a and A (Figure 2) responded to low concentrations of aliphatic acids largely independently of hydrocarbon structure or carbon number (Figure 1, top and middle rows, large arrows). Much higher concentrations of pentanol were needed to activate these modules, and even higher concentrations of methyl valerate and pentanal were required (Figure 2). 2-Hexanone did not activate these modules. They were activated by ethyl esters, but not by isoamyl esters (Johnson et al., 1998). These findings are consistent with the recognition of a hydrogen bond acceptor by the receptors associated with modules a and A. Modules e and E were activated by all compounds that shared a four-carbon, straight-chained hydrocarbon structure in the functional group study (Figure 2). However, 2-methyl-butyric acid stimulated these modules robustly, while the very closely related structural isomer, isovaleric (3-methylbutyric) acid, caused only low activity in these modules when presented at the same concentration (Figure 1, middle row, small arrows). This difference suggests that modules e and E may recognize some steric feature present in many aliphatic odorant molecules (i.e., a specific geometric arrangement of hydrocarbon hydrogens) (Johnson and Leon, 2000b).

By identifying glomerular modules activated by one odorant that are not activated by other odorants differing only slightly in chemical structure, we therefore can generate specific hypotheses concerning the specificity of individual modules. These hypotheses then can be tested through the use of another odorant set. This systematic approach to odor coding should accelerate our understanding of what odorant chemical features are compartmentalized within the olfactory bulb. Furthermore, by studying a wider range of functional groups and hydrocarbon structures, we should be able to approach a more complete stimulus map of bulbar activity.

3. CHEMOTOPIC ORGANIZATION OF GLOMERULI WITHIN MODULES

Most of the modules we have identified are comprised of multiple, adjacent glomeruli. For example, at a concentration of 7.2 parts per million, valeric acid stimulated about 30 glomeruli each within modules a and A (Johnson et al., 1999). The 2-DG technique is capable of resolving the activation of as few as one or two glomeruli (Johnson et al., 1998). Therefore, the activation of larger numbers of adjacent glomeruli by a single odorant likely indicates that these glomeruli are all activated directly by the odorant, and that the specificities of these glomeruli may be closely related.

To explain tuning of individual bulbar projection neurons to aliphatic acids of a particular carbon number, Mori and coworkers suggested that adjacent glomeruli in the rostral, dorsomedial bulb may respond optimally, but broadly, to acids of slightly distinct carbon number (Yokoi et al., 1995). Lateral inhibition between neighboring glomeruli or between projection neurons via granule cell interneurons was proposed to accomplish the observed tuning. To test this hypothesis, we analyzed centroids of 2-DG uptake within module A.

The centroids differed significantly across different straight-chained aliphatic acids (Johnson et al., 1999), and the dorsal-ventral position of the module was correlated with carbon number (Figure 3A). Carbon number in straight-chained, saturated compounds is exactly correlated with hydrophobicity, molecular volume, and molecular length, three properties that could affect an odorant ligand's interaction with receptors. To determine if only one of these properties dictated the dorsal-ventral position of module A, we calculated centroids of 2-DG uptake for five- and six-carbon aliphatic acid odorants possessing different hydrocarbon structures (Johnson and Leon, 2000b). The straight-chained, branched, cyclic, and double-bonded molecules we chose differed independently in hydrophobicity, volume, and length. Indeed, centroids within module A differed across these odorants, and the dorsal-ventral position of the module was significantly correlated only with molecular length (Figure 3B). Therefore, longer aliphatic acid odorants appear to activate glomeruli located progressively more ventrally within the module.

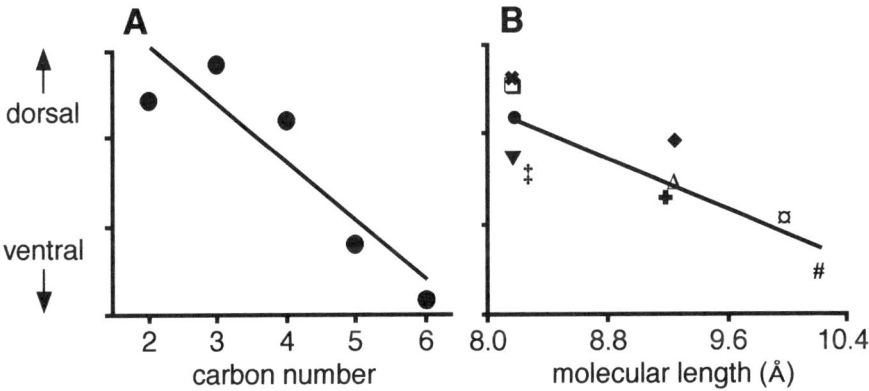

Figure 3. Dorsal-ventral positions of module A are significantly correlated with carbon number in straight-chained, saturated aliphatic acids (**A**: $r = 0.91$, $P < 0.05$), and with molecular length in five- and six-carbon aliphatic acids of different hydrocarbon structure (**B**: $r = 0.85$, $P < 0.005$). Lines are the results of linear regression. The y-axes are not comparable between the two plots due to differences in the boundaries of the modules used for centroid determination. The odorants in **B** were 2-methybutyric acid (✖), cyclobutane-carboxylic acid (❑), isovaleric acid (●), tert-butylacetic acid (▼), cyclopentanecarboxylic acid (‡), isocaproic acid (◆), valeric acid (Δ), trans-2-pentenoic acid (✚), trans-3-hexenoic acid (▫), and caproic acid (#).

The spatial organization of glomeruli within module A is an example of chemotopic organization, wherein glomeruli that are nearest neighbors respond to the most closely related stimuli. Such arrangements may typify the organization of glomeruli within and between functional modules of the olfactory bulb. For example, module a of the lateral bulb shifts rostrally with increasing carbon number for aliphatic acids, and centroids within the caudo-lateral and caudo-medial aspects of the bulb shift both rostrally and ventrally with increasing carbon number for both acids and esters (Johnson et al., 1998; 1999). Also, the modules that distinguish low concentrations of odorants with related functional groups are clustered together in the caudal bulb (Figure 1, bottom row). Lateral inhibition may enhance contrast between projection neurons associated with these nearby glomeruli, thereby allowing them to discriminate subtly distinct chemical properties that may not be distinguished adequately by an individual odorant receptor.

4. PARALLEL REPRESENTATIONS OF ODORANTS WITHIN THE LATERAL AND MEDIAL ASPECTS OF THE OLFACTORY BULB

For every module in the lateral aspect of the bulb, there appears to be a module of similar specificity in the medial aspect of the bulb (Johnson *et al.*, 1998, 1999; Johnson and Leon, 2000a, b). This phenomenon is perhaps best illustrated in Figure 2, where lateral modules are assigned lower case letters and the corresponding medial modules are assigned upper case letters. The similarities in the relative amounts of 2-DG uptake between lateral and medial modules across both odorant concentration and odorant functional groups results in a similar pattern of circle size between the left and right sides of the figure. Therefore, there appear to be two similar representations of aliphatic odorants within each olfactory bulb, one in the lateral aspect and one in the medial aspect.

The medial modules are located about 1.7 mm more caudally than the corresponding lateral modules, and the medial modules also are situated more ventrally. There are similar spatial relationships between lateral and medial glomeruli that receive projections from sensory neurons expressing the same odorant receptor gene (Ressler *et al.*, 1994). It is likely that our paired modules reflect this sensory neuron projection pattern (Johnson *et al.*, 1998, 1999). Possible reasons for two representations in each bulb include redundancy to insure odor perception after damage to part of the bulb or epithelium, coincidence detection to resolve stimulus-evoked activity from spontaneous activity, and/or separate cortical projections underlying different odor-influenced behaviors (Johnson *et al.*, 1999).

5. DIFFERENT REPRESENTATIONS AT DIFFERENT ODORANT CONCENTRATIONS

If bulbar spatial activity patterns are involved in the coding of perceived odor quality, then different odors should be associated with different activity patterns. Indeed, odorants that differ only slightly in chemical structure evoke both different perceived odors and different patterns of 2-DG uptake (Figure 1). Certain odorants also evoke different perceived odors at different concentrations. In our functional group study, we included two odorants that humans report to change in odor quality with concentration (pentanal and 2-hexanone) and three odorants with constant odors (valeric acid, methyl valerate, and pentanol). We found that the patterns evoked by pentanal and 2-hexanone were significantly different at different concentrations (Johnson and Leon, 2000a). New glomerular modules were evoked at higher concentrations, and these modules were located far away from those evoked at lower concentrations (Figure 2). Quantitative comparisons indicated that the patterns evoked by certain concentrations of pentanal or 2-hexanone were more similar to the pattern evoked by methyl valerate than to patterns evoked by different concentrations of the same odorant (Johnson and Leon, 2000a). Patterns evoked by valeric acid, methyl valerate, and pentanol did not differ significantly across different odorant concentrations. Thus, our data are consistent with a relationship between olfactory bulb spatial activity patterns and odor quality perception. The data further suggest that rats may better discriminate between different concentrations of pentanal and 2-hexanone than between certain concentrations of these odorants and methyl valerate.

6. CONCLUSIONS

By mapping activity across the entire glomerular layer in rats exposed to systematically different odorant chemicals, we find that aspects of odorant chemistry are represented spatially in the olfactory bulb. Responses to particular odorant molecular features are compartmentalized into glomerular modules, where responses may be tuned by using chemotopic glomerular arrangements and lateral inhibition. Further research, using odorants that differ in other aspects of chemical structure, as well as behavioral studies correlating differences in odor perception with quantitative differences in spatial patterns of bulbar activity, should greatly increase our understanding of olfactory bulb function.

REFERENCES

Johnson, B. A., and Leon, M., 2000a, Modular representations of odorants in the glomerular layer of the rat olfactory bulb and the effects of stimulus concentration, *J. Comp. Neurol.* **409**:495-509.

Johnson, B. A., and Leon, M., 2000b, Odorant molecular length: one aspect of the olfactory code, *J. Comp. Neurol.* **426**:330-338.

Johnson, B. A., Woo, C. C., and Leon, M., 1998, Spatial coding of odorant features in the glomerular layer of the rat olfactory bulb, *J. Comp. Neurol.* **393**:457-471.

Johnson, B. A., Woo, C. C., Hingco, E. E., Pham, K. L., and Leon, M., 1999, Multidimensional chemotopic responses to n-aliphatic acid odorants in the rat olfactory bulb, *J. Comp. Neurol.* **409**:529-548.

Ressler, K. J., Sullivan, S. L., and Buck, L. B., 1994, Information coding in the olfactory system: Evidence for a stereotyped and highly organized epitope map in the olfactory bulb, *Cell* **79**:1245-1255.

Slotnick, B. M., Bell, G. A., Panhuber, H., and Laing, D. G., 1997, Detection and discrimination of propionic acid after removal of its 2-DG identified major focus in the olfactory bulb: a psychophysical analysis, *Brain Res.* **762**:89-96.

Yokoi, M., Mori, K., and Nakanishi, S., 1995, Refinement of odor molecule tuning by dendrodendritic synaptic inhibition in the olfactory bulb, *Proc. Natl. Acad. Sci. USA* **92**:3371-3375.

PRENATAL GROWTH AND ADULT SIZE OF THE VOMERONASAL ORGAN IN MOUSE LEMURS AND HUMANS

Timothy D. Smith[1,4], Mark P. Mooney[2,4], Annie M. Burrows[1,4], Kunwar P. Bhatnagar[3], and Michael I. Siegel[4]

[1] School of Physical Therapy, Slippery Rock University
Slippery Rock, PA 16057
[2] Department of Oral Medicine and Pathology, University of Pittsburgh,
Pittsburgh, PA 15261
[3] Department of Anatomical Sciences and Neurobiology
University of Louisville, Louisville, KY 40292, USA
[4] Department of Anthropology, University of Pittsburgh
Pittsburgh, PA 15260

INTRODUCTION

Previous studies of VNO size have primarily sought to compare the magnitude of VNO function (via a presumed link between VNO size and receptor population) among vertebrate species (Dawley, 1998) and sexes (Dawley and Crowder, 1995; Weiler *et al.*, 1999). A few have sought to determine whether the structure degenerates or persists in humans. These studies have described a continuous increase in prenatal VNO anteroposterior length and epithelial volume (Smith *et al.*, 1996, 1997; Sherwood *et al.*, 1999), especially during the late second and third trimesters (Smith *et al.*, 1997). A comparison of prenatal VNO data with the same measures in adults indicates that some postnatal growth, of variable magnitude, occurs during postnatal human ontogeny (Smith *et al.*, 1998; Bhatnagar and Smith, in preparation). Although these studies have established that the human VNO does not degenerate (prenatally or postnatally), no studies have attempted to determine the magnitude of human VNO growth via a comparison to other species.

The chemosensory potential of the human VNO is highly debated (e.g., Takami *et al.*, 1993; Monti-Bloch *et al.*, 1998; Preti and Wysocki, 1999; Wysocki and Preti, 2000; Bhatnagar and Smith, in preparation). A comparison of VNO growth between humans and a primate having a demonstrably functional VNO may shed light on the likelihood of a functional human VNO. In the prosimian primate *Microcebus murinus*, VNO function has been experimentally demonstrated (Aujard, 1997) and the organ is morphologically well-developed in this species (Schilling, 1970). We examined size of the VNO in two species of Microcebus (*M. murinus* and *M. myoxinus*) for a comparison to previous data sets on the human VNO (Smith et al., 1997, 1998; Smith and Bhatnagar, 2000; Bhatnagar and Smith, in preparation).

MATERIALS AND METHODS

We examined histologically sectioned embryos and fetuses of *M. myoxinus* (N = 13) and *M. murinus* (N = 6) and adults of *M. murinus* (N = 4). Embryos and fetuses were part of the Bluntschli collection, Department of Mammalogy, Division of Vertebrate Zoology, American Museum of Natural History, New York, NY. Adult mouse lemur tissues were harvested from cadavers purchased from the Duke University Primate Center. Specimens in the Bluntschli collection had been previously sectioned at 10- 40 µm, and were stained using various procedures. These specimens ranged from 7.5 to 37 mm in total overall length (which corresponded to crown-rump length (CRL), based on archived photographs). Using archived photographs, it was found that the earliest *Microcebus* embryo (7.5 CRL) corresponded to a stage 16 embryo of the Carnegie stages. Adult prosimians were decalcified using a formic acid solution (Cal-Ex II or formic acid-sodium citrate), embedded in paraffin, and sectioned at 10-16 µm.

Data on human VNOs were previously published (Smith et al., 1997, 1998; Smith and Bhatnagar, 2000; Bhatnagar and Smith, in preparation) based on a sample of 50 embryos/fetuses and 19 adults. Quantification of VNO length in pre- and postnatal *Microcebus* specimens was accomplished as in previous studies (see Smith et al., 1997). Briefly, the number of slides in which the VNO could be found was counted. This number was multiplied by the sectional thickness for each specimen. Since a clear sensory epithelium could be detected for *Microcebus*, both the length of the neuroepithelium (VNNE), and the length of the entire VNO (including a cranially positioned, short, non-sensory duct) were quantified. Since a more complete series of *M. myoxinus* was available, these measures were compared to previously published VNO data from *Homo sapiens* (Smith et al., 1997, 1998; Smith and Bhatnagar, 2000; Bhatnagar and Smith, in preparation) at comparable stages of development. The growth curves generated by regression equations were compared between species using a t_s test for the homogeneity of regression line slopes (Sokal and Rohlf, 1981)[1]. Differences were considered significant at $p < 0.05$.

RESULTS

The VNO was present in all *Microcebus* embryos and fetuses. In *M. myoxinus* embryos from 7.5 to 9 mm CRL, the VNO was similar to that described for human embryos (Smith and Bhatnagar, 2000) - tubular (oval or round) in shape with no recognizable duct cranially (Figure 1). In *M. myoxinus* that were 12 mm CRL or larger, the VNO was preceded by a short duct, of stratified epithelium, that communicated the VNO lumen to the nasopalatine duct. In specimens that were 13 mm CRL and larger, the VNO was crescent-shaped with a distinct neuroepithelium (Figure 1). VNO development in *M. murinus* appeared to precede similarly, with smaller specimens in which a distinct duct was not clearly recognizable.

---

[1] There is currently no published data on perinatal CRL of *Microcebus myoxinus*. Therefore, for the purposes of comparison, we necessarily assume that the prenatal size ranges of our samples are analogous, although heterochronic differences may certainly exist between species.

The range of right VNO length was similar between prenatal *M. murinus* and *Homo*, but lower for *M. myoxinus* (Table 1). Relative to crown-rump length, VNO length, VNNE length, and vomeronasal cartilage (VNC) length all increased in a sigmoidal manner in *M. myoxinus*, with the steepest slope in the VNC. In contrast, VNO length increased in a linear fashion in *Homo* (Table 2; Figure 2). Comparison of regression line slopes revealed a significant ($p < 0.05$) difference between *M. myoxinus* and *Homo* (Table 2). VNO length increase occurred more rapidly and at an earlier time period (*i.e.*, late embryonic, early fetal) in *Microcebus* compared to *Homo*.

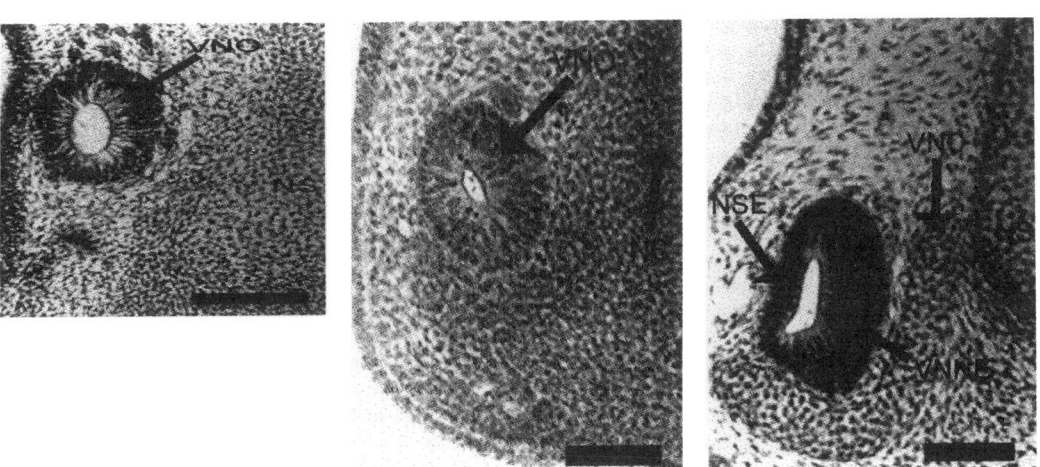

Figure 1. The vomeronasal organs (VNO) of mouse lemur embryos (middle) showed similarities to that of human embryos (left), but fetal mouse lemurs (right) had distinct neuroepithelia (VNNE), and non-sensory epithelia (NSE). Left: 22 mm *Homo* embryo. Middle: 9 mm *Microcebus myoxinus* embryo. Right: 13 mm *M. myoxinus* fetus. NS = nasal septum; VNC = vomeronasal cartilage; scale bars = 100 µm.

Table 1. Prenatal length of the VNO in *Microcebus* and *Homo*.

	VNO length range (µm)	VNE length range (µm)
Microcebus myoxinus	140-1515	140-1395
Microcebus murinus	350-2620	350-2380
Homo sapiens	228-3015	U

U = VNE length range unknown in humans since receptor cells most likely disappear during prenatal development (Ortmann, 1989; Smith and Bhatnagar, 2000); data on humans from Smith *et al.*, 1997 and Smith and Bhatnagar, 2000

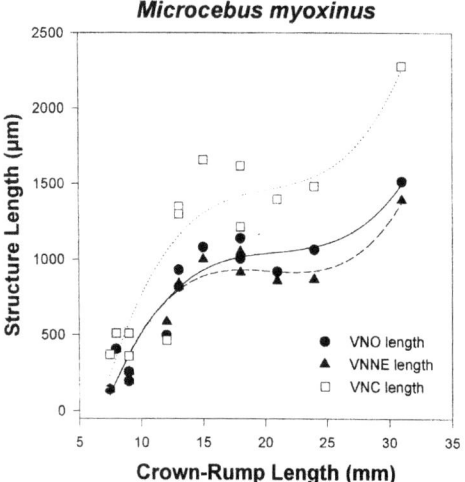

Figure 2. Left: individual specimen's distribution and regression lines for right vomeronasal organ (VNO) length (solid regression line), vomeronasal neuroepithelium (VNNE) length (long dashed regression line), and vomeronasal cartilage (VNC) length (short dashed regression line) in *Microcebus myoxinus*. Right: superimposed specimen's distributions and regression lines for right VNO length in *Microcebus myoxinus* (solid regression line) *versus Homo* (long dashed regression line).

Table 2. Regression equations* of right vomeronasal organ length against crown-rump length for *Microcebus myoxinus* and *Homo sapiens* and results of *t*-test for homogeneity of slopes.

Microcebus myoxinus

$Y = -2330.0 + (470.28 \times CRL) + (-21.19 \times CRL^2) + (0.36 \times CRL^3)$
$R^2 = 0.90$
$SE = 146.7$

Homo sapiens

$Y = 53.01 + (7.7 \times CRL)$
$R^2 = 0.65$
$SE = 1.25$
$t_s = 3.68; p < 0.05$

*Linear regression equations based on the equation: $Y = a + (b \times CRL)$; polynomial regression equations based on the equation: $Y = a + (b \times CRL) + (c \times CRL^2) + (d \times CRL^3)$; CRL = crown-rump length; regression equation on human VNO length calculated from data of Smith and Bhatnagar, 2000.

VNO length was greater and less variable in adult *M. murinus* than adult *Homo* (Table 3). Right VNO lengths ranged from 6100 to 7200 μm in *Microcebus* and from 1450 to 9620 μm in *Homo*. Proportionately, adult VNO length ranged from 16.4 to 19.7% of head length (rhinion-inion) in *Microcebus*, but only 0.5 to 5.6% of head length in *Homo*.

Table 3. Adult VNO length in *Microcebus* and *Homo*.

	VNO length mean (μm) +/- SD	VNE length mean (μm) +/- SD
Microcebus murinus	6700 +/- 592	5280 +/- 678
Homo sapiens	5077 +/- 2270	U

data on humans from Smith *et al.*, 1998 and from Bhatnagar and Smith, in preparation; U = VNE length unknown in adult humans since the presence of receptor cells is controversial (Takami *et al.*, 1993; Bhatnagar and Smith, in preparation)

DISCUSSION

It has only recently been acknowledged that the human vomeronasal organ (VNO) is a consistent anatomical structure, both prenatally and postnatally (e.g., Johnson *et al.*, 1985; Smith *et al.*, 1996), after decades in which the VNO has been thought to disappear before birth (Crosby and Humphrey, 1939; Nakashima *et al.*, 1985; Kjaer and Fisher-Hansen, 1996). Although it derives from the same embryonic tissue as in other mammals (Smith and Bhatnagar, 2000), the human VNO is atypical among mammals in location and microscopic anatomy (Johnson *et al.*, 1985; Smith *et al.*, 1998). Based on conflicting data among physiological or anatomical studies (e.g., Ortmann, 1989; Takami *et al.*, 1993; Monti-Bloch *et al.*, 1998; Preti and Wysocki, 1999; Wysocki and Preti, 2000; Smith and Bhatnagar, 2000; Bhatnagar and Smith, in preparation), its functionality as a chemoreceptive organ remains in question.

The utility of chemosensory organ size as a measure corresponds to its relationship to receptor population (Dawley, 1998). Using this relationship as an underlying assumption, previous studies have compared VNO size between male and female vertebrates (Segovia and Guillamon, 1982; Dawley and Crowder, 1995) or between vertebrate species (Dawley, 1998). Although there is no consensus that the human VNO functions as a chemosensory apparatus, the present study allows a size comparison between the human structure (vestigial or otherwise) and the VNO of a primate for which VNO function has been experimentally confirmed (Aujard, 1997). Our results indicate that the human VNO is more highly variable and proportionately smaller compared to those of mouse lemurs, both prenatally and postnatally. It is possible that this reflects a different ontogeny of the two tissues, the chemosensory VNNE of *Microcebus* (Schilling, 1970; Aujard, 1997), and the non-chemosensory VNO[2] of *Homo* (Roslinski *et al.*, 2000; Bhatnagar and Smith, in preparation). It should also be emphasized that the earliest (embryonic) stages of VNO development are very similar between these species, and begin to diverge in morphology through heterochronic or heterotropic mechanisms during fetal development (Smith and Bhatnagar, 2000).

The prenatal growth curves generated for *Microcebus myoxinus* were comparable to those seen in other prenatal facial growth studies (Siegel *et al.*, 1987; Mooney *et al.* 1994; Smith *et al.*, 1996). Specifically, the third order growth curves suggested multiple accelerations („spurts") in VNO length increase in *Microcebus*. It is noteworthy that the human VNO is much more variable in this regard, even more so than other structures in the

---

[2] The morphological evidence for receptors in the vomeronasal organ is in debate. The present study utilizes descriptions of the human VNO in Johnson *et al.* (1985), Roslinski *et al.* (2000), and Bhatnagar and Smith (in preparation) for purposes of discussion.

midfacial region (Smith *et al.*, 1996). It is clear that adult *Microcebus* has a proportionately larger VNO size than adult humans, and that this discrepancy may begin to be manifested during fetal growth.

This proportional size difference is in keeping with the assertion that strepsirhine primates possess relatively well developed VNOs compared to haplorhines (Maier, 1980), especially some Old World primates[3]. Since the VNO and olfactory epithelium both arise from the invaginating olfactory pit (Bossy, 1980), this difference may also relate to relative reduction of the olfactory portion of the nasal fossa in humans compared to prosimians (Loo, 1973; and see Negus, 1958). At a purely functional level, these differences may reflect a different level or even mode of VNO function. Based on previous histological descriptions, the VNNE of *Microcebus* and other prosimians exhibits marked nuclear stratification (Schilling, 1970; Evans and Schilling, 1995) similar to olfactory epithelium, whereas the human VNO epithelium is more similar to the respiratory type in this regard (Smith and Bhatnagar, 2000; Bhatnagar and Smith, in preparation). Therefore, differences in growth may well reflect profound differences in epithelial thickness, which is much greater in chemosensory types compared to respiratory epithelium. The acquisition of volumetric data, which has been suggested to provide a more accurate consideration of epithelial thickness (Smith *et al.*, 1997; Dawley, 1998) is the focus of our ongoing study.

ACKNOWLEDGMENTS

The authors wish to thank E. Brothers and R. Randall for access to the Bluntschli collection of primate embryos at the Department of Mammalogy, Division of Vertebrate Zoology, American Museum of Natural History.

REFERENCES

Aujard, F., 1997, Effect of vomeronasal organ removal on male socio-sexual responses to female in a prosimian primate (*Microcebus murinus*), *Physiol Behav.* **62**:1003-1008.
Bossy, J., 1980, Development of olfactory and related structures in staged human embryos, *Anat. Embryol.* **161**:225-236.
Crosby, E. C., and Humphrey, T., 1939, Studies of the vertebrate telencephalon. I. The nuclear configuration of the olfactory and accessory olfactory formation and the nucleus olfactorius anterior of certain reptiles, birds, and mammals, *J. Comp. Neurol.* **71**:121-213.
Dawley, E. M., 1998, Species, sex, and seasonal differences in VNO size, *Microsc. Res. Techn.* **41**:506-518.
Dawley, E. M., and Crowder, J., 1995, Sexual and seasonal differences in the vomeronasal epithelium of the red-backed salamander (*Plethodon cinereus*), *J. Comp. Neurol.* **359**:382-390.
Evans, C., and Schilling, A., 1995, The accessory (vomeronasal) chemoreceptor system in some prosimians, in: *Creatures of the Dark: The Nocturnal Prosimians,* (L. Alterman, G. A. Doyle, and M. K. Izard, eds.,), Plenum Press, New York pp. 393-411.
Johnson, A., Josephson, R., and Hawke, M., 1985, Clinical and histological evidence for the presence of the vomeronasal (Jacobson's) organ in adult humans, *J. Otolaryngol.* **14**:71-79.
Kjaer, I., and Fischer Hansen, B., 1996, The human vomeronasal organ: prenatal developmental stages and distribution of luteinizing hormone-releasing hormone, *Eur. J. Oral Sci.* **104**:34-40.
Loo, S. K., 1973, A comparative study of the nasal fossa of four nonhuman primates, *Folia Primatol.* **20**:410-422.
Maier, W., 1980, Nasal structures in Old and New World primates, in: *Evolutionary Biology of the New World monkeys and Continental Drift*, (R. L. Ciochon, and A. B. Chiarelli, eds.,), Plenum Press, New York, pp. 219-241.

---

[3] Our recent work has shown that the juvenile and adult chimpanzee (*Pan troglodytes*) has a VNO that is virtually identical to that of adult humans (Smith *et al.*, in press), showing that humans are not the only Old World primate to possess a VNO.

Monti-Bloch, L., Diaz-Sanchez, V., Jennings-White, C., and Berliner, D. L., 1998, Modulation of serum testosterone and autonomic function through stimulation of the male human vomeronasal organ (VNO) with pregna-4,20-diene-3,6-dione, *J. Steroid Biochem. Molec. Biol.* **65**:237-242.

Mooney, M. P., Siegel, M. I., Kimes, K., Todhunter, J., and Smith, T. D., 1994, Development of the paraseptal cartilages in normal and cleft lip and palate human fetal specimens, *Cleft Palate-Craniofac. J.* **31**:239-245.

Nakashima, T., Kimmelman, C. P., and Snow, J. B., 1985. Vomeronasal organs and nerves of Jacobson in the human fetus, *Acta Otolaryngol. (Stockh)* **99**:266-271.

Negus, V., 1958, *Comparative Anatomy and Physiology of the Nose and Paranasal Sinuses*, E. and S. Livingstone Ltd., Edinburgh.

Ortmann, R., 1989, Über Sinneszellen am fetalen vomeronasalen Organ des Menschen, *HNO* **37**:191-197.

Preti, G., and Wysocki, C. J., 1999, Human pheromones: Releaser or primers. Fact or myth., in: *Advances in Chemical Signals in Vertebrates*, (R. E. Johnston, D. Müller-Schwarze, P. W. Sorensen, eds.), Kluwer Academic/ Plenum Publishers, New York pp. 315-331.

Roslinski, D. L., Bhatnagar, K. P., Burrows, A. M., and Smith, T. D., 2000, Comparative morphology and histochemistry of glands associated with the vomeronasal organ in humans, mouse lemurs, and voles, *Anat. Rec.* **260**:92-101.

Schilling, A., 1970, L'organe de Jacobson du lemurien malgache *Microcebus murinus* (Miller, 1977), *Mem. Mus. d'Hist. Nat. (Serie A)* **61**:203-280.

Segovia, S., and Guillamón, A., 1982, Effects of sex steroids on the development of the vomeronasal organ in the rat, *Develop. Brain Res.* **5**:209-212.

Sherwood, R. J., McClachlan, J. C., Aiton, J. F., and Scarborough, J., 1999, The vomeronasal organ in the human embryo, studied by means of three-dimensional computer reconstruction, *J. Anat.* **195**:413-418.

Siegel, M. I., Mooney, M. P., Kimes, K. R., and Todhunter, J. S., 1987, Analysis of the size variability of the human normal and cleft palate fetal nasal capsule by means of three-dimensional computer reconstruction of histological preparations, *Cleft Palate J.* **24**:190-199.

Smith, T. D., and Bhatnagar, K. P., 2000, The human vomeronasal organ. Part II: prenatal development, *J. Anat.* **197**:421-436.

Smith, T. D., Siegel, M. I., Mooney, M. P., Burdi, A. R., and Todhunter, J. S., 1996, Vomeronasal organ +growth and development in normal and cleft lip and palate human fetuses, *Cleft Palate-Craniofac. J.* **33**:385-394.

Smith, T. D., Siegel, M. I., Mooney, M. P., Burdi, A. R., Burrows, A. M., and Todhunter, J. S., 1997, Prenatal growth of the human vomeronasal organ, *Anat. Rec.* **248**:447-455.

Smith, T. D., Siegel, M. I., Burrows, A. M., Mooney, M. P., Burdi, A. R., Fabrizio, P. A., and Clemente, F. R., 1998, Searching for the vomeronasal organ of adult humans: Preliminary findings on location, structure, and size, *Microsc. Res. Techn.* **41**:483-491.

Smith, T. D., Siegel, M. I., Bonar, C. J., Bhatnagar, K. P., Mooney, M. P., Burrows, A. M., Smith, M. A., and Maico, L. M., The existence of the vomeronasal organ in postnatal chimpanzees and evidence for its homology to that of humans, *J. Anat.* (in press).

Sokal, R. R., and Rohlf, F. J., 1981, *Biometry*, 2nd Ed., W.H. Freeman and Co., San Fransisco.

Takami, S., Getchell, M. L., Chen, Y., Monti-Bloch, L., Berliner, D. L., Stensaas, L. J., and Getchell, T. V., 1993, Vomeronasal epithelial cells of the adult human express neuron-specific molecules, *NeuroReport* **4**:375-378.

Weiler, E., McCulloch, M. A., and Farbman, A. I., 1999. Proliferation in the vomeronasal organ of the rat during postnatal development, *Euro. J. Neurosc.* **11**:700-711.

Wysocki, C. J., and Preti, G., 2000, Human body odors and their perception, *Japan. J. Smell Taste Res.* **7**:19-42.

SIZE OF THE VOMERONASAL ORGAN IN WILD *MICROTUS*

Lisette M. Maico[1], Dana L. Roslinski[1], Annie M. Burrows[1,2,5], Mark P. Mooney[2,3], Michael I. Siegel[2], Kunwar P. Bhatnagar[4], and Timothy D. Smith[1,2,5]

[1] School of Physical Therapy, Slippery Rock University
Slippery Rock, PA 16057
[2] Department of Anthropology, University of Pittsburgh
Pittsburgh, PA 15260
[3] Department of Oral Medicine and Pathology, University of Pittsburgh
Pittsburgh, PA 15260
[4] Department of Anatomical Sciences and Neurobiology
University of Louisville, School of Medicine
Louisville, KE 40292
[5] Section of Mammals, Carnegie Museum of Natural History
Pittsburgh, PA 15206

INTRODUCTION

The structure and function of the mammalian vomeronasal organ (VNO) have been the focus of numerous studies (e.g., Adams and Weikamp, 1984; Vaccarezza *et al.*, 1981). Functionally, this chemosensory epithelial structure has been linked to various behaviors such as mate-finding and aggressive interactions (Powers and Winans, 1975; Wysocki and Lepri, 1991). In mammals, the evidence that VNOs function as a pheromone receptor has derived from experimental data. Determinants of VNO size are less well-understood, but have been hypothesized to include hormonal factors (Segovia and Guillamón, 1982, 1993.). It is therefore possible that morphology of the VNO may vary with gonadal development.

Most VNO studies have focused on laboratory-bred animals (e.g., Segovia and Guillamón, 1982; Weiler *et al.*, 1999). The present study utilizes wild-caught rodents in which reproductive behavior varies considerably. Furthermore, this study uses two species of voles (*Microtus pennsylvanicus* and *Microtus ochrogaster*) that vary in mating system and degree of paternal tendencies (Gruder-Adams and Getz, 1985; Oliveras and Novak, 1986). It has been hypothesized that degree of parental care, mate finding behavior and hormonal levels may correlate with the size of the accessory olfactory structures in vertebrates (Dawley and Crowder, 1995; Segovia and Guillamón, 1993). The purpose of this study is to investigate whether there is a relationship between VNO size and sex, reproductive behavior or reproductive status based on histological findings.

METHODS

Microtus pennsylvanicus (8 males, 8 females) and *M. ochrogaster* (8 males, 7 females) were captured between June and August 1998. Specimens were live-trapped in Pittsburgh or Slippery Rock, Pennsylvania *(M. pennsylvanicus)* and Bloomington, Indiana or Effingham, Illinois *(M. ochrogaster)*. More detailed explanations of how the voles were trapped, selected, and prepared for histological assessment are provided in Smith *et al.* (in press). Prior to histological preparation, selected craniometric measurements were taken. Heads and gonadal tissues were processed at the Basic Science Research Laboratory in the Graduate School of Physical Therapy, Slippery Rock University. Tissues were embedded in paraffin and sectioned at 10-16 µm. Every tenth section was mounted on numbered glass slides, stained with hemotoxylin-eosin, and examined under light microscope. Selected sections were stained with Gomori trichrome.

Evidence for „pregnant" or „non-pregnant" status was grossly and histologically examined. Initially, gross swelling of the abdomen was observed. When the voles were dissected, swellings in the uterine horns were considered as signs of pregnancy. Swellings were dissected open and sectioned for embryos. The male group was subdivided based on sexual maturity level as evidenced by the number of primary spermatocytes and late spermatids seen in the seminiferous tubules. Primary spermatocytes and late spermatids were identified based on Gartner and Hiatt's (1994) description. These cells were counted at 630x magnification in an 80x80 µm grid. The slide selected for the count was from the approximate middle section through the testis. Seminiferous tubules (ST) used for the count showed a distinct tubular appearance, clearly making the two cell types in question identifiable. Cell counts were made within three sites (cranial, midsection and caudal) in which STs were randomly chosen. Cells that were not completely part of the designated working area were not included. The mean of the number of late spermatids (A) and number of primary spermatocytes (B) was taken for each male specimen. A spermatogenic index (A/B) was calculated and rounded off to the nearest whole number, with a higher number representing a specimen that had more late spermatids than primary spermatocytes. One M. ochrogaster testis was unusable due to histological damage.

Data analysis of each specimen was performed at the University of Pittsburgh in the Image Processing Laboratory, Department of Anthropology. For all measurements, we arbitrarily used the right VNO since an earlier study indicated that there were no significant differences between right and left sides (Smith *et al.*, in press). For volume quantification, the thick vomeronasal neuroepithelium (VNNE) was identified and digitized using a well-documented three dimensional reconstruction technique (Siegel *et al*, 1983). The validity of this method has been established (Smith *et al.*, 1996). VNNE length (VNNEL) was calculated by summing up the recorded thickness of all the sections in which the VNNE was found. Ratios of VNNE lengths and volumes to palatal length were calculated to control for any differences that may result solely from body size differences. These ratios were converted using the arcsine transformation prior to statistical analysis (Sokal and Rohlf, 1981). Absolute lengths of the VNOs were calculated in an antero-posterior direction.

A two-way analysis of variance (ANOVA, sex by species) was performed on palatal length and VNNE measurements. A one-way ANOVA was used to compare VNNE size between male voles (both species) according to the spermatogenic index. A Student's t-test was used to compare female voles (both species) according to pregnancy status.

RESULTS

The VNNE was easily identified in all voles (Figure 1), exhibiting a contrasting thickness with the medial non-sensory, or receptor-free epithelium. Anteroposteriorly, the VNNEL was from 2.8 to 3.5 mm in *M. pennsylvanicus* and 2.7 to 3.8 mm in *M. ochrogaster*. The VNNEV ranged from 0.95 to 2.78 mm in *M. pennsylvanicus* and 0.92 to 2.08 mm in *M. ochrogaster*. Means and standard deviations of absolute VNO measurements are shown on Table 1. Results show significant ($p < 0.05$) differences in the palatal length between the species and in the species by sex interaction. In VNO size comparison, the only significant ($p < 0.05$) difference found was in the ratio of VNNE volume to palatal length (VNNEL/PL). No significant differences ($p > 0.05$) were noted in the analysis of absolute VNNE length or volume. Results also revealed no significant ($p > 0.05$) differences between the two species, the sexes, and sex by species interaction in absolute or proportional measures of VNNE size. Post hoc Student's *t*-tests revealed that significant ($p < 0.05$) differences existed in palatal length and VNNEV/PL ratio between sexes of *M. pennsylvanicus* but not between sexes in *M. ochrogaster*.

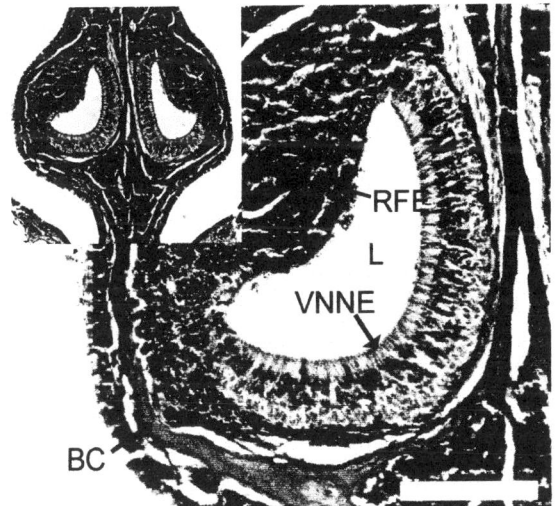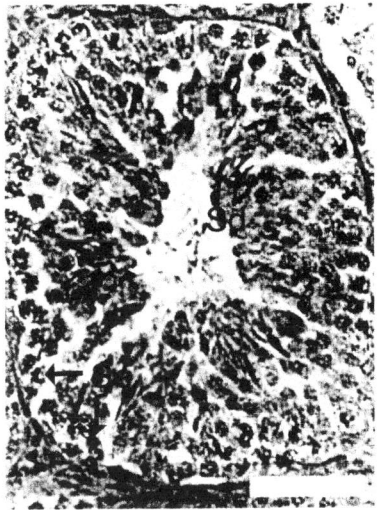

Figure 1. The right vomeronasal organ (above left), is surrounded by a bony capsule (BC). Bilateral organs are also shown (inset). The Vomeronasal neuroepithelium (VNNE) was easily differentiated from the thinner, more superomedial receptor-free epithelium (RFE) during quantification. The seminiferous tubules (above right) contained the primary spermatocytes (Se) and late spermatids (Sd) in all male voles.

All male voles exhibited both primary spermatocytes and late spermatids that were easily differentiated during cell counts (Figure 1), although ratios varied (Table 2). The two species were combined for analysis of VNNE size and gonadal development (Tables 2 and 3). No significant ($p > 0.05$) differences were found between VNNE size in groups according to the spermatogenic index (one-way ANOVA) and pregnancy status (t-test).

Table 1: Descriptive statistics and analysis of variance (ANOVA) of palatal length and vomeronasal organ (VNO) measurements.

	M. pennsylvanicus mean (SD)		*M. ochrogaster,* mean (SD)		*F*-values		
	Females	Males	Females	Males	Species	Sex	Spec. × Sex
PL (mm)	14.20 § (0.56)	15.26 § (0.78)	14.23 (1.14)	13.87 (0.47)	5.89 *	1.51 ns	6.48 *
RVL (mm)	3.21 (0.20)	3.20 (0.26)	3.14 (0.39)	3.16 (0.21)	0.31 ns	0.0001 ns	0.02 ns
RVV (cc × 10^{-4})	1.94 (0.50)	1.56 (0.29)	1.55 (0.44)	1.56 (0.22)	1.85 ns	1.74 ns	1.83 ns
RVL/PL (0.01)	0.23 (0.02)	0.21 (0.03)	0.22 (0.02)	0.23	0.85 ns	0.43 ns	2.02 ns
crRVV/PL	0.0041 § (0.0004)	0.0035 § (0.0004)	0.0038 (0.0004)	0.0039 (0.0003)	0.12 ns	3.06 ns	5.55 *

PL = palatal length; RVL = right vomeronasal neuroepithelium length; RVV = right vomeronasal neuroepithelium volume; crRVV = cube root of RVV; § = paired means that were significantly ($p < 0.05$) different from each other; * = significant difference, p-value < 0.05; ns = no significant difference, p-value > 0.05.

Table 2: Comparison of vomeronasal neuroepithelium (VNNE) measurements and maturity level of *Microtus* (both species combined).

	SPERMATOGENIC INDEX, mean (std dev)				*F* values[†]
	I	II	III	IV	
RVL (mm)	3.17 (0.32)	3.20 (0.21)	3.15 (0.14)	3.20 (Ł)	0.03 ns
RVV (cc)	1.61×10^{-4} (2.14×10^{-5})	1.56×10^{-4} (2.33×10^{-5})	1.50×10^{-4} (3.88×10^{-5})	1.82×10^{-4} (Ł)	0.32 ns

[†] = based on one-way *ANOVA*; RVL = right vomeronasal neuroepithelium length; RVV = right vomeronasal neuroepithelium volume; ns = no significant difference, p-value > 0.05; Ł = analysis not done

Table 3: Comparison of vomeronasal neuroepithelium (VNNE) measurements and pregnancy status of *Microtus* (both species combined).

	PREGNANCY STATUS, mean (std dev)		*t* values[‡]
	Pregnant	Non-pregnant	
RVL (mm)	3.29 (0.27)	3.03 (0.23)	0.15 ns
RVV (cc)	1.82×10^{-4} (3.89×10^{-5})	1.71×10^{-4} (6.54×10^{-5})	1.13 ns

[‡] = based on Student's *t*-tests; RVL = right vomeronasal neuroepithelium length; RVV = right vomeronasal neuroepithelium volume; ns = no significant difference, p-value > 0.05

DISCUSSION

Previous work on various species of vertebrates has suggested that a relationship may exist between VNNE size and sex, reproductive behavior or reproductive status (Dawley, 1998; Segovia and Guillamón, 1982). Our results showed no significant difference in VNO size, i.e. VNNE length and volume, between the sexes in either species. This contradicts Segovia and Guillamón's (1982) study that suggests that there is sexual dimorphism in rat VNO. The authors showed a male > female pattern on the VNO volume and receptor number. Neonatal male orchidectomy resulted in decreased VNO volume, neuroepithelial volume, and number of receptors while neonatal female androgenization resulted in increased values for the same measurements. In contrast to *M. pennsylvanicus*, *M. ochrogaster* males bond with one partner and contribute more parental care than the latter species (Gruder-Adams and Getz, 1985). Thus, if VNO size is negatively related to parental behavior we would expect to see a significant difference in the VNO size between males of the species. However, results only showed significant difference between the palatal lengths of the two species. Also, since it has been hypothesized that VNO size is related to hormonal level (Segovia and Guillamón, 1982, 1993), it was somewhat of a surprise to find no significant difference between the pregnant and non-pregnant voles and between the males who were grouped into four maturity levels according to the spermatogenic index. One can argue that our sample may have been biased toward sexually mature males to begin with. Hence, such a study may be more appropriately carried out using wider age-ranging of animals in a laboratory setting.

Although our study indicates a proportional VNO size difference for one measure (i.e., RVNNEV/palatal length), it is unclear that even this difference was functionally important. Even though proportional differences existed where sexes differed somatically, the receptor population may have indeed been similar. Proportional differences may be more meaningful in comparisons of species that differ greatly in size.

As mentioned earlier, other VNO studies have been on laboratory-bred animals. Our results were based on two species of wild *Microtus*. One might have expected different findings between wild and captive animals because external environmental factors are expected to affect wild animals more so than the latter. Factors such as pheromones, which are important in reproductive behavior and reproductive status (e.g., Schilling *et al.*, 1984), can be more easily controlled in laboratory settings than in the wild.

On the other hand, our results might not be surprising in view of other studies that did not find a relationship between reproductive physiology and the postnatal size of the VNO (Weiler *et al.*, 1999), suggesting that multiple factors may affect VNO size (Smith *et al.*, in press). It has been established that climatic conditions and habitats play important roles in influencing plasma testosterone level. In fact, several internal and external environmental factors can affect testicular activity (Grocock and Clarke, 1974; Ko *et al.*, 1998). With this in mind, it seems reasonable to expect that the window of opportunity to relate VNO to the reproductive system may be small. Data from specimens collected during a different time of the season may have shown more significant relationships with VNO size.

In conclusion, our results did not corroborate current literature regarding the relationship of VNO size to mammalian characteristics, i.e. sex, reproductive behavior and reproductive status. We did not find a male > female pattern in VNO size in either *Microtus* species. Reproductive behavior and reproductive status also did not show a clear relationship with VNO size. Since multiple factors (e.g., climatic conditions) are suggested to play an important role in reproductive behavior and development, future studies may be to collect *Microtus* at different seasons.

ACKNOWLEDGMENTS

The authors are grateful to the State System of Higher Education of Pennsylvania for funding this study. We also thank S.J.C. Gaulin for his loan of live traps and S.B. McLaren for help in selecting trapping sites.

REFERENCES

Adams, D. R., and Wiekamp, M. D., 1984, The canine vomeronasal organ, *J. Anat.* **138**(4):171-187.
Dawley, E. M., 1998, Species, sex, and seasonal differences in VNO size, *Microsc. Res. Techn.* **41**:506-518.
Dawley, E. M., and Crowder, J., 1995, Sexual and seasonal differences in the vomeronasal epithelium of the red-backed salamander (*Plethodon cinereus*), *J. Comp. Neurol.* **359**:382-390.
Gartner, L. P., and Hiatt, J. L., 1994, *Color Atlas of Histology*, 2nd. Ed., Williams and Wilkins, Baltimore.
Grocock, C. A., and Clarke, J. R., 1974, Photoperiodic control of testis activity in the vole, *Microtus agrestis*, *J. Reprod. Fert.* **39**:337-347.
Gruder-Adams, S., and Getz, L. L., 1985, Comparison of the mating system and paternal behavior in *Microtus ochrogaster* and *M. pennsylvanicus*, *J. Mamm.* **66**:165-167.
Ko, S. K., Kang, H. M., Im, W. B., and Kwon, H. B, 1998, Testicular cycles in three species of Korean frogs: *Rana nigromaculata, Rana rugosa*, and *Rana dybowskii*, *Gen. Comp. Endocrinol.* **111**:347-358.
Oliveras, D., and Novak, M., 1986, A comparison of paternal behavior in the meadow vole *Microtus pennsylvanicus*, the pine vole *M. pinetorum* and the prairie vole *M. ochrogaster*, *Anim. Behav.* **34**:519-526.
Powers, J. B., and Winans, S. S., 1975, Vomeronasal organ: critical role in mediating sexual behavior in the male hamster, *Science* **187**:961-963.
Schilling, A., Perret, M., and Predine, J., 1984, Sexual inhibition in a prosimian primate: a pheromone-like effect, *J. Endocrinol.* **102**:143-151.
Segovia, S., and Guillamón, A., 1982, Effects of sex steroids on the development of the vomeronasal organ in the rat, *Dev. Brain Res.* **5**:209-212.
Segovia, S., and Guillamón, A., 1993, Sexual dimorphism in the vomeronasal pathway and sex differences in reproductive behaviors, *Brain Res. Rev.* **18**:51-74.
Siegel, M. I., Todhunter, J. S., Doyle W. J., and Rood, S. R., 1983, Computer reconstruction of eustachian tube Anatomy, *Ann. Otol. Rhinol. Laryngol.* **92**:10-14.
Smith, T. D., Siegel, M. I., Mooney, M. P., Burdi, A. R., and Todhunter, J. S., 1996, Vomeronasal organ growth and development in normal and cleft lip and palate fetuses, *Cleft Palate-Craniofac. J.* **33**:385-394.
Smith, T. D., Roslinski, D. L., Burrows, A. M., Bhatnagar, K. P., Mooney, M. P., and Siegel, M. I., Size of the vomeronasal neuroepithelium in two species of *Microtus* with differing levels of paternal behavior, *J. Mamm.* (in press).
Sokal, R. R., and Rohlf, F. J., 1981, *Biometry*, W. H. Freeman and Co., New York.
Vaccarezza, O. L., Sepich, L. N., and Tramezzani, J. H., 1981, The vomeronasal organ of the rat, *J. Anat.* **132**:167-185.
Weiler, E., Apfelbach, R., and Farbman, A. I., 1999, The vomeronasal organ of the male ferret, *Chem. Senses* **24**:127-136.
Wysocki, C. J., and Lepri, J. J. L., 1991, Consequences of removing the vomeronasal organ, *J. Steroid Biochem. Molec. Biol.* **39**(4B):661-669.

OXYTOCIN, NOREPINEPHRINE AND OLFACTORY BULB MEDIATED RECOGNITION

Dean E. Dluzen,[1] Yili Shang,[1] and Rainer Landgraf[2]

[1]Department of Anatomy
Northeastern Ohio Universities College of Medicine
Rootstown, OH 44272
[2]Max-Planck Institute of Psychiatry
Munich, Germany

INTRODUCTION

In addition to its putative role in parturition and lactation, the neuropeptide oxytocin (OXT), exerts a myriad of appended actions within the central nervous system (Argiolas and Gessa, 1991). In the present report we focus upon the capacity for OXT to enhance memory/recognition responses. The beneficial effects of OXT upon memory/recognition have been demonstrated in a variety of paradigms (Engelmann et al., 1996). Since many of these paradigms consist of tests involving social recognition, which rely heavily upon olfaction (Sawyer et al., 1984; Dantzer et al., 1990), we were interested in the extent to which the olfactory bulb (OB) was involved with mediating these recognition responses. Accordingly, we addressed two fundamental questions which comprised the basis for this report: 1) Will OXT affect social recognition when applied within the OB? and 2) How does this neuropeptide function at this site? In an attempt to answer these questions, we combined a behavioral assay of social recognition with infusions/measurements of agents within the OB to evaluate whether the localized application of neuromodulators at this site would alter social recognition.

OXYTOCIN AND OLFACTORY BULB INFUSION

Our first venture in this endeavor consisted of infusing OXT directly into the OB of adult male rats to determine whether social recognition responses would be altered. Twenty-one gauge guide cannulae were implanted bilaterally above the dorsal center surface of the OB. At 5-7 days post-implant of guide cannulae, two 27 gauge infusion cannulae were inserted into the OB to permit bilateral infusion of test agents. Within one-minute following infusion, a stimulus animal (21-30 day old juvenile rat) was placed within the cage of the male rat and the amount of behavioral investigation directed to the stimulus animal was recorded. At 120-minutes following this initial exposure, this same juvenile along with a novel juvenile were placed within the male's cage and the amount of investigation directed to the two juveniles was recorded. Under ordinary (control) conditions, similar amounts of investigation would be directed to the two stimulus animals at this 120-minute test period and this lack of discriminatory responses would suggest an absence of recognition. A significantly greater amount of investigation directed to the novel versus same (original) juvenile, however, is interpreted to indicate that a social recognition response was present (Engelmann *et al.*, 1995). In this way, the display of discriminatory investigatory responses at this 120-minute inter-exposure interval (IEI) indicates that recognition has been preserved or prolonged to periods beyond the normal capacity for the display of responses indicative of recognition.

When OXT was infused bilaterally within the OB, clear recognition responses were present when tested at the 120-minute interval as revealed by statistically greater amounts of investigation directed to the novel stimulus animal (Figure 1B). Such data show that the OB, like the septal (Dantzer *et al.*, 1987; Popik *et al.*, 1992) and medial preoptic (Popik *et al.*, 1991) areas, represents an important target site where OXT can function to enhance social recognition responses. Since OXT is activated under social interactions (Hughes *et al.*, 1987) and reported to be present within the OB (Halasz and Shepherd, 1983; Sofroniew, 1985) and/or transported to the OB following release from the paraventricular nucleus (Yu *et al.*, 1996) our findings suggest a potential new role for this neuropeptide at this site.

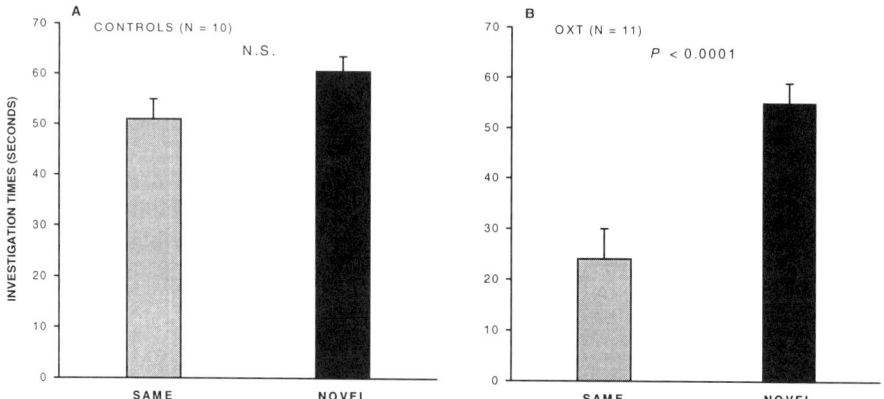

Figure 1. Investigation times (Mean+SEM in Seconds) directed to the same versus novel stimulus animal as tested at a 120-minute IEI for animals infused with either Ringer's solution (A. Controls, N=10) or Ringer's solution with oxytocin (B. OXT @ 0.5 ng, N=11) into the OB. Animals receiving OXT showed significantly greater amounts of investigation directed toward the novel stimulus animals indicating a recognition response, that was not present for Control animals. Data from Dluzen *et al.* (1998a).

To verify these effects and establish that OXT is functioning through an OXT receptor mediated process we co-infused OXT with either an OXT or arginine vasopressin receptor antagonist into the OB. Discriminatory social recognition responses were abolished under conditions of OXT+OXT- receptor antagonist infusion (Figure 2A), but remained in animals co-infused with the OXT + arginine vasopressin receptor antagonist (Figure 2B). These results confirm that OB OXT infusion preserves recognition responses and this process involves activation of an OXT receptor within this site. In support of this functional assay for OXT receptors within the OB are data indicating the expression of OB OXT receptors as identified in the rat with *in situ* hybridization (Yoshimura *et al.*, 1993) and in the guinea pig with immunocytochemistry (Tribollet *et al.*, 1992).

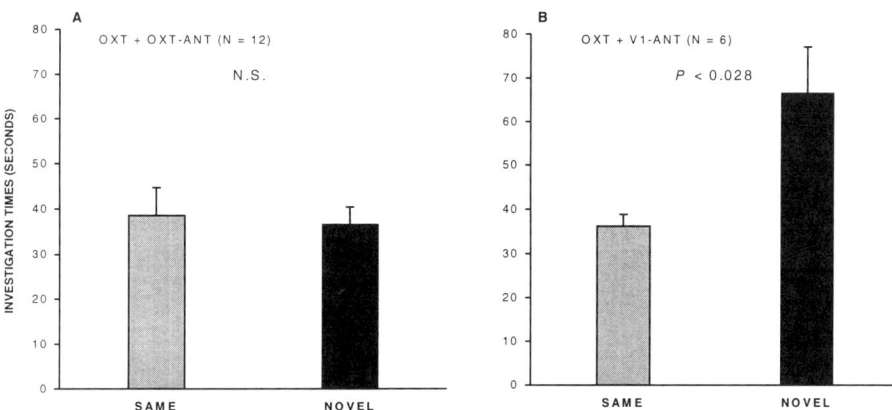

Figure 2. Investigation times (Mean and SEM in Seconds) as described in Figure 1 for animals infused with either oxytocin + oxytocin-receptor antagonist (**A.** OXT@0.5 ng+OXT-ANT@ 5.0 ng, N = 12) or oxytocin + arginine vasopressin-receptor antagonist (**B.** OXT+V1-ANT, N = 6) into the OB. Co-infusion of OXT+OXT-ANT abolished recognition responses obtained with OXT infusion (Figure 1B), but recognition responses remained in animals receiving a co-infusion of OXT+V1-ANT. Data from Dluzen *et al.* (2000).

OXYTOCIN AND NOREPINEPHRINE INTERACTIONS

Our next goal was to consider some of the means by which OXT may function within the OB to preserve social recognition responses. Since there existed considerable evidence linking OXT with norepinephrine (NE) as related to recognition (Kendrick *et al.*, 1997), our efforts were channeled toward this catecholamine. A microdialysis probe was inserted within the OB which allowed us to not only measure changes in NE levels but also infuse OXT within this site. One action resulting from OXT infusion was to produce a significant increase in NE output within the OB (Figure 3A). No such changes in OB NE output were obtained in response to an infusion of the OXT antagonist (Figure 3B) or under basal conditions of Ringer's solution infusion (Figure 3C). Similar findings were reported in the ewe where it was demonstrated that OB NE output can be activated by OXT under physiological conditions (Kendrick *et al.*, 1988; Levy *et al.*, 1993), as well as in responses to OXT infusion (Levy *et al.*, 1995).

The significance of this OB OXT/NE relationship for recognition was highlighted in a follow-up experiment in which OB OXT infusion was tested in rats where OB NE was depleted by the neurotoxin, 6-hydroxydopamine. In animals receiving the vehicle for 6-hydroxydopamine into the OB and subsequently tested with OXT, clear recognition responses were present (Figure 4A). By contrast, rats infused with 6-hydroxydopamine followed by testing with OXT showed no indication of a recognition response (Figure 4B). The relevance of this OXT-NE relationship within the OB is that a number of reports have demonstrated the importance of the OB NE system for memory/recognition responses (Gervais *et al.*, 1988; Brennan *et al.*, 1990; Levy *et al.*, 1990; Guan *et al.*, 1993a; b; Kaba and Nakanishi, 1995; Dluzen and Kreutzberg, 1993).

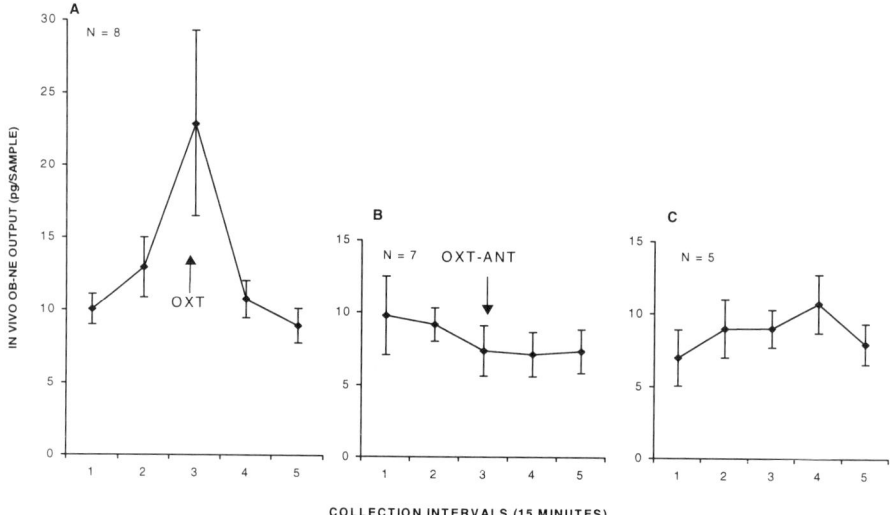

Figure 3. OB NE output (pg/sample) from sequential microdialysis samples (15-minute intervals @ 3.5 ul/minute) from male rats in which either oxytocin (**A.** OXT@2-3 ng, N=8), oxytocin-antagonist (**B.** OXT-ANT, N=7) or Ringer's solution (**C.** N=5) was infused through the probe during interval 3 of the dialysis. Infusion of OXT produced a significant increase in OB NE output that was not observed for either the OXT-ANT or Ringer's treated rats. Data from Dluzen *et al.* (2000).

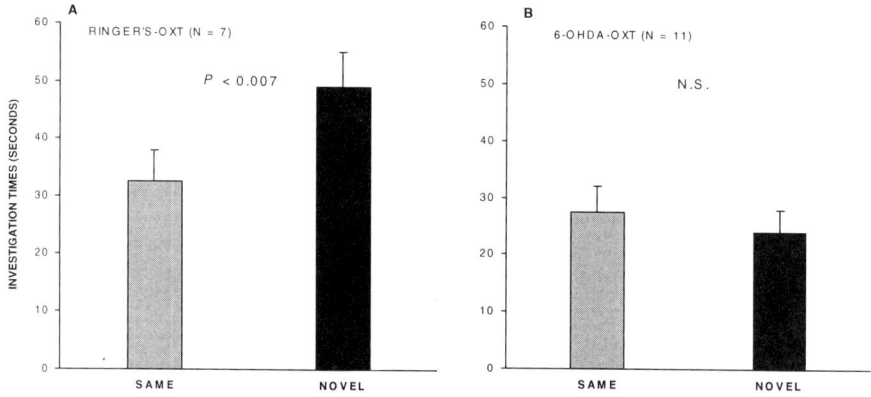

Figure 4. Investigation times (Mean+SEM in Seconds) as described in Figure 1 for animals infused with oxytocin into the OB under conditions of an intact OB (**A.** Ringer's - OXT, N=7) or a NE depleted OB (**B.** 6-OHDA-OXT, N=11). Depletion of OB NE abolished the capacity for OXT to enhance recognition responses. Data from Dluzen *et al.* (1998b).

NOREPINEPHRINE AND SOCIAL RECOGNITION

The fact that depletion of OB NE with 6-hydroxydopamine abolishes the capacity for OXT to preserve social recognition responses along with the ability for OXT to increase OB NE output have two salient implications. First, it establishes a close relationship between these two neuroactive agents. Second, it suggests that one means by which OXT may enhance recognition responses is through activating the OB NE system. Consequently, we initiated a series of experiments in which relatively specific adrenoceptor agonists and antagonists of NE were employed to test the significance of their roles in this recognition process. In effect, our working hypothesis was that OB OXT was functioning as an agent to evoke OB NE as the critical step involved with enhancing recognition. The ability for a neuropeptide to modulate NE neurotransmission to enhance memory has been proposed for the related neuropeptide, arginine vasopressin, over 20 years ago (Kovacs *et al.*, 1979) and a generalized review of OXT/neurotransmitter interactions has been presented recently (Kendrick, 2000). To test this hypothesis, we bypassed OXT and directly infused the NE agonists, clonidine or isoproterenol, into the OB. An infusion of the α-adrenoceptor agonist, clonidine, within the OB in a manner identical to that performed with OXT produced a preservation in social recognition like that observed with OXT (Figure 5A). When the β-adrenoceptor agonist, isoproterenol, was similarly tested, no beneficial effects upon social recognition were obtained (Figure 5B).

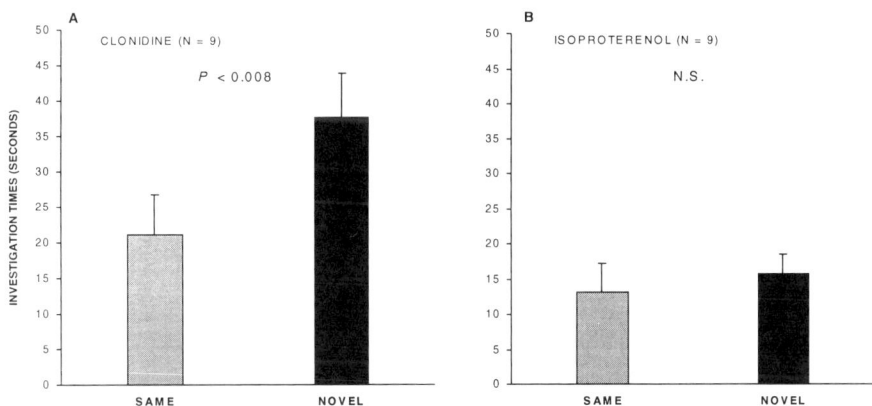

Figure 5. Investigation times (Mean+SEM in Seconds) as described in Figure 1 for animals infused with either the α-adrenoceptor, clonidine (**A.** 10 uM, N=9) or the β-adrenoceptor, isoproterenol (**B.** 10 uM, N=9) into the OB. Infusion of the clonidine, but not isoproterenol, resulted in the display of significant recognition responses. Data from Dluzen *et al.* (2000).

As an approach to further test our hypothesis, we examined the effects of co-infusion of OXT with relatively selective NE adrenoceptor antagonists. The logic behind these experiments was that OXT would evoke NE output (Figure 3A), but the function of this OB NE output would be negated by the presence of the NE adrenoceptor antagonists present within the infusion cocktail. When OXT was co-infused with the α-adrenoceptor antagonist, phentolamine, no evidence for social recognition was present, i.e., the amount of behavioral investigation directed to the same and novel stimulus animals was virtually identical (Figure 6A). However, significantly greater amounts of investigation were directed to the novel stimulus animal, suggestive of recognition, under conditions in which OXT was co-infused with the β-adrenoceptor antagonist, timolol (Figure 6B). Such results imply that preventing the OXT-evoked NE output from activating α-adrenoceptors abolishes the display of social recognition responses and are in concert with the data presented in Figure 5A illustrating the importance for α-adrenoceptor involvement in social recognition. This selectivity for the OB α-adrenoceptor was also reported for recognition responses as assessed for the pregnancy block paradigm in mice (Kaba and Keverne, 1998), however, the β-adrenoceptor may be more critical for recognition responses in ewes (Levy et al., 1990).

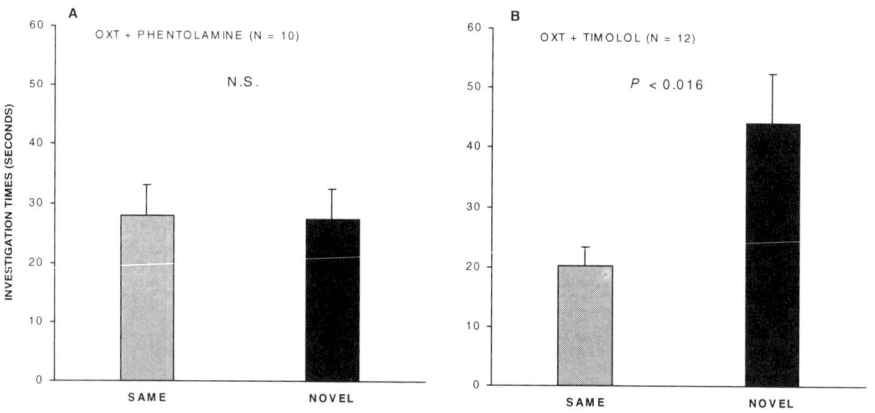

Figure 6. Investigation times (Mean+SEM in Seconds) as described in Figure 1 for animals co-infused with oxytocin+the α-adrenoceptor antagonist, phentolamine (**A.** 40 nM, N=10) or oxytocin+the β-adrenoceptor antagonist, timolol (**B.** 40 nM, N =12) into the OB. The addition of phentolamine, but not timolol, to the OXT infusion abolished recognition responses. Data from Dluzen et al. (2000).

In the final series of experiments, we combined neurochemical and behavioral assays as a means to achieve a more comprehensive profile of the relationship between OB NE and the display of recognition. In these experiments, male rats were subjected to microdialysis under conditions in which OB NE levels were manipulated while testing simultaneously for behavioral recognition responses. Following a basal period of effluent sample collections, either vehicle or nisoxetine was infused through the microdialysis probe for a 40-minute interval. A stimulus animal (ovariectomized female) was then placed into the dialyzed animal's cage and the amount of investigation directed to this stimulus animal was recorded. The dialysis was continued and at 120-minutes later the amount of investigation to this stimulus animal was again recorded. Since the NE uptake blocker, nisoxetine, increases extracellular levels of NE, we were interested in determining whether this treatment would alter social recognition. Under control (vehicle infusion) conditions, the amount of OB NE recorded was low and stable throughout the entire period of the experiment (Figure 7A) and the behavioral measures provided no evidence for a recognition responses in these animals (Figure 7B). In animals receiving nisoxetine infusion marked and relatively prolonged increases in NE levels were detected (Figure 7C), results similar to that reported following OB infusion of the NE uptake inhibitor desipramine (El-Etri *et al.*, 1999). Accompanying this augmented OB NE were significantly reduced amounts of behavioral investigation obtained during the second exposure to the stimulus female indicative of a recognition response (Figure 7D) (Thor and Holloway, 1982). These data reinforce previous findings indicating the importance for the increase in OB NE output levels as associated with enhanced recognition responses.

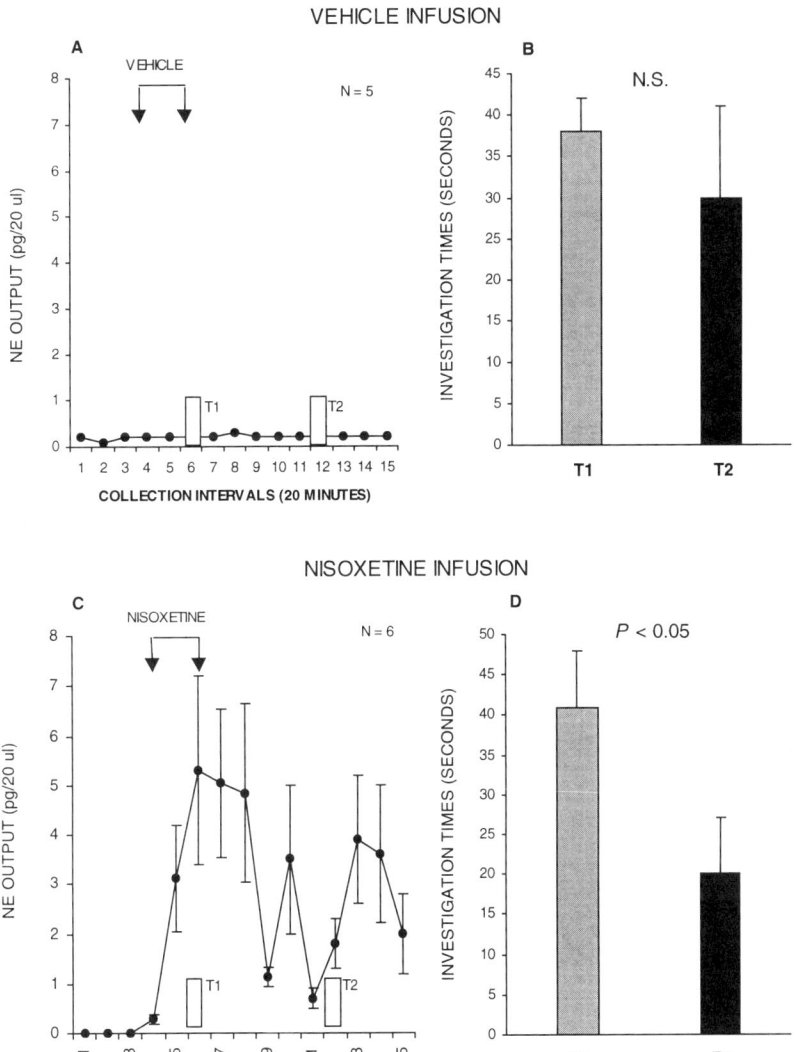

Figure 7. OB NE output (pg/20 ul) and simultaneous behavioral investigatory recordings from male rats during microdialysis. In rats receiving a vehicle infusion into the OB, there is no change in OB NE output (**A**) nor any indication of a recognition responses (**B**). Rats receiving an infusion of the NE uptake blocker, nisoxetine (1 mM), show increased OB NE output (**C**) and evidence of enhanced recognition responses (**D**). Data from Shang and Dluzen (1999).

OVERVIEW

From the data it is clear that OXT infusion into the OB can preserve or prolong social recognition responses to periods that are typically not observed under normal (control) conditions. This process represents an OXT receptor mediated mechanism within the OB; however, our data do not allow us to identify the specific point of contact for these OXT receptors. The capacity for OXT to enhance social recognition involves a critical interaction with NE as evidenced by the ability for OXT infusion to increase OB NE output and the abolition of OXT's effects in animals with depleted OB NE concentrations. More specifically, it appears that the increase in OB NE output, as achieved with either a neuropeptide, like OXT, or through an artificial agent, like nisoxetine, selectively activates post-synaptic α-adrenoceptors to result in this display of prolonged recognition.

REFERENCES

Argiolas, A., and Gessa, G. L., 1991, Central functions of oxytocin, *Neurosci. Biobehav. Rev.* **15**:217-231.

Brennan, P., Kaba, H., and Keverne, E. B., 1990, Olfactory recognition: A simple memory system, *Science* **250**:1223-1225.

Dantzer, R., Bluthe, R.-M., Koob, G. F., and LeMoal, M., 1987, Modulation of social memory in male rats by hypophyseal peptides, *Psychopharmacology* **91**:363-368.

Dantzer, R., Tazi, A., and Bluthe, R.-M., 1990, Cerebral lateralization of olfactory-mediated affective processes in rats, *Behav. Brain Res.* **40**:53-60.

Dluzen, D. E., and Kreutzberg, J. D., 1993, 1-methyl-4-phenyl-1,2,3,6-tetrahydropyridine (MPTP) disrupts social memory/recognition responses in the male mouse, *Brain Res.* **609**:98-102.

Dluzen, D. E., Muraoka, S., Engelmann, M., and Landgraf, R., 1998a, The effects of infusion of arginine vasopressin, oxytocin or their antagonists into the olfactory bulb upon social recognition responses in male rats, *Peptides* **19**:999-1005.

Dluzen, D. E., Muraoka, S., and Landgraf, R., 1998b, Olfactory bulb norepinephrine depletion abolishes vasopressin and oxytocin preservation of social recognition responses, *Neurosci. Lett.* **254**:161-164.

Dluzen, D. E., Muraoka, S., Engelmann, E., Ebner, K., and Landgraf, R., 2000, Oxytocin induces preservation of social recognition in male rats by activating α-adrenoceptors of the olfactory bulb, *Eur. J. Neurosci.* **12**:760-766.

El-Etri, M. M., Ennis, M., Griff, E. R., and Shipley, M. T., 1999, Evidence for cholinergic regulation of basal norepinephrine release in the rat olfactory bulb, *Neuroscience* **93**:611-617.

Engelmann, M., Wotjak, C., and Landgraf, R., 1995, Social discrimination procedure: An alternative method to investigate juvenile recognition abilities in rats, *Physiol. Behav.* **58**:315-321.

Engelmann, M., Wotjak, C. T., Neumann, I., Ludwig, M., and Landgraf, R., 1996, Behavioral consequences of intracerebral vasopressin and oxytocin: Focus on learning and memory, *Neurosci. Biobehav. Rev.* **20**:341-358.

Gervais, R., Holley, A., and Keverne, B., 1988, The importance of central noradrenergic influences on the olfactory bulb in the processing of learned olfactory cues, *Chem. Sens.* **13**:3-12.

Guan, X., Blank, J., and Dluzen, D. E., 1993a, Depletion of olfactory bulb norepinephrine by 6-OHDA disrupts chemical cue but not social recognition responses in male rats, *Brain Res.* **622**:51-57.

Guan, X., Blank, J., and Dluzen, D. E., 1993b, Role of the olfactory bulb norepinephrine in the identification and recognition of chemical cues, *Physiol. Behav.* **53**: 437-441.

Halasz, H., and Shepherd, G. M., 1983, Neurochemistry of the vertebrate olfactory bulb, *Neuroscience* **10**:579-619.

Hughes, A. M., Everitt, B. J., Lightman, S. L., and Todd, K., 1987, Oxytocin in the central nervous system and sexual behaviour in male rats, *Brain Res.* **414**: 133-137

Kaba, H., and Keverne, E. B., 1988, The effect of microinfusions of drugs into the accessory olfactory bulb on the olfactory block to pregnancy, *Neuroscience* **25**: 1007-1011.

Kaba, H., and Nakanishi, S., 1995, Synaptic mechanisms of olfactory recognition memory, *Rev. Neurosci.* **6**:125-141.

Kendrick, K. M., 2000, Oxytocin, motherhood and bonding, *Exp. Physiol.* **85S**:111S-124S.

Kendrick, K. M., Da Costa, A. P. C., Broad, K. D., Ohkura, S., Guevara, R., Levy, F., and Keverne, E. B., 1997, Neural control of maternal behaviour and olfactory recognition of offspring, *Brain Res. Bull.* **44**:383-395.

Kendrick, K. M., Keverne, E. B., Chapman, C., and Baldwin, B. A., 1988, Intracranial dialysis measurements of oxytocin, monoamine and uric acid release from the olfactory bulb and substantia nigra of sheep during parturition, suckling, separation from lambs and eating, *Brain Res.* **439**:1-10.

Kovacs, G. L., Bohus, B., and Versteeg, D. H. G., 1979, Facilitation of memory consolidation by vasopressin: Mediation by terminals of the dorsal noradrenergic bundle? *Brain Res.* **172**:73-85.

Levy, F., Gervais, R., Kinderman, U., Orgeur, P., and Piketty, V., 1990, Importance of β-noradrenergic receptors in the olfactory bulb for recognition of lambs, *Behav. Neurosci.* **104**:464-469.

Levy, F., Geuvara-Guzman, R., Hinton, M. R., Kendrick, K. M., and Keverne, E. B., 1993, Effects of parturition and maternal experience on noradrenaline and acetylcholine release in the olfactory bulb of sheep, *Behav. Neurosci.* **107**:662-668.

Levy, F., Kendrick, K. M., Goode, J. A., Geuvara-Guzman, R., and Kendrick, E. B., 1995, Oxytocin and vasopressin release in the olfactory bulb of parturient ewes: Changes with maternal experience and effects on acetylcholine, γ-aminobutyric acid and noradrenaline release, *Brain Res.* **669**:197-206.

Popik, P., and van Ree, J. M., 1991, Oxytocin, but not vasopressin facilitates social recognition following injection into the medial preoptic area of the rat brain, *Eur. Neuropsychopharmacol.* **1**:555-560.

Popik, P., Vos, P. E. and van Ree, J. M., 1992, Neurohypophyseal hormone receptors in the septum are implicated in social recognition in the rat, *Behav. Pharmacol.* **3**:351-358.

Sawyer, T. F., Hengehold, A. K., and Perez, W. A., 1984, Chemosensory and hormonal mediation of social memory in male rats, *Behav. Neurosci.* **98**:908-913.

Shang, Y., and Dluzen, D. E., 1999, The selective noradrenergic uptake blocker, nisoxetine, increases norepinephrine output from the olfactory bulb and enhances recognition responses, *Soc. Neurosci.* **25**:82 (Abst. # 37.6).

Sofroniew, M. V., 1985, Vasopressin, oxytocin and their related neurohypophysins, in *Handbook of Chemical Neuroanatomy, Volume 4, GABA and neuropeptides in the CNS - Part 1* (A. Bjorkund, and T. Hokfelt, eds.), pp. 93-165.

Thor, D. H., and Holloway, W. R., 1982, Social investigation in laboratory rats, *J. Comp. Physiol. Psychol.* **96**:1000-1006.

Tribollet, E., Barberis, C., Dubois-Dauphin, M., and Dreifuss, J. J., 1992, Localization and characterization of binding sites for vasopressin and oxytocin in the brain of the guinea pig, *Brain Res.* **589**:15-23.

Yoshimura, R., Kiyama, H., Kimura, T., Araki, T., Maeno, H., Tanizawa, O., and Tohyama, M., 1993, Localization of oxytocin receptor messenger ribonucleic acid in the rat brain, *Endocrinology* **133**:1239-1246.

Yu, G. Z., Kaba, H., Okutani, F., Takahashi, S., Higuchi, T., and Seto, K., 1996, The action of oxytocin originating in the hypothalamic paraventricular nucleus on mitral and granule cells in the rat main olfactory bulb, *Neuroscience* **72**:1073-1082.

A POSSIBLE HUMORAL PATHWAY FOR THE PRIMING ACTION OF THE MALE PHEROMONE ANDROSTENOL ON FEMALE PIGS

Tadeusz Krzymowski, Stanisława Stefańczyk-Krzymowska,
Waldemar Grzegorzewski, Janina Skipor, and Barbara Wąsowska

Division of Reproductive Endocrinology and Pathophysiology
Institute of Animal Reproduction and Food Research
The Polish Academy of Sciences
Tuwima 10, 10-747 Olsztyn, Poland

INTRODUCTION

Intraspecific communication by chemical signals (pheromones) plays important behavioral and physiological roles in coordinating reproduction in mammals. Females of a variety of domestic species respond to pheromones from males by undergoing an earlier onset of puberty (Brooks and Cole, 1970; Kirkwood et al., 1983; Kalbrom, 1982; Pearce and Paterson, 1992), alterations in ovarian cycles (Oldham et al., 1979; Booth and Baldwin, 1983; Booth and Signoret, 1992) and estrous behavior (Signoret, 1970; Dorries et al., 1997). The precocious attainment of puberty in gilts was studied after contact with the boar (Kirkwood et al., 1983; Kalbrom, 1982) or following the treatment with the male pheromone androstenone, commercially available as an aerosol spray called Suidor (Glei et al., 1989). According to Pearce and Paterson (1992), physical contact with the boar is essential for the maximal pubertal acceleration as it allows the direct transfer of the priming pheromone from the boar to the snout of the recipient gilt; among adult females, nosing and sniffing of the genitalia corresponded to 47% of interactions between the boar and anestrous gilts, while the remaining 43% of interactions involved head-to-head contacts (Signoret, 1970).

The mechanism of olfaction seems to be well adapted for the transmission of information by pheromones responsible for changes in behavioral and sexual activities. Pheromones act by stimulating the dendritic receptors of chemosensory neurons in the olfactory neuroepithelium located in the dorsal-caudal region of the nasal cavity, and in the vomeronasal organ (Rhein and Cagan, 1981; Chen and Lancet, 1984; Hancock et al., 1985; Buck and Axel, 1991; Read, 1992). Neurons of the olfactory and vomeronasal organs, as well as their targets in the main and accessory olfactory bulbs, play key roles in evoking behavioral responses to pheromones. For example, removal of the olfactory bulbs induced anestrus in gilts (Kirkwood et al., 1983; Booth and Baldwin, 1983). In rodents, the vomeronasal chemosensory system appears to be the primary route for information processing of signals influencing neuroendocrine function (Marchlewska-Koj, 1984). However, this pathway is not necessary for the neuroendocrine response of ewes to males, since neither vomeronasal organ lesions nor olfactory nerve cuts inhibited LH responses to males (Cohen-Tannoudji et al., 1989). In female domestic pigs, the vomeronasal organ was not necessary for androstenone-evoked sexual behaviors in estrous females, which persisted followed ablation of the organ (Dorries et al., 1997). It is possible that odorant binding protein and vomeromodulin function as stimulus transporters in olfactory and vomeronasal organs (Krishna et al., 1994). The application of molecular genetic techniques led to the identification of the components of a second-messenger response cascade in chemoreception (Read, 1992). The interaction of pheromones with their specific receptors on the membrane of chemosensory neurons leads to the release of the GTP-coupled-α subunit of G-proteins linked to adenyl cyclase, thus opening cAMP-gated cation channels and altering the membrane potential (Buck and Axel, 1991).

Prolonged contact of receptors on olfactory membranes with their ligands quickly induces chemosensory adaptation, which occurs as the membrane hyperpolarizes despite the continued presence of the chemical stimulus. This adaptation suggests that perireceptor events in olfaction might be terminated long before neuroendocrine responses such as puberty onset or estrus induction are completed. While it is possible that the stimulation of the chemosensory pathway by the pheromones set in motion self-sustaining neuroendocrine responses, we suggest that it is possible that the pheromones at the nasal mucosa enter the circulation via the vasculature in the nasal cavity and then act in a hormone-like fashion over the long time needed for puberty onset or estrus induction. To that end, we have demonstrated that: (1) the venous effluent from the nasal cavity flows through the angularis oculi vein to the cavernous sinus located in the perihypophyseal vascular complex known as the cavernous sinus-carotid rete, (2) hormones, including progesterone (Krzymowski et al., 1992), testosterone (Skipor et al., 2000) and neuropeptides (LH-RH, oxytocin, beta-endorphin), can pass from the venous blood in the cavernous sinus into the arterial blood of the carotid rete supplying the brain, and, (3) efficacy of this local system of hormone exchange is dependent on the phase of the estrous cycle (Grzegorzewski et. al, 1995; 1997; Skipor et al., 1997).

The perihypophyseal cavernous sinus-carotid rete mirabile complex is very well developed in pigs, sheep, goats, and many other members of the Artiodactyla order (Gillian, 1974). This vascular complex is located intracranially below and on both sides of the hypophysis, under the dura mater. The carotid rete mirabile lies inside the cavernous sinus along the route of the venous blood stream (Gillian, 1974; Daniel et al., 1973). Circulation of blood in the perihypophyseal vascular complex is presented in Figure 1. Several different functions have been suggested for the vascular complex, including its possible role in the regulation of brain temperature (Baker and Hayward, 1968). Recently, we have demonstrated retrograde transfer of hormones from the venous outflow of the hypophysis and brain into the arterial blood supplying the hypophysis and brain (Krzymowski et al., 1992; Grzegorzewski et al., 1995; 1997; Skipor et al., 1997). In rabbits, rats, and humans, the carotid rete mirabile is not anatomically similar to that of Artiodactyla species, yet passage of the carotid artery throughout the cavernous sinus (Daniel et al., 1973) could provide a counter-current transfer route for the transfer of steroids in rabbits (Krzymowski et al., 1990), and tritiated water, tyrosine and propanol in rats (Einer-Jensen and Larsen, 2000).

The superficial veins of the nose and face, as shown in Figure 1, which carry blood from the nasal mucosa into the cavernous sinus of the perihypophyseal vascular complex (Ghoshal and Khamas, 1984; Mitchel et al., 1998), may alternatively direct the venous effluent from the nose through the cavernous sinus to the jugular vein. Until recently, the superficial veins of the nose and face were thought to function only in the cooling system of the brain (Maloney and Mitchel, 1997). Our research efforts have begun to explore the relationship between morphology of the nasal veins and reproductive status of the pig to determine whether or not the hormonal status of the animal influences the structure and function of these vessels.

Pheromones in urine, saliva, excrement, vaginal and skin secretions, can have a number of behavioral effects. The mechanism by which pheromones influence the hormonal control of reproduction is still unclear. To study this, the chemical identity of the pheromone molecule and the synthesis of radiolabeled ligands were useful. Two steroid pheromones, 5α-androstenone (5α-androst-16-en-3-one) and 5α-androstenol (5α-androst-16-en-3-ol), produced by boars effect lordosis responses in female pigs. We conducted the work described here using radiolabeled 5α-androstenol, since it is the predominant pheromone in the maxillary salivary gland and its concentration in boar saliva is 10-20 times higher than that of 5α-androstenone (Booth, 1982). To address our hypothesis that male pheromones influence females not only by stimulation of chemosensory neurons in the nasal cavity but also by arterial transport directly to the limbic system (humoral pathway) *in vitro*, *ex vivo* and *in vivo* experiments were performed on female pigs.

LOCAL TRANSFER OF ^{3}H-ANDROSTENOL (^{3}H-A) WITH BLOOD FROM THE NASAL CAVITY TO THE BRAIN AND HYPOPHYSIS

Experiments *ex vivo* on Isolated Heads of Pigs

In these *ex vivo* experiments, the isolated heads of pigs were used to study ^{3}H-androstenol (^{3}H-A) transfer from the nasal cavity to the blood perfusing the brain and hypophysis. Sexually mature gilts at days 16-21 of the estrous cycle were slaughtered and their heads were separated from the bodies. The heads were attached to a perfusion system using heated, oxygenated and heparinized, autologous blood. A total amount of 10^8 dpm (758 ng) of radiolabeled androstenol (^{3}H-5α-androst-16-en-3-ol) was either infused over a 5 min period into the angularis oculi veins, which drain the nasal cavities, or was sprayed through intranasal catheters onto the nasal mucosa surface for 2 min. Before and after both treatments, blood samples were frequently collected from the carotid rete and from venous effluent. In both groups of pigs, we found that the concentration of ^{3}H-A in the arterial blood of the carotid rete, supplying the brain and hypophysis, was significantly higher than the background radioactivity of blood samples collected before ^{3}H-A administration ($p<0.001$ and $p<0.01$, respectively; Krzymowski *et al.*, 1999). Moreover, after perfusion, the brain and hypophysis were dissected and analysis of tissue samples revealed significantly greater accumulation of ^{3}H-A in neurohypophysis ($p<0.001$), adenohypophysis ($p<0.01$), ventromedial hypothalamus ($p<0.05$), corpus mammillare ($p<0.01$) and perihypophyseal vascular complex ($p<0.001$), than in tissues taken from other locations in the brain.

Experiments *in vivo* on Anaesthetized Gilts

The experiment was performed in anesthetized gilts on days 18-21 of the estrous cycle (Stefańczyk-Krzymowska *et al.*, 2000). ^{3}H-A in total amount of 10^8 dpm (758 ng) was applied for 1 min onto the respiratory part of the nasal mucosa, 4-6 cm from the nares. Arterial blood samples from the aorta and from the carotid rete were collected every two min over a 60-min interval following the administration of ^{3}H-A. Total radioactive venous effluent from the head was removed and replacement volume of homologous blood was transfused into the carotid external vein. We found that the concentration of ^{3}H-A was significantly higher in the arterial blood of the carotid rete than in the aorta over entire experiment (60 min), with highly significant differences appearing 20 min from the beginning of ^{3}H-A administration (p<0.001). The mean rate of ^{3}H-A transfer from venous to arterial blood in the perihypophyseal vascular complex, defined as a ratio of ^{3}H-A concentration in the carotid rete blood to that simultaneously collected from the aorta, was 1.96 ± 0.1. We also observed that ^{3}H-A selectively accumulated in olfactory bulb, amygdala, septum, adenohypophysis, neurohypophysis, and in the perihypophyseal vascular complex (Stefańczyk-Krzymowska et al., 2000). ^{3}H-A also accumulated in the anterior hypothalamic areas of the periventricular and supraoptic nuclei, sites of GnRH, oxytocin and vasopressin synthesis, but not in the mediobasal hypothalamic areas of the infundibular and ventromedial nuclei, where transport and release of these peptides take place (Stefańczyk-Krzymowska et al., 2000).

Uptake, Accumulation, and Release of Androstenol by Nasal Mucosa

In *ex vivo* and *in vitro* experiments (Krzymowski et al., 1999), active uptake of ^{3}H-A in the epithelium of the nasal mucosa located 3-5 cm from the nares was greater than uptake in the rostral regions of the nasal cavities, i.e., the neuroepithelium of the nasal cavity. The uptake was significantly lower in more distal segments of the epithelium and in the neuroepithelium. In ex vivo experiments (Krzymowski et al., 1999) and *in vivo* on anaesthetized gilts (Stefańczyk-Krzymowska et al., 2000) it was been found that large amounts of labeled androstenol accumulated in the nasal mucosa. The presence of odorant binding protein (OBP) and vomeromodulin in the nasal mucosa (Krishna et al., 1994), suggest the existence of a mechanism that enables binding of the priming pheromone, perhaps followed by its gradual release to the venous blood.

EFFECT OF INTRAMUSCULAR INJECTION OF ANDROSTENOL ON ATTAINMENT OF PUBERTY IN GILTS

Table 1. Effect of intramuscular injection of androstenol from 160 to 220 days of age on the puberty attainment in gilts.

Group	Number of gilts	Injection (160 - 220 days of age)	Age of puberty attainment (day of the first ovulation)
A	20	Intramuscular injection (3 times a week): 10 μg androstenol in 0.5 ml saline	224.5 ± 2.9
B	20	Intramuscular injection (3 times a week): 0.5 ml saline	240.7 ± 2.8
0	20	No injection	244.5 ± 2.4

A vs. 0: - P < 0.001; A vs. B: - P< 0.001 B vs. 0: - P > 0.05

This study was performed to ascertain whether the boar pheromone – androstenol (5α-androst-16-en-3-ol) injected intramuscularly may act via the humoral pathway in gilts to accelerate puberty attainment. Sixty female piglets born within a one-week span of time were individually marked after weaning and randomly assigned to three groups of twenty:

0 - control without any injection, A - for androstenol injection and B - for saline injection. All gilts were housed on the farm, in a building without any boars. The group 0 was located in a distant part of the building, over 100 m from groups A and B. Gilts of groups A and B were placed in two adjacent pens, with the possibility of direct contact, through the partition. Gilts of the group A were intramuscularly injected with 10 µg androstenol dissolved in 0.5 ml of saline, and gilts of group B with 0.5 ml saline, three times a week, from 160 to 220 days of age.

We interpret the data in Table 1 to demonstrate that injections of androstenol can accelerate attainment of the puberty in gilts.

CONCLUSION

Numerous experiments, performed *in vitro*, *ex vivo*, *in vivo* on anaesthetized gilts and with prepubertal gilts in farm habitat, enabled us to present the concept that in addition to the standard neural pathway, typical for signaling pheromones, androstenol as a priming pheromone may influence the brain and hypophysis by a humoral pathway (Figure 1).

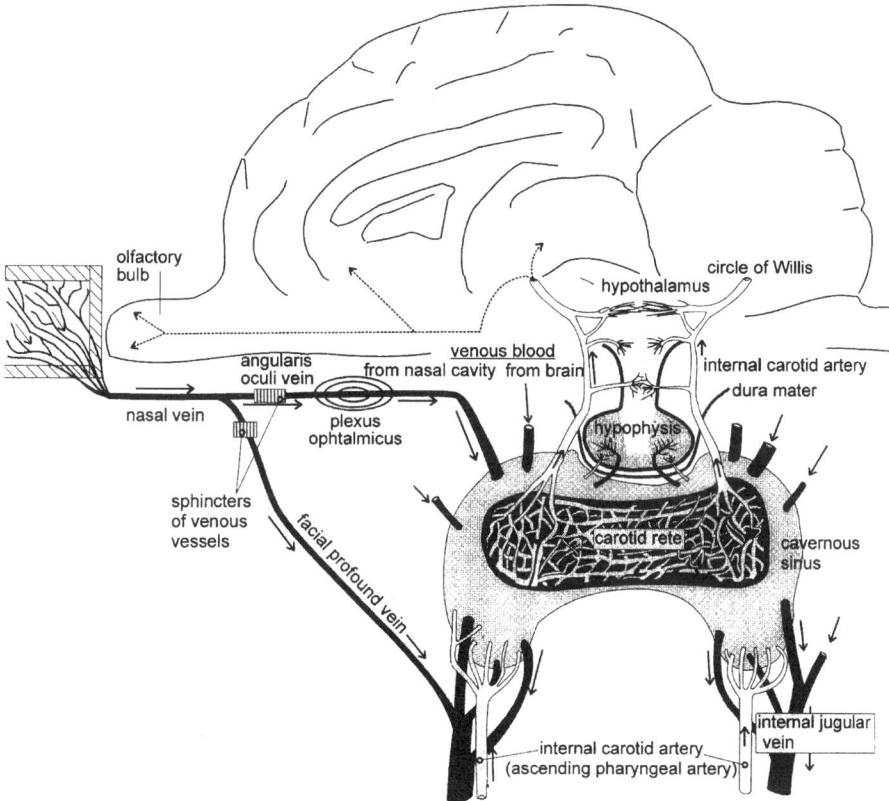

Figure 1. Schematic diagram presenting the putative humoral pathway for the delivery of a priming pheromone from the nasal cavity to the hypophysis and brain. For better understanding the part of the scheme with the perihypophyseal vascular complex was magnified, but the proportional representation of the hypophysis and vascular complex was maintained.

We demonstrated that androstenol can be absorbed into the nasal mucosa of the respiratory part of the nasal cavity, transported by the venous blood to the cavernous sinus and then locally transferred within the perihypophyseal vascular complex to the arterial blood of carotid rete supplying the brain and hypophysis. This local mechanism, by avoiding systemic blood circulation in its early stages, protects androstenol against metabolic degradation by the liver. Since endogenous androstenol does not occur in the systemic blood plasma of gilts, we propose that even a very small amount of this priming pheromone delivered to the brain and hypophysis by the local humoral transfer may evoke an endocrine response resulting in puberty acceleration. Our future experiments will address this mechanism further.

ACKNOWLEDGMENTS

This work was supported by Polish State Committee on Scientific Research (project No 5P06D 005 17). Authors are grateful to unknown referee who proposed constructive changes that made significant contribution to the quality of this manuscript.

REFERENCES

Baker, M. A. and Hayward, J. N., 1968, The influence of the nasal mucosa and the carotid rete upon hypothalamic temperature in sheep, *J. Physiol. (Lond).* **198**:561-579.

Booth, W. D. and Signoret, J. P., 1992, Olfaction and reproduction in ungulates, *Oxford Rev. Reprod.* **14**:265-301.

Booth, W. D. and Baldwin, B. A., 1983, Changes in estrous cyclicity following olfactory bulbectomy in post-pubertal pigs, *J. Reprod. Fertil.* **67**:143-150.

Booth, W. D., 1982, Steroid hormone and pheromone in the submaxillary gland and saliva of pig, in: *Olfaction and Endocrine Regulation*, (W. Breipohl, ed.), IRL Press, London, pp. 353-356.

Brooks, P. H. and Colle, D. J. A., 1970, Effect of the presence of a boar pheromone on attainment of puberty in the gilts, *J. Reprod. Fertil.* **23**:425-440.

Buck, L. L. and Axel, R. A., 1991, A novel multigene family may encode odorant receptors: a molecular basis for odor recognition, *Cell* **65**:175-187.

Chen, Z. and Lancet, D., 1984, Membrane proteins unique to vertebrate olfactory cilia: Candidates for sensory receptor molecules, *Proc. Nat. Acad. Sci. USA.* **81**:1859-1863.

Cohen-Tannoudji, J., Lavenet, C., Locatelli, A., Tillet, Y and Signoret, J. P., 1989, Non-involvement of the accessory olfactory system in the LH response of anoestrous ewes to male odour, *J. Reprod. Fertil.* **86**:135-144.

Daniel, P. M., Dawes, D. K. and Prichard, M. M. L., 1973, Studies of the carotid and associated arteries, *Phil. Trans. Roy. Soc. London.* B **237**:173-208.

Dorries, K. M., Adkins-Regan, E. and Halpern, B. P., 1997, Sensitivity and behavioral responses to the pheromone androstenone are not mediated by the vomeronasal organ in domestic pigs, *Brain Behav. Evol.* **49**:53-62.

Einer-Jensen, N. and Larsen, L., 2000, Transfer of tritiated water, tyrosine and propanol from the nasal cavity to cranial arterial blood in rates, *Exp. Brain Res.* **130**:216-220.

Ghoshal, N. G. and Khamas, W. A. H., 1984, Light microscopic study of blood vessels of the nasal cavity of the pig. *Acta Anat.* **120**:202-206.

Gillian, L. A., 1974, Blood supply to brains of ungulates with and without a rete mirabile caroticum, *J Com. Neurol.* **153**:275-290.

Glei, M., Schlegel, W., Straube, D. and Blankenger, J., 1989, Untersuchungen zur Beeinflussung des Pubertätseintrittes von Jungsauen mittels maskuliner Stimuli, *Archiv. Tierzucht.* **32**:173-179.

Grzegorzewski, W., Skipor, J., Wąsowska, B and Krzymowski, T., 1997, Countercurrent transfer of ^{125}I-LH-RH in the perihypophyseal cavernous sinus-carotid rete vascular complex, demonstrated on isolated pig heads perfused with autologous blood, *Dom. Anim. Endocrinol.* **14**:149-160.

Grzegorzewski, W., Skipor, J., Wąsowska, B and Krzymowski, T., 1995, Counter current transfer of oxytocin from the venous blood of the perihypophyseal cavernous sinus to the arterial blood of carotid rete supplying the hypophysis and brain depends on the phase of the estrous cycle in pigs, *Biol Reprod.* **52**:139-144.

Hancock, M. R., Gennings, J. N. and. Gower, D. B., 1985, On the existence of the receptor to the pheromonal steroid 5α-androst-16-en-3-one in porcine nasal epithelium, *FEBS Lett.* **181**:328.-334

Karlbom, I., 1982, Attainment of puberty in female pigs: influence of boar stimulation, *Anim. Reprod. Sci.* **4**:313-319.

Kirkwood, R. N., Hughes, P. E., and Both, W. D., 1983, The infuence of boar-related odours on puberty attainment in gilts, *Anim. Prod.* **36**:131-136.

Krishna, N. S. R., Getchell, M. L. and Getchell, T. V., 1994, Expression of the putative pheromone and odorant transfer vomeromodulin mRNA and protein in nasal chemosensory mucose,. *Neurosci. Res.* **39**:243-259.

Krzymowski, T., Bartlewski, P., Skipor, J., Grzegorzewski, W., Ziemińska, A., 1990, Counter current transfer of hormones from venous into arterial blood vessels in the base of brain in rabbit, *Acta Physiol. Pol.* Supl. **34**:172-173.

Krzymowski, T., Grzegorzewski, W., Stefańczyk-Krzymowska, S., Skipor, J. and Wąsowska, B., 1999, Humoral pathway for transfer of the boar pheromone, androstenol, from the nasal cavity to the brain and hypophysis, *Theriogenology* **52**:1225-1240.

Krzymowski, T., Skipor, J. and Grzegorzewski, W., 1992, Cavernous sinus and carotid rete of sheep and sows as a possible place for counter current exchange of some neuropeptides and steroid hormones, *Anim. Reprod. Sci.* **29**:225-240.

Maloney, S. K. and Mitchell, G., 1997, Selective brain coolong: The role of angularis oculi vein and nasal thermoreception, *Amer. J. Physiol.*, **273** (Reg. Integ. Comp. Physiol. 42):R1108-R1116.

Marchlewska-Koj A., 1984, Pheromones and mammalian reproduction, *Oxford Rev. Reprod. Biol.* **6**:66-302.

Mitchell, J., Thomalla, L. and Mitchell, G., 1998, Histological studies of the dorsal nasal, angularis oculi and facial veins of sheep (*Ovis aries*), *J. Morphol.* **237**:275-281.

Oldham, C. M., Martin, G. B. and Knight, T. W., 1979, Stimulation of seasonally anovular Merino ewes by rams: I. Time from introduction of the rams to the preovulatory LH surge and ovulation, *Anim. Reprod. Sci.* **1**:283-290.

Pearce, G. T. and Paterson, A. M., 1992, Physical contact with the boar is required for maximum stimulation of puberty in the gilt because it allows transfer of boar pheromone and not because it induces cortisol release, *Anim. Reprod. Sci.* **27**:209-224.

Read, R. R., 1992, Signalling pathways in odorant detection, *Neuron* **8**:205-209.

Rhein, L. D. and Cagan, R. H., 1981, Role of cilia in olfactory recognition, in: *Biochemistry of Taste and Olfaction*, R. H. (Cagan and M. R. Kare, eds., Acad. Press, New York, pp. 47-68.

Signoret, J.P., 1970, Reproductive behaviour of pigs. *J. Reprod. Fertil.* Suppl. **11**:5-117.

Skipor, J., Grzegorzewski, W., Krzymowski, T. and Einer-Jensen, N., 2000, Local transport of testosterone from nasal mucosa to the carotid blood and brain in the pig, *Polish J. Vet. Sci.* **3**:19-22.

Skipor, J., Grzegorzewski, W., Wąsowska, B and Krzymowski, T., 1997, Counter current transfer of β-endorphin in the perihypophyseal cavernous sinus-carotid rete vascular complex in sheep, *Exp. Clin. Endocrinol. Diabetes.* **105**:308-313.

Stefańczyk-Krzymowska, S., Krzymowski, T., Grzegorzewski, W., Wąsowska, B. and Skipor, J., 2000, Humoral pathway for the priming pheromone (^{3}H-androstenol) local transfer from the nasal cavity to the brain and hypophysis in anaesthetized gilts. *Exp. Physiol.* **85-6**:801-809.

DEMONSTRATION OF VOLATILE C_{19}-STEROIDS IN THE URINE OF FEMALE ASIAN ELEPHANTS, *ELEPHAS MAXIMUS*

M. Dehnhard,[1] M. Heistermann,[2] F. Göritz,[1] R. Hermes,[1]
T. Hildebrandt,[1] G. Strauss,[3] C. Weisgerber,[4] and H. Haber[5]

[1]Institute for Zoo Biology and Wildlife Research
PF 601103, D-10252 Berlin, Germany
[2]Department of Reproductive Biology, German Primate Center
Kellnerweg 4, D-37077 Göttingen, Germany
[3]Sigma-Aldrich GmbH, 30926 Seelze, Germany
[4]Tierpark Friedrichsfelde, 10307 Berlin, Germany
[5]Institute of Pharmacy, Humboldt University, Goethestr. 54
D-13086 Berlin, Germany

INTRODUCTION

In vertebrates behavioral as well as non-behavioral cues regulate many aspects of social interactions related to reproduction. Among the non-behavioral cues visual and chemical stimuli (pheromones; Karlson and Lüscher, 1959) release specific behavioral or endocrine reactions in the recipient. Pheromones may consist of a single substance or a specific mixture of substances. Mechanism of their release into the environment are manifold and secretion with urine represents one principle. Behavioral effects of estrous-related pheromones have been described in several species, including mouse (Schwende *et al.*, 1984), hamster (Singer *et al.*, 1976), sheep (Blisset *et al.*, 1994), cow (Dehnhard *et al.*, 1991), cotton-top tamarin (Ziegler *et al.*, 1993) and the Asian elephant (Rasmussen *et al.*, 1997). Their secretion depends on ovarian estrogens (Dehnhard *et al.*, 1991) and their function is primarily to inform males about the female's reproductive state. The Asian elephant is one of the few vertebrate species in which the identity of such a pheromonal compound, (Z)-7-dodecenyl acetate, is known (Rasmussen *et al.*, 1997).

There are also evidences for the existence of gestagen dependent pheromones in fish, reptiles and mammals. In fish pheromonally active estrogen and gestagen metabolites are released into the water and enhance reproductive fitness of the male goldfish (Defraipont and Sorensen, 1993). In the rabbit, progesterone effectively reinstates the emission of the nipple pheromone in ovariectomized animals (Gonzales-Mariscal *et al.*, 1994).

Moreover, examples of pheromones from males triggering female reproductive activity have been described (Claus *et al.*, 1990). The recently identified puberty-accelerating pheromone of the house mouse (E,E-α-farnesene and E-β-farnesene) was as effective as the homogenate of its synthesizing preputial gland (Ma et al., 1999). Adult male pigs produce large quantities of 16-unsaturated C_{19}-steroids (Patterson, 1968), which are sequestered in the submaxillary glands. When aroused, boars champ vigorously, producing copious amounts of saliva and releasing the volatile steroids into the air facilitating the expression of a rigid mating posture in estrous females (Melrose et al., 1971). The substance 5α-androst-16-en-3-one is the most effective pheromonal component of boar saliva. Additionally, 5α-androst-16-en-3-one and their corresponding 3α/β-hydroxy steroids have been demonstrated in humans (Claus, 1994), moose (*Alces alces*; Claus, 1994) and camel (*Camelus dromedarius*; Claus et al., 1999). Previously, these steroids were considered to have a pheromonal function only in domestic pigs.

The objective of this study was to analyze urinary volatiles of Asian elephants throughout the estrous cycle. Using a combined approach of hormonal analysis and solid-phase microextraction (SPME; Arthur and Pawliszyn, 1990) of urinary volatiles, reproductive stage related substances should been detected. Based on these results the study aimed to identify estrous and luteal activity related substances and to assess their reliability for monitoring female ovarian function.

MATERIAL AND METHODS

Animals and Sample Collection

Nine adult female Asian elephants (12 to 29 years) maintained in mixed social groups in the zoos of Leipzig and Berlin were involved in the study. Urine samples were collected 1 - 3 times a week. Additionally blood samples were collected weekly from 4 elephants. One of these elephants was treated with estradiol implants (down regulation of the ovarian function). All samples were frozen immediately after collection and stored at -20°C until analyzed. Freezing at –20°C for 4 months did not result in a significant loss of 2-unsaturated C_{19}-steroids.

Determination of Female Reproductive Status

The concentration of immunoreactive 5β-pregnanetriol, the most abundant urinary progesterone metabolite in the Asian elephant, was determined following the method of Niemüller et al. (1993). Samples were diluted 1 : 200 in assay buffer (0.04 mol/l PBS, containing 0.1% BSA, pH 7.0) and 50 µl aliquots were analyzed directly using a microtitreplate enzyme-immunoassay (EIA). Urinary hormone concentrations are expressed as mass/mg creatinine. 5α-pregnane-3,20-dione (DHP) was measured with an EIA in 5 µl plasma aliquots using a monoclonal antibody (rat) against progesterone-7-BSA, progesterone-peroxidase as label and DHP as standard.

5α-androst-16-en-3-one (androstenone) Immunoreactivity

For the analysis of androstenone-like immunoreactivity an ELISA-System (Sigma-Aldrich) developed for the analysis of androstenone (boar pheromone) in pork/lard was adapted. Calibration standards (0.01 – 5 ng/ml androstenone) were prepared in assay buffer (0.04 M Na2HPO4, 0.15 M NaCl, 0.1% BSA, pH 7.2). Urine samples were diluted 1: 40 in assay buffer and analyzed according to the instructions of the manufacture. Cross-reactivity studies were performed with 5α-androst-2-en-17-one and –17β-ol (1 – 500 ng/ml) and serial dilutions of luteal phase urine (1:10 to 1:640) and fiber elute (methanol elute dissolved in assay buffer), respectively.

Solid-phase Micro Extraction (SPME) Sampling

SPME sampling was done in the headspace above the surface of 5 ml Tris-buffered urine (pH 7.0, containing 1.83 g NaCl and 500 ng 5α-androstan as internal standard) in 20 ml headspace vials (Perkin Elmer). A fiber with an 85 μm polyacrylate coating was used. The vial were heated to 100°C for 60 min and agitated with a magnetic stirring bar. Analytes were desorbed at 300°C in the injector of the gas chromatograph, and analysis was started.

Analyses were carried out in a Perkin Elmer (PE) AutoSystem XL gas chromatograph provided with a flame ionization detector (FID). A 60 m x 0.25 mm ID (0.25 μm film thickness) PE-5 coating fused-silica capillary column (PE) was used. The GC oven was kept at 45°C for 5 min, increased to 240°C at 5°C min^{-1} and then from 240 to 310°C at 10°C min^{-1}. The carrier gas was nitrogen at 140 kPa. The detector was kept at 300°C.

Gas Chromatography-Mass Spectrometry (GC-MS)

GC-MS determinations were conducted with a FISONS TRIO 1000 GC-MS data system. The samples were analyzed using a 30 m DB-5MS capillary column (0.32 mm ID and 0.12 μm film thickness, J & W Scientific, USA). Ultra pure helium was used as carrier gas, with a head pressure setting of 15 psig. Injector temperature was 300 °C; the interface and ion source temperature were maintained at 300 °C and 200 °C, respectively. Splitless injection mode was used; the purge valve was turned on 3 min after injection, with a split flow of 25 ml/min during the GC run. The GC oven was kept at 50°C for 5 min, increased to 300°C at 10°C min^{-1}. Electron-impact ionization mass spectra were recorded in the full scan mode. The mass spectra were identified by computer MS library research and compared with those of the authentic standards.

Quantification of 2-unsaturated C_{19}-steroids

A calibration line was prepared by adding various amounts of 5α-androst-2-en-17β-ol (Steraloids: 10, 25, 50, 75, 100, 250, 500, 750 ng/ml) and a constant amount of 100 ng/ml 5α-androstan to pooled urines. The peak areas of the calibration samples were determined, and the ratio was used to evaluate the biological samples. The calibration curve showed a good linearity in the range of 10 to 750 ng/ml. The linear regression equation and variance explained were $y = 0.236 - 0.015x$ and $r^2 = 0.999$. Urinary concentrations were calculated from the peak ratios. To evaluate the concentrations of the minor compound 5α-androst-2-en-17-one in Figure 2, the ratio of its peak area against an external standard (100 ng/ml) added to follicular phase urine was used.

Data Analysis

The date of ovulation was determined by a rise in urinary pregnanetriol levels above a given threshold value (mean + 2SD) of the proceeding follicular phase values. The same criteria were applied to 5α-androst-2-en-17β-ol levels. Spearman's rank correlation coefficient was used to determine the association between levels of urinary pregnanetriol and 5α-androst-2-en-17β-ol in each individual and 5α-androst-2-en-17-one for one individual.

RESULTS

Comparison of Volatile Profiles from Urines of Different Reproductive Stages

The comparison of profiles of volatiles between estrous and diestrous urines revealed (data not shown):
1. a number of unknown follicular phase-specific, probably estrous related compounds with high inter- and intra-individual variance
2. two unknown substances appearing occasionally but synchronous in all animals of a group. Maximal concentrations were reached in an animal on the day before parturition, suggesting that these substances might be related to adrenocortical activity
3. two luteal phase specific compounds occurring in all animals investigated so far.

Identification of Two Luteal Phase Dependent Compounds

According to the mass spectra obtained (see Figure 1a,b) the major compound was identified as 5α-androst-2-en-17β-ol. The minor compound eluting 0.2 min earlier was identified as 5 α-androst-2-en-17-one. 5 α-androst-2-en-17 β-ol and –17-one showed a base peak at m/z 220 and m/z 218, respectively, which arises from the presence of the double bond in the C-2 position. This double bond directs a retro Diels-Alder fragmentation of the A ring with charge retention on the remaining steroid moiety (Labows *et al.*, 1979).

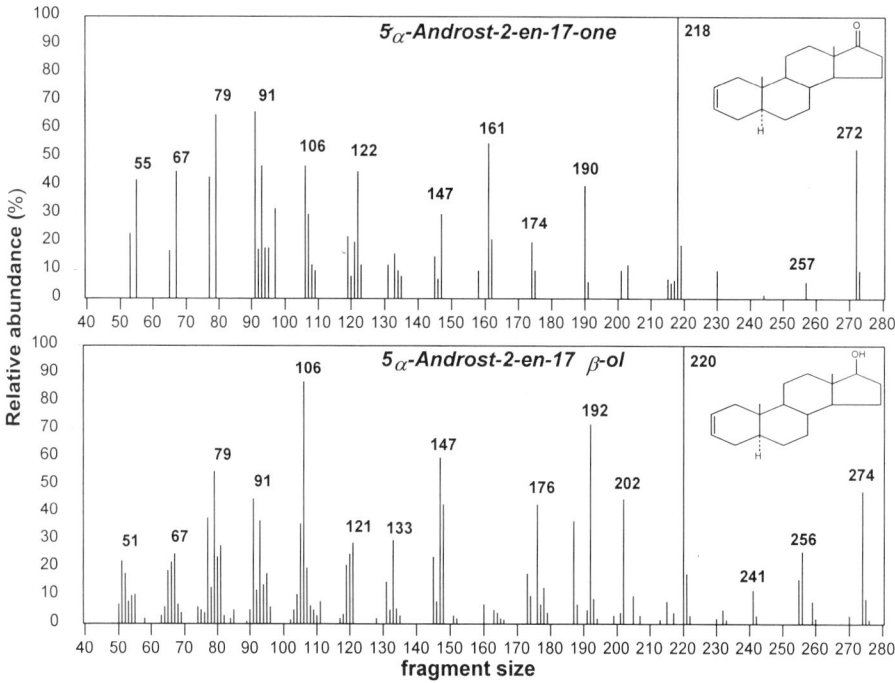

Figure 1. Mass spectra of urinary 5α-androst-2-en-17-one (**A**) and 5α-androst-2-en-17β-ol (**B**).

Concentrations of 5α-androst-2-en-17 β-ol, -17-one and a 5α-androst-16-en-3-one immunoreactivity during the Ovarian Cycle

The profiles of pregnanetriol and the luteal phase-specific substances 5 α -androst-2-en-17 β-ol and 5 α -androst-2-en-17-one over an 72-week period encompassing 4 ovarian cycles in an individual female are shown in Figure 2a-c.

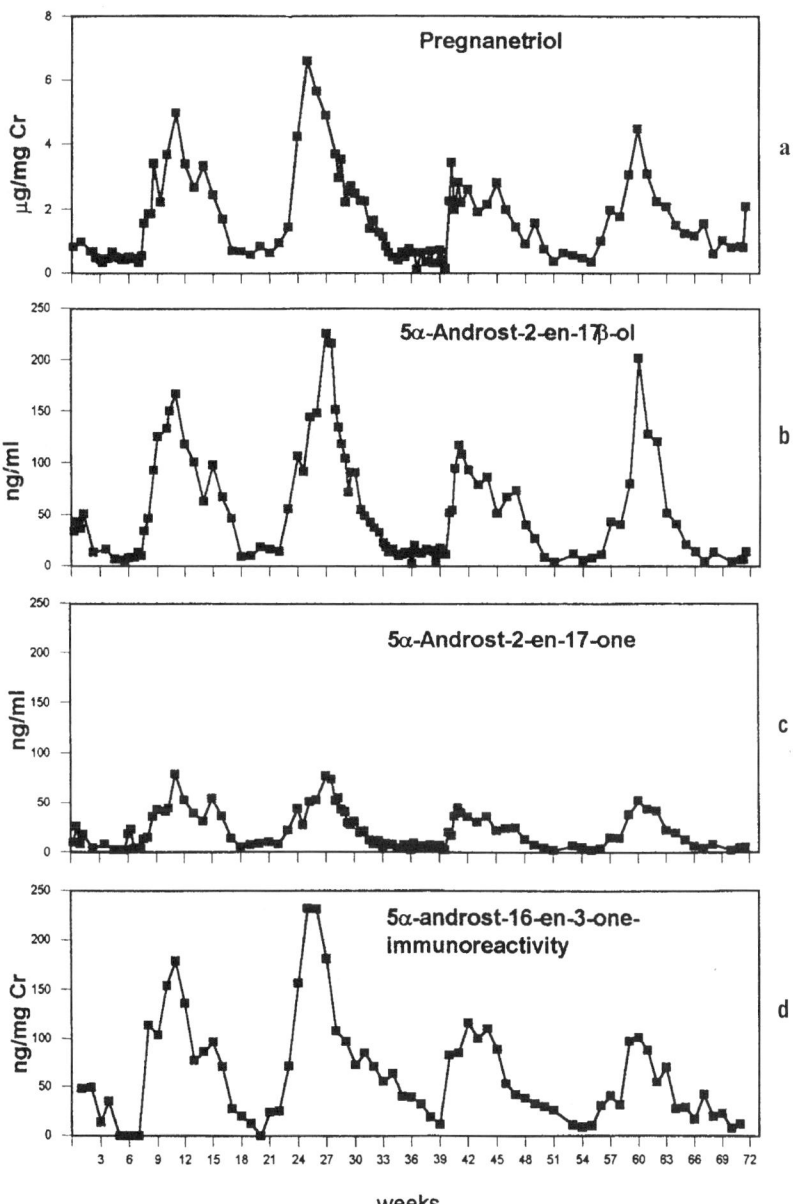

Figure 2. Profiles of urinary pregnanetriol, 5α-androst-2-en-17β-ol, 5α-androst-2-en-17-one, and 5α-androst-16-en-3-one-immunoreactivity throughout four complete ovarian cycles in an individual female.

The concentrations of pregnanetriol followed a cyclic pattern in which the follicular and luteal phases could be clearly distinguished (a). The estrous cycle lengths based on the time between successive increases in pregnanetriol levels varied from 14 to 19 weeks (mean 15.9 ± 2.0). Levels of the gas chromatographically determined 5 α -androst-2-en-17 β-ol followed a similar pattern and increased with those of pregnanetriol ($\rho_S = 0.86$, n = 97, $P < 0.01$). Concentrations of 5 α -androst-2-en-17 β-ol up to 225 ng/ml were obtained during luteal phases, whereas concentrations decreased below 10 ng/ml during follicular phases (b). The levels of the minor compound 5α-androst-2-en-17-one are shown in Figure 2c. They did not exceed 50 ng/ml during the luteal phase and decreased to 5 ng/ml during the follicular phase. Correlations of $\rho_S = 0.85$ ($P < 0.01$) were obtained when 5α-androst-2-en-17-one was compared with pregnanetriol. Similar profiles for all analytes were obtained for another three cycling females (not shown). The course of urinary 5α-androst-2-en-17β-ol also reflects the course of 5α-pregnane-3,20-dione (DHP) in blood plasma, which is the predominant gestagen in the African elephant (Figure 3).

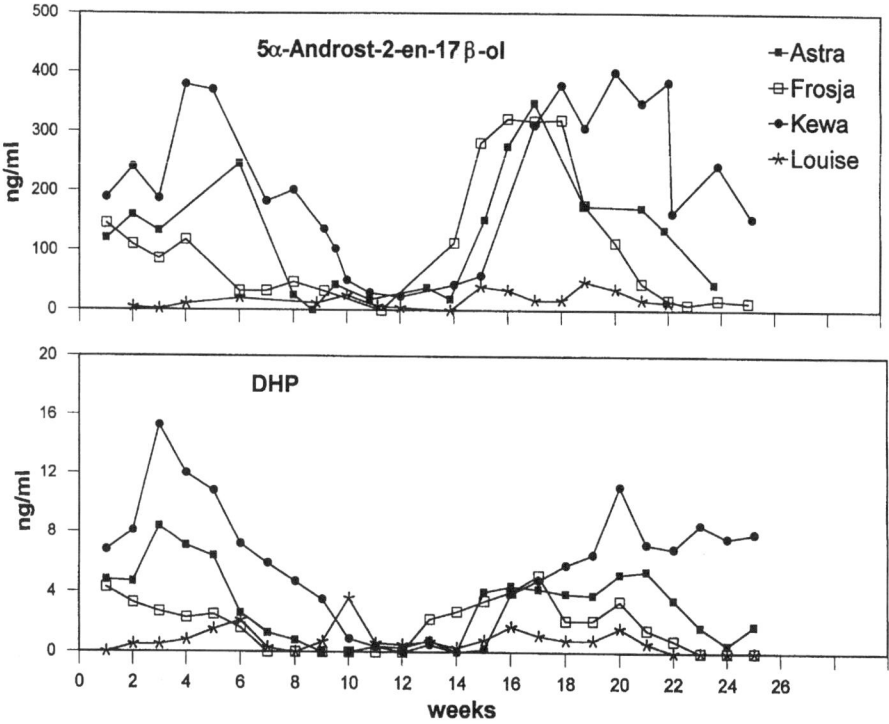

Figure 3. Profiles of DHP in blood plasma and urinary 5α-androst-2-en-17β-ol in four elephant cows.

Due to the boar pheromone-like odor emitted for the injection port of the GC and the binding of urinary substances to the antibody directed against 5α-androst-16-en-3-one, a commercially available EIA (developed to measure androstenone in pork/lard from pigs) was used for continuous measurements in urines from one elephant cow. The course of the androstenone immunoreactivity fits with the course of urinary pregnantriol (Figure 2d).

DISCUSSION

We have identified two androgens, 5α-androst-2-en-17-one and the corresponding alcoholic compound 5α-androst-2-en-17β-ol in the headspace volatiles of urine of female Asian elephants. The concentrations of 5α-androst-2-en-17β-ol and 17-one were positively associated with luteal activity and reflected cyclic ovarian activity. The results suggest that measurement of 5α-androst-2-en-17β-ol should provide a useful non-invasive method for monitoring reproductive status in this species.

Our findings of two unsaturated androgens in the elephant with either a 17-oxo- or 17-hydroxy-group were surprising. So far, one of the substances, 5α-androst-2-en-17-one has been demonstrated in human auxiliary bacterial isolates (Gower et al., 1997). Our results are similar to results in domestic pigs. In the pig three steroidal compounds with either a 3-oxo- or 3-hydroxy- group (5α-androst-16-en-3-one, 5α-androst-16-en-3β-ol and –3β-ol; Patterson, 1968) contribute to the boar pheromone, where they stimulate both sexual behavior and oxytocin release (Mattioli et al., 1986). In addition, we also demonstrated a urinary androstenone-like immunoreactivity in the elephant. Its measurement in concentrations up to 250 ng/mg Cr (corresponding to 75 ng/ml) cannot be explained with a cross-reactivity of 5α-androst-2-en-17-one and –17β-ol with the androstenone-specific antibody which was determined to be below 1%. Therefore additional urinary androgens are likely to occur.

The strong luteal dependence of urinary 5α-androst-2-en-17-one /-17β-ol in Asian elephants suggest that they are gestagen metabolites. However, this is unlikely because metabolism studies in several species (Schwarzenberger et al., 1997) and GC-MS analysis of urinary progesterone metabolites of Asian elephants (Niemüller et al., 1993) indicated that gestagens are generally metabolized to hydroxylated pregnanes.

Alternatively, 5α-androst-2-en-17-one and-17β-ol might both be androgen metabolites. Gestagens are precursors for androgen biosynthesis and high quantities of conjugated androgens, e.g. androsterone (5α-androstan-3β-ol-17-one), have been reported in urine of man (Choi et al., 2000). Enzymatic removal of H_2O across C-1,2 has been observed (Dorfman and Ungar, 1965) which might convert androsterone into 5α-androst-2-en-17-one. To our knowledge, however, no previous study demonstrated a luteal-dependent androgen secretion and own measurements of testosterone in urines of female elephants obtained throughout the cycle did not reveal luteal-dependent courses (Heistermann and Dehnhard, unpublished data). Thus, the patterns of 5α-androst-2-en-17-one and-17β-ol excretion do not reflect androgen metabolism.

The measurements of urinary 5α-androst-2-en-17-one and-17β-ol in substantial concentrations suggest that they are not merely a metabolic by-product. Thus, we assume a separate way of biosynthesis of 2-unsaturated androgens in the ovary of the Asian elephant. 5α-androst-2-en-17β-ol and -17-one may be part of a multidirectional chemical communicative system. There is a distinct possibility that both 2-unsaturated androgens might serve as female-emitted, female-received pheromones to regulate ovarian function in Asian elephants, which does not preclude that they are a signal for males. Longitudinal data from pregnanetriol analyses revealed that females with close social relationships tend to synchronize their timing of estrus (Oerke et al., 1999). However, until these substances may be portrayed as pheromones, studies on their behavioral relevance are urgently required.

REFERENCES

Arthur, C. L., and Pawliszyn, J., 1990, Solid-phase microextraction with thermal desorption used fused silica optical fibres, *Anal. Chem.* **62**:2145-2148.

Blisset, M. J., Bland, K. P., and Cottrell, D. F., 1994, Detection of oestrous-related odour in ewe urine by rams, *J. Reprod. Fert.* **101**:189-191.

Choi, M. H., Kim, K. R., and Chung, B. C., 2000, Simultaneous determination of urinary androgen glucuronides by high temperature gas chromatography-mass spectrometry with selected ion monitoring, *Steroids* **65**:54-59.

Claus, R., 1994, *Pheromone. Veterinärmedizinische Endokrinologie*, Gustav Fischer Verlag, Jena, pp. 691-712.

Claus, R., Over, R., and Dehnhard, M., 1990, Effect of male odour on LH secretion and the induction of ovulation in seasonally anoestrus goats, *Anim. Reprod. Sci.* **22**:27-38.

Claus, R., Kaufmann, B., Dehnhard, M., and Spitzer, V., 1999, Demonstration of 16-unsaturated C-19 steroids ('boar pheromones') in tissues of the male camel *(Camelus dromedarius)*, *Reprod. Dom. Anim.* **34**:455-458.

Defraipont, M., and Sorensen, P. W., 1993, Exposure to the pheromone 17a,20ß-dihydroxy-4-pregnen-3-one enhances the behavioural spawning success, sperm production and sperm motility of male goldfish, *Anim. Behav.* **46**:245-256.

Dehnhard, M., Claus, R., Pfeiffer, S., and Schopper, D., 1991, Variation in estrus-related odors in the cow and its dependency on the ovary, *Theriogenology* **35**:645-652.

Dorfman, R. I., and Ungar, F., 1965, *Metabolism of steroid hormones*,. Academic Press, New York, pp. 224-283.

Gonzalez-Mariscal, G., Chirino, R., and Hudson, R., 1994, Prolactin stimulates emission of nipple pheromone in ovariectomized new zealand white rabbits, *Biol. Reprod.* **50**:373-376.

Gower, D. B., Mallet, A. I., Watkins, W. J., Wallace, L. M., and Calame, J.-P., 1997, Capillary gas chromatography with chemical ionization negative ion mass spectrometry in the identification of odorous steroids formed in metabolic studies of the sulphates of androsterone, DHA and 5α-androst-16-en-3β-ol with human axillary bacterial isolates, *J. Steroid Biochem. Mol. Biol.* **63**:81-89.

Karlson, P., and Lüscher, M., 1959, „Pheromones": a new term for a class of biologically active substances, *Nature* **183**:55-56.

Labows, J. N., Preti, G., Hoelzle, E., Leyden, J., and Kligman, A., 1979, Steroid analysis of human apocrine secretion, *Steroids* **34**:249-258.

Ma, W., Miao, Z., and Novotny, M., 1999, Induction of estrus in grouped female mice (Mus domesticus) by synthetic analogues of preputial gland constituents, *Chem. Senses* **24**:289-93.

Mattioli, M., Geleati, G., Conte, F., and Seren, E., 1986, Effect of 5α-androst-16-en-3-one on oxytocin release in oestrous sows, *Theriogenology* **25**:399-403.

Melrose, R., Patterson, R. L. S., and Reed, H. C. B., 1971, Androgen steroids associated with boar odour as an aid to the detection of oestrus in pig artificial insemination, *Br. Vet. J.* **127**:497-501.

Niemüller, C. A., Shaw, H. J., and Hodges, J. K., 1993, Non-invasive monitoring of ovarian function in Asian elephants *(Elephas maximus)* by measurement of urinary 5ß-pregnanetriol, *J. Reprod. Fert.* **99**:617-25.

Oerke, A.-K., Heistermann, M., Hodges, K., 1999, Elephant-service: non-invasive methods to monitor the reproductive status in the Asian and African elephant, *EEP Research Group Newsletter* **6**:8.

Patterson, R. L. S., 1968, Identification of 3-hydroxy-5α-androst-16-ene as the musk odour component of boar submaxillary salivary gland and its relationship to the sex odour taint in boar fat, *J. Sci. Food Agric.* **19**:434-438.

Rasmussen, L. E. L., Lee, T. D., Zhang, A., Roelofs, W. L., and Daves, G. D. Jr., 1997, Purification, identification, concentration and bioactivity of (Z)-7-dodecen-1-yl acetate: sex pheromone of the female Asian Elephant, *Elephas maximus*, *Chem. Senses* **22**:417-437.

Schwarzenberger, F., Palme, R., Bamberg, E., and Möstl, E., 1997, A review of faecal progesterone metabolite analysis for non-invasive monitoring of reproductive function in mammals, *Z. Säugetierk.* **62**: (Suppl. II):214-221.

Schwende, F. J., Wiesler, D., and Novotny, M., 1984, Volatile compounds associated with estrus in mouse urine: potential pheromones, *Experientia* **40**:213-215.

Singer, A.G., Agosta, W. C., O'Connell, R. J., Pfaffmann, C., Bowen, D. V., and Field, F., 1976, Dimethyl disulfide: an attractant pheromone in hamster vaginal secretion, *Science* **191**:948-950.

Ziegler, T. E., Epple, G., Snowdon, C. T., Porter ,T. A., Becher, A. M., and Kuderling, I., 1993, Detection of the chemical signals of ovulation in the cotton-top tamarin, *Saginus oedipus*, *Anim. Behav.* **45**:313-322.

THE PHEROMONE OF THE MALE GOAT: FUNCTION, SOURCES, ANDROGEN DEPENDENCY AND PARTIAL CHEMICAL CHARACTERIZATION

Rolf Claus[1], Martin Dehnhard[2], Ursula Götz[1], and Markus Lacorn[1]

[1]University of Hohenheim
Institute for Animal Husbandry and Animal Breeding
Section for Animal Husbandry and Physiology
Garbenstrasse 17, 70599 Stuttgart, Germany
[2]Institute for Zoo Biology and Wildlife Research
Postfach 601103, 10252 Berlin, Germany

FUNCTION OF MALE PHEROMONES AS A FINE TUNING MECHANISM OF SEASONAL OVARIAN ACTIVITY

The cyclic ovarian activity in sheep and goats of most breeds is limited to autumn and winter (Shelton, 1960; Claus et al., 1985; Martin et al., 1986). The photoperiod is the main environmental factor regulating the ovarian activity so that the decreasing daylength stimulates the slow onset of cyclicity in autumn (Legan and Karsch, 1980; Mori et al., 1984). The introduction of males to flocks of light-conditioned females additionally induces the ovarian function in acyclic does and leads to a synchronized resumption of the cycle. Thus a highly synchronized lambing in a group of females is ensured (Shelton, 1960; Chemineau, 1987; Walkden-Brown et al., 1999). Such a birth synchrony was also reported from other species and is regarded to improve survival of offspring in the presence of predators (Lindsay, 1988).

In the goat, this "male effect" can be released in the absence of males by presenting the odor of buckhairs or buckhair extracts, demonstrating the existence of pheromones (Claus et al., 1990; Over et al., 1990). In goats, the typical odor of the buck is most pronounced during the rutting season and was demonstrated to originate from the sebaceous glands in the parietal region (McEwan-Jenkinson et al., 1967). By the mere presentation of buckhair odor to anestrous goats for 3 days follicular maturation could be induced in 4 out of 5 females (Claus et al., 1990).

This follicle maturation is explained by an increased pulse frequency of luteinizing hormone (LH). A first rise of LH can be observed within 20 min after application (Claus et al., 1990; Chemineau, 1987) and an increased pulse frequency is maintained furtheron.

Under experimental conditions, it increased from 1.6 ± 0.55 pulses per hour during the 6 h control period to 2.6 ± 1.14 pulses per hour during the first 6 h following onset of stimulation (Claus et al., 1990). The first rise of LH after odor stimulation is preceded by characteristic increases in the multiple-unit activity of the medial basal hypothalamus (Hamada et al., 1996) within a few minutes. The activity of the medial basal hypothalamus is the interface between the odor stimulus and the reaction of gonadotropin releasing hormone (GnRH) producing neurons, which regulate LH discharge by the anterior pituitary.

Based on the short-term changes in the LH secretion pattern a bioassay was developed to monitor the biological activity in odor emitting samples. The LH pulses were evaluated either as the percentage of does reacting with an immediate rise after odor presentation or according to changes of the pulse frequency (Claus et al., 1990). Recently, the recording of electrophysiological events (Iwata et al., 2000) was used as a bioassay for monitoring the presence of the pheromone. The latter method avoids time-consuming LH-assays but may find its limitation due to animal welfare aspects.

Based on the changes of LH secretion it could be demonstrated that the reaction is highly specific. Androstenone e.g., which is the pheromone of the boar, did not lead to a reaction and similarly 4-ethyl-branched fatty acids which had been discussed as pheromones in the goat (Sasada et al., 1983) had no effect on LH secretion (Dehnhard et al., 1988; Claus et al., 1990). In addition to hairs from the parietal area also urine of bucks was shown to be pheromone-positive and the activity increased after stimulation of the endocrine testicular function by human chorionic gonadotropin application (hCG; Götz, 1993). In other studies, only buckhairs were found to be positive (Walkden-Brown et al., 1993). Saliva, which is a source of pheromone in the boar, was also tested but did not lead to an altered LH pattern (Götz, 1993). The presence of the pheromone in urine is also supported by the buck's behavior, because he impregnates his beard and forehand coat with urine thus producing a high surface, which facilitates odor emission. The availability of two different sources of the pheromone might be valuable for pheromone isolation because the spectrum of contaminating substances differs so that only identical peaks in the GC-analysis are possible candidates.

ANDROGEN-DEPENDENCY OF THE PHEROMONES

Three mature male goats were castrated. They had been pretreated with hCG at days – 4 and - 1 to elevate the endocrine testicular function. Portions of parietal hair were shorn before and 2 and 4 weeks after castration and urine was collected. Before castration both materials were highly active in the bioassay. Already 2 weeks after castration both hair and urine were pheromone-negative.

Because testosterone substitution to castrated male sheep led to pheromone activity in this species (Knight and Lynch, 1980; Fulkerson et al., 1981), also male goats, which had been castrated six month earlier, were supplemented with i.m. depots of 1 mg testosterone/kg bodyweight in sesam oil three times in weekly intervals. Four weeks after onset of treatment the pheromone activity was determined using both hair and urine but no significant reaction could be demonstrated. These results contrast with the recent demonstration of a change of electrical activity in the hypothalamus after presentation of buckhair from testosterone treated castrated males (Iwata et al., 2000). The discrepancy cannot be definitely explained, but probably might be related to the different bioassays. The occurrence of the pheromone in urine, however, largely excludes the possibility that androgen-dependent sebaceous glands synthesize the pheromone.

ATTEMPTS TO ISOLATE THE PHEROMONE

Figure 1 shows a schematic outline of the extraction- and purification procedure. Volumes of 200 ml urine were extracted with three portions of 200 ml dichloromethane (DCM) after adding 40 g NaCl and adjusting the pH to 7.5. Amounts of 50 g hairs were extracted with 2 l DCM. The extracts were dried over anhydrous sodium sulfate and concentrated in a rotary evaporator. Hair extracts were taken up in 50 ml petroleumether and successive solvent distributions were performed against 90 % methanol (50 ml, 25 ml, 25 ml). The petroleumether was discarded and the bulk of methanol in the combined portions evaporated in a rotary evaporator. The remaining methanol-water mixture was filled up with water to a volume of 50 ml and reextracted with three portions of DCM (50 ml, 25 ml, 25 ml). Urine extracts were processed without a solvent distribution.

An aliquot portion was used for the confirmation of the pheromone presence with the bioassay. Volumes of the DCM extracts corresponding to 15 g hairs were taken up in 1 ml methanol and 0.2 ml portions of the methanol were chromatographed consecutively by HPLC to avoid overloading of the column. A semipreparative reversed phase column (Lichrosorb 100, RP 18, 250 x 8 mm I.D., 7 µm, Knauer Berlin) was used at 40 °C and flow rate of 2.5 ml/min. Each run was carried out with a methanol/water gradient containing 1 % acetic acid: 84 % methanol (4 min), then up to 90 % methanol (10 min), followed by an increase up to 100 % methanol (10 min) which was maintained for another 10 min.

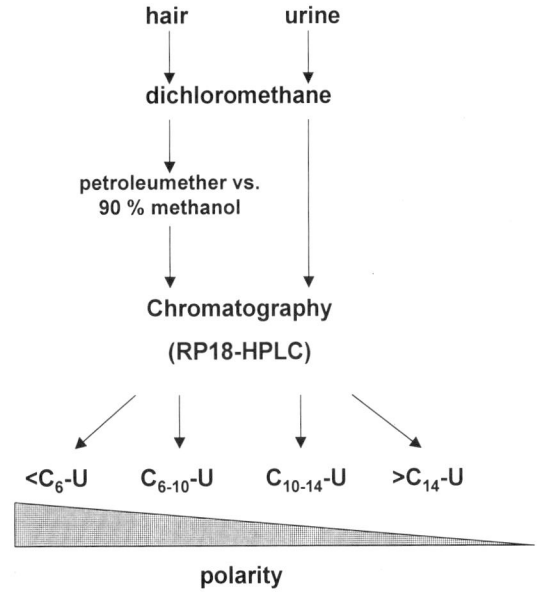

Figure 1. Schematic outline of the extraction procedure and HPLC-clean-up of buckhair and –urine.

For collection of reproducible fractions, the samples had been spiked with the phenylurethanes of hexanol (C_6-U), decanol (C_{10}-U) and tetradecanol (C_{14}-U) respectively. These markers do not occur naturally and can be easily detected with the UV-detector (Björkqvist and Toivonen, 1978; Wintersteiger and Wenniger-Weinzierl, 1982).

Four different fractions were collected from each run. Fraction 1 represented the eluates from the solvent peak up to the C_6-U-peak, fraction 2 contained the combined eluates between C_6-U and C_{10}-U, fraction 3, those between C_{10}-U and C_{14}-U and the eluates following C_{14}-U were fraction 4. The corresponding fractions from five runs were combined, the methanol was evaporated and the remaining water-phase was filled up with water to a volume of 25 ml. After extraction with 3 portions of DCM (25, 12.5,12.5 ml) the extracts were combined and concentrated to 10 ml by evaporation.

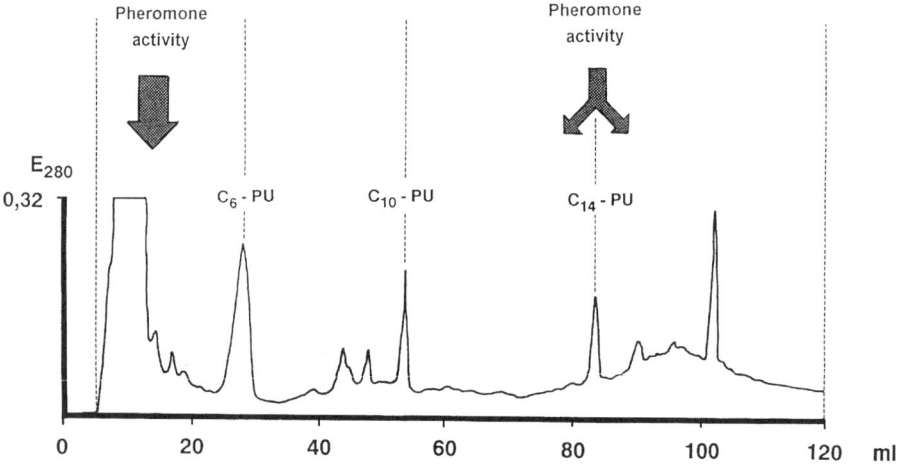

Figure 2. Elution profile of buckhair extracts on RP18-HPLC. Vertical lines indicate the fractions 1 to 4. The arrows signal the location of bioactive substances.

In case of urine a similar chromatographic system was used. Urine extracts (corresponding to 100 mL urine) were spiked with U-standards and separated on an analytical RP 18 column (RP 18 250 x 4.6 mm Bischoff, Leonberg). The injection volumes were 100 µl, the column temperature 40°C and the flow rate 1 ml/min. The following methanol/water gradient was used: 75 % methanol increased to 82.5 % in 1 min and then up to 92.5 % in 1 min. 92.5 % methanol was kept for 4 min and then increased up to 95 % in 2 min. The final increase to 100 % was performed in 11 min and then kept for 6 min. Aliquot portions of the fractions corresponding to 1.5 g of hairs (average amounts shorn from the parietal area of individual bucks) or 10 ml of urine were used for the bioassay.

The results of the bioassay both for purified hair and urine extracts are shown in Table 1. Positive fractions of buckhair are additionally attributed to the elution profile in Figure 2. Fraction 1 ($< C_6$-U) revealed nearly the same biological activity as the crude extract both for hair and urine. In the latter, none of the other fractions was positive. In contrast, hair-extract eluates did not lead to a response in fraction 2 (C_{6-10}-U) but again, even if less pronounced, in fraction 3 and 4. These results indicate that the pheromone activity is mainly attributable to the most polar of the HPLC-fractions (compare Figure 2). Data from hairs suggest 2 other compounds with decreased activity in fraction 3 and 4. Based on the results of derivatization reactions (see below) rather one second substance has to be assumed which is found with a polarity similar to C_{14}-U so that it was split into two fractions. Nevertheless, the second substance is less active than the first one.

Table 1. Biological activity of RP-18 fractions compared to native materials of hairs and urine.

Fraction	Does reacting [%]	
	Hairs	Urine
Native material	88	88
<C_6-U	86	86
C_{6-10}-U	0	13
C_{10-14}-U	43	13
>C_{14}-U	25	0

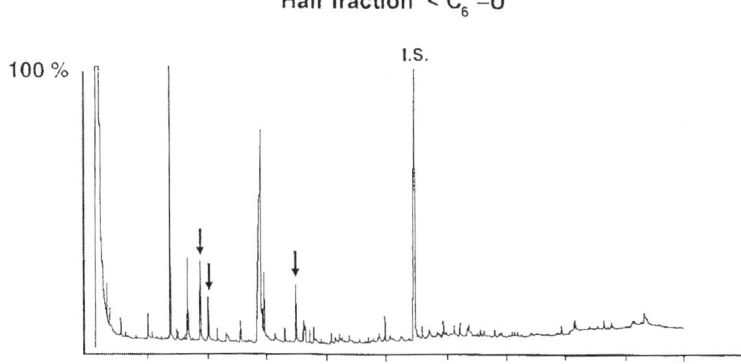

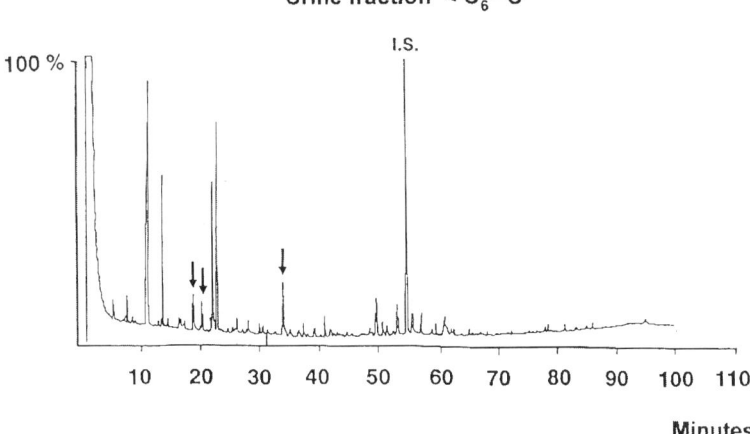

Figure 3. Comparison of gas chromatographic separation of HPLC-purified extracts from buckhair and urine. Arrows mark major peaks in both materials (I.S.: Internal Standard C_6-U).

Gaschromatographic comparisons of the crude extracts and HPLC-fraction 1 ($< C_6$-U) revealed a highly efficient purification due to the solvent distribution and HPLC-clean up. In addition, when comparing fraction 1 from hairs and urine by GC only 3 major peaks coincided (Figure 3: arrows). Because there is no knowledge about the concentrations of the pheromone, however, the compound may as well be represented by one of the minor peaks. Therefore additional data on the chemical characteristics were collected by derivatization reactions.

PARTIAL CHEMICAL CHARACTERIZATION OF THE PHEROMONE BY DERIVATIZATION REACTIONS

Information on the chemical characteristics should be obtained by derivatization of functional groups, which might be important for the biological activity. Two different reactions were carried out to clarify the presence of carbonyl compounds and functional groups with active H-atoms such as alcohols or amines. For the reaction of carbonyl compounds (Shriner et al., 1966) a crude extract corresponding to 12 g buckhairs, was taken up in 100 ml of a 2,4-dinitrophenylhydrazine (DNPH) solution (Merck; 3 g in 15 ml hydrochloric acid, 67 ml ethanol and 23 ml water) and stirred for 24 h at 22°C. The sample was concentrated to dryness in a rotary evaporator, dissolved in 20 ml DCM and washed with 3 portions of 10 ml water to remove excessive DNPH. Aliquot portions of the DCM phase were tested in the bioassay. Additionally the carbonyl compounds were regenerated from their DNPH-derivatives by heating the sample for 3 h at 65°C with a 200-fold excess of laevulinic acid (Keeney, 1957) and again tested for pheromone activity.

In addition, the carbonyl compounds in the extracts were also reduced to alcohols with $NaBH_4$. The extracts were dissolved in 1 ml of a 1.5 mM solution of $NaBH_4$ in isopropanol and allowed to react for 24 h at room temperature. Thereafter, 1 N hydrochloric acid was added dropwise, until the formation of hydrogen had disappeared. This sample again was evaluated in the bioassay.

The silylation of the extract (24 mg) was carried out by 11 ml N-methyl-N-tri-methylsilyl-trifluoroacetamide (MSTFA; MN, Düren) for 2 h at 60°C, and the sample again checked for pheromone activity after removing excess MSTFA.

As summarized in Table 2, the biological activity was reduced but not abolished after derivatization with DNPH pointing to a carbonyl substance whereas the remaining activity can be attributed to a second non-carbonyl compound. This inactivation could be reversed by laevulinic acid. Reduction with $NaBH_4$ largely maintained the biological activity, whereas the additional blockage of active H-atoms with MSTFA abolished the pheromone activity. The remaining activity when the extract was silylated by MSTFA without preceding reduction with $NaBH_4$ probably can be attributed to the presence of the second substance, which under these conditions should be the carbonyl compound.

Table 2. Pheromone activity after derivatization of hair extracts.

Sample/derivative	Does reacting [%]	Interpretation
Extract	88	-
DNPH	33	(reduced activity)
Regenerated after DNPH	73	Carbonyl
MSTFA	38	(reduced activity)
$NaBH_4$	75	Alcohol
$NaBH_4$ / MSTFA	25	(inactive)

In regard to the known function of ketones and alcohols for perfumes and pheromones as well (Van Toller and Dodd, 1992) the results suggest, that the pheromone of the male goat may be active, both as a ketone (or aldehyde) and an alcohol. A similar situation is known for the steroidal pheromones of the boar, which also occur as 5α-androst-16-en-3-one and 5α-androst-16-en-3α-ol and –3ß-ol (Patterson 1968a, b; Claus, 1979). The results of the derivatization reactions also suggest that the main part of the two molecules is identical.

ACKNOWLEDGMENTS

The investigations were supported by the German Research Organization (DFG).

REFERENCES

Björkqvist, B., and Toivonen, H., 1978, Separation and determination of aliphatic alcohols by high-performance liquid chromatography with UV detection, *J. Chromatogr.* **153**:265-270.
Chemineau, P., 1987, Possibilities for using bucks to stimulate ovarian and oestrous cycles in anovulatory goats – a review, *Livestock Prod. Sci.* **17**:135-147.
Claus, R., 1979, Pheromone bei Säugetieren unter besonderer Berücksichtigung des Ebergeruchsstoffes und seiner Beziehung zu anderen Hodensteroiden, *Fortsch. Tierphysiol. Tierernähr.* **10**:1-136.
Claus, R., Over, R., and Dehnhard, M., 1990, Effect of male odour on LH secretion and the induction of ovulation in seasonally anoestrous goats, *Anim. Reprod. Sci.* **22**:27-38.
Claus, R., Schopper, D., and Thume, O., 1985, Evidence for different types of seasonal anoestrus in the dairy goat as revealed by progesterone determination in milk fat, *Livestock Prod. Sci.* **13**:71-77.
Dehnhard, M., Over, R., Hoppen, H. J., Schams, D., and Claus, R., 1988, LH release in the goat in response to male pheromones, *Acta Endocrinol.* **117**(Supp. 287):58-59.
Fulkerson, W. J., Adams, V. R., and Gherardi, P. B., 1981, Ability of castrate male sheep treated with oestrogen or testosterone to induce and detect oestrus in ewes, *Appl. Anim. Ethol.* **7**:57-66.
Götz, U., 1993, Abhängigkeit der Pheromonaktivität in Haaren und Urin beim Ziegenbock vom Hoden, Diss. vet med. München.
Hamada, T., Nakajima, M., Takeuchi, Y., and Mori, Y., 1996, Pheromone-induced stimulation of hypothalamic gonadotropin-releasing hormone pulse generator in ovariectomized, estrogen-primed goats, *Neuroendocrinol.* **64**:313-319.
Iwata, E., Wakabayashi, Y., Kukuma, Y., Kikusui, T., Takeuchi, Y., and Mori, Y., 2000, Testosterone-dependent primer pheromone production in the sebaceous gland of male goats, *Biol. Reprod.* **62**:806-810.
Keeney, M., 1957, Regeneration of carbonyls from 2,4-dinitrophenylhydrazones with levulinic acid, *Anal. Chem.* **29**:149-191.
Knight, T. W., and Lynch, P. R., 1980, Source of ram pheromones that stimulate ovulation in the ewe, *Anim. Reprod. Sci.* **3**:133-136.
Legan, S. J., and Karsch, F. J., 1980, Photoperiodic control of seasonal breeding in ewes: modulation of negative feedback action of estradiol, *Biol. Reprod.* **23**:1061-68.
Lindsay, D. R., 1988, Reproductive behaviour in survival: a comparison between wild and domestic sheep, *Aust. J. Biol. Sci.* **41**:97-102.
Martin, G. B., Oldham, C. M., Cognie, Y., and Pearce, D. T., 1986, The physiological responses of anovulatory ewes to the introduction of rams – a review, *Livestock Prod. Sci.* **15**:219-247.
McEwan-Jenkinson, D., Blackburn, P. S., and Proudfoot, R., 1967, Seasonal changes in the skin glands of the goat, *Br. Vet. J.* **123**:541-549.
Mori, Y., Maeda, K., Sawasaki, T., and Kano, Y., 1984, Effect of long days and short days on oestrus cyclicity in two breeds of goats with different seasonality, *Jpn. J. Anim. Reprod.* **30**:239-245.
Over, R., Cohen-Tannoudji, J., Dehnhard, M., Claus, R., and Signoret, J. P., 1990, Effect of pheromones from male goats on LH-secretion in anoestrous ewes, *Physiol. Behav.* **48**:665-668.
Patterson, R. L. S., 1968a, 5α-androst-16-en-3-one: compound responsible for taint in boar fat, *J. Sci. Food. Agric.* **19**:31-37.

Patterson, R. L. S., 1968b, Identification of 3α-hydroxy-5α-androst-16-en as the musk odor component of boar submaxillary salivary gland and its relationship to the sex odor taint in pork meat, *J. Sci. Food. Agric.* **19**:434-438.

Sasada, H., Sugiyama, T., Yamashita, K., and Masaki, J., 1983, Identification of specific odor components in mature male goat during the breeding season, *Jpn. J. Zootech. Sci.* **54**:401-408.

Shelton, M., 1960, The influence of the presence of the male on initiation of oestrus cycle activity in Angora does, *J. Anim. Sci.* **19**:368-375.

Shriner, R. L., Fuson, R. C., and Curtin, D. Y., 1966, *Systematic identification of organic compounds*. A Wiley and Sons Inc., London, p. 458.

Van Toller, S., and Dodd, G. H., 1992, *Fragrance: the psychology and biology of perfume*, Elsevier, London.

Walkden-Brown, S. W., Restall, B. J., and Henniawati, 1993, The male effect in Australian cashmere goats. 3. Enhancement with buck nutrition and use of oestrous females, *Anim. Reprod. Sci.* **32**:69-84.

Wintersteiger, R., and Wenniger-Weinzierl, G., 1982, High performance liquid chromato-graphy of napthyl-urethanes with fluorescence detection, *J. Chromatogr.* **237**:399-406.

ANALYSIS OF VOLATILE COMPOUNDS IN SCENT-MARKS OF SPOTTED HYENAS (*CROCUTA CROCUTA*) AND THEIR POSSIBLE FUNCTION IN OLFACTORY COMMUNICATION

Heribert Hofer[1,2], Marion East[1,2], Ina Sämmang[1], and Martin Dehnhard[1]

[1]Institute of Zoo and Wildlife Research
Alfred-Kowalke-Str. 17, D-10315 Berlin, Germany
Hofer@izw-berlin.de
[2]Max-Planck-Institut für Verhaltensphysiologie
Postfach 1564, D-82305 Seewiesen, Germany

INTRODUCTION

The function of scent-marks and scent-marking behavior in carnivore societies has traditionally concentrated on territorial behavior, although scent-marking is known to serve a variety of functions (Albone, 1984). The four (Mills and Hofer, 1998) extant members of the family *Hyaenidae*, the spotted hyena (*Crocuta crocuta*), the striped hyena (*Hyaena hyaena*), the brown hyena (*Hyaena brunnea*) and the aardwolf (*Proteles cristatus*) have long been recognized as possessing sophisticated olfactory communication systems using the secretion of anal or subcaudal scent-glands (Kruuk 1972; Mills *et al.*, 1980; Nel and Bothma, 1983; Gorman and Mills, 1984; Mills and Gorman, 1987; Richardson, 1990, 1991; Sliwa, 1998). *Hyaenidae* also show a form of scent-marking behavior unique amongst carnivores known as "pasting". During pasting, individuals assume a more or less crouching position, partly or fully evert the anal scent-gland, and deposit the secretion by dragging or swiping the gland over an exposed vegetation structure such as a grass stalk. Previous analyses of the chemical constitution of carnivores' anal gland secretions have revealed unusual and interesting volatile and non-volatile components in all species (Wheeler *et al.*, 1975; Mills *et al.*, 1980; Apps *et al.*, 1989; Buglass *et al.*, 1990).

Functions of Scent-marking in *Hyaenidae*

Previous studies of scent-marking in the four hyena species have documented and emphasized its role in territorial behavior. Several studies have documented how hyenas change their strategy of territory marking in response to the size of their territory in order to maximize the probability that an intruder encounters the scent-mark of a territory owner. For instance, in the aardwolf, small territories are maintained by pasting secretions from the anal gland on grass stalks (Nel and Bothma, 1983; Richardson, 1990, 1991). Both sexes paste, although males mark more than females. Scent marks are concentrated along the territory boundary (Richardson, 1991). Gorman and Mills (1984) distinguished such a strategy of scent-marking suitable for small territories from strategies of 'hinterland' marking in the large territories of brown and spotted hyenas in the Kalahari, where territory owners distribute scent-marks throughout the territory (Gorman and Mills, 1984; Mills and Gorman, 1987).

Apart from territorial defense, scent-marking may have other functions, an issue rarely addressed in these studies. It could well be that the location of scent-marks and the composition of scent-gland secretions provide additional information about the originator. There are at least two ways by which such additional functions might be identified.

Chemical analysis of the composition of a scent-mark could analyze the variation between classes of individuals if scent gland secretions are collected from individually recognized study animals. Brown hyenas, for instance, produce two different types of secretion containing a rich cocktail of volatile compounds including short-chain and long-chain fatty acids (Mills *et al.*, 1980; Buglass *et al.*, 1990), and they can distinguish scent-marks of their own group from those of neighbors (Mills *et al.*, 1980). Spotted hyenas also produce secretions rich in short-chain and long-chain fatty acids (Buglass *et al.*, 1990), and Mills (1990) suggested that spotted hyenas may be able to distinguish the scent-marks of clan members from those of non-clan members. This idea has not yet been tested.

Secondly, detailed behavioral observations may reveal the active use of scent glands and scent-marking in more than one behavioral or social context. For instance, Nel and Bothma (1983) documented high rates of pasting by aardwolves during foraging and suggested that these scent-marks marked food patches of termites consumed during a particular night and thus prevent the aardwolf from wasting time by searching depleted food patches. During social interactions, striped hyenas lift their tails and protrude slit-like scent glands on either side of the anus, a behavior most frequently displayed by subordinates (Fox, 1971).

In spotted hyenas, pasting is used in the context of territory defense both along the periphery of the territory and in the 'hinterland' (Mills and Gorman, 1987). However, there are at least two further contexts in which pastings and anal gland secretions play an important role. In East Africa, spotted hyenas live in large clans in a highly structured society dominated by females (Kruuk, 1972; Frank, 1986; Hofer and East, 1993a). A clan is a fission-fusion society where members often move solitarily or in small groups, even if total clan membership is large and comprises on average 45 adults and subadults (Hofer and East, 1993a). In the vicinity of the communal den, the social center of a clan (Hofer and East, 1993a) and a location usually avoided by intruders, clan members frequently paste at a high rate and inspect pasting sites of other individuals (pers. obs.). This includes immigrant, potentially reproductively active males that inspect scent-marks of females and those of reproductive competitors (pers. obs.). Secondly, spotted hyenas frequently engage in ritualized meetings known as greeting ceremonies in which one or both participants may lift their tails and thus expose their sub-caudal scent-gland and have it inspected by their greeting partner. East *et al.* (1993) proposed that greeting partners present their scent-gland to advertise by olfactory cues their individual identity and/or class properties such as sex or clan membership. The subordinate of the two greeting partners communicates the acceptance of its relative social status by several gestures of submission, including the lifting of a hind leg, the erection of the penis or clitoris, the lifting of the tail and thereby the exposure of the anal scent-gland (East *et al.*, 1993).

In this study, we tested the ideas proposed by Mills (1990) and East *et al.* (1993). We asked whether or not volatile constituents of spotted hyena scent-marks can provide clues to class characteristics such as sex or clan membership. We also compared individuals in fatty acid composition of their scent-marks to determine whether or not such variation provides sufficient scope for individual recognition.

MATERIAL AND METHODS

Scent-marks were collected from three study groups ("clans") in a population of several hundred known individuals in the Serengeti National Park, Tanzania continuously studied since 1987. We also obtained scent-gland secretion from one captive adult female ("Patze") from Berlin Zoo of West African origin (Albert *et al.*, 2000) and one adult male from the Northern woodlands of the Serengeti National Park. Individuals in all three study clans were recognized by their spot patterns, scars and other natural features such as ear "notches". All clan members were individually known (Hofer and East, 1993a). Individuals were sexed using the shape of the phallic glans (Frank *et al.*, 1990) and reproductive status (lactation). Females were classified as adults when they reached two years of age (Hofer and East, 1993a). Males were classified as adults when they reached two years of age, and most males dispersed from their natal clan after this age (Hofer and East, 1993a). All adult males that provided scent-marks for this study were potentially reproductively active immigrant males in our study clans. The three study clans held territories at the woodland/plains boundary in the Serengeti National Park. For details on the social and spatial behavior of clans in the Serengeti, see Hofer and East (1993 a, b, c). All study clans were habituated to the presence of a vehicle at the den.

All scent-marks collected during the interval from October 1997 to April 2000 were from adult immigrant males and adult females from three study clans in the vicinity of the communal den of each study clan or elsewhere in the "hinterland" of the territory (Gorman and Mills, 1984). Samples were collected within minutes after individuals pasted on an exposed vegetation feature, usually a tall grass stalk."

We took care to collect scent-marks from grass stalks or other kinds of vegetation only if (1) we could unambiguously identify the relevant vegetation feature, and (2) it had not been previously scent-marked by another individual. Scent-marks were also collected directly from the anal scent-gland using cotton wool sticks from one immobilized captive adult female hyena ("Patze") in the Berlin Zoo, and from one immobilized male from a clan in the Northern woodlands of the Serengeti. Grass-stalks and cotton wool sticks with secretion were placed in headspace glasses (20 ml), sealed with a headspace aluminium cap with a PTFE-coated silicone rubber septa (Perkin Elmer), stored at $-20°C$ and transported to Europe in an uninterrupted cold chain until processing. As controls for field samples, we cut one piece out of vegetation in the vicinity of the scent-marked vegetation at approximately the same height as the vegetation feature with scent-mark. Controls served to check to detect any volatile substances that may arise from the vegetation.

The fatty acid content of anal scent secretions was analyzed with headspace solid-phase microextraction (SPME). In principle, this method uses a fiber coated with an adsorbent that can extract organic compounds from the headspace above a liquid or solid sample or from the surface of a biological material. SPME has the advantage that no solvent-consuming extraction procedures are required, there are no losses of volatile substances if evaporation steps are required, and extracted compounds are only desorbed upon exposure of the SPME fiber in the heated injector port of a gas chromatograph (GC). Thus, for processing, volatiles were released by heating the sample for 60 min at 60°C or 100°C, respectively. During this time, a polyacrelate fiber was exposed to the headspace above the sample. The fiber was then inserted into the injector of a Perkin Elmer (PE) AutoSystem XL gaschromatograph provided with a flame ionization detector (FID) and desorbed at 300°C. Analytes were separated on a SPB^{TM}-1000 column (Supelco,

Bellefonte, Pennsylvania; USA) with dimensions of 30m by 0.25 mm by 0.25μm film thickness at 50°C for 5 min and subsequently increased to 200°C at 5°C min^{-1}. Identification of fatty acids was performed in two ways. Peaks from scent samples were compared with synthetic standards and showed, as expected, identical GC retention times. Secondly, augmentation of scent-marks with synthetic fatty acids increased GC profiles in height, as expected, but did not broaden peak widths.

For the current study we analyzed the chemical composition of fatty acids of one adult captive female, and that of five adult males and seven adult females. Statistical analyses were performed using SYSTAT 9.0 (Wilkinson 1999). All probabilities were two-tailed. Because of the large variations and the violation of the assumption of normality, we compared males and females using non-parametric tests (Mann-Whitney U-test, Kruskal-Wallis test, and Wilcoxon signed ranks test; see Sokal and Rohlf, 1981).

RESULTS

As Figure 1 illustrates, scent-marks from free-ranging spotted hyenas contained both short-chain and long-chain fatty acids (C2 to C12 at least). Unlike many felids and other carnivore families, this sample of spotted hyena scent did not contain any indol.

There were no significant differences between the sexes in the relative amount each fatty acid contributed to the secretion (Table 1).

There was substantial variation between individuals of the same sex indicated by the sex-specific large coefficients of variation (CV) for the relative contribution of each fatty acid to the total amount indicate (Table 1). Fatty acids where CV exceeded 100% were, in males, butyric and nonanoic acid, and in females acetic, proprionic, butyric, and pentanoic acid (Table 1).

Table 1. Relative amounts of volatiles (fatty acids) in scent-marks of male ($n=5$) and female ($n=7$) Serengeti spotted hyenas.*

Fatty acid	Males		Females		Comparison	
	Mean ± St. Dev.	C.V. (%)	Mean ± St. Dev.	C.V. (%)	U	P
Acetic (C2)	0 ± 0	-	0.006 ± 0.016	264.6	20.0	Ns
Proprionic (C3)	0 ± 0	-	0.007 ± 0.020	264.6	20.0	Ns
Butyric (C4)	0.025 ± 0.027	108.3	0.016 ± 0.036	219.1	7.5	Ns
Pentanoic (C5)	0.028 ± 0.016	59.0	0.040 ± 0.061	151.2	10.0	Ns
Hexanoic (C6)	0.129 ± 0.106	82.2	0.073 ± 0.071	96.8	11.0	Ns
Heptanoic (C7)	0.047 ± 0.035	74.4	0.050 ± 0.044	87.2	17.0	Ns
Octanoic (C8)	0.127 ± 0.056	44.1	0.086 ± 0.051	59.5	11.0	Ns
Nonanoic (C9)	0.056 ± 0.060	107.2	0.036 ± 0.014	39.0	17.0	Ns
Decanoic (C10)	0.220 ± 0.125	56.7	0.222 ± 0.119	53.4	19.0	Ns
Undecanoic (C11)	0.090 ± 0.047	52.0	0.085 ± 0.042	49.1	15.0	Ns
Dodecanoic (C12)	0.279 ± 0.132	47.3	0.313 ± 0.185	59.2	15.0	Ns

* St. Dev.: standard deviation; C.V.: coefficient of variation (in %); U: Mann-Whitney U-test; Ns: not significant.

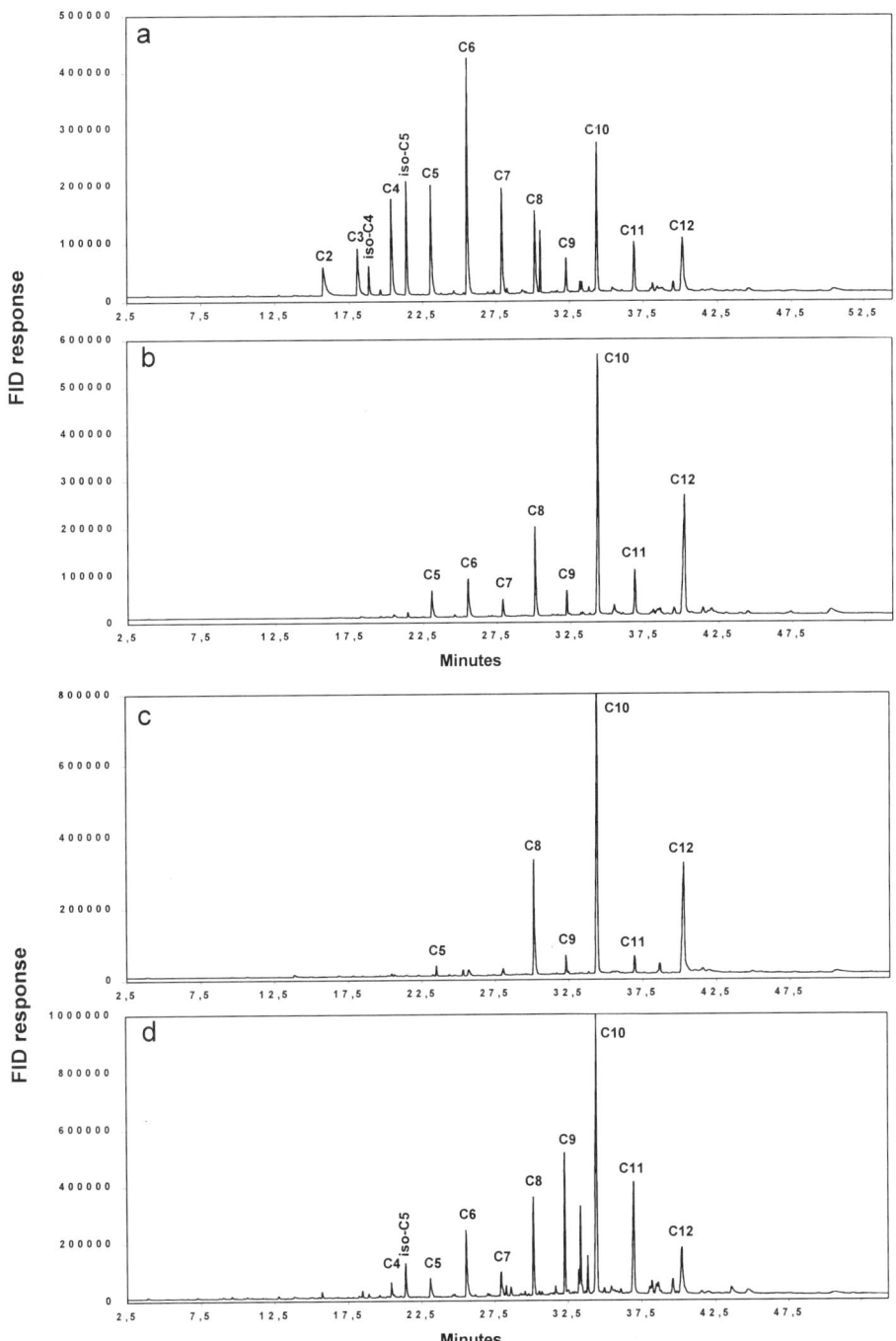

Figure 1. GC profile of the (**a**) scent-mark of the Serengeti adult female P163 (Pool clan); (**b**) scent-mark of the Serengeti adult male M301 (Mamba clan); (**c**) scent-mark of the Serengeti adult female I201 (Isiaka clan), (**d**) anal gland secretion of a captive adult female ("Patze") from Berlin Zoo. Abbreviations (C2 – C12) see Table 1.

Overall, females had a more variable composition of scent-marks than did males (Wilcoxon signed-ranks test on differences in the coefficients of variation between male and females, Z=2.48, p=0.013). Preliminary results suggest that there were significant differences between individuals from different clans in the relative amount of decanoic acid but not for other fatty acids (Table 2).

Table 2. Relative amounts of volatiles (fatty acids) in scent-marks of Serengeti spotted hyenas from different clans.*

Fatty acid	Isiaka Clan ($n=2$)		Mamba Clan ($n=4$)		Pool Clan ($n=5$)		Comparison	
	Mean ± St. Dev.	C.V. %	Mean ± St. Dev.	C.V. %	Mean ± St.Dev.	C.V. %	H	p
Acetic	0 ± 0	-	0 ± 0	-	0.008 ± 0.019	223.6	1.20	Ns
Proprionic	0 ± 0	-	0 ± 0	-	0.010 ± 0.023	223.6	1.20	Ns
Butyric	0.008 ± 0.009	110.8	0.001 ± 0.001	115.5	0.030 ± 0.039	130.2	3.39	Ns
Pentanoic	0.012 ± 0.004	30.7	0.048 ± 0.069	143.9	0.031 ± 0.043	139.4	0.53	Ns
Hexanoic	0.049 ± 0.051	103.9	0.066 ± 0.027	41.6	0.099 ± 0.087	87.2	0.72	Ns
Heptanoic	0.030 ± 0.030	99.0	0.054 ± 0.042	78.4	0.041 ± 0.038	92.0	0.40	Ns
Octanoic	0.143 ± 0.030	20.8	0.088 ± 0.026	29.3	0.076 ± 0.046	60.3	4.66	0.097
Nonanoic	0.037 ± 0.023	61.2	0.039 ± 0.012	31.8	0.054 ± 0.061	112.1	0.62	Ns
Decanoic	0.101 ± 0.077	75.8	0.341 ± 0.056	16.5	0.200 ± 0.081	40.7	7.57	0.023
Undecanoic	0.074 ± 0.069	93.3	0.100 ± 0.034	34.2	0.093 ± 0.043	46.6	0.61	Ns
Dodecanoic	0.336 ± 0.067	20.0	0.259 ± 0.041	16.0	0.357 ± 0.224	62.6	1.45	Ns

* St. Dev.: standard deviation; C.V.: coefficient of variation (in %); H: Kruskal-Wallis test for differences between clans with 2 df; Ns: not significant.

DISCUSSION

The anal gland secretions of Serengeti spotted hyenas contained short-chain and long-chain fatty acids from C_2 to at least C_{12}, confirming previous results from captive spotted hyenas (Buglass et al., 1990). There were no differences between the sexes in the relative amount that each fatty acid contributed to the total pool of volatiles. Relative amounts of decanoic acid significantly varied between clans whereas other fatty acids did not. Variation between individuals was substantial as documented by large coefficients of variation for several fatty acids (Table 1 and 2).

These results are consistent with the idea of Mills (1990) and East et al. (1993) that spotted hyenas produce clan-specific identity cues, and thus provide information by which intruders can recognize an individual as territory owner by matching (Gosling, 1982). The results are also consistent with the idea (East et al., 1993) that spotted hyenas use secretions of the anal scent-gland for individual recognition. We illustrate this with a re-assessment of the results of the detailed behavioral study of more than 3,000 greeting ceremonies by East et al. (1993) during which greeting partners may lift their tail and permit access to their scent-gland.

If, as Mills (1990) suggested, spotted hyenas use anal gland secretion to distinguish clan members from non-clan members, then we would predict that individuals that encounter each other infrequently could, by permitting the inspection of their scent, minimize the chance of aggression that could arise from mistaken identity. In this case, greeting partners that meet infrequently should lift their tails and expose their scent-glands more often than partners that regularly meet. Adult females, in particular nursing females, that are present at the communal den more frequently than adult males (Hofer and East, 1993c), should therefore expose their scent-gland less frequently in all-female greetings than should adult males in all-male greetings. This was highly significantly so (East et al., 1993).

If scent-glands provided only information about clan membership, but no additional information about individual identity, social status or other class characteristics, then there should be no difference in the behavior of dominant and subordinate greeting partners in terms of the likelihood by which they present their scent-gland. This, however, was not the case. During greetings between adult females, each female was more likely to provide access to its caudal scent-gland when it was the subordinate participant than when it was dominant. Furthermore, as the rank difference between two greeting females increased, the proportion of subordinates that held their tail up and exposed their scent-gland also increased. With decreasing status of a cub's mother, the proportion of cubs that exposed their scent-gland increased when greeting adult females (East et al., 1993). All of these results suggest that anal gland secretions provide cues to individual identity, perhaps similar to the way in which individual voices provide cues for individual recognition in the long-distance call of spotted hyenas (East and Hofer, 1991).

If volatile compounds do provide cues for individual recognition, then experiments with artificially created "skeleton" scent-marks made of synthetic fatty acids dissolved in a carrier substance could test whether fatty acids do indeed play a role in olfactory communication. We have already created such artificial scent-marks consisting of a fatty acid composition that mimics the scent of specific Serengeti individuals and a study is underway that uses them in behavioral tests with spotted hyenas.

ACKNOWLEDGMENTS

We are grateful to the Tanzanian Commission for Science and Technology (COSTECH) for permission to conduct the study and the Director Generals of the Tanzania Wildlife Research Institute (formerly the Serengeti Wildlife Research Institute) and Tanzania National Parks, and the Conservator of the Ngorongoro Conservation Area for cooperation and support. We thank A. Francis, F. Göritz, A. Lema, M. Faßbender, P. Göltenboth, G. Orio, D. Thierer, L. Trout, and K. Wilhelm for assistance. The Fritz-Thyssen-Stiftung, the Stifterverband der deutschen Wissenschaft, the Max-Planck-Gesellschaft and the Institut für Zoo- und Wildtierforschung funded the study.

REFERENCES

Albert, R., Hofer, H., East, M. L., and Pitra, C., 2000, Genetische Identifizierung der geographischen Herkunft von Tüpfelhyänen (*Crocuta crocuta* Erxleben 1777), *Zool. Garten N.F.* **69**:1-10.

Albone, E. S., 1984, *Mammalian semiochemistry: the investigation of chemical signals between mammals*, John Wiley, Chichester.

Apps, P., Viljoen, H. W., Richardson, P. R. K., and Pretorius, V., 1989, Volatile components of the anal gland secretion of the aardwolf (*Proteles cristatus*). *J. Chem. Ecol.* **15**:1681-1688.

Buglass, A. J., Darling, F. M. C., and Waterhouse, J. S., 1990, Analysis of the anal sac secretion of the Hyaenidae, in: *Chemical Signals in Vertebrates 5*, (D. W. Macdonald, D. Müller-Schwarze, and S. E. Natynczuk, eds.), Oxford University Press, Oxford, pp. 65-69.

East, M. L., and Hofer, H., 1991, Loud calling in a female-dominated mammalian society: I. Structure and composition of whooping bouts of spotted hyaenas, (*Crocuta crocuta*), *Anim. Behav.* **42**:637-649.

East, M. L., Hofer, H., and Wickler, W., 1993, The erect penis is a flag of submission in a female-dominated society: greeting ceremonies in Serengeti spotted hyenas, *Behav. Ecol. Sociobiol.* **33**:355-370.

Fox, M. W., 1971, Ontogeny of a social display in *Hyaena hyaena*: anal protrusion, *J. Mammal.* **52**:467-469.

Frank, L. G., 1986, Social organisation of the spotted hyaena (*Crocuta crocuta*). I. Demography, *Anim. Behav.* **35**:1500-1509.

Frank, L. G., Glickman, S. E., and Powch, I., 1990, Sexual dimorphism in the spotted hyaena (*Crocuta crocuta*), *J. Zool. Lond.* **221**:308-313.

Gorman, M. L., and Mills, M. G. L., 1984, Scent marking strategies in hyaenas (Mammalia), *J. Zool. Lond.* **202**:535-547.

Gosling, L. M., 1982, A reassessment of the function of scent marking territories. *Z. Tierpsychol.* **60**:89-118.

Hofer, H., and East, M. L., 1993a, The commuting system of Serengeti spotted hyaenas: how a predator copes with migratory prey. I. Social organization, *Anim. Behav.* **46**:547-557.

Hofer, H., and East, M. L., 1993b, The commuting system of Serengeti spotted hyaenas: how a predator copes with migratory prey. II. Intrusion pressure and commuters' space use, *Anim. Behav.* **46**:559-574.

Hofer, H., and East, M. L., 1993c, The commuting system of Serengeti spotted hyaenas: how a predator copes with migratory prey. III. Attendance and maternal care, *Anim. Behav.* **46**: 575-589.

Kruuk, H., 1972, *The spotted hyena*, University of Chicago Press, Chicago.

Mills, M. G. L., 1990, *Kalahari hyaenas. The comparative behavioural ecology of two species*, Unwin Hyman, London.

Mills, M. G. L., and Gorman, M. L., 1987, The scent-marking behaviour of the spotted hyaena *Crocuta crocuta* in the southern Kalahari. *J. Zool. Lond.* **212**:483-497.

Mills, M. G. L., and Hofer, H., 1998, *Hyaenas of the world: status survey and action plan*, IUCN, Gland.

Mills, M. G. L., Gorman, M. L., and Mills, M. E. J., 1980, The scent-marking behaviour of the brown hyaena *Hyaena brunnea*, *S. Afr. J. Zool.* **15**:240-248.

Nel, J. A. J., and Bothma, J. du P., 1983, Scent marking and midden use by aardwolves (*Proteles cristatus*) in the Namib Desert, *Afr. J. Ecol.* **21**:25-39.

Richardson, P. R. K., 1990, Scent marking and territoriality in the aardwolf, in: *Chemical Signals in Vertebrates 5*, (D. W. Macdonald, D. Müller-Schwarze, and S. E. Natynczuk, eds.), Oxford University Press, Oxford, pp. 378-387.

Richardson, P. R. K., 1991, Territorial significance of scent marking during the nonmating season in the aardwolf *Proteles cristatus* (Carnivora, Protelidae), *Ethology* **87**:9-27.

Sliwa, A., 1998, Responses of aardwolves, *Proteles cristatus*, Sparrman 1783, to translocated scent marks, *Anim. Behav.* **56**:137-146.

Sokal, R. R., and Rohlf, F. J., 1981, *Biometry*, 2nd edn., Freeman, San Francisco.

Wheeler, J. W., Endt, D. W., and Wemmer, C., 1975, 5-thiomethylpentane-2,3-dione. A unique natural compound from the striped hyena, *J. Am. Chem. Soc.* **97**:441-442.

Wilkinson, L., 1999, *SYSTAT 9: the system for statistics*, SPSS Inc, Chicago.

MICE, MUPS AND MYTHS: STRUCTURE-FUNCTION RELATIONSHIPS OF THE MAJOR URINARY PROTEINS

Robert J. Beynon,[1] Jane L. Hurst,[2] Simon J. Gaskell,[3] Simon J. Hubbard,[4] Rick E. Humphries,[2] Nick Malone,[2] Amr Darwish Marie,[1] Line Martinsen,[4] Charlotte M. Nevison,[2] Caroline E. Payne,[2] Duncan H. L. Robertson,[1] and Christina Veggerby[1]

[1]Protein Function Group and [2]Animal Behaviour Group
Faculty of Veterinary Science
University of Liverpool, PO Box 147, Liverpool L69 3BX, UK
[3]Michael Barber Centre for Mass Spectrometry and
[4]Department of Biomolecular Sciences
UMIST, PO Box 88, Manchester M60 1QD, UK

1. INTRODUCTION

The presence of high concentrations of protein in the urine of many mammals is often a pathological event reflecting a failure of the renal mechanisms to prevent passage of plasma proteins through the glomerular filter, or for recovery of protein in the proximal tubule. In adult humans, for example, the normal urinary protein concentration is of the order of 50µg/ml, a value that serves to emphasise the remarkable protein output in the urine of mice, and to a lesser extent rats. The average protein concentration in the house mouse (*Mus domesticus*) reaches between 200 and 2000 times that of a normal human, at concentrations of the order of 30mg/ml. Over 99% of this protein comprises the Major Urinary Proteins, a group of 18-20kDa proteins, synthesised in the liver, secreted into plasma and subsequently passed through the glomerular filter into the urine as it is elaborated.

Thus, the daily loss of protein from a mouse may be estimated in the tens of milligrams per day. Moreover, production of MUPs is an irreversible protein loss. This is a substantial investment of energy in protein synthesis, although perhaps not as dramatic as might be thought in the context of hepatic protein metabolism, which in the mouse amounts to a turnover flux of several hundred milligrams a day. The mouse, in effect, replaces nearly all of its liver protein every 24h. Nonetheless, the cost of production of this protein implies a critical functional role. The MUPs are lipocalins, 8-stranded beta-barrel structures enclosing a central cavity or 'calyx' that is lined with apolar, hydrophobic residues and which therefore favours biding of lipophilic molecules. In the mouse, it has been clearly demonstrated that the MUPs bind several pheromones, including 2-*sec* butyl-4,5-dihydrothiazole and 3,4-dehydro-

exo-brevicomin (Bacchini *et al.*, 1992; Robertson *et al.*, 1993). A combined biochemical and behavioural study, which used a competitive displacer molecule (menadione, discovered as a MUP ligand in a semi-natural environment, Robertson *et al.*, 1998) to demonstrate co-ordination in time of pheromone release and behavioural response proved unambiguously that one consequence of binding of these pheromones to the MUPs was a prolonged release pattern, extending to 24h rather than the few minutes in which the pheromones would evaporate if not protein bound. The kinetics of release of the pheromones is therefore as much a function of affinity for the MUPs as of their volatility.

A simple role for MUPs as slow release mechanism does not explain satisfactorily the notable molecular heterogeneity of this group of proteins. In this chapter, we develop the hypotheses that MUPs have additional functions, for which molecular heterogeneity is an essential prerequisite. These ideas are based on the data we have derived on aspects of mechanism (Robertson *et al.*, this volume), structure (Veggerby *et al.*, this volume) and extent of variation (Payne *et al.*, this volume) of MUPs. Such structural and mechanistic information is an essential component of models that place urinary chemical signalling in an ecological and evolutionary context (Hurst *et al.*, this volume).

2. MOLECULAR HETEROGENEITY OF MUPS

The early studies of MUPs, using inbred laboratory strains, provided clear evidence for heterogeneity in the net charge on MUPs, assessed by isoelectric focussing (Groen and Lagerwerf, 1979; Hainey and Bishop, 1982; Hayakawa *et al.*, 1983; Sampsell and Held, 1985; Duncan *et al.*, 1988). However, charge variation can be created by post-translational modifications, such as deamidation (conversion of a neutral amino acid residue to an anionic residue), proteolysis (potential removal of a segment of polypeptide chain carrying charged residues), glycosylation, phosphorylation and similar. It was not until cDNA sequences, and subsequently the gene structure were elucidated that there was clear evidence for multiple genes, giving rise to distinct gene products, each recognisable as a member of the MUP family, but distinct in primary sequence. At the time of writing, there are a total of over 1800 entries in Genbank indexed with the term 'major urinary protein', of which about 20 are full length cDNA sequence, the remainder being fragments or EST (expressed sequence tag) sequences. A note of caution should be interjected in relation to the EST clones – they include, for example, entries that suggest that mouse diaphragm is a source of MUP mRNA – this is at variance with data for other skeletal muscle, and may reflect trace contamination of a diaphragm preparation with liver mRNA – at 5% of the total liver messenger mRNA population, the contamination would not have to be excessive for MUP to crop up in an EST search.

The source of most MUP cDNA or EST clones is inbred mouse strains, notable BALB/C and C57BL/6. Inasmuch as each inbred strain represents one 'snapshot' of the wild-type genome, we have concentrated on the expressed, urinary MUPs from wild caught mice. Early surveys had suggested that the wild population was a rich source of genetic diversity in the MUP gene cluster (Held and Sampsell, 1986). The MUPs of wild mice (Robertson *et al.*, 1997; Pes *et al.*, 1999) demonstrated a level of heterogeneity that was much greater than had been anticipated, and the number of unique MUPs, even when defined simply by accurate mass and charge state now exceeds 60. Detailed analysis of single MUPs from individuals confirmed that these are true gene products (Veggerby *et al.*, this volume).

3. THE RELATIONSHIP BETWEEN SEQUENCE VARIATION AND STRUCTURE

From the full length sequences that have been cloned and sequenced, and from those proteins defined by mass spectrometry and Edman degradation (Robertson *et al.*, 1996; Veggerby *et al.*, this volume) it is possible to map sequence variants onto the three dimensional structure of the MUP, as originally solved by Bocskei *et al.* (1992), and supported by the more recent structural analyses (Lucke *et al.*, 1999; Zidek *et al.*, 1999a). When the sequence variant positions (see for example, the alignment in Robertson *et al.*, 1996) are superimposed onto the three dimensional structure (Figure 1a), most of these are solvent-exposed, and relatively few are located within the calyx, which creates a hydrophobic cavity of volume $475Å^3$. The recent NMR study of Lucke *et al.*, (1999) has highlighted a difference in the rigidity of the cavity, inasmuch as one wall of the cavity shows much greater side chain flexibility than the other (Figure 1b). This flexible face may be able to undergo side chain rearrangements to accommodate the different ligands, consistent with a role for MUPs in binding and releasing multiple pheromones (Novotny *et al.*, 1999). Certainly, the MUP calyx is able to accommodate much larger molecules than are normally associated with this protein in nature (Robertson *et al.*, 1998; Robertson *et al.*, this volume; Marie *et al.*, submitted).

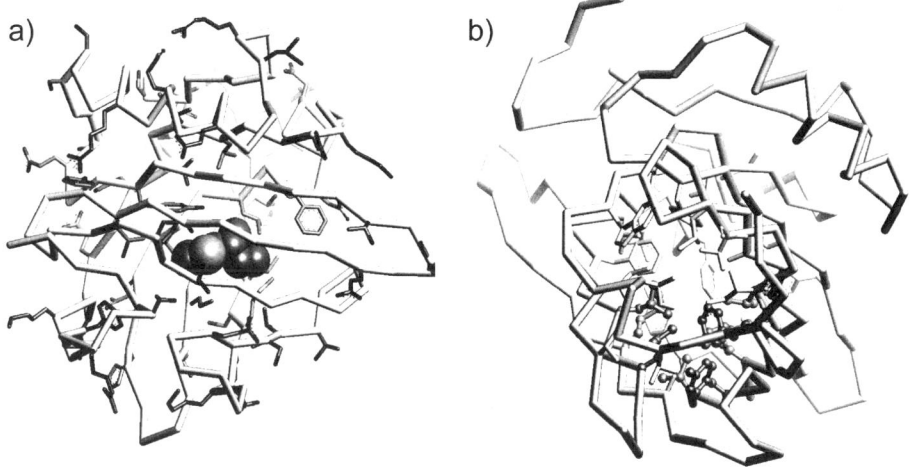

Figure 1. (a) The three dimensional structure of MUP, based on one of the X-ray crystal structure of Bocskei *et al.*, (1992), highlighting sequence positions known to be variable from cDNA sequence data, or protein chemistry characterisation of individual proteins. No data from EST sequences has been included, since these sequences are often unreliable or of poor quality. **(b)** The three dimensional structure of MUP, based on one of the NMR ensemble described by Lucke *et al.*, (1999). The residues lining the cavity are highlighted, and those residues that show a degree of backbone flexibility, and which therefore might confer flexibility on the cavity are highlighted as 'ball and stick' side chains.

The characterisation of MUPs from wild-caught mice has identified many more variant proteins than were previously known from inbred laboratory animals that are clearly new gene products. The precise masses (determined to within ±2Da by electrospray ionisation mass spectrometry) are not reconciled by consideration of post-translational changes such as proteolysis. Further, when these proteins are digested by the endopeptidase

LysC (which digests the MUP at each lysine residue within the sequence) and the masses of the peptides are determined by matrix assisted laser desorption ionisation time of flight (MALDI-TOF) mass spectrometry, the N-terminal and C-terminal peptides are nearly always correctly identified. By contrast, we commonly encounter variant masses for LysC peptides L4 and L5 that differ from known peptides, suggesting that there is a focus of genetic variation in this part of the sequence (residues 32 to 75). These residues form beta sheet strands B, C and D (Lucke et al., 1999) and thus, constitute an external patch on one side of the molecule, and one face of the internal cavity. This region is distributed between the rigid and flexible faces of the cavity.

Further work on MUP characterisation is required to define the sequence differences of additional MUP variants. However, an approach based on tandem mass spectrometry of those peptides not identified and reconciled by MALDI-TOF mass spectrometry will probably be more efficient than cDNA library construction and DNA sequencing, and is of course, non-invasive.

4. THE RELATIONSHIP BETWEEN STRUCTURAL VARIATION AND FUNCTION

The sequence variation in MUPs can be differentiated on the basis of whether it leads to changes that are solvent exposed or internally facing into the cavity. Cavity polymorphisms could manifest as differences in the rate of binding (k_{on}) or release (k_{off}) for a range of ligands, leading to the possibility that variant proteins might exhibit different affinities for specific pheromones. Some evidence for this behaviour was obtained by Robertson et al., (1993) who showed variation in the thiazole: brevicomin ratio for different MUPs resolved by anion exchange chromatography.

The analysis of polymorphic MUPs variants has been limited by two factors; the need for knowledge of protein sequence (whether determined directly or inferred from the cDNA sequence) and the ability to isolate the protein in a pure state. Many of the MUPs co-elute on high-resolution anion exchange chromatography (Robertson et al., 1996) but we have succeeded in purifying to homogeneity (assessed by mass spectrometry) several MUPs as a prelude to detailed binding studies. For some of these proteins, we have used a fluorescent probe, N-phenyl-naphthylamine (NPN), as a ligand that reports the environment of the MUP calyx. NPN is non-fluorescent in aqueous solution, but when transferred to an apolar or hydrophobic environment shows strong fluorescence (Robertson et al., this volume; Marie et al., submitted). The binding of NPN is stoichiometric, and NPN is displaced by high concentrations of menadione, the competitive displacer used to demonstrate the role of MUPs in providing slow release of ligands (Hurst et al., 1998). It is thus clearly resident in the central cavity.

We purified three MUPs that exhibit variation in surface residues, and one that was variant in a cavity residue, residue 56, which has been demonstrated to be one of Phe, Val (Robertson et al., 1996) or most recently, Leu/Ile (Veggerby et al., this volume). NPN binds to all variants thus purified, but the binding to the Val56 variant was substantially weaker and showed a lower fluorescent yield (Marie et al., submitted). The insertion of a different residue into the calyx was therefore able to modulate binding of a ligand.

The availability of the three-dimensional structure of MUP allowed homology modelling and molecular dynamics studies of the binding of NPN to the Phe and Val variants (Hubbard et al., in preparation). The molecular dynamics analysis indicated that NPN potentially bound to MUP-Phe56 and MUP-Val56 in very different configurations (Figure 2). Moreover, there was some suggestion of a degree of ring stacking between NPN and Phe56, the outcome of which might be expected to be an enhanced binding and fluorescence yield, in keeping with the experimental observations.

All of these ligands for MUPs have the common property of substantial apolar characteristics, and the hierarchy of ligand displacement is consistent with their relative hydrophobicity (Table 1). The volume of each of these ligands is much less than the cavity volume, and each can be accommodated with relative ease, particular with flexible structures such as NPN, which, in the modelling studies adopts different configurations in the two MUP variants. Moreover, the recent work of Zidek et al., (1999b) has discovered an enhanced flexibility of the MUP backbone upon ligand binding, which brings about an increase in conformational entropy. This backbone flexibility, coupled with the side chain flexibility on one face of the calyx (Lucke et al., 1999) may assist binding of a range of ligands. In wild mice however, the predominant MUP-associated ligands are 2-sec butyl-4,5-dihydrothiazole and 3,4-dehydro-exo-brevicomin (Robertson et al., 1998) and it is clear that these play an important role in wild populations as well as in inbred animals.

Figure 2. The binding of N-phenyl-naphthylamine (NPN) to two MUP variants, differing a cavity residue at sequence position 56, which is either a valyl or phenylalanyl residue. The simulation was performed by Dr S J Hubbard using the ICM program. The single amino acid substitution elicits a substantial change in NPN conformation and orientation.

Table 1. Properties of four MUP ligands.

Ligand	Volume ($Å^3$)	Accessible area ($Å^2$)	Accessible hydrophobicity (%)
2-sec butyl-4,5-dihydrothiazole	136	314	96
3,4-dehydro-exo-brevicomin	142	309	92
menadione	152	325	80
N-phenyl-naphthylamine	212	409	97

Although such studies provide some supporting evidence for a role of MUP variants in differential binding and release of pheromones, the true test of such a hypothesis lies in direct observation of the release of the natural ligands. The pheromones, being volatile, are difficult to analyse once released, and we have developed an approach based on determination of residual ligands when urine is deposited on an inert surface (Robertson et *al.*, this volume). This approach allows ligand loss to be monitored over extended periods, generates data of sufficient high quality for subsequent kinetic analysis and should allow us to assess the effect of polymorphism on natural ligand release. Modelling studies of each of the MUP variants allowed the size of the calyx to be measured, which varied from 430 to 620Å^3, all large volumes in relation to the size of the natural ligands. However, different binding modes between the ligands and the variants might lead to dramatic variation in the on and off rate constants, and direct determination of such parameters will be informative.

5. MUPS AS INFORMATION MOLECULES

There is good evidence to support the contention that MUPs act as semiochemicals in their own right. The first such study to provide information for the role of MUP-derived peptide sequence used an N-terminal hexapeptide derived from the sequence of a MUP Class 2 pseudogene (Mucignat-Caretta *et al.*, 1995). It is unlikely that such a peptide would ever be expressed and excreted, and since the effect (on puberty acceleration) has since been questioned in a related study (Novotny *et al.*, 1999) the true effect of such a peptide has yet to be clarified. More recent studies have provided further data to support a semiochemical role for MUPs. Using countermarking as a behavioural, rather than a physiological response, Humphries *et al.*, (1999) demonstrated that a protein fraction from mouse urine, even when depleted of volatiles by time or competitive displacement, was countermarked preferentially. Further information for the role of MUPs derives from the work of Brennan *et al.*, (1999) who provide evidence that the MUPs not only mediate the pregnancy blocking effects of male urine (through bound ligands) but also that they conveyed the strain recognition signal. In this context, strain recognition reflects their studies with laboratory mice and, extrapolated to wild mice, refers to individuality signalling. The countermarking response, which used wild mice, is also consistent with a role in individuality signalling – competitive countermarking requires de facto that the countermarking individual recognises the first donor as non-self (Hurst *et al.*, this volume). Finally, there is now evidence for a receptor for the rat equivalent to MUPs (Krieger *et al.*, 1999).

If the MUPs act as information molecules, communicating individual identity, it might be expected that they are expressed in both sexes. Previous data has been ambiguous about the level of expression of MUPs in female laboratory mice (Finlayson *et al.*, 1963; Duncan *et al.*, 1988; Held and Sampsell, 1986) but the generally held view, often repeated in reviews, is that the numbers of MUPs, as well as their concentration, is much reduced in females (e.g. Flower, 1996). The urinary MUP protein concentration has thus been reported as being between 3% and 20% of the male expression level, but we find that the differences between the sexes are much less pronounced, with the mean concentration in MUPs in female urine being between one third and one half that in males (see also Held and Sampsell, 1986). Moreover, there is considerable individual variation in urinary MUP protein concentration in both sexes, to the extent that some females produce more MUP than some males (unpublished data).

6. CONCLUSIONS

MUPs are more than simple delivery agents for pheromone molecules. Key to these additional roles is the notable molecular heterogeneity that these proteins evince. Cavity mutants have the potential to modulate the delivery of pheromones in the time domain, manipulating the ratio of molecules released and remaining in the sample. Surface mutants might also alter ligand affinity through manipulation of backbone conformational entropy. More importantly, the discovery of a MUP receptor introduces the possibility of manipulation of surface mediated interactions at the vomeronasal organ, and the key datum delivered by this interaction would be individual identity.

Previous emphasis on urine-mediated individuality signals has focussed on either volatile metabolites released in urine according to haplotype (Singer *et al.*, 1997; Yamazaki *et al.*, 1999) or the secretion of the soluble histocompatibility (MHC) molecules (or fragments thereof) into mouse or rat urine (reviewed in Singh, 1999). In the latter model, the high degree of allelic variation in MHC molecules is a key component of individuality signalling. Further, it has been postulated that the binding groove of the MHC proteins, normally a binding site for an antigenic peptide, might also function as a binding site for a volatile molecule. This has evolved to the notion of the 'carrier hypothesis' for communication of individuality. Whilst this may be attractive in principle, we suggest that MUPs are far better candidates for such carriers than MHC molecules. They are highly polymorphic, have evolved to be secreted in urine and posses a large, flexible binding pocket for lipophilic (and thus, volatile) molecules. It has to be acknowledged that in all studies using urine-derived protein fractions, these fractions contain both MHC molecules and MUPs. It has not been possible to assess the role of MUPs in the absence of MHC molecules, or vice versa. Studies with recombinant MUPs should eliminate these problems, provided it can be demonstrated that the proteins are not contaminated with proteins derived from the expression host (usually bacteria or yeast). Alternatively, techniques will need to be developed for selective purification of the two components. Further tests should explore the magnitude of the MHC effect on a variant MUP background, and vice versa.

MHC molecules and MUPs share many properties that make them ideal as mediators of individual identity; they are 'hard-coded' into the genome, are expressed in both sexes, exhibit similar degrees of polymorphism and are largely independent of status. The plethora of MUP polymorphic variants hints at a matching family of MUP receptors, and the matching of two polymorphic proteins (MUP and MUP receptor) would have the combinatorial potential to encode individuality data. In such a model, the role of volatile ligands may be subsumed into different functions, such as communication of sex or status or acting as 'alerter' molecules.

7. ACKNOWLEDGMENTS

This work was supported by BBSRC grants to RJB and JLH.

8. REFERENCES

Bacchini, A., Gaetani, E., and Cavaggioni, A., 1992, Pheromone binding proteins of the mouse, *Mus musculus*, *Experientia* **48**:419-421.

Bocskei, Z., Groom, C. R., Flower, D. R., Wright, C. E., Phillips, S. E., Cavaggioni, A., Findlay, J. B., and North, A. C., 1992, Pheromone binding to two rodent urinary proteins revealed by X-ray crystallography, *Nature* **360**:186-188.

Brennan, P. A., Schellinck, H. M., and Keverne, E. B., 1999, Patterns of expression of the immediate-early gene egr-1 in the accessory olfactory bulb of female mice exposed to pheromonal constituents of male urine, *Neuroscience* **90**:1463-1470.

Duncan, R., Matthai, R., Huppi, K., Roderick, T., and Potter, M., 1988, Genes that modify expression of major urinary proteins in mice, *Mol. Cell. Biol.* **8**:2705-2712.

Finlayson, J. S., Potter, M., and Runner, R. C., 1963, Electrophoretic variation and sex dimorphism of the major urinary protein complex in inbred mice: a new genetic marker, *J. Natl. Cancer Inst.* **31**:91-107.

Flower, D. R., 1996, The lipocalin protein family: structure and function, *Biochem. J.* **318**:1-14.

Groen, A., and Lagerwerf, A. J., 1979, Genetically determined electrophoretic variants of the major urinary protein (Mup) complex in mouse urine, *Anim. Blood Groups Biochem. Genet.* **10**:107-114.

Hainey, S., and Bishop, J. O., 1982, Allelic variation at several different genetic loci determines the major urinary protein phenotype of inbred mouse strains, *Genet. Res.* **39**:31-39.

Hayakawa, J., Nikaido, H., and Koizumi, T., 1983, Components of major urinary proteins (MUP's) in the mouse. Sex, strain, and subspecies differences, *J. Hered.* **74**:453-456.

Held, W. A., and Sampsell, B. M., 1986, Genetic variation in major urinary proteins in wild mice. *Curr. Top. Microbio. Immunol.* **127**: 124-130.

Humphries, R. E., Robertson, D. H. L., Beynon, R. J., and Hurst, J. L., 1999, Unravelling the chemical basis of competitive scent marking in house mice., *Anim. Behav.* **58**:1177-1190.

Hurst, J. L., Robertson, D. H. L., Tolladay, U., and Beynon, R. J., 1998, Proteins in urine scent marks of male house mice extend the longevity of olfactory signals, *Anim. Behav.* **55**:1289-1297.

Krieger, J., Schmitt, A., Lobel, D., Gudermann, T., Schultz, G., Breer, H., and Boekhoff, I., 1999, Selective activation of G protein subtypes in the vomeronasal organ upon stimulation with urine-derived compounds, *J. Biol. Chem.* **274**:4655-4662.

Lucke, C., Franzoni, L., Abbate, F., Lohr, F., Ferrari, E., Sorbi, R. T., Ruterjans, H., and Spisni, A., 1999, Solution structure of a recombinant mouse major urinary protein, *Eur. J. Biochem.* **266**:1210-1218.

Mucignat-Caretta, C., Caretta, A., and Cavaggioni, A., 1995, Acceleration of puberty onset in female mice by male urinary proteins, *J. Physiol. (Lond)* **486**:517-522.

Novotny, M. V., Ma, W., Wiesler, D., and Zidek, L., 1999, Positive identification of the puberty-accelerating pheromone of the house mouse: the volatile ligands associating with the major urinary protein, *Proc. R Soc. Lond. B. Biol. Sci.* **266**:2017-2022.

Pes, D., Robertson, D. H. L., Hurst, J. L., Gaskell, S. J., and Beynon, R. J., 1999, How many major urinary proteins are produced by the house mouse *Mus domesticus*?, in: *Advances in Chemical Communication in Vertebrates* (R. E. Johnston, D. Müller-Schwarze, and P.W. Sorensen, eds.) Plenum Press, New York, pp. 149-161.

Robertson, D. H. L., Beynon, R. J., and Evershed, R. P., 1993, Extraction, characterization and binding analysis of two pheromonally active ligands associated with major urinary protein of house mouse (Mus musculus), *J. Chem. Ecol.* **19**:1405-1416.

Robertson, D. H. L., Gaskell, S. J., Cox, K., Evershed, R. P., and Beynon, R. J., 1996, Analysis of molecular heterogeneity in Major Urinary Proteins of the house mouse, *Mus musculus, Biochem. J.* **316**: 265-272.

Robertson, D. H. L., Hurst, J. L., Bolgar, M. S., Gaskell, S. J., and Beynon, R. J., 1997, Molecular heterogeneity of urinary proteins in wild house mouse populations, *Rapid Communications in Mass Spectrometry* **11**:786-790.

Robertson, D. H. L., Hurst, J. L., Hubbard, S. J., Gaskell, S. J., and Beynon, R. J., 1998, Ligands of urinary lipocalins from the mouse: uptake of environmentally derived chemicals, *J. Chem. Ecol.* **24**:1127-1140.

Sampsell, B. M., and Held, W. A., 1985, Variation in the major urinary protein multigene family in wild-derived mice, *Genetics* **109**:549-568.

Singer, A. G., Beauchamp, G. K., and Yamazaki, K., 1997, Volatile signals of the major histocompatibility complex in male mouse urine, *Proc. Natl. Acad. Sci. U. S. A.* **94**:2210-2214.

Singh, P. B., 1999, The present status of the 'carrier hypothesis' for chemosensory recognition of genetic individuality, *Genetica* **104**:231-233.

Yamazaki, K., Beauchamp, G. K., Singer, A., Bard, J., and Boyse, E. A., 1999, Odortypes: their origin and composition, *Proc. Natl. Acad. Sci. U. S. A.* **96**:1522-1525.

Zidek, L., Stone, M. J., Lato, S. M., Pagel, M. D., Miao, Z., Ellington, A. D., and Novotny, M. V., 1999a, NMR mapping of the recombinant mouse major urinary protein I binding site occupied by the pheromone 2-sec-butyl-4,5-dihydrothiazole, *Biochemistry* **38**:9850-9861.

Zidek, L., Novotny, M. V., and Stone, M. J., 1999b, Increased protein backbone conformational entropy upon hydrophobic ligand binding [see comments], *Nat. Struct. Biol.* **6**:1118-1121.

POLYMORPHIC VARIANTS OF MOUSE MAJOR URINARY PROTEINS

Christina Veggerby[1], Caroline E. Payne[2], Simon J. Gaskell[3], Duncan H.L. Robertson[1], Jane L. Hurst[2], and Robert J. Beynon[1]

[1]Protein Function Group and [2]Animal Behaviour Group
Faculty of Veterinary Science
University of Liverpool, Liverpool L69 3BX, UK
[3]Michael Barber Centre for Mass Spectrometry
UMIST, Manchester M60 1QD, UK

INTRODUCTION

The components of mouse urine include a group of proteins known as major urinary proteins (MUPs). These are small (18-19 kDa) proteins belonging to the lipocalin superfamily (Flower *et al.*, 1993). The members of this family bind small hydrophobic ligands that are possible chemosignals (Böcskei *et al.*, 1992; Bacchini *et al.*, 1992; Robertson *et al.*, 1993). Five of the known MUP ligands, 2-*sec*-butyl-4,5-dihydrothiazole and 3,4-dehydro-*exo*-brevicomin, α- and β-farnesene, and 6-hydroxy-6-methyl-3-heptanone, elicit primer pheromonal effects (Jemiolo *et al.*, 1985; Jemiolo *et al.*, 1986; Novotny *et al.*, 1999) and behavioural responses (Hurst *et al.*, 1998; Mucignat-Caretta *et al.*, 1998; Humphries *et al.*, 1999) in mice, suggesting a role for these proteins in olfactory communication. One function of the proteins as a facilitator of slow release of ligand has been shown (Hurst *et al.*, 1998). In addition, it has been suggested that the proteins interact directly with receptors in the vomeronasal organ, exerting its effect on puberty onset in females, even when free of ligands (Mucignat-Caretta *et al.*, 1995), and that the proteins play a role in conveying the strain recognition signal of the male pheromone (Brennan *et al.*, 1999).

The function of the heterogeneity of this group of proteins is not known. Our recent work on MUPs from inbred mice (Robertson *et al.*, this volume) has studied the possibility that the heterogeneity accomplishes the differential release of the ligands, facilitated by the polymorphs having different ligand binding affinity. A prerequisite for understanding the function of the observed polymorphism is defining the polymorphism. Most of the work to date has concentrated on MUPs from inbred laboratory mice strains (typically BALB/C and/or C57BL/6) (Kuhn *et al.*, 1984; Shahan *et al.*, 1987; Robertson *et al.*, 1996). Comparing two different populations of wild mice, we found a much wider degree of MUP heterogeneity in mass and charge than was seen in inbred mice (Pes *et al.*, 1999).

We report here our further characterisation of the polymorphic variation in one of these populations, namely the population of feral mice on the Isle of May, a small island off the east coast of Scotland. The heterogeneity of individual urine samples was analysed by isoelectric focusing, and the major MUPs from five of these samples were purified and further characterised by electrospray ionisation mass spectrometry (ESI/MS) and peptide fingerprinting using a restricted specificity endoprotease (Lys-C) and subsequent analysis of the peptide mixture by matrix-assisted laser desorption/ionisation mass spectrometry (MALDI-TOF/MS).

METHODS

Male Wild House Mice (*Mus domesticus*). Mice were captured from an isolated population living ferally on the Isle of May off the east coast of Scotland. Each mouse was given a unique code for identification (e.g. M9, M17). Urine was collected in response to being held by the scruff of the neck and samples were frozen immediately and stored at -20°C until analysis.

Isoelectric Focusing (IEF). Urine samples were analysed by IEF using Immobiline dryplates pH 4.2-4.9 (Pharmacia) on the Multiphor flatbed system cooled to 10°C with a Multitemp III thermostatic circulator (Pharmacia). Samples were diluted 1:5 in water and run into the gel at 200V, 5mA, and 15W for 200vh. Sample applicators were then removed and the gel run at 3500V, 5mA, and 15W for 14.8Kvh. The gel was stained with Coomassie Brilliant Blue in a Hoefer automated gel stainer.

Ion Exchange Chromatography (IEC). Urinary proteins were purified by high-resolution anion exchange chromatography on the SMART chromatography platform (Pharmacia). Prior to chromatography, 10 µl of urine was desalted on micro Sephadex G25 spin columns and filtered through a 0.22 µm filter. Urinary protein (15 µg) was applied to a MiniQ PC 3.2/3 column (240 µl bed volume, Pharmacia) equilibrated with 20 mM Bistris buffer, pH 5.5, at a flow rate of 120 µl/min. Following sample application, the column was washed with 5 column volumes of 20 mM Bistris buffer, pH 5.5, and bound protein eluted with a linear gradient of NaCl from 0-150 mM in 20 column volumes. Protein peaks, identified by absorbance at 280 nm, were collected. Peak fraction sizes were typically 60 µl, containing 1-3 µg of protein.

Electrospray Ionisation Mass Spectrometry (ESI/MS). Peak fractions from IE chromatography were desalted into 0.2 % formic acid on Microcon-10 concentrators (Millipore) and diluted 1:10 in 50 % (v/v) acetonitrile/0.1 % (v/v) formic acid. All analyses were performed on a VG Quattro mass spectrometer, upgraded to Quattro-II specifications, fitted with an electrospray ionisation source. Raw data, collected between m/z 700 and 1400, were subsequently refined and transformed to a true mass scale using maximum entropy software (MaxEnt 2, Micromass). Raw data were processed between 18400 and 19000 Da. After each sample the instrument was calibrated using a 2 pmol/µl solution of horse heart myoglobin in 50 % (v/v) acetonitrile/0.1 % (v/v) formic acid. The calibration spectrum also served to determine the peak width parameter (0.75 Da) during MaxEnt processing.

Peptide Fingerprinting. The protein of the major peaks from IE of M22 urine and of the uMUP-I peak from BALB/C was denatured in 4 M guanidine thiocyanate for 1 hour at room temperature and desalted into Lys-C digestion buffer (25 mM Tris, 1 mM EDTA,

2 mM β-mercaptoethanol, pH 8.5). Sequencing grade Lys-C (0.1 mg/ml, Roche) was added in a 1:100 (w/w) ratio. Digestion took place overnight at 37°C and the reaction was stopped by the addition of 1 μl formic acid. The peptides were analysed on a matrix-assisted laser desorption ionisation time-of-flight mass spectrometer (MALDI-TOF/MS) (PE Biosystems Voyager Elite). A pulsed nitrogen laser (337 nm) was used as a desorption/ionisation source and mass spectrometry was performed in reflectron mode with positive ion detection. External mass calibration was performed with a mixture of des-Arg bradykinin, neurotensin, ACTH, and insulin β-chain. To 1 μl of digested protein or peptide calibrants, 1 μl of α-cyano-4-hydroxy cinnamic acid (saturated in 50 % acetonitrile/0.1 % triflouroacetic acid) was added, and 1 μl was spotted onto the target. Signal averaged mass spectra for 80-256 laser shots were collected and data analysed using the PE Biosystems GRAMS/386 software. The spectra for the wild mice MUP digests were compared to that of the uMUP-I digest and theoretical digests (using the BioLynx software, Micromass) of other published MUP sequences.

RESULTS

To assess the level of heterogeneity in the Isle of May population of feral house mice, urine samples collected from 18 male mice were analysed by isoelectric focusing (IEF) (Figure 1), whereby proteins are separated in a gel according to their isoelectric point. These proteins differ only slightly in charge and accordingly a narrow pH-range, high-resolution gel was used. A urine sample from inbred BALB/C was included for comparison. The May samples displayed greater variation in individual MUP profiles than seen in inbred strains, expressing up to nine distinguishable bands. The pattern varied between samples, although the same bands occurred in most samples. The greatest difference between individuals was in the relative amounts of each protein band expressed.

The 18 wild mouse urine samples were also analysed by IEC, an analytical as well as preparative chromatographic separation of proteins according to overall charge. Figure 2 shows results for one of the May samples (M22) compared to BALB/C urine. Eleven distinct peaks were seen in the profile for M22, whereas only six bands were discernible on the IEF gel. Seven peaks were resolved in the BALB/C sample, five of which had similar elution volumes to the M22 May sample (shaded peaks) and must thus contain proteins of the same net charge.

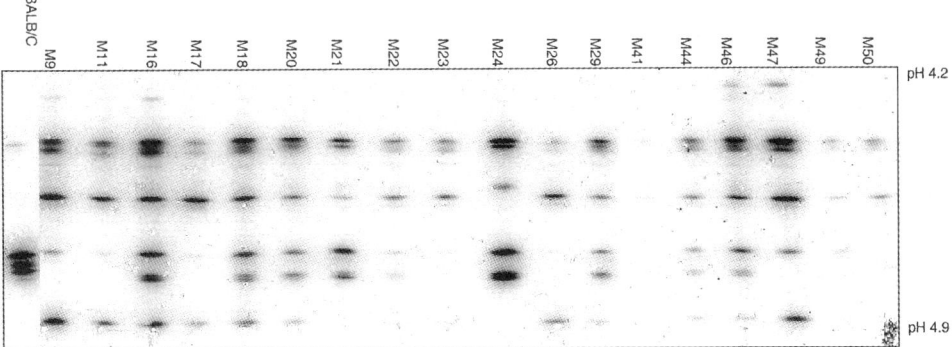

Figure 1. Analysis of urine samples from May mice by isoelectric focusing.
Samples were diluted 1:5 in water and applied to an Immobiline dryplate pH 4.2-4.9 on the Multiphor flatbed system. The gel was run at 3500V, 5mA, and 15W for 14.8Kvh and stained with Coomassie Brilliant Blue.

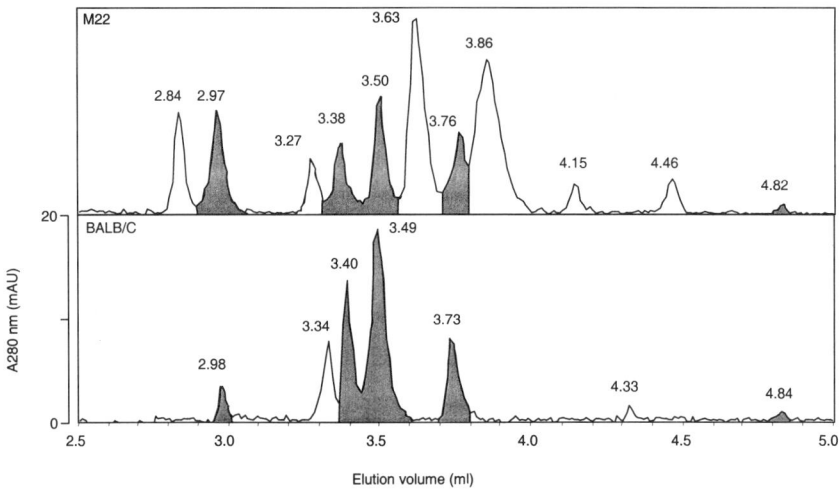

Figure 2. Purification of MUPs from M22 urine compared with the BALB/C MUP profile
Urine samples were desalted on micro Sephadex G25 spin columns and 15 µg of protein was applied to a MiniQ PC 3.2/3 column (Pharmacia) equilibrated with 20 mM Bistris buffer, pH 5.5. The column was washed with 5 column volumes of 20 mM Bistris buffer, pH 5.5 and bound protein eluted with a linear gradient of NaCl from 0-150 mM in 20 column volumes. Peaks of similar elution volume present in both are shaded.

To further characterise the May MUPs, the proteins of the five main peaks (2.84, 2.97, 3.38, 3.50, and 3.60 ml) and the cluster of peaks between 3.76 and 3.86 ml (shoulders indicates the presence of more than 1 protein) were analysed by ESI/MS, which yields accurate masses (Table 1). The accuracy and precision of this mass measurement on ESI/MS is a powerful tool for protein characterisation. For example, the masses of MUPs are determined typically to ±2 Da, well within the error introduced by single amino acid differences and readily matched against masses predicted from cDNA data.

Table 1. Electrospray ionisation mass spectrometric analysis of purified M22 MUPs

Peak 1: 2.84 ml	Peak 2: 2.97 ml	Peak 4: 3.38 ml	Peak 5: 3.50 ml	Peak 6: 3.63 ml	Peak 7+8: 3.79-3.86 ml
					18624.0 Da
18682.2 Da	18669.0 Da	18658.2 Da	18693.1 Da	18707.3 Da	18692.1 Da
					18712.6 Da
New	New	New	Found in both BALB/C and C57BL/6J urine		

The five main peaks contained singly purified proteins, the first three of these masses identify new proteins not expressed in any inbred strain (Robertson *et al.*, 1996). Peak 2 and 4 both eluted at positions similar to BALB/C MUPs in IEC, but the ESI/MS results established that the similarity was only at the level of overall charge. The proteins of peak 5 and 6 may correspond to proteins expressed in BALB/C and C57BL/6J, peak 5 eluted at the same volume on IEC as the 18693 Da uMUP-I from BALB/C (Robertson *et al.*, 1996) and peak 6 eluted at the same position as the 18707 Da uMUP-IX from C57BL/6J (Robertson *et al.*, 1996) (we have used the nomenclature suggested by Robertson *et al.*, 1996, where the prefix u is used to define these MUPs as being expressed in urine rather than other tissues). Correspondence of mass and charge, at this precision, is an excellent indicator of a protein match.

The five proteins were subjected to peptide fingerprinting analysis. The five purified proteins were reduced, denatured, and digested with the endoprotease Lys-C, which cleaves specifically after every lysyl residue, and the peptide mixture were analysed by MALDI-TOF/MS. The peptide masses thus obtained can be compared with those of known MUPs to identify peptides containing sites of heterogeneity, by a mass shift corresponding to one or more amino acid substitutions (Figure 3).

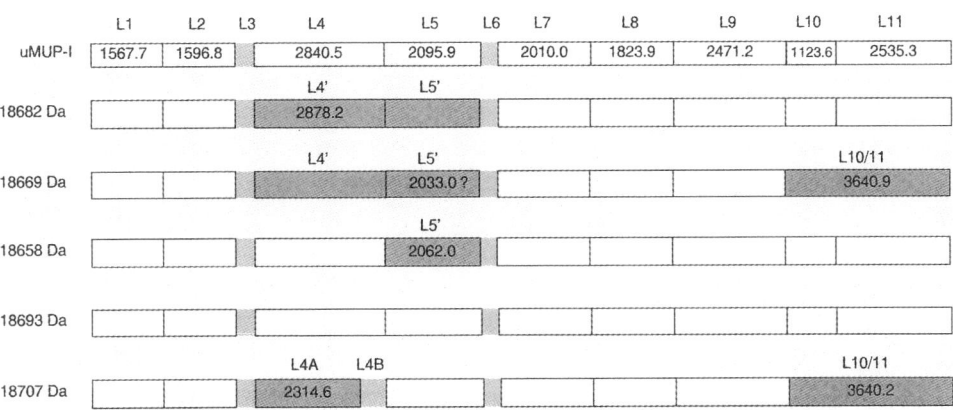

Figure 3. Peptide mapping of the major MUPs from M22. The proteins in the major IEC peaks were digested with Lys-C. The peptides were analysed on a matrix-assisted laser desorption ionisation time-of-flight mass spectrometer (MALDI-TOF/MS) (PE Biosystems Voyager Elite). The spectra for the wild mice MUP digests were compared to that of the uMUP-I digest and theoretical digests (using the BioLynx software) of other published MUP sequences. The light grey boxes represent the peptides too small to be detected. The darker grey boxes represent peptides differing from the uMUP-I sequence.

Extensive sequence identity to uMUP-I was found, fragments L1, L2, L7, L8, and L9 were found in all spectra. The two tripeptides L3 and L6 are too small for detection on the MALDI-TOF/MS. Three fragments corresponding to known mutations were found: L5': 2047.8 Da fragment, $F_{56} \rightarrow V_{56}$ mutation in uMUP-II (Kuhn *et al.*, 1984, clone p1057), L4': 2314.6 Da fragment, $N_{50} \rightarrow K_{50}$ mutation in uMUP-IX (Kuhn *et al.*, 1984, clone p499), and L10/11: 3640.2/3640.9 Da fragment, $K_{140} \rightarrow E_{140}$ mutation in uMUP-VIII (Robertson *et al.*, 1996) and uMUP-IX (Kuhn *et al.*, 1984, clone p499). For the 18682 Da protein the 2878.2 Da L4 fragment would account for the remaining protein mass difference. No single base change can cause the observed mass shift, and as two base changes in different codons is more likely than two in the same codon, this fragment probably contains two amino acid changes. More sophisticated analyses by tandem mass spectrometry would reveal these substitutions. The same fragment was seen in the spectrum for the 18669 Da protein. The remaining protein mass difference must reside in the L5 fragment, but no peak of the appropriate size (2033 Da) were seen.

For the 18658 Da protein only one fragment (L5) differed from uMUP-I. A 2062 Da fragment is found in the spectrum, which would account for the protein mass difference. The only single base change possible for this mass shift results in an $F_{56} \rightarrow I/L_{56}$ substitution (isoleucine and leucine are isobaric amino acids and therefore indistinguishable). The peptide map for the 18693 Da protein matched that of uMUP-I (Robertson et al., 1996) from BALB/C while the 18707 Da protein matched that of uMUP-IX from C57BL/6J (Kuhn et al., 1984, clone p499).

4. DISCUSSION

MUP heterogeneity arises from the expression of about 35 related genes on chromosome 4, inherited in a co-dominant fashion (Hainey and Bishop, 1982).

Inbred mice excrete multiple MUPs in the urine and the urines of all individuals have the same patterns in IEF and IEC. The wild mice exhibit greater complexity, although the extent of the difference between individuals is less than in other wild populations (Payne et al., this volume). This may be due to a degree of inbreeding in an isolated island population or it might be a consequence of dramatic population changes during the year, leading to 'pinch points' (Triggs, 1991).

Detailed analysis of MUPs from single May mice has allowed identification of several proteins. Two proteins (uMUP-I and uMUP-IX) were previously observed in inbred mice (Kuhn et al., 1984; Robertson et al., 1996). A further three proteins representing new gene products have been identified. In one instance, it was possible that a phenylalanine residue had substituted for either leucine or isoleucine in position 56. This amino acid residue resides in the MUP cavity, with the side chain protruding into the cavity space (Zidek et al., 1999) and may have an effect on ligand binding. Confirmation of this change and identification of the changes in these new proteins require peptide sequencing, possibly by tandem mass spectrometry.

Having obtained a more detailed picture of the current polymorphic variants of one population of wild mice, the next step will be to investigate the function of these polymorphisms in olfactory communication. It is striking that only one of the mutations seen in inbred mice and possibly one of the mutations of the new MUPs characterised here are mutations in the cavity of the MUPs. The majority of the known mutations are of surface residues (Beynon et al., this volume), suggesting a function additional to control of ligand release.

ACKNOWLEDGMENTS

This work was supported by BBSRC (to RJB and JLH) and NERC (to JLH). The help of Samantha Gray in collecting wild mice is gratefully acknowledged.

REFERENCES

Bacchini, A., Gaetani, E., and Cavaggioni, A., 1992, Pheromone binding proteins of the mouse, *Mus musculus*, *Experientia*. **48**:419-421.

Böcskei, Zs., Groom, C. R., Flower, D. R., Wright, C. E., Phillips, S. E. V., Cavaggioni, A., Findalay, J. B. C., and North, A. C. T., 1992, Pheromone binding to two rodent urinary proteins revealed by X-ray crystallography, *Nature* **360**:186-188.

Brennan, P. A., Schellinck, H. M., and Keverne, E. B., 1999, Patterns of expression of the immediate-early gene egr-1 in the accessory olfactory bulb of female mice exposed to pheromonal constituents of male urine, *Neuroscience* **90**:1463-1470.

Flower, D. R., North, A. C. T., and Attwood, T. K., 1993, Structure and sequence relationships in the lipocalins and related proteins, *Protein Science* **2**:753-761.

Hainey, S., and Bishop, J. O., 1982, Allelic variation at several different genetic loci determines the major urinary protein phenotype of inbred mouse strains, *Genet. Res. Camb.* **39**:31-39.

Humphries, R. E., Robertson, D. H. L., Beynon, R. J., and Hurst, J. L., 1999, Unravelling the chemical basis of competitive scent marking in house mice, *Anim. Behav.* **58**:1177-1190.

Hurst, J., Robertson, D. H. L., Tolladay, U., and Beynon, R. J., 1998, Proteins in urine scent marks of male house mice extend the longevity of olfactory signals, *Anim. Behav.* **55**:1289-1297.

Jemiolo, B., Alberts, S., Sochinski-Wiggins, S., Harvey, S., and Novotny, M. V., 1985, Behavioural and endocrine responses of female mice to synthetic analogs of volatile compounds in male urine, *Anim. Behav.* **33**:1114-1118.

Jemiolo, B., Harvey, S., and Novotny, M., 1986, Promotion of the Whitten effect in female mice by synthetic analogues of male urinary constituents, *Proc. Natl. Acad. Sci. USA.* **83**:4576-4579.

Kuhn, N. J., Woodworth-Gutai, M., Gross, K. W., and Held, W. A., 1984, Subfamilies of the mouse major urinary protein (MUP) multi-gene family: sequence analysis of cDNA clones and differential regulation in the liver, *Nucleic Acids Res*, **12**:6073-6090.

Mucignat-Caretta, C., Caretta, A., and Cavaggioni, A., 1995, Acceleration of puberty onset in female mice by male urinary proteins, *J. Physiol.* **486**:517-522.

Mucignat-Caretta, C., Caretta, A., and Baldini, E., 1998, Protein-bound male urinary pheromones: differential responses according to age and gender, *Chem. Senses* **23**:67-70.

Novotny, M. V., Ma, W., Wiesler, D., and Zidek, L., 1999, Positive identification of the puberty-accelerating pheromone of the house mouse: the volatile ligands associating with the major urinary protein, *Proc. R. Soc. Lond.* **266**:2017-2022.

Pes, D., Robertson, D. H. L., Hurst, J. L., Gaskell, S., and Beynon, R. J., 1999, How many Major Urinary Proteins are produced by the house mouse *Mus domesticus*? in *Advances in Chemical Signals in Vertebrates*, (R. E. Johnston, D. Müller-Schwarze, and P. W. Sorensen, eds.), Plenum Publishers, New York, pp. 149-162.

Robertson, D. H. L., Beynon, R. J., and Evershed, R. P., 1993, Extraction, characterisation, and binding analysis of two pheromonally active ligands associated with major urinary protein of house mouse (*Mus musculus*), *J. Chem. Ecol.* **19**:1405-1416.

Robertson, D. H. L., Cox, K. A., Gaskell, S. J., Evershed, R. P., and Beynon, R. J., 1996, Molecular heterogeneity in the Major Urinary Proteins of the house mouse *Mus domesticus*, *Biochem. J.* **316**:265-272.

Shahan, K., Gilmartin, M., and Derman, E., 1987, Nucleotide sequences of liver, lachrymal, and submaxillary gland mouse major urinary protein mRNAs: mosaic structure and construction of panels of gene-specific synthetic oligonucleotide probes, *Mol. Cell. Biol.* **7**:1938-1946.

Triggs, G. S., 1991, The population ecology of house mice (*Mus domesticus*) on the Isle of May, Scotland, *J. Zool. Lond.* **225**:449-468.

Zidek, L., Stone, M. J., Lato, S. M., Pagel, M. D., Miao, Z., Ellington, A. D., and Novotny, M. V., 1999, NMR mapping of the recombinant mouse major urinary protein I binding site occupied by the pheromone 2-*sec*-butyl-4,5-dihydrothiazole, *Biochemistry*, **38**:9850-9861.

DIFFERENTIAL RESPONSES ELICITED IN MALE MICE BY MUP-BORNE ODORANTS

Andrea Cavaggioni, Carla Mucignat-Caretta, and Antonio Caretta

Dipartimento di Anatomia e Fisiologia Umana
Universita' di Padova, 35131 Padova, Italy

INTRODUCTION

Pheromones transmit information on sex, age, social rank and hormonal status among mice (Halpern, 1987), are responsible for attraction and mate recognition, and modulate intraspecific interactions as well as reproductive physiology (Keverne, 1983). Most male pheromones are mainly androgen-dependent molecules that are excreted in urine of adult males (Reynolds, 1971). They modulate both behavioral and neuroendocrine outcomes through the activation of chemosensory receptors of the main olfactory epithelium and vomeronasal organ (Brennan and Keverne, 1997; Tirindelli *et al.*, 1998).

Different types of molecules may play a pheromonal role. The Major Urinary Proteins (MUPs) are 18 kDa androgen-dependent lipocalins secreted in urine (Szoka and Paigen, 1978) and characterized by a hydrophobic binding site for odorants (Cavaggioni *et al.*, 1987; 1990). The most abundant odorants in male urine are 2-sec-butyl-4,5dihydrothiazole (SBT) and 3,4-dehydro-exo-brevicomin (DHB) (Schwende *et al.*, 1986), they are secreted bound to MUPs and then released in air (Bacchini *et al.*, 1992, Böcskei *et al.*, 1992).

We have shown that the odorants released by MUPs attract female, while repelling male mice (Mucignat-Caretta *et al.*, 1998), and modify exploration of different environments (Mucignat-Caretta and Caretta, 1999a). Moreover, when odorants are present on the fur of a sexually receptive female mouse they turn the normal sexual approach of a male into an aggressive one (Mucignat-Caretta and Caretta, 1999b). Noteworthy, the simultaneous presence of DHB and SBT on the fur is sufficient to enhance aggression of experienced fighter males towards castrated mice (Novotny *et al.*, 1985). Interestingly, the amygdala, which mediates aggressive behaviors in mice, receives inputs both from the main and accessory olfactory systems. Aggressive behaviors towards adult and also towards newborn mice are reduced following vomeronasal ablation (Bean and Wysocki, 1989). On the other hand, aggression towards females was not elicited by MUPs stripped of their odorants, excluding thus a direct role of the protein in aggressive behaviors (Mucignat-Caretta and Caretta, 1999b).

We therefore assume that MUP-borne volatiles are a trigger for aggressive behaviors in male mice, and tested their effect on the aggressive behavior displayed by an adult isolated male mouse in the presence of a pup, an example of noncompetitive intraspecific aggression.

2. METHODS

2.1. Animals

Albino Swiss mice were housed in plastic cages (42x26x15 cm) with food and water freely available, on a 12 hours light cycle, lights on at 5:00 a.m. The temperature was 24± 1°C. Experiments conformed to the EEC 86/609 guidelines. Nine adult males were tested. Care was taken to avoid severe injury to the pups (Elwood et al., 1991). Pups were promptly returned to the dams at the end of the test.

2.2. Biochemical procedure

Urine collected from 60 adult male mice was chromatographed through a molecular sieve column (G50, Pharmacia) and fractions containing MUPs were chromatographed through an ion-exchange column (DE-52, Pharmacia) (Mucignat-Caretta et al., 1995).

2.3. Behavioral test

We observed male mice, previously isolated for 15 days, during 5min tests. In the test each mouse was presented at once with two newborn pups (0-48 hours after birth), one of which had been painted with MUPs releasing the odorants (50 µl, 10 mg ml^{-1}) and the other was a control. We measured the latency to the first attack (seconds) to each of the pups. In the absence of an attack, it was assigned the maximal latency. Data were subjected to ANOVA, completely randomized design.

RESULTS AND DISCUSSION

Male mice isolated for 15 days became aggressive toward mouse pups and displayed infanticide behavior; odorants released by MUPs painted on a pup, however, prevented the attacks. This was shown by the latency to the first attack. The mean value ± SEM for control pups was 99.4 ± 33.2 sec., while for MUPs-painted pups 223.9 ± 37.5 seconds, significantly longer for MUP-painted pups than for controls, $F (1,8) = 23.633$, $p < 0.002$. Moreover, while all nine control pups were attacked within 190 seconds, only 4 out of 9 painted pups were attacked in the same time frame.

Thus, odorants released by MUPs do inhibit infanticide, although in the case of receptive female mice they are known to trigger aggressive behavior, overcoming sexually appealing olfactory, visual and tactile stimuli. This means that the odorants released by MUPs act in concert with other stimuli for the perceptual identification of conspecifics, and the complex relationship linking stimuli to aggressive behavior of mice cannot be restricted to pheromones only.

REFERENCES

Bacchini, A., Gaetani, E., and Cavaggioni, A., 1992, Pheromone binding proteins of the mouse, Mus musculus, *Experientia* **48**:419-421.

Bean, N. J., and Wysocki, C. J., 1989, Vomeronasal organ removal and female mouse aggression: the role of experience, *Physiol. Behav.* **45**:875-882.

Böcskei, Zs., Groom, C. R., Flower, D. R., Wright, C. E., Phillips, S. E. V., Cavaggioni, A., Findlay, J. B. C., and North, A. C. T., 1992, Pheromone binding to two rodent urinary proteins revealed by X-rays crystallography, *Nature* **360**:186-188.

Brennan, P. A., and Keverne, E. B., 1997, Neural mechanisms of mammalian olfactory learning, *Progress Neurobiol.* **51**:457-481.

Cavaggioni, A., Findlay, J. B. C., and Tirindelli, R., 1990, Ligand binding characteristics of homologous rat and mouse urinary proteins and pyrazine-binding protein of calf, *Comp. Biochem. Physiol.* **96B**:513-520.

Cavaggioni, A., Sorbi, R. T., Keen, J. N., Pappin, D. J. C., and Findlay, J. B. C., 1987, Homology between the pyrazine-binding protein from nasal mucosa and Major Urinary Proteins, *FEBS Letters* **212**:225-228.

Elwood, R. W., Masterson, D., and O'Neill, C., 1991, Protecting pups in tests for infanticidal responsiveness in mice, Mus domesticus, *Anim. Behav.* **40**:778-780.

Halpern, M., 1987, The organization and function of the vomeronasal system, *Ann. Rev. Neurosci.* **10**:352-362.

Keverne, E. B., 1983, Pheromonal influences on the endocrine regulation of reproduction, *Trends Neurosci.* **6**:381-384.

Mucignat-Caretta, C., and Caretta, A., 1999a, Urinary chemical cues affect light-avoidance behaviour in male laboratory mice, Mus musculus, *Anim. Behav.* **57**:765-769.

Mucignat-Caretta, C., and Caretta, A., 1999b, Chemical signals in male house mice urine: protein-bound molecules modulate interactions between sexes, *Behaviour* **136**:331-343.

Mucignat-Caretta, C., Caretta, A., and Baldini, E., 1998, Protein-bound male urinary pheromones: differential responses according to age and gender, *Chem. Senses* **23**:67-70.

Mucignat-Caretta, C., Caretta, A., and Cavaggioni, A., 1995, Acceleration of puberty onset in female mice by male urinary proteins, *J. Physiol. (Lond.)* **486**:517-522.

Novotny, M., Harvey, S., Jemiolo, B., and Alberts, J., 1985, Synthetic pheromones that promote inter-male aggression in mice, *Proc. Natl. Acad. Sci. USA* **82**:2059-2061.

Reynolds, E., 1971, Urination as a social response in mice, *Nature* **234**:481-483.

Schwende, F. J., Wiesler, D., Jorgenson, M., Carmack, M., and Novotny, M., 1986, Urinary volatile constituents of the house mouse, *Mus musculus*, and their endocrine dependency, *J. Chem. Ecol.* **12**:277-269.

Szoka, P., and Paigen, K., 1978, Regulation of mouse Major Urinary Protein production by the Mup-a gene, *Genetics* **90**:597-612.

Tirindelli, R., Mucignat-Caretta, C., and Ryba, N. J. P., 1998, Molecular aspects of pheromonal communication via the vomeronasal organ of mammals, *Trends Neurosci.* **21**:482-486.

CHARACTERISTICS OF LIGAND BINDING AND RELEASE BY MAJOR URINARY PROTEINS

Duncan H. L. Robertson,[1] Amr Darwish Marie,[1] Christina Veggerby,[1] Jane L. Hurst,[2] and Robert J. Beynon[1]

[1]Protein Function Group, Department of Veterinary Preclinical Sciences
[2]Animal Behaviour Group, Department of Clinical Veterinary Science and Animal Husbandry
Faculty of Veterinary Science
University of Liverpool
Liverpool L69 3BX, UK

1. INTRODUCTION

The Major Urinary Proteins (MUPs) are abundant in the urine of house mice (Finlayson and Baumann 1958). They have a characteristic eight-stranded β-barrel structure, which incorporates a central, hydrophobic pocket or calyx (Böcskei *et al.*, 1992; Zidek *et al.*, 1999 a, b). This distinctive structure is common to a number of other proteins, collectively called the Lipocalin Superfamily (Flower *et al.*, 1993). The central hydrophobic pocket of the lipocalins allows the binding and transport of hydrophobic ligands in aqueous solutions (Flower 1994). When found in urine, MUPs are associated with a number of endogenous ligands (Bacchini *et al.*, 1992; Robertson *et al.*, 1993; Novotny *et al.*, 1999 a, b). A number of these compounds including, brevicomin (3,4-dehydro-exo-brevicomin) and thiazole (2-sec-butyl-4,5-dihydrothiazole) are primer pheromones, which exhibit a variety of behavioural and physiological effects (Jemiolo *et al.*, 1985, Jemiolo *et al.*, 1986). The association of MUPs with these pheromones and their abundant presence in urine has led to the suggestion that MUPs play an important role in olfactory communication.

MUPs are produced by a family of closely related genes (Bishop *et al.*, 1982) and consequently, the mature proteins have slight differences in their amino acid sequences (Robertson *et al.*, 1995). By surveying MUPs in the urine of wild mice from a number of populations, we have identified more than 50 different MUPs (Robertson *et al.*, 1997, Pes *et al.*, 1999, Veggerby *et al.*, this volume).

One established function of MUPs is to prolong the release of the associated volatile pheromones, allowing an otherwise short-lived signal to be extended in time (Hurst *et al.*, 1998). Other studies have also suggested that MUPs themselves can act as pheromones

Chemical Signals in Vertebrates 9, edited by Marchlewska-Koj *et al.*
Kluwer Academic/Plenum Publishers, New York, 2001.

(Mucignat-Caretta et al., 1995, Brennan et al., 1999), although this still remains the subject of some conjecture (Novotny el al. 1999b). The precise role of the MUP heterogeneity however remains unclear. The following study attempts to address this by characterising the effect of MUP heterogeneity on the characteristics of the ligand-binding cavity.

Two approaches have been taken to investigate the effect of heterogeneity on the MUP cavity. A fluorescent probe of the MUP cavity, N-phenyl napthylamine (NPN) has been identified. When NPN acts as a MUP ligand, it undergoes a marked fluorescence enhancement. This has been used to measure the affinity of NPN for the cavity in a number of purified MUPs, with known sites of heterogeneity. Additionally, the release of two endogenous ligands, thiazole and brevicomin, from urine and MUP deposits has been characterised.

2. METHODS AND MATERIALS

Assessment of Putative Fluorescent Reporter Ligands Four potential MUP ligands were assessed for fluorescence enhancement in the presence of MUPs; 1-anilinonapthalene-8-sulphonic acid (1-8 ANS), 2-anilinonapthalene-6-sulphonic acid (1-6 ANS), 2-(p-toluidinyl) naphthalene-6-sulphonic acid (2-6 TNS) and N-phenyl-1-napthylamine (NPN). Each was mixed with unfractionated MUPs and the fluorescence spectra recorded.

Measurement of NPN Affinity for Individual MUPs Individual MUPs were purified from BALB/C (uMUP-I and uMUP-VII) and C57BL/6 (uMUP-IX and uMUP-X) urine, using anion exchange chromatography as recorded previously (Robertson et al., 1995). A solution (50 nM) of purified MUP was titrated with increasing additions of NPN, the range of final NPN concentrations was 0-1500 nM. The fluorescence (λ_{ex} = 337 nm, λ_{em} = 395 nm) was measured and plotted as a function of NPN concentration. Non-linear least squares analysis was used to calculate the dissociation constants (K_d) of the individual MUPs. The same data were also processed by Scatchard analysis (not shown) to confirm that the stoichiometry of the association was 1:1.

Characterisation of Endogenous Ligand Release from MUPs and Urine Whole urine and desalted MUPs were obtained from C57BL/6 mice. To make a chloroform extract of free ligands from C57BL/6 urine, equal volumes of chloroform and urine were vortexed for 10 seconds, allowed to settle for 15 minutes, prior to the chloroform being removed with an automatic pipette. Aliquots of each preparation (30 µl) were deposited onto circular (2.5 cm diameter), GF/C glass microfibre filters (Whatman, U.K.). Each GF/C filter was suspended 5 mm above a polystyrene base by means of a needle inserted through it, 3-4 mm from the edge. To minimise variations in airflow and light intensity, the whole apparatus was housed in a Perspex chamber, open to the atmosphere and lined with aluminium foil. All experiments were conducted in an air-conditioned room, where the temperature was maintained between 20 and 22 °C. GF/C filters treated in this manner were removed at times 0 h, 15 min, 30 min, 45 min, 1, 2, 3, 4, 5 and 21h and, along with a 500 µl aliquot of deionised water, placed in a glass vial that was sealed with an airtight cap. The zero time sample varied slightly from this protocol, in that an unused GF/C filter was placed in a vial followed by 470 µl of deionised water and 30 µl of urine, immediately before capping.

All vials were analysed by Gas Chromatography/Mass Spectrometry (GC/MS) with headspace sampling, as stated elsewhere (Humphries et al., 1999). The samples were incubated at 100°C for 20 minutes to completely dissociate the MUP-ligand complex, prior to removal of the headspace sample. The GC/MS was used in selected ion mode (SIM) to detect ions of m/z 95 (brevicomin) and 60 (thiazole). The amount of residual ligand contained in each sample was determined from the integrated peak area in the total ion chromatograph. This was then expressed as a percentage of the zero time sample and plotted as a function of incubation time. Finally, the data were analysed as a bi-exponential function by non-linear least squares curve fitting.

3. RESULTS AND DISCUSSION

3.1 Binding of Fluorescent Probes

In considering a function for MUPs, we have entertained the possibility that the rates at which pheromones are released from a urine mark could reflect information about its age or donor. If individual MUPs were to loose a particular ligand at different rates, they would be expected to vary in their affinity for that ligand. As MUP ligands are housed in a cavity of the molecule, differences in binding affinity would result from changed characteristics of this cavity. In inbred mice, the MUP amino acid chain is known to vary at positions 50, 56, 136 and 140. The purpose of this study was to ascertain the effect of these variations on the ligand-binding cavity, in order to test the above hypothesis of ligand release rates.

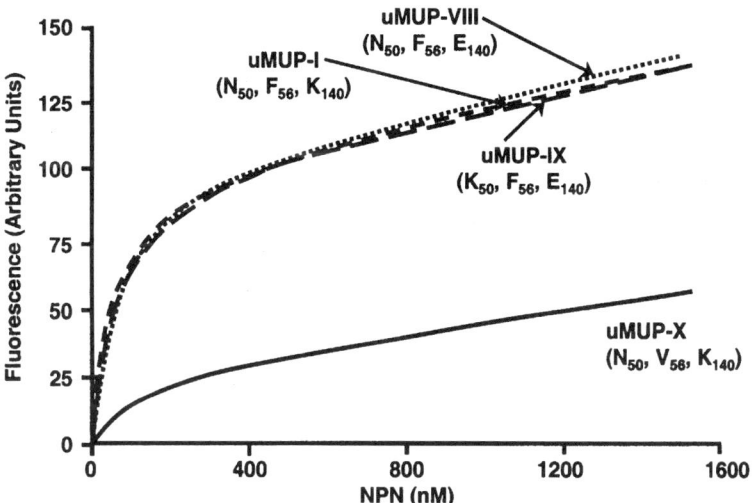

Figure 1. NPN binding to purified MUPs. NPN (0-1500nM) was titrated against 50nM solutions of four different purified MUPs, from BALB/C (uMUP-I and uMUP-VII) and C57BL/6 (uMUP-IX and uMUP-X). The resulting data were analysed by non-linear least squares curve fitting. The identity of the heterogeneous amino acids at positions 50, 56 and 140 is annotated.

The first approach taken in this characterisation was to identify a molecule that could be used as a fluorescent probe of the ligand-binding cavity. This was achieved by measuring the fluorescence spectra of four molecules, 1-anilinonapthalene-8-sulphonic acid (1-8 ANS), 2-anilinonapthalene-6-sulphonic acid (2-6 ANS) and 2-(p-toluidinyl) naphthalene-6-sulphonic acid (2-6 TNS) and N-phenyl-1-napthylamine (NPN), in the presence of MUPs. The fluorescence spectra of three of these molecules (1-8 ANS, 2-

6ANS and 2-6 TNS), were unchanged. One molecule, NPN, had a greatly enhanced fluorescence spectrum, with a fluorescence maximum at 395 nm under excitation at 337 nm. This is consistent with transfer of NPN from aqueous solution, to the non-polar environment of the MUP cavity. The stoichiometry of the reaction was determined by Scatchard analysis (data not shown), to be 0.96 ± 0.018, (mean ± S.E., n = 4) confirming that NPN was acting as a MUP ligand and causing displacement of the endogenous ligands. It is not known whether the lack of fluorescence enhancement in the other three molecules is due to their inability to act as MUP ligands or their inability to fluoresce upon doing so.

In a previous study (Robertson *et al.*, 1995), we determined the amino acid sequence of MUPs from inbred mice using mass spectrometry. In this study, the affinity of purified preparations of such MUPs for NPN was assessed by incubation with increasing amounts of NPN and measurement of fluorescence intensity. Non-linear least squares analysis of such data allows the dissociation constant (K_d) to be determined according to the analysis of Epps *et al.*, (1995). In three of the four MUPs we measured; uMUPs-I, VIII and IX (Robertson *et al.*, 1995) the K_d values were very similar, 60, 61 and 56 nM respectively (Figure 1). One MUP (uMUP-X), showed lower fluorescence enhancement, with a measured K_d of 122 nM. The unique difference between uMUP-X and the other MUPs measured is a V/F substitution at position 56.

From published crystallographic co-ordinates (Böcskei *et al.*, 1992) it is possible to determine the position of the variant residues in the three dimensional MUP structure. Position 56 is an inward facing residue in one of the β-strands that define the calyx. The effect of this substitution on NPN binding suggests that variations at residue 56 significantly alter the properties of the ligand-binding cavity. This might influence the ligand binding specificity and/or the rate at which ligands are released. The remaining variant sites are all remote from the calyx and outward facing.

3.2 Release of Endogenous Ligands

In addition to NPN binding, the manner in which the endogenous ligands thiazole and brevicomin are released from urine has also been investigated. Observations that mice liberally cover their home environment with urine (Hurst, 1987), lead us to the conclusion that ligand release was likely to happen from dry or semi-dry MUPs, rather than in aqueous solution. We have therefore attempted to investigate this by depositing urine or MUPs on a glass fibre membrane, in an attempt to mimic natural deposition. Previous experiments using laboratory Benchkote as a deposition medium were unsuccessful due to non-specific adsorption of thiazole and brevicomin.

In the first instance, we examined thiazole and brevicomin loss from whole BALB/c urine (Figure 2). The residual thiazole and brevicomin were plotted as a function of time and analysed by bi-exponential curve fitting, according to the general equation:

$$L_t = L_0 \cdot [P_f \exp(-k_f \cdot t) + P_s \exp(k_s \cdot t)]$$

Where P_f and P_s are the proportions of each function and k_f and k_s are the first order rate constants for loss. In the case of thiazole and brevicomin loss from urine, P_f and P_s represent fractional pool sizes of the two components and k_f and k_s represent the rate at which they are lost from the deposit. The thiazole and brevicomin contained in urine are comprised of two distinct pools; one (P_f) which is lost rapidly from the sample (the fast pool) and another, (P_s) which prevails for longer (the slow pool). The curve fitting data for thiazole and brevicomin loss from BALB/c urine is summarised in Table 1. The proportions of the pools, and the rates at which each is lost from the sample are different for each ligand.

Table 1. Bi-exponential, non-linear least squares curve fitting parameters for thiazole and brevicomin loss from BALB/C urine.

Fitted Parameter	Thiazole (± SE)	Brevicomin (± SE)
P_f(%)	75.4 ± 6.3	62.5 ± 3.5
k_f 99(h^{-1})	1.36 ± 0.1	10.4 ± 2.8
P_s(%)	24.2 ± 6.5	37.5 ± 2.5
k_s (h^{-1})	0.1 ± 0.06	0.3 ± 0.04

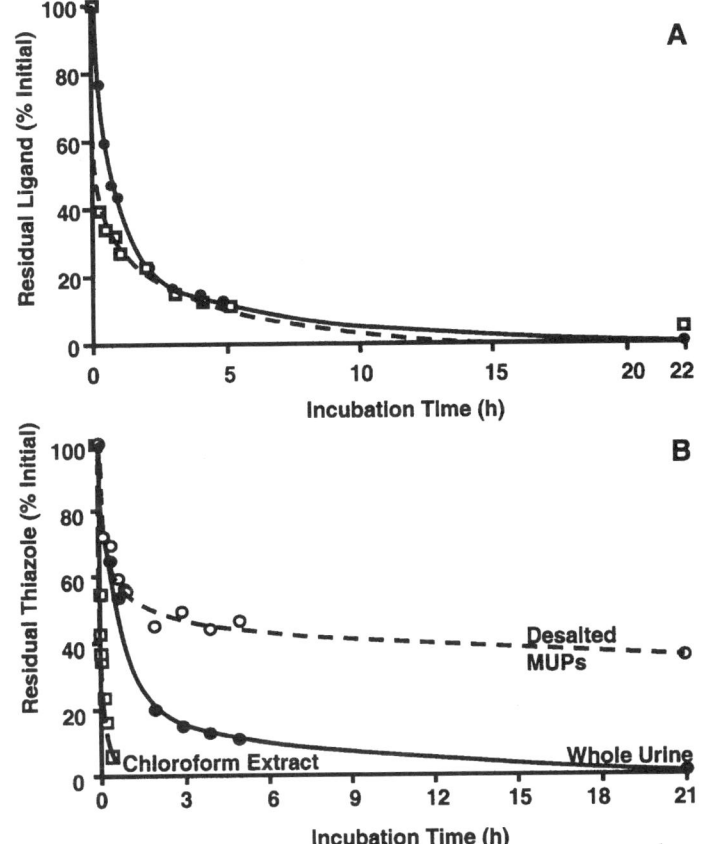

Figure 2. A. Loss of thiazole ●— and brevicomin ···□··· from whole urine. Aliquots (30μl) of whole BALB/C urine were deposited onto circular glass fibre filters, which were incubated at room temperature for 0h, 15, 30, 45mins, 1, 2, 3, 4, 5 and 21h. At the end of the allotted time, the filters were analysed by GC/MS with headspace sampling. The amount of residual ligand was analysed by bi-exponential non-linear least squares curve fitting. B. Loss of free and MUP associated thiazole. Aliquots (30μl) of whole urine ●—, desalted MUPs, ···○···· and chloroform extract of urine (free thiazole) —□—, were analysed using the above method.

One possible interpretation of these results is that the fast pool represents loss of free ligands from the urine and the slow pool represents loss of MUP associated ligands. To test this hypothesis, we measured the rate of free thiazole loss from a chloroform extract of whole urine and the rate of MUP associated thiazole loss from desalted urine (Figure 2). The free ligands were lost very rapidly from the GF/C membrane ($k = 8.8h^{-1}$) in a monophasic manner. The loss of thiazole from desalted MUPs was more complex. In this instance, the loss followed the same biphasic pattern as whole urine, with the rates of the two pools largely unchanged ($k_f = 2.1 \pm 0.4h^{-1}$ and $k_s = 0.02 \pm 0.006h^{-1}$). The proportion of the two pools however, was significantly different ($P_f = 49.7\%$ and $P_s = 48.5\%$). The MUP-ligand complex had probably dissociated between desalting and deposition, as a consequence of the removal of the free thiazole, forming a small amount of free thiazole, which was lost rapidly from the sample. The large amount of residual thiazole associated with desalted MUPs after 21 hours may reflect MUP species with particularly tight binding.

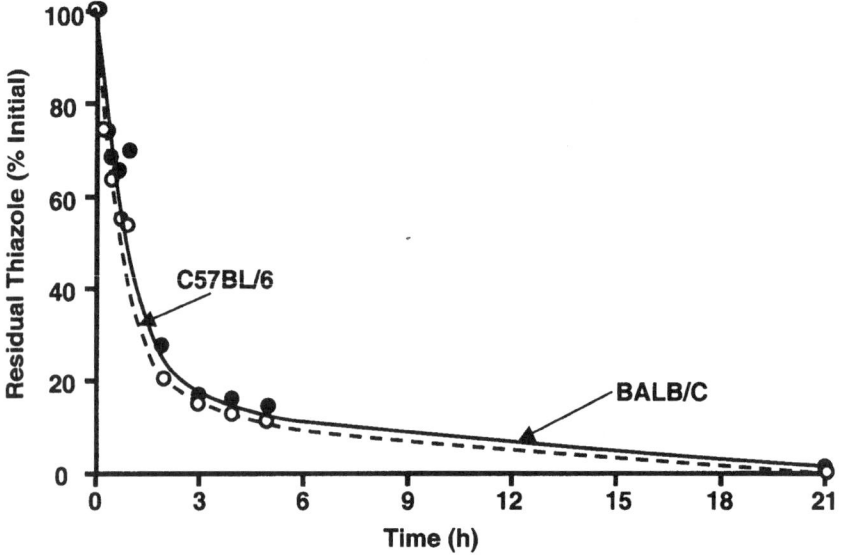

Figure 3. Loss of urinary thiazole from two different strains of mouse. Urine was collected from group housed BALB/C and C57BL/6 mice. Aliquots (30μl) of each were deposited onto glass fibre membranes and allowed to stand for 0h, 15, 30 and 45mins, 1, 2, 3, 4, 5 and 21h. The amount of residual thiazole was then analysed by GC/MS.

The thiazole loss characteristics of urine from C57BL/6 and BALB/C mice were very similar (Figure. 3). Each strain contains a complex mixture of MUPs. In wild mice, the heterogeneity is much greater than in inbred animals (Robertson et al., 1997, Pes et al., 1999). Variation in wild mice is observed in both the type of MUPs present and the relative amounts of common MUPs in individual urines. In a related study, Veggerby et al., (this volume) have identified MUPs in wild mice with mass and charge characteristics not hitherto seen in inbred strains. We have also tentatively identified a second variation at position 56. Urine from individual wild mice could thus reflect even greater variability in its characteristics of ligand release.

5. ACKNOWLEDGMENTS

This work is supported by the BBSRC and the Skeath-Hughes travel fund (University of Liverpool).

6. REFERENCES

Bacchini, A., Gaetani, E., and Cavaggioni, A., 1992, Pheromone binding proteins of the mouse *Mus musculus*, *Experentia*, **48**:419-421.

Bishop, J. O., Clark, A. J., Clissold, P. M., Hainey, S., and Francke, U., 1982, Two main groups of mouse urinary protein genes both largely located on chromosome 4, *EMBO J.* **1**:615-620.

Böcskei, Zs., Groom, C. R., Flower, D. R., Wright, C. E., Phillips, S. E. V., Cavaggioni, A., Findlay, J. B. C., and North, A. C. T., 1992, Ligand binding of two rodent pheromone transport proteins revealed by X-ray crystallography, *Nature* **360**:186-188.

Brennan, P. A., Schellinck, H. M., and Keverne, E. B., 1999, Patterns of expression of the immediate-early gene egr-1 in the accessory olfactory bulb of female mice exposed to pheromonal constituents of male urine, *Neuroscience* **90**:1463-1470.

Epps, D. E., Raub, T. J., and Kedzy, F. J., 1995, A general, wide range spectrofluorimetric method for measuring site specific affinities of drugs toward human serum albumin, *Anal. Biochem.* **227**:342-349.

Finlayson, J. S., and Baumann, C. A., 1958, Mouse proteinuria, *Amer. J. Physiol.* **192**:69-72.

Flower, D. R., 1994, The lipocalin protein family: a role in cell regulation, *FEBS letters* **345**:7-11.

Flower, D. R., North, A. C. T., and Attwood, T. K., 1993, Structure and sequence relationships in the lipocalins and related proteins, *Protein Science* **2**:753-761.

Humphries, R. E., Robertson, D. H. L., Beynon, R. J., and Hurst, J. L., 1999, Unravelling the chemical basis of competitive scent marking in house mice, *Anim. Behav.* **58**:1177-1190.

Hurst, J. L., 1987, The functions of urine marking in a free living population of house mice, *Anim. Behav.* **35**:1422-1433.

Hurst, J. L., Robertson, D. H. L., Tolladay, U., and Beynon, R. J., 1998, Lipocalins in mouse urine provide a slow release mechanism for olfactory signals, *Anim. Behav.* **55**:12891297.

Jemiolo, B., Alberts, S., Sochinski-Wiggins, S., Harvey, S., and Novotny, M., 1985, Behavioural and endocrine responses of female mice to synthetic analogs of volatile compounds in male urine, *Anim. Behav.* **33**:1114-1118.

Jemiolo, B., Harvey, S., and Novotny, M., 1986, Promotion of the Whitten effect in female mouse by synthetic analogs of male urinary constituents, *Proc. Nat. Acad. Sci. USA*, **83**:4576-4579.

Mucignat-Caretta, C., Caretta, A., and Cavaggioni, A., 1995, Acceleration of puberty onset in female mice by male urinary proteins, *J. Physiol.* **486**:517-522.

Novotny, M. V., Jemiolo, B., Wiesler, D., Ma, W., Harvey, S., Xu, F., Xie, T. M., and Carmack, M., 1999a, A unique urinary constituent, 6-hydroxy-6-methyl-c-heptanone, is a pheromone that accelerates puberty in female mice, *Chem. Biol.* **6**:377-383.

Novotny, M. V., Ma, W., Wiesler, D., and Zidek, L., 1999b, Positive identification of the puberty-accelerating pheromone of the house mouse: the volatile ligands associating with the major urinary protein, *Proc R. Soc. Lond. B Biol. Sci.* **266**:2017-2022.

Pes, D., Robertson, D. H. L., Hurst, J. L., Gaskell, S., and Beynon, R. J., 1999, How many major urinary proteins are produced by the house mouse *Mus domesticus?* in: *Advances in Chemical Signals in Vertebrates* (R. E. Johnston, D. Müller-Schwarze, and P. W. Sorensen eds.), Plenum Press, New York, pp. 149-161.

Robertson, D. H. L., Beynon, R. J., and Evershed, R. P., 1993, Extraction, characterisation and binding analysis of two pheromonally active ligands associated with major urinary protein of the house mouse, *J. Chem. Ecol.* **19**:1405:1415.

Robertson, D. H. L., Evershed, R. P., Cox, K., Gaskell, S., and Beynon, R. J., 1995, Characterisation of molecular heterogeneity in the major urinary proteins of the house mouse *Mus musculus*. *Biochem. J.* **316**:265-272.

Robertson, D. H. L., Hurst, J. L., Bolgar, M. S., Gaskell, S. J., and Beynon, R. J., 1997, Molecular Heterogeneity of Urinary Proteins in Wild House Mouse Populations. *Rap. Comm. Mass Spec.* **11**:786-790.

Zidek, L., Novotny, M. V., and Stone, M. J., 1999a, Increased protein backbone conformational entropy upon hydrophobic ligand binding, *Nat. Struct. Biol.*, **6**:1118-1124.

Zidek, L., Stone, M. J., Lato, S. M., Pagel, M. D., Miao, Z., Ellington, A. D., and Novotny, M. V., 1999b, NMR mapping of the recombinant mouse major urinary protein I binding site occupied by the pheromone 2-sec-butyl-4,5-dihydrothiazole, *Biochemistry*, **38**:9850-9861.

CHEMICAL COMMUNICATION IN THE PIG

Dietrich Loebel,[1] Andrea Scaloni,[2,3] Sara Paolini,[4] Silvana Marchese,[5] Carlo Fini,[4] Lino Ferrara,[3] Heinz Breer,[1] and Paolo Pelosi[5]

[1] Institut für Physiologie, University of Hohenheim
Garbenstrasse, 30, 70599 Stuttgart, Germany
[2] Centro Internazionale Servizi di Spettrometria di Massa
National Research Council
via Pansini 5, 80131 Napoli, Italy
[3] IABBAM., Natl. Research Council, via Argine 1085, 80100 Napoli, Italy
[4] Dipartimento di Medicina Interna e Sezione INFM, University of Perugia
via del Giochetto, 6, 06126 Perugia, Italy
[5] Dipartimento di Chimica e Biotecnologie Agrarie, University of Pisa
via S. Michele, 4, 56124 Pisa, Italy

INTRODUCTION

Chemical communication in mammals often utilises soluble binding proteins, both to deliver the specific volatile pheromones in the environment and to detect them (Pelosi, 1994, 1996). These proteins belong to the large family of lipocalins, polypeptides of 150-200 residues, sharing a compact three-dimensional structure and a carrier function for hydrophobic ligands in aqueous biological fluids (Flower, 1995).

Sexual communication has been best studied in mice and rats. It has long been known that sexually mature males of both species excrete in their urine exceptionally large amounts of protein, up to several mg per ml of urine. These proteins, called major urinary proteins (MUPs) in house mice and α2-u in Norway rats, reversibly bind a large number of compounds (Cavaggioni *et al.*, 1990), including volatile pheromones in the urine (Novotny *et al.*, 1985; Jemiolo *et al.*, 1992). In house mice, it is particularly interesting that the volatile pheromones were found tightly bound to MUPs, even after several purification steps (Bacchini *et al.*, 1992). Proteins nearly identical to the urinary lipocalins in amino acid sequences are expressed by glands of the nasal mucosa in both sexes of diverse species (Pelosi, 1994, 1996; Pes and Pelosi, 1995; Utsumi *et al.*, 1999). Traditionally, they have been labelled as odorant-binding proteins (OBPs) and represent an intranasal subtype of lipocalins that has the ability to bind reversibly with volatile odorants. The expression of nasal OBPs is not under hormonal control nor is it sex specific, in contrast to MUPs, but in some cases, an interesting temporal pattern has been reported. In Norway rats, for instance, the synthesis of one of the two lipocalins of the vomeronasal organ increases soon after birth and again at puberty; a pattern that has been related to the importance of chemical communication during these two periods of life, between mother and newborn and between sexes, respectively (Miyawaki *et al.*, 1997).

Another protein-rich fluid is the saliva, both in rats and mice. These lipocalins in saliva have been named „urinary" by analogy to those proteins found in urine, and the salivary lipocalins likely perform the same carrier function of MUPs (Shaw *et al.*, 1983; Shahan *et al.*, 1987). The role of saliva in sexual communication among mice has been demonstrated (Marchlewska-Koj *et al.*, 1990).

In golden hamster, a lipocalin with structural similarities to MUPs and OBPs has been labelled aphrodisin, and is a component of hamster vaginal discharge. A search for endogenous ligands for these proteins has not been successful, though a pheromonal role of the polypeptide itself was suggested (Singer *et al.*, 1986; Henzel *et al.*, 1988).

Information on the pheromone-binding proteins of mammals is presently limited to few species, although it seems reasonable that similar mechanisms might be more widespread. We have focused our attention on pigs, whose steroidal sex pheromones, 5α-androst-16-en-3-one and 5α-androst-16-en-3-ol had been identified long ago (Katkov *et al.*, 1972).

PHEROMONE-BINDING PROTEINS IN THE PIG

In domestic pigs, sex pheromones are produced by males and secreted into the saliva. The pattern of soluble proteins synthesised in the three salivary glands has revealed a striking sex specificity only in the submaxillary gland, while no major differences were associated with the sublingual or parotid glands. The boar's submaxillary glands are much larger than those of the sow, and express an abundant protein of about 20 kDa, that is completely absent in females, prepubertal young, and castrated males (Marchese *et al.*, 1998). This lipocalin protein, which we have labelled SAL, is significantly different than mouse MUPs (Figure 1), but there are similarities. As reported for MUPs, porcine SAL has been found to bind endogenous ligands, namely, both of the porcine steroidal pheromones, 5α-androst-16-en-3-one and 5α-androst-16-en-3-ol. These compounds were tightly bound to the protein even after several purification steps (Marchese *et al.*, 1998).

Two isoforms of SAL are expressed in the same individual, differing by these three amino acid substitutions: Val45→Ala, Ile48→Val, Ala73→Val. Interestingly, a three-dimensional model, built on the similarity of this protein with other lipocalins of known structure, has indicated that two of these substitutions are in residues located inside the binding pocket. This observation might suggest the hypothesis that the two isoforms of SAL could each bind a different component of the pheromone mixture (Loebel *et al.*, 2000).

Both isoforms are N-glycosilated at Asn53. The glycosilation pattern consists of chains containing up to four N-acetyl-glucosamine residues bound to a fucosylated pentasaccharide core linked to Asn53.

MALDI-TOF mass spectrometry, performed on the peptide mixture after controlled enzymatic hydrolysis, revealed that only two of the four cysteines (Cys68 and Cys160) are linked by a disulphide bridge, while the others (Cys75 and Cys141) are present in the reduced form.

pig SAL	HKEAGQDVVTSNFDASKIAGEWYSILLASDAKENIEENGSMRVFV	45
mus MUP6	EEASSTGRNFNVEKINGEWHTIILASDKREKIEDNGNFRLFL	42
rat α2-uG	EEASSTRGNLDVAKLNGDWFSIVVASNKREKIEENGSMRVFM	42
pig OBP-I	QEPQPEQDPFELSGKWITSYIGSSDLEKIGENAPFQVFM	39
	* * * * * ** * ** ** * *	

pig SAL	-QRKVNGECTDFYAVCDKVG-DGVYTVAYYEHIRVLDNSSLAFKF	88
mus MUP6	-HTVRDEECSELSMVADKTEKAGEYSVTYDEQIHVLEN-SLVLKF	85
rat α2-uG	-RIKENGECRELYLVAYKTPEDGEYFVEYDQHIDVLEN-SLGFKF	85
pig OBP-I	FFSKENGICEEFSLIGTKQ-EGNTYDVNYARSIEFDDKESKVYLN	83
	** * * * * * * * * ** ** **	

pig SAL	GENKFRLLEVNYSDYVILHLVNVNGDKTFQLMEFYGRKPDVEPKL	133
mus MUP6	GFNTFTIPKTDYDNFLMAHLINEKDGETFQLMGLYGREPDLMSDI	130
rat α2-uG	GGNTFTILKTDYDRYVMFHLINFKNGETFQLMVLYGRTKDLSSDI	130
pig OBP-I	GNNKFVVSYASETALIISNINVDEEGDKTIMTGLLGKGTDIEDQD	128
	* * * * ** * ***** *** *	

pig SAL	KDKFVEICQQYGIIKENIIDLTKIDRCFQLRGSGGVQESSAE	175
mus MUP6	KERFAQLCEEHGILRENIIDLSNANRCLQARE	162
rat α2-uG	KEKFAKLCEAHGITRDNIIDLTKTDRCLQARG	162
pig OBP-I	LEKFKEVTRENGIPEENIVNIIERDDCPA	157
	* * * ** ** ** ** *	

Figure 1. Amino acid sequence of SAL compared with urinary protein of mouse (*mus* MUP6) and rat (*rat* α2-uG) and with pig OBP-I. Residues identical with urinary proteins are indicate by asterisks. Asn53 is the only glycosilation site. Cysteines 68 and 160 are connected by a disulphide bridge, while the other two cysteines (75 and 141) are present in the native protein in their reduced form. A second isoform of SAL contains three substitutions (Val45→Ala, Ile48→Val and Ala73→Val).

Ligand-binding experiments, performed with both the radioactive odorant 2-isobutyl-3-methoxypyrazine and the fluorescent probe 1-aminoanthracene, indicated that porcine SAL binds a variety of volatile molecules, with dissociation constants in the micromolar range. The native steroidal pheromones, however, are better ligands than the synthetic ligands by at least one order of magnitude (Loebel et al., 2000).

One of the isoforms of SAL has been expressed in high yield in a bacterial system. CD spectra have shown that the recombinant protein is folded as the native SAL, while MALDI-TOF mass spectrometry confirmed the same redox state of the four cysteines. The recombinant protein also binds the same ligands as the native one with similar binding constants, thus indicating that the glycane moiety in the native protein is not required for its structural or for binding properties (Loebel et al., 2000).

Electrophoretic analysis of crude extracts of olfactory, nasal respiratory, and vomeronasal tissues of pigs, followed by Western blot, has revealed bands of about 29 kDa which cross-react with the antiserum against SAL. These bands, present only in the respiratory sample of both sexes, have been purified by immunoaffinity chromatography. The N-terminal sequence is identical to that of SAL. Cloning and sequencing of the corresponding cDNA has revealed the presence of isoforms identical to those of SAL. The major difference between these proteins, which we have classified as OBP-III, and SAL lies in the much larger size of the N-linked glycan region at Asn53, accounting for its higher molecular weight.

The affinity and specificity of OBP-III, purified as a mixture of isoforms, was measured with the fluorescent ligand 1-aminoanthracene in competitive binding assays with a series of odorants, as well as with the steroidal pheromones. The results are very similar to those obtained with the salivary lipocalin SAL, confirming once again the negligible effect of the glycan region in the function of these proteins.

CONCLUSIONS

Chemical communication in the pig utilises lipocalins that bind and solubilise the steroidal sex pheromones. In the salivary glands, these proteins are involved in delivering volatile, hydrophobic, pheromones, while in the nose these same polypeptides capture those volatiles and ferry them to specific membrane-bound receptors. It has been recently shown in mice that volatile pheromones are able to stimulate the neurons of the vomeronasal organ at extremely low concentrations and with very high specificity (Leinders-Zufall *et al.*, 2000). In addition, there is strong evidence that pheromone-binding proteins, including MUPs, are endowed with pheromonal activity, being able, even when devoid of their ligands, to accelerate sexual maturation in young female mice (Mucignat-Caretta *et al.*, 1995). This hypothesis is supported by the discovery of two alternative pathways of signal transduction in the vomeronasal organs, using different types of G-proteins. The first pathway is activated by volatile pheromones, the second by lipocalins, such as the urinary proteins (Krieger *et al.*, 1999).

Two classes of membrane-bound receptors establish the links between the two classes of pheromones (volatiles and proteins) and their corresponding transduction pathways. Receptor of class V1R (Dulac and Axel, 1995) are structurally similar to olfactory receptors and could be the targets of volatile pheromones, whose chemical nature is not different from that of common odorants. V2R receptor, instead, contain, in addition to the seven transmembrane domains of V1R and olfactory receptors, a large extracellular region of about 600 amino acids (Herrada and Dulac, 1997; Matsunami and Buck, 1997; Ryba and Tirindelli, 1997). This part of the molecule has been suggested to represent a binding site for polypeptide ligands, such as urinary proteins and OBPs. This model will receive a more definitive test when reversible interactions between pheromone-binding proteins and the extracellular domain of V2R receptor can be measured.

The above system, using two classes of pheromones (volatile compounds and proteins) and two classes of vomeronasal receptors (V1R and V2R) could be active also in other species of mammals. We are further investigating the presence of these biochemical and physiological mechanisms in pigs.

ACKNOWLEDGMENTS

This work was supported by EEC grant no. BIO4-CT98-0420 to H.B. and P.P., by the Deutsche Forsuchgemeinschaft and the BMBF to H.B., by Italian National Research Council grants to L.F. and A.S., by MURST Ricerca Scientifica 40% grant to P.P. and by EEC Training and Mobility Program "Access to Large-Scale Facilities" grant no. ERB FMGECT95 0061 to A.S.

REFERENCES

Bacchini, A., Gaetani, E., and Cavaggioni, A., 1992, Pheromone-binding proteins in the mouse *mus musculus*, *Experientia* **48**:419-421.

Cavaggioni, A., Findlay, J. B. C., and Tirindelli, R., 1990, Ligand binding characteristics of homologous rat and mouse urinary proteins and pyrazine binding protein of the calf, *Comp. Biochem. Physiol.* **96B**:513-520.

Dulac, C., and Axel, R., 1995, A novel family of genes encoding putative pheromone receptors in mammals, *Cell* **83**:195-206.

Flower, D. R., 1995, Multiple molecular recognition properties of the lipocalin protein family, *J. Mol. Recognit.* **8**:185-195.

Henzel, W. J., Rodriguez, H., Singer, A. G., Stults, J. T., Macrides, F., Agosta, W. C., and Niall, H., 1988, The primary structure of aphrodisin, *J. Biol. Chem.* **263**:16682-16687.

Herrada, G., and Dulac, C., 1997, A novel family of putative pheromone receptors in mammals with a topographically organized and sexually dimorphic distribution, *Cell* **90**:763-773.

Jemiolo, B., Xie, T. M., and Novotny, M., 1992, Urine marking in male mice: responses to natural and synthetic chemosignals, *Physiol. Behav.* **52**:521-526.

Katkov, T., Booth, W. D., and Gower, D. B., 1972, The metabolism of 16-androstenes in boar salivary glands, *Biochim. Biophys. Acta* **270**:546-56.

Krieger, J., Schmitt, A., Lobel, D., Gudermann, T., Schultz, G., Breer, H., and Boekhoff, I., 1999, Selective activation of G protein subtypes in the vomeronasal organ upon stimulation with urine-derived compounds, *J. Biol. Chem.* **274**:4655-4662.

Leinders-Zufall, T., Lane, A. P., Puche, A. C., Ma, W., Novotny, M. V., Shipley, M. T., and Zufall, F., 2000, Ultrasensitive pheromone detection by mammalian vomeronasal neurons, *Nature* **405**:792-796.

Loebel, D., Scaloni, A., Paolini, S., Fini, C., Ferrara, L., Breer, H., and Pelosi, P., 2000, Cloning, post-translational modifications, heterologous expression and ligand-binding of boar salivary lipocalin, *Biochem. J.*, (in press).

Marchese, S., Pes, D., Scaloni, A., Carbone, V., and Pelosi, P., 1998, Lipocalins of boar salivary glands binding odours and pheromones, *Eur. J. Biochem.* **252**:563-568.

Marchlewska-Koj, A., Pochron, E., and Sliwowska, A., 1990, Salivary glands and preputial glands of males as source of estrus-stimulating pheromone in female mice, *J. Chem. Ecol.* **16**:2817-2822.

Matsunami, H., and Buck, L. B., 1997, A multigene family encoding a diverse array of putative pheromone receptors in mammals, *Cell* **90**:775-784.

Miyawaki, A., Matsushita, F., Ryo, Y., and Mikoshiba, K., 1997, Possible pheromone-carrier function of two lipocalin proteins in the vomeronasal organ, *EMBO J.* **13**:5835-5842.

Mucignat-Caretta, C., Caretta, A., and Cavaggioni, A., 1995, Acceleration of puberty onset in female mice by male urinary proteins, *J. Physiol.* **486**:517-522.

Novotny, M., Harvey, S., Jemiolo, B., Alberts, J., 1985, Synthetic pheromones that promote inter-male aggression in mice, *Proc. Natl. Acad. Sci. U. S. A.* **82**:2059-2061.

Pelosi, P., 1994, Odorant-binding proteins, *Crit. Rev. Biochem. Molec. Biol.* **29**:199-228.

Pelosi, P., 1996, Perireceptor events in olfaction, *J. Neurobiol.* **30**:3-19.

Pes, D., and Pelosi, P., 1995, Odorant-binding proteins of the mouse, *Comp. Biochem. Physiol.*, **112B**:471-479.

Ryba, N. J., and Tirindelli, R., 1997, A new multigene family of putative pheromone receptors, *Neuron* **19**:371-379.

Shaw, P. H., Held, W. A., and Hastie, N. D., 1983, The gene family for major urinary proteins:expression in several secretory tissues of the mouse, *Cell* **32**:755-761.

Shahan, K. M., Denaro, M., Gilmartin, M., Shi, Y., and Derman, E., 1987, Expression of six mouse major urinary protein genes in the mammary parotid sublingual submaxillary and lachrymal glands and in the liver, *Mol. Cell. Biol.* **7**:1947-1954.

Singer, A. G., Macrides, F., Clancy, A. N., and Agosta, W. C., 1986, Purification and analysis of a proteinaceous aphrodisiac pheromone from hamster vaginal discharge, *J. Biol. Chem.* **261**:13323-13326.

Utsumi, M., Ohno, K., Kawasaki, Y., Tamura, M., Kubo, T., and Tohyama, M., 1999, Expression of major urinary protein genes in the nasal glands associated with general olfaction, *J. Neurobiol.* **39**:227-236.

INTER-FETAL COMMUNICATION AND ADULT PHENOTYPE IN MICE

John G. Vandenbergh and Andrew K. Hotchkiss

Department of Zoology
North Carolina State University
Raleigh, NC 27695-7617

INTRODUCTION

Chemical signals can influence mammalian reproduction at several levels, ranging from molecular signals within the cell derived from the genome to pheromones being communicated among individuals. Among adult mammals, chemical signals can serve as conspecific attractants to bring mates into proximity (Johnston, 1983) or result in regulation of the onset of puberty and adult ovarian cyclicity (Vandenbergh, 1983; 1994). We have become accustomed to calling such signals pheromones when they are emitted by one organism, transmitted through the environment, and have a behavioral or physiological effect on another organism. Here, we propose to include the chemical agents that are transmitted from one fetus to another in the intrauterine environment as priming pheromones (Vandenbergh, 1991).

The phenomenon of inter-fetal communication grew out of the early work by vom Saal and Bronson (1978) and Clemens *et al.* (1978) showing that, in mice and rats respectively, neighboring males have a masculinizing effect on their sisters. This effect is due to the transfer of androgens to adjacent fetuses either through amniotic fluid and membranes (vom Saal, 1989; vom Saal and Bronson, 1980) or through the maternal uterine vasculature (Meisel and Ward, 1981; vom Saal and Dhar, 1992). In mice, for example, males begin to produce testosterone in significant quantities about day 12 of gestation. Testosterone is then available to masculinize adjacent female siblings. Differences due to the proximity of males to females during development can alter the anatomy, physiology and behavior of females. These alterations are organizational in nature and have long-term effects, often extending into adulthood. A recent review of this phenomenon is available (vom Saal *et al.*, 1999).

Determination of the position of males relative to females in utero has been done by performing Caesarian sections on term females. More recently, procedures have been developed to utilize the anogenital distance index to detect the prior intrauterine position of the female mouse (Vandenbergh and Huggett, 1995). Measuring the distance from the anus to the genital papilla in either newborn or weanling female mice and adjusting the length for the body weight yields an index which is statistically representative of the females prior position relative to males in utero. We will show later in this paper how intrauterine position can influence a number of physiological events which can be predicted by the anogenital distance index (AGDI).

The purpose of this paper is to expand our concept of priming pheromones affecting reproduction. There is ample evidence of priming pheromones being important during adulthood (Vandenbergh, 1983; 1994) and juveniles (growth in voles, maturation) but now we must consider that these priming effects begin in utero. While such priming effects require us to consider hormones, such as testosterone, as a pheromone we think they do fit the classical definition for a priming pheromone (Bronson, 1968). That is, a chemical substance that is emitted by one organism, transmitted through the environment and has a developmental or physiological effect on another organism of the same species. The environment we will discuss here is that found inside the uterus.

ANATOMY

The initial discovery that the intrauterine position (IUP) could influence adult phenotype was an anatomical observation that the ano-genital distance of females was variable and that the variability was due to the intrauterine position of the female fetus in relation to males (Clemens *et al.*, 1978). Among the reproductive-related anatomical changes are relatively larger sexually dimorphic nuclei in the preoptic area of the hypothalamus of female rats if they are adjacent to male siblings in utero (Faber and Hughes, 1992). Reproductive organs have been influenced by the presence of adjacent siblings in the male gerbil (Clark *et al.*, 1993). Here we will focus on effects on the uterine weight of juvenile female mice. Females with different AGDI's have similar body weight but significantly different uterine weights at thirty days of age after receiving estradiol benzoate injections. The uterine mass is approximately twenty-six percent heavier in females with short AGDI's representing an intrauterine position without adjacent males. It is not known whether this difference is due to a difference in adult endocrine levels or to differences in the number of hormone receptors. It is quite likely the difference is due to receptors because there do not seem to be significant differences in adult hormone levels. Research is under way in our laboratory to identify differences in receptor content of the adult uterus.

In addition to the reproductive organ weights, body mass and fat composition have also shown to be different between females with males and without males adjacent in the uterus. Body mass and percent fat content were measured in mice and shown to be significantly different with male-exposed females being leaner. Females with males on both sides have only about six percent body fat compared to sixteen percent body fat in the females without males adjacent (Table 1).

Table 1. Body mass and fat composition of adult CD-1 female mice from an intrauterine position without males adjacent (0M) and with males on both sides (2M).

	0M		2M	
	Mean	±S.E.	Mean	±S.E.
Body Mass (g)	37.5	0.99	33.1	0.94
% Fat	16.0	1.40	5.9*	0.72
* p = <0.01				

PHYSIOLOGY

An array of physiological events have been shown to be influenced by intrauterine position. Puberty onset is later and estrous cycles are longer and more irregular among 2M females. We monitored the estrous cycles of twelve month old female mice with known AGDI's and found that more than twice as many of the females exposed to low androgen levels in utero were continuing to cycle compared to females that were exposed to males in utero. It thus seems that the irregularity and cessation of ovarian cycles in older laboratory mice is related to their prior intrauterine position.

One of the more interesting physiological findings is that intrauterine position can result in the alterations of the sex ratio. This was first discovered by Clark *et al.* (1993) in the gerbil and confirmed in the house mouse (Vandenbergh and Huggett, 1994). In both these studies females in utero adjacent to males produce approximately sixty percent male pups upon reaching adulthood whereas females that are in utero without adjacent male pups produce approximately forty percent male pups in their litters. The intermediate form, females with a male adjacent on one side, produce the expected fifty percent sex ratio. Thus, in a population the overall sex ratio remains approximately fifty percent. However, the variability among litters can be at least partly ascribed to prior intrauterine exposure to an adjacent male. We have recently begun to explore the mechanism of this sex ratio alteration and have found that testosterone, at a dose of 0.5 or 0.75 milligrams per kilogram body weight, injected into females from day 11 to day 17 of gestation produced female pups that, upon adulthood produced male biased litters (Konzelmann, Hotchkiss and Vandenbergh, submitted). The finding that the hormone testosterone either naturally from adjacent males or artificially through injection of the mother can alter subsequent sex ratio may have important implications for understanding population dynamics in rapidly breeding rodents. Sex ratio alterations have been found in a number of species (Clutton-Brock and Iason, 1986; Krakow, 1995) but there is little experimental evidence providing a mechanism for such alteration. It is possible that pregnant females could influence the sex ratio of their grand offspring if they were either exposed to androgenic compounds or produced it naturally, perhaps from the adrenal gland.

RELEVANCE TO HUMANS

The finding in rodents (vom Saal *et al.*, 1999) that exposure to hormones in utero can have profound and lasting effects stimulates the question whether such effects can occur in a species that typically has a single offspring, such as the human. Two approaches to this question are possible. We can examine human male/female twins or we can study humans that were in utero when the mother had high levels of testosterone. Both of these conditions were studied by Berenbaum (1998) and I will briefly review her findings here. The frequency of twin births in humans in western nations is becoming greater due to assisted reproduction. An outcome of this is the availability of higher numbers of heterozygous twins for study. Such studies have begun to reveal interesting cross-twin effects. Characteristics that have long been associated with masculinity such as spatial ability and sensation seeking are both higher among females who had been in utero with a male co-twin. Another characteristic, which is less well known as a masculine trait, is that spontaneous otic emissions in the inner ear occur at a lower rate in males than in females. Females that were in utero with a male co-twin have an intermediate rate of emissions. While neither of these characteristics relate directly to reproduction they are all associated with masculine rather than feminine characteristics.

Another opportunity to study the influence of prenatal exposure on human females occurs as a result of congenital adrenal hyperplasia (CAH). In this condition the adrenal glands of the female produce high levels of androgens during pregnancy. These androgens can cross the placenta and, in a series of studies, have been found to have significant effects on the female offspring. Spatial ability is influenced in a masculine direction and interestingly, the selection of toys is diverted to a more masculine direction in offspring of CAH mothers. There is evidence from studies of CAH daughters that sexual orientation is more directed in a masculine direction and fantasies have also been reported to be more related to lesbian or bisexual lifestyles.

While these observations are relatively preliminary there is a good deal of consistency from study to study suggesting that testosterone from a male sibling in utero or from a maternal source can have influences on human behavioral outcomes. The finding that hormones can act as priming pheromones in humans as well as in animals opens the unsettling possibility that endocrine disruptors in the environment may serve a similar role. Several compounds have been identified as estrogenic and anti-androgenic. To date there is only a small amount of information suggesting that endocrine disrupters exist with androgenic activity (Parks *et al.*, 2000; Bauer *et al.*, 2000; Schiffer *et al.*, 2000). This area of inquiry awaits investigation.

ACKNOWLEDGMENTS

We thank the North Carolina Agricultural Research Service (#NC 95087) and the W. M. Keck Foundation for financial support.

REFERENCES

Bauer, E. R. S., Daxenberger, A., Petri, T., Sauerwein, H., and Meyer, H. H .D., 2000, Characterisation of the affinity of different anabolics and synthetic hormones to human sex hormone binding globulin and to the bovine gestagen receptor *Workshop, Hormones and Endocrine Disrupters in Food and Water*, Copenhagen, (abstract).

Berenbaum, S. A., 1998, How hormones affect behavioral and neural development: Introduction to the special issue on "Gonadal Hormones and Sex Differences in Behavior", *Dev. Neuropsych.* **14**(2/3), 175-196.

Bronson, F. H., 1968, Pheromonal influences on mammalian reproduction. in: *Reproduction and Sexual Behavior*, (M. Diamond, ed.) Indiana Univ. Press, Bloomington, pp. 344-365.

Clark, M. M., Karpiuk, P., and Galef, Jr., B. G., 1993, Hormonally mediated inheritance of acquired characteristics in Mongolian gerbils, *Nature* **364**:712.

Clemens, L., Gladue, B., and Coniglio, L., 1978, Prenatal endogenous androgenic influences on masculine sexual behavior and genital morphology in male and female rats, *Horm. Behav.* **10**:40-53.

Clutton-Brock, T. H., and Iason, G. R., 1986, Sex ratio variation in mammals. *Quart. Rev. Biol.* **61**:339-374.

Faber, K. A., and Hughes, C. L., 1992, The effect of neonatal exposure to diethylstilbestrol, genistein and zearalenone on pituitary responsiveness and sexually dimorphic nucleus volume in the castrated adult rat, *Biol. Reprod.* **45**:649-653.

Johnston, R. E., 1983, Chemical signals and reproductive behavior, in: *Pheromones and Reproduction in Mammals*, (J. G. Vandenbergh, ed.), Academic Press, NY, pp. 3-38.

Krakow, S., 1995, Potential mechanisms for sex ratio adjustment in mammals and birds, *Biol. Rev.* **70**:225-241.

Meisel, R. L., and Ward, I. L., 1981, Fetal female rats are masculinized by male littermates located caudally in the uterus, *Science* **213**:239-242.

Parks, L. G., Lambright, C., Orlando, E., Guillette, L., and Gray, Jr., L. Earl, 2000, In vitro confirmation of androgenic activity in Kraft Mill effluent known to masculinize female fish, in *Soc. Study Reprod.*, (abstract).

Schiffer, B., Daxenberger, A., and Meyer, H. H. D., 2000, Possible transition of the anabolic growth promoter trenbolone into soil via liquid manure and solid dung, in *Workshop, Hormones and Endocrine Disruptors in Food and Water*, (abstract).

Vandenbergh, J. G., ed., 1983, *Pheromones and Reproduction in Mammals*, Academic Press, New York.

Vandenbergh, J. G., 1991, Chemosignal influences on mammalian reproduction, *Verh. Dtsch. Zool. Ges.* **84**:195-199.

Vandenbergh, J. G., 1994, Pheromones and mammalian reproduction, in: *The Physiology Reproduction*, (E. Knobil, and J. Neill, eds.), 2nd ed., Raven Press, New York, pp. 343-362.

Vandenbergh, J. G., and Huggett, C. L., 1994, Mother's prior intrauterine position affects the sex ratio of her offspring in house mice, *Proc. Natl. Acad. Sci. USA* **91**:1155-1159.

Vandenbergh, J. G., and Huggett, C. L., 1995, The anogenital distance index, a predictor of the intrauterine position effects on reproduction in female house mice, *Lab. Anim. Sci.* **45**:567-573.

vom Saal, F. S., 1989, Sexual differentiation in litter bearing mammals: Influence of sex of adjacent fetuses in utero, *J. Anim. Sci.* **67**:1824-1840.

vom Saal, F. S., and Bronson, F., 1978, In utero proximity of female house mouse fetuses to males: effect on reproductive performance during later life. *Biol. Reprod.* **19**:842-853.

vom Saal, F. S., and Bronson, F., 1980, Sexual characteristics of adult female mice are correlated with their blood testosterone levels during prenatal development, *Science* **208**:597-599.

vom Saal, F. S., and Dhar, M. G., 1992, Blood flow in the uterine loop artery and loop vein is bidirectional in the mouse: implications for transport of steroids between fetuses. *Physiol. Behav.* **52**:163-171.

vom Saal, F. S., Clark, M. M., Galef, Jr., B. G., Drickamer, L. C., and Vandenbergh, J. G., 1999, Intrauterine position phenomenon , *Encyclopedia Reprod.* **2**:893-900.

THE ROLE OF THE MAIN AND ACCESSORY OLFACTORY SYSTEMS IN PRENATAL OLFACTION

David M. Coppola

Neuroscience Program
Centenary College
Shreveport, LA 71134

INTRODUCTION

Olfaction, an „early" sensory system both phylogenetically and ontogenetically, is required for survival in the newborn mammal. For example, odor cues orient neonates to their mother, her nipples and the nest; neonatal house mice whose olfactory bulbs are lesioned subsequently undergo fatal malnutrition (Coppola et al., 1994). Altricial rodents, like the mouse and rat, are not only born deaf and blind but are denied vision and hearing for some time after birth.

Almost twenty years ago, developmental scientists began to ask whether olfaction, so necessary immediately after birth, might be functioning prenatally (Pedersen and Blass, 1982; Smotherman, 1982). In the intervening years, we have learned that some mammals detect chemical cues in the womb, and that prenatal olfactory learning subsequently influences postnatal behavior (reviewed by Schaal and Orgeur, 1992).

In this chapter I will neither review the now numerous behavioral studies done by others on prenatal olfaction nor will I give a detailed general account of the anatomical and physiological development of the olfactory system. Both topics are too broad in scope to treat here. I refer the interested reader to excellent reviews by others (Pedersen et al., 1985, Breipohl, 1986, Schaal and Orgeur, 1992). Instead, I will rather myopically review work done by my collaborators and I, with a mention of but a few other investigators whose work has been particularly influential in guiding our thinking. The primary question we have pursued is one of developmental staging. We have asked when the various chemosensory systems become functional and what is their pattern of obtaining adult-like functional competence. In pursuing these questions we have focused only on the main and accessory olfactory systems. Our work, taken together, sheds considerable light on which olfactory system is functioning prenatally in the only two species in which prenatal olfaction has been studied in detail: the house mouse and Norway rat.

VOMERONASAL ORGAN DEVELOPMENT

A paradox of olfactory development sparked our initial interest in the vomeronasal organ (VNO) as a potential mediator of prenatal olfaction. While a functional olfactory system is necessary for survival at birth, the main olfactory system (MOS) undergoes dramatic postnatal development in size and in circuitry (Figure 1). Perhaps an alternate system, such as that including the VNO, mediates pre- and peri-natal olfaction while the MOS matures in this interval. In favor of the idea that the VNO mediates prenatal olfaction is its structure and function which are specialized for detecting chemicals in a liquid environment like the amniotic fluid (Meredith and O'Connell, 1979). More direct evidence of this idea came from a study on fetal rats showing that the metabolic marker 2-deoxyglucose (2-DG) is taken up more actively by the accessory olfactory bulb, the neural target of the VNO, than by the main olfactory bulb (Pederson et al., 1983).

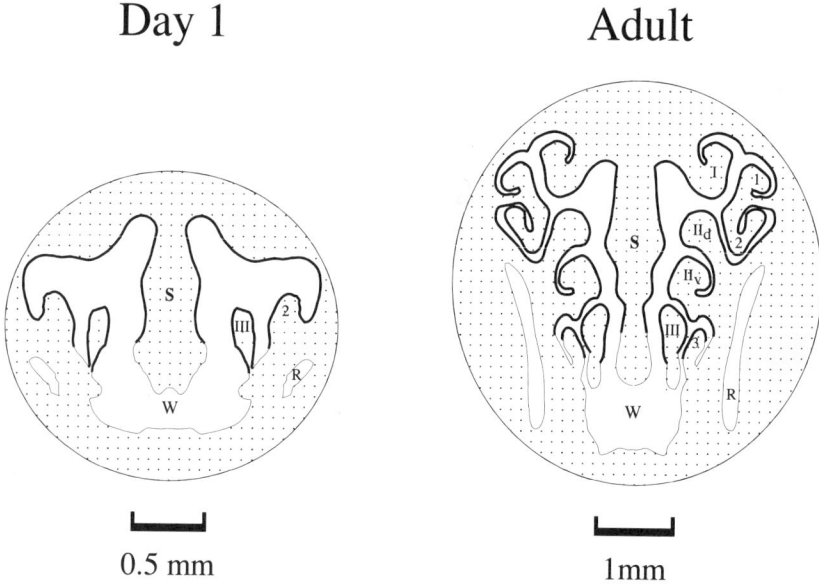

Figure 1. Schematics of coronal sections through the nasal cavity of one-day old and adult mouse. Diagrams were drawn from sections taken at the position of the septal window in the rostrocaudal axis. Only the olfactory cavities are shown. Thick lines represent the olfactory receptor epithelium lining the nasal cavity. R = Maxillary Recess; S = Septum; W = Septal Window. Roman numerals represent the endoturbinates; arabic numerals represent the ectoturbinates; d = dorsal; v = ventral. Note different scales. Adapted with permission from Coppola et al., 1994.

The Mouse

We initially tested the hypothesis that the VNO mediates fetal olfaction in the mouse, the species in which the function of the VNO has been most thoroughly studied (reviewed by Wysocki, 1989). Receptor neurons of the VNO gain access to chemical signals dissolved in the mucous of the nasal cavity by a vascular pumping mechanism. Robert O'Connell and I reasoned if the fetal VNO functions prenatally, it ought to be actively sampling the amniotic fluid with its vascular „pump." To test this we injected fluorescent microspheres or „beads" into the amniotic sacs of late-term mouse fetuses. Despite finding beads throughout the nasal cavity, we never found beads inside the VNO lumen (Coppola and O'Connell, 1989). Subsequent anatomical analysis of the developing mouse nasal cavity explained the lack of beads in the VNO. Not only is the narrow canal which connects the VNO lumen with the nasal cavity not patent on the day of birth, but it undergoes a protracted period of postnatal development (Figure 2; Coppola and O'Connell, 1989; Coppola et al., 1993). Thus, the VNO can not mediate olfaction in the womb.

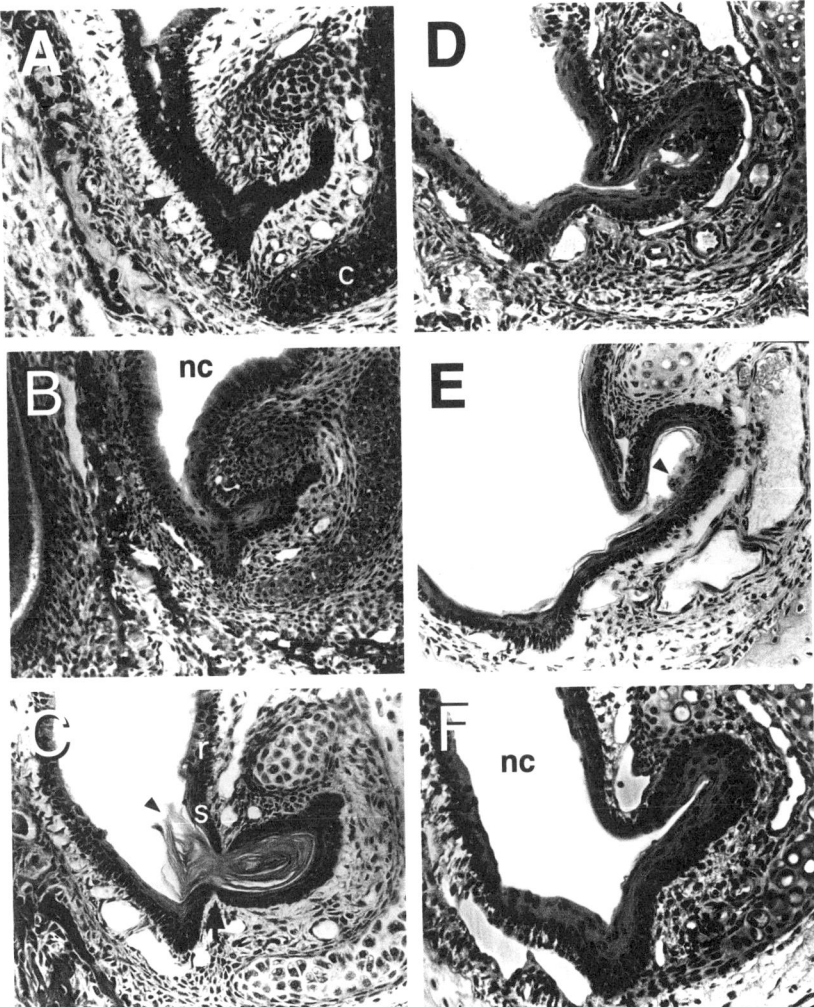

Figure 2. Coronal sections through the base of the mouse nasal cavity in the region of the left vomeronasal duct. For all sections, dorsal is toward top; medial is toward the right. Abbreviations: c = cartilage; nc = nasal cavity; r = respiratory epithelium; s = squamous epithelium. A. The vomeronasal duct at embryonic day 19. Arrow points to the duct precursor. B. The duct at one day after birth. C. The duct at postnatal day 5. Note the layers of sloughed epithelium in the lumen of the canal. D. The duct at postnatal day 10. E. The duct at postnatal day 15. Note the persistence of what appears to be rapidly desquamating tissue inside the duct. Small arrows point to pycnotic nuclei. F. The duct at postnatal day 25. Note lack of cellular debris in the lumen of the duct at this age. Adapted with permission from Coppola et al., 1993.

The Rat

Despite the situation in the mouse, the rat potentially has a different developmental schedule. Leah Millar and I addressed this possibility by examining the VNO canal and vascular pump of the late-term rat fetus. Interestingly, unlike the mouse, the rat canal is patent before birth (Figure 3A). Also, microspheres injected into the amniotic sac can subsequently be found in the VNO lumen, suggesting the organ's vascular pump, which samples chemical cues dissolved in the fluid of the nasal cavity, is active before birth (Figure 3B). Based on these findings, one could conclude, consistent with the 2-DG results of Pedersen *et al.* (1983), that the rat accessory olfactory system is functional in the fetus and could even be the dominant chemosensory system at this stage of development.

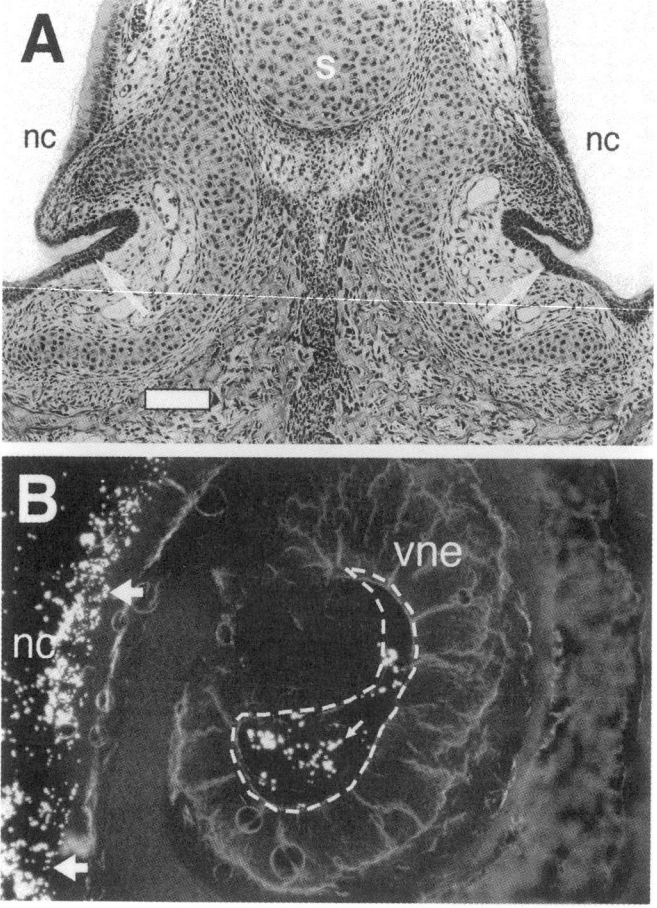

Figure 3. A. Coronal section through the base of the nasal cavity in the embryonic day 20 rat. The already patent vomeronasal ducts can be seen lying mediolaterally (arrows). Abbreviations: nc = nasal cavity; s = nasal septum; scale bar = 100μm. **B.** Coronal section through the right vomeronasal organ (VNO) and adjacent nasal cavity of embryonic day 20 rat. Under fluorescent illumination, fluorescent microparticles, injected 24 hrs before into the amniotic sac, can be seen in the nasal cavity (large arrows) and in the VNO (small arrow). The dashed line follows border of the VNO lumen. Abbreviation: nc = nasal cavity; vne = vomeronasal epithelium. Adapted with permission from Coppola and Millar, 1994.

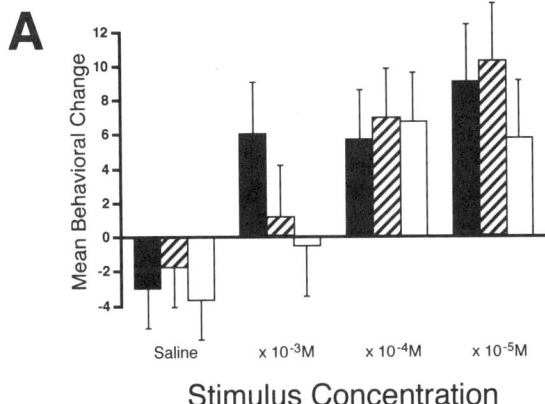

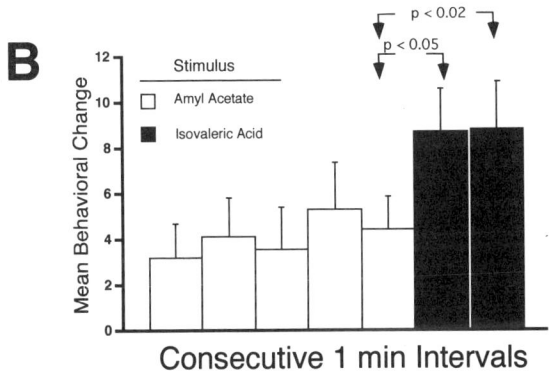

Figure 4. A. Plotted is the mean (+ SEM) increase in fetal behavior after a stimulus was delivered. Pretest activity (before stimulus was delivered) was subtracted from the total behaviors in the test period (after stimulus was delivered). At the beginning of each 2-min. test period, five µl of stimulus solution was added to a constant flow of saline purging the nasal cavity. The stimuli were saline or one of three dilutions of iso-amyl acetate. Sample sizes ranged between 22 and 24 fetuses for each stimulus concentration. Note the significant increase in behavior at all concentrations of iso-amyl acetate. **B.** Plotted is the mean (+ SEM) increase in fetal behavior after a stimulus was delivered (see part **A**). Seven consecutive one-min. stimulus trials are shown. Stimulus protocol was identical to part **A**. Iso-amyl acetate was the stimulus for the first five trials and iso-valeric acid was the stimulus for the last two. Note lack of adaptation to Iso-amyl acetate and increase in behavior to iso-valeric acid. . Used with permission from Coppola and Millar, 1997.

FETAL BEHAVIORAL RESPONSES TO ODORS

I'll come back to the rat in the last section, but I first want to return to the story in the mouse. Our results in the mouse provided an interesting test of fetal olfactory function. Around the time of our work on fetal olfaction in the mouse, William Smotherman and his colleagues had perfected a technique for direct observation of the unanesthetized rat fetus. They used this technique to show that the rat fetus can undergo chemically-cued aversive conditioning (possibly mediated by gustation, olfaction or the trigeminal system) and the rat fetus can respond to chemicals in the gas phase by increasing behavioral activity (Smotherman, 1982; Smotherman and Robinson, 1989). Leah Millar and I further adapted the technique to study the behavioral responses of the unanesthetized mouse fetus to odors delivered into the nasal cavity through a tiny cannula attached to a perfusion pump (Coppola and Millar, 1997). There have been surprising few studies of the development of behavioral responses to odors and those available fail to differentiate the contributions of the various chemosensory subsystems in the nasal cavity and the mouth. As we have seen, the mouse is a species in which the VNO is clearly not contributing to fetal or early postnatal olfaction due to the organ's immature access canal. Could the mouse fetus still detect odors? To test this question, we chose purified odorants that would not stimulate the gustatory system and used concentrations below the trigeminal system threshold. Not only did we find that the late-term mouse fetus can detect odors such as iso-amyl acetate and iso-valeric acid, but they do so at surprisingly low concentrations (Figure 4A). Moreover, the mouse fetus can consistently discriminate between these two odors and shows very slow adaptation to repeated odor presentations (Figure 4B). Through a process of elimination (VNO and trigeminal), our results strongly implicate the main olfactory system in fetal olfaction. More importantly, they reveal an unexpectedly high level of olfactory ability in a system destined for major postnatal growth (Figure 1).

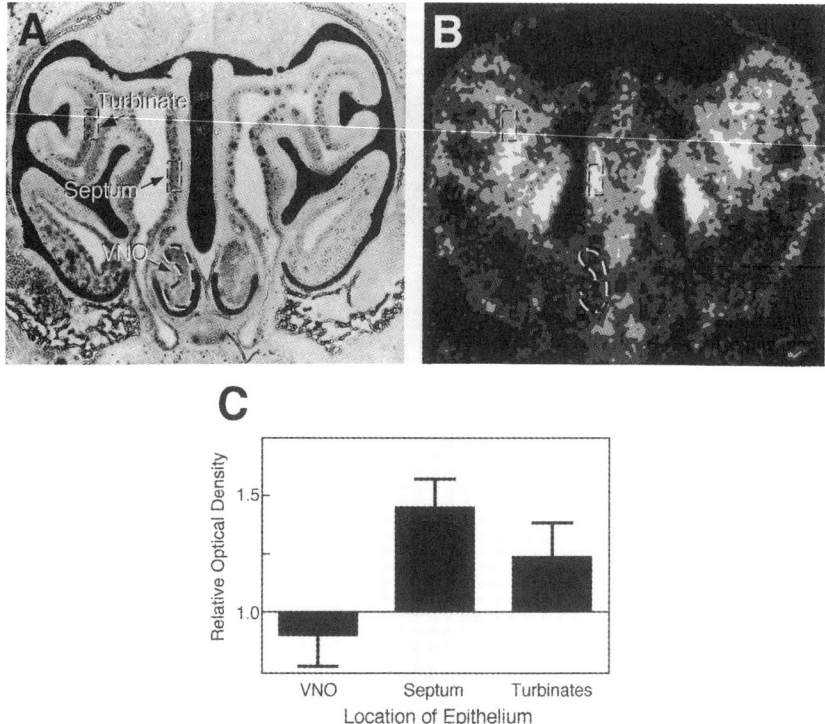

Figure 5. A. Unstained frozen coronal section of embryonic day 20 rat. Two-deoxyglucose was injected into maternal circulation 45 min. prior to harvesting the tissue. Note regions of interest (dashed lines) in main olfactory mucosa and VNO where optical density of autoradiographs was systematically measured in 4 animals (see part B). **B.** False gray level assignments from a digitized autoradiograph made from section in **A**. White corresponds to the highest 2-DG uptake and black the lowest. Regions of interest are outlined with dashed line as in **A**. Note the low level of 2-DG uptake throughout the VNO relative to the multiple foci of high 2-DG uptake in the main olfactory mucosa. **C.** Mean (+ SEM) relative optical density at three regions of interest measured across four rat fetuses. Data from Coppola and O'Connell, unpublished observations.

METABOLIC ACTIVITY MAPPING

With conflicting results in the mouse and the rat, it is not clear which, if either, is the more typical condition for mammals. The work described thus far leaves open the possibility the accessory olfactory system mediates prenatal olfaction in the rat. However, data collected several years ago by Robert O'Connell and I cast doubt on this conclusion. We used the 2-DG method in the fetal rat as Pederson *et al.* (1983) had done before us except that we focused on the receptor epithelium instead of the olfactory bulbs (Coppola and O'Connell, unpublished observations). Our results, consistent across four animals, revealed foci of high metabolic activity at several locations in the main olfactory mucosa (Figure 5). High 2-DG uptake was never found in the VNO epithelium, which had measurements consistently below the average of the entire tissue sample (Figure 5C.) These data strongly suggest that the main olfactory mucosa is more active functionally than the accessory olfactory system in the later periods of gestation. They do not, of course, eliminate the possibility that the VNO is operating. However, that the measurements were near the normal maximum for olfactory tissue was born out by our inability to increase 2-DG uptake in the main olfactory epithelium (or VNO) when we infused odorants into the nasal cavity during the entire period of tracer incorporation (data not shown). If our earlier experiment had merely measured differences in basal metabolic rate between the accessory and main olfactory tissue, we might have expected that odor infusions would have caused higher levels of 2-DG incorporation. Collectively, our 2-DG results suggest, as in the mouse, that the rat main olfactory system, not the accessory, is mediating prenatal olfaction.

CONCLUSIONS

It can be concluded with considerable confidence that the accessory olfactory system is not functioning in the womb. The work I have outline by my collaborators and I has been further supported by recent studies demonstrating the perinatal immaturity of accessory olfactory circuits and neuropeptide systems (Horowitz *et al.*, 1999; Zancanaro *et al.*, 1999). The intriguing idea that the olfactory system develops in stages through different subsystems, beginning with the accessory olfactory system (while having more than a little circumstantial support at the time it was proffered) turns out to be incorrect (reviewed by Pedersen *et al.*, 1985). Modern molecular data make clear that the seven-transmembrane receptors of the VNO and main olfactory system are quite distantly related, pointing to their interaction with very different ligands (Dulac and Axel, 1995; Matsunami and Buck, 1997). These data are consistent with the conventional view that in mammals the accessory olfactory system is specialized for detecting pheromones which help coordinate reproduction (reviewed by Keverne, 1999; also see Vandenbergh and Coppola, 1986 for an ecological perspective).

A common metaphor for development, in general, is that of gradual unfolding. And, in particular, we have seen how dramatic the postnatal increase in size and circuitry is in the main olfactory system (Figure 1). Nevertheless, our behavioral data in the mouse fetus challenge this view by documenting an uncanny ability to detect and discriminate odors at this stage. Given this precociousness, it is a puzzle what all that postnatal development is adding. I speculate it might not be adding as much additional functionality to the olfactory system as has been assumed but rather adds a large measure of redundancy to a system so critical for survival. Whatever the resolution of this mystery, it is clear that the mammalian fetus has the chemosensory capabilities to take advantage of the abundance of chemical information available in the amniotic fluid about kin, food, and other qualities of the world it is about to be born into.

REFERENCES

Breipohl, W. (ed.), 1986, *Ontogeny of Olfaction,* Springer-Verlag, Berlin.
Coppola, D. M., and Millar, L. C., 1994, Stimulus access to the accessory olfactory system in the prenatal and perinatal rat, *Neurosci.* 60:463-468.
Coppola, D. M., and Millar, L. C., 1997, Olfaction in utero: Behavioral studies of the mouse fetus, *Behav. Process.* 39:53-68.
Coppola, D. M., and O'Connell, R. J., 1989, Stimulus access to olfactory and vomeronasal receptors in utero, *Neurosci. Let.* 106:241-248.
Coppola, D. M., Budde, J., and Millar, L., 1993, The Vomeronasal duct has a protracted postnatal development in the mouse, *J. Morph.* 218:59-64.
Coppola, D. M., Coltrane, J. A., and Arsov, I., 1994, Retronasal or internasal olfaction can mediate odor-guided behaviors in newborn mice, *Physiol. Behav.* 56:729-736.
Dulac, C., and Axel, R., 1995, A novel family of genes encoding putative pheromone receptors in mammals, *Cell* 83:195-206.
Horowitz, L. F., Montmayeur, J., Echelard, Y., and Buck, L. B., 1999, A genetic approach to trace neural circuits, *P.N.A.S.* 96:3194-3199.
Keverne, E. B., 1999, The vomeronasal organ, *Science* 286:716-720.
Matsunami, H., and Buck, L. B., 1997, A multigene family encoding a diverse array of putative pheromone receptors in mammals, *Cell* 90:775-784.
Meredith, M., and O'Connell, R. J., 1979, Efferent control of stimulus access to hamster vomeronasal organ, *J. Physiol.* 289:301-316.
Pedersen, P. E., and Blass, E. M., 1982, Prenatal and postnatal determinants of the first suckling episode in the albino rat, *Dev. Psychobiol.* 15:349-356.
Pedersen, P. E., Stewart, W. B., Greer, C. A., and Shepherd, G. M., 1983, Evidence for olfactory function in utero, *Science* 221:478-480.
Pedersen, P. E., Greer, C. A., and Shepherd, G. M., 1985, Early development of olfactory function, in: *Handbook of Behavioral Neurology,* Vol.8. (E. M. Blass, ed.), Plenum Press, New York, pp. 163-203.
Smotherman, W. P., 1982, Odor aversion learning by the rat fetus, *Physiol. Behav.* 29:769-771.
Smotherman, W. P., and Robinson, S. R., 1989, Rat fetuses respond to chemical stimuli in gas phase, *Physiol. Behav.* 47:863-868.
Schaal, B., and Orgeur, P., 1992, Olfaction in utero: Can the rodent model be generalized, *Q. J. Exp. Psych.* 44B:245-278.
Vandenbergh, J. G., and Coppola, D. M., 1986, The physiology and ecology of puberty modulation by primer pheromones, in: *Advances in the Study of Animal Behavior* Vol 16 . (J. S. Rosenblatt *et al.*, eds.), Academic Press, Orlando, pp. 71-107.
Wysocki, C. J., 1989, Vomeronasal chemoreception: its role in reproductive fitness and physiology, in: *Neural Control of Reproductive Function,* (J. M. Lakoski, J. R. Perez-Polo and D. K. Rassin, eds.), Alan R. Liss, New York, pp. 545-566.
Zancanaro, C., Mucignat-Caretta, C., Merigo, F., and Osculati, F., 1999, Neuropeptide expression in the mouse vomeronasal organ during postnatal development, *Neuroreport* 10:2023-2027.

FETAL OLFACTORY COGNITION PREADAPTS NEONATAL BEHAVIOR IN MAMMALS

Benoist Schaal,[1,2] Gérard Coureaud,[1] Luc Marlier,[2] and Robert Soussignan[3]

[1]Unité de Physiologie de la Reproduction et des Comportements, CNRS, Inra, 37380 Nouzilly, France
[2]Centre Européen des Sciences du Goût, CNRS
15, rue Picardet, 21000 Dijon, France
[3]Laboratoire 'Personnalité et Conduites Adaptatives', CNRS Paris, France

1. INTRODUCTION

Several evolutionary adaptations have appeared in mammals that allow them to cope with the physical challenges presented during the early postnatal period. One set of general adaptations facilitates the maternal transmission of sensory information to the fetal and neonatal niches. The other set of adaptations results in fetuses and neonates developing the necessary sensory, cognitive, and motor abilities to be fully prepared to utilize maternally-provided information, even during the traumatic interval surrounding birth.

Fetal perception extracts certain cues from a complex multimodal information flow. The reception of these sensory inputs by newborn offspring elicits responses that attenuate distress responses, evoke active searching and positive orientation responses, and optimize the metabolic processes that support survival and growth. We will focus here on the olfactory 'arrowing' that the mammalian mother establishes, and which favors the adaptive responses of neonates, especially in their first attempts to locate the mammae and initiate nursing.

2. EVIDENCE FOR CHEMOSENSORY CONTINUITY IN THE ECOLOGICAL NICHES TO WHICH MAMMALIAN 'PERINATES' ARE EXPOSED

The initial evidence for chemosensory continuity between the prenatal and postnatal niche was based on studies conducted using Norway rats (Teicher and Blass, 1977). These authors first removed the natural odor cues from the females' abdomen by thoroughly washing it with organic solvents. This treatment resulted in the pups' refusal to orally grasp the nipple to which they were directed. If such prewashed nipples were, in a second test, painted with amniotic fluid, saliva from a parturient female or from a newborn, the oral seizing behavior of the pups was apparently normal. Thus emerged the hypothesis that an odor cue from the prenatal environment could be responsible for the triggering of the initial nipple attachment by neonatal rat pups. During parturition, amniotic odors are assumed to be spread on the abdominal surface by the female's intense self-licking activities at birth. More amniotic odors are spread over the mother's abdominal surface by the newly born pups in their first searching movements.

Positive responsiveness by neonates to the odors of birth fluids has been confirmed in a wide array of mammals, including rodents (Hepper, 1987; Kodama and Smotherman, 1996), lagomorphs (Coureaud et al., 1998), ungulates (Schaal et al., 1995a; Parfet and Gonyou, 1991) and primates (Schaal et al., 1995b). In parallel, empirical evidence was also obtained on the positive responsiveness of neonates to volatile compounds in lacteal secretions (including colostrum, and transitional and mature milks) in lagomorphs (Keil et al., 1990), ungulates (Schaal and Orgeur, unpublished data) and primates (Marlier and Schaal, 1997). Regardless of their origin, both types of species-specific fluids must include odors that elicit attraction or positive pre-feeding behavior of the newborn organisms. Presumably, the development of these adaptive responses to the odors begins when the odors first contact the nasal chemoreceptors at either the fetal or neonatal stage.

We measured the responses of newborn humans when they were exposed to a simultaneous choice between the odor of their amniotic environment and the odor of their mother's colostrum or milk. We observed that 2-day old, breast-fed, human infants oriented their faces (noses) equivalently to both stimuli (Marlier et al., 1998a). This lack of a discriminatory preference was not based on the infants' inability to detect either odor substrate or to orient, because they readily oriented toward either odorous stimulus when the alternative choice was an unodorized control stimulus.

Comparative studies have shown that the human neonates' lack of a discriminatory preference for amniotic fluid or mother's milk can be generalized to other mammalian newborns. For example, newborn rabbits introduced into a 2-choice arena presenting a choice between the odors of placenta and rabbit milk did not display a discriminatory preference (Coureaud et al., 1998). Likewise, ingestion-naïve lambs tested shortly after birth did not display a discriminatory preference between two pieces of cloth, one impregnated with their own amniotic fluid and the other with their mother's colostrum (Schaal and Orgeur, unpublished). Finally, newly born piglets exposed to a 3-choice test between water, colostrum and birth fluids, exhibited higher, but equal, durations of exploration and nosing toward the colostrum and birth fluids (Parfet and Gonyou, 1991).

The forced-choice procedures used in the above experiments are inadequate to determine whether or not the equivalent responses of newborns towards birth and lacteal fluids are caused by an early inability to discriminate these two sources of fluid or by a simple lack of preference. Regardless of the underlying perceptual mechanisms responsible for this outcome, it is our assumption that newborns treat both stimuli as equivalent.

Our "transnatal chemosensory continuity" hypothesis based on the observations described above proposes that the mammalian 'perinate' is exposed to a relative continuity in the chemosensory and/or motivational properties of the salient biological substrates they normally encounter during the late prenatal and early postnatal developmental phases of life. This continuity hypothesis has two tenets: 1) there is a continuity in the processing of odor information between the fetal and neonatal stages of development, and, 2) there is a chemical continuity linking the late amniotic stages and the early postnatal stages. The concept of continuity in chemosensory processing during development has been discussed at length elsewhere (e.g., Smotherman and Robinson, 1987, 1995; Schaal and Orgeur, 1992; Schaal et al., 1995c). Regarding the continuity in odor information, several non-exclusive mechanisms may be at work to promote the qualitative similarity in perinatal chemical ecologies. The most evident pathway to achieve chemical continuity is the contemporaneous transfer of odorous compounds into the fetal (including the fetal blood stream and the amniotic pool) and lacteal compartments. Exogenic compounds introduced to the maternal blood by the mother's dietary choices, e.g., food aromas, or by her local environment, e.g., tobacco smoke, have been shown to pass to the fetus (Schaal et al., 1995d; Mennella et al., 1995), into milk (Galef and Henderson, 1972; Bilko et al., 1994; Schaal et al., 1994; Desage et al., 1996), or into both (Lambers and Clark, 1996). Metabolic processes associated with physical exertion or stress in mothers can also exert a simultaneous influence on the chemosensory properties of amniotic and lacteal fluids. Finally, volatile compounds originating in the fetal and maternal immunogenetic system are reciprocally transferred (Beauchamp et al., 1994, 1995; Yamazaki et al., 2000), and may thus provide odor cues that are present in both perinatal fluids.

Another possibility for chemosensory overlap between amniotic and milk substrates is that they both contain specialized odorants that function as pheromones, such as those, which elicit teat-searching behavior. Such a specialized odorant to which rabbit pups react by attraction, and searching and oral grasping has recently been characterized in rabbit milk (Coureaud et al., 1999, Schaal et al., in preparation).

Our research to date has mainly focused on those biological processes we can control experimentally, i.e., the synchronous transfer of odorants from the pregnant and lactating mother's diet into both the amniotic and lacteal compartments. Several instrumental analyses of the transfer of dietary aromas into both biological compartments have been conducted (Schaal et al., 1995d; Desage et al., 1996), but we will summarize here only those data based on the chemosensory competence of neonatal organisms.

3. EMPIRICAL TESTS OF SOME PREDICTIONS DERIVED FROM THE TRANSNATAL CHEMOSENSORY CONTINUITY HYPOTHESIS

3.1. Transnatal Chemosensory Similarity Should Decrease in the First Postnatal Days

The olfactory similarities between amniotic and lacteal fluids should be maximal right after birth, when both media are influenced by the same source of variation. Since colostrum is already present in the mammae in the last days or weeks of pregnancy, these lacteal fluids are as much exposed to the entry of exogenous aromatic compounds as the amniotic fluids. Accordingly, the higher degree of shared odorants and chemosensory continuity between both fluids should result in an absence of a discriminatory preference in newborns that tested before their first colostrum intake. But as time goes on, the amniotic fluid should become more distinct from the lacteal secretion, since even by postpartum days 3 there is a substantial alteration in the composition of the expressed milk.

These predictions were examined in humans. Within the first two days postpartum, the composition of colostrum remains relatively stable, it enters a transitional phase at day 3, and by days 4-5 the onset of mature milk excretion increasingly replaces the colostrum. To assess how infants differentiate their amniotic fluid odor from the changing odors of their mothers' lacteal secretion, five groups of infants were tested between day 1 and day 5 of life to a 2-choice test between the odor of their amniotic fluid and of their mother's colostrum/milk (Marlier et al., 1997). Two phases in the pattern of neonatal response became apparent in these conditions: 1) between days 1-3, infants did not show evidence of a reliable differentiation of both odors; however, 2) by day 4, infants not only discriminated the two fluids, they displayed greater attraction, in terms of longer duration of head orientation, to their mother's milk odor than to their own amniotic fluid odor. This pattern of results is in support of a progressive shift in the infant's perception of the relative quality of milk odors relative to those of amniotic fluid.

3.2. Transnatal Chemosensory Similarity Cannot Exist between Amniotic Fluid and Artificial Formula Milks

In theory, the degree of chemosensory continuity between amniotic fluid and formulas derived from bovine milk cannot rival that between the amniotic and lacteal fluids of a same woman. Consequently, bottle-fed infants should be able to differentiate the olfactory qualities of their amniotic fluid and of their formula. This prediction was empirically confirmed: 2-day old bottle-fed infants exposed to a two-choice test opposing these odors clearly discriminated them in favor of the amniotic odors (Marlier *et al.*, 1998b), in contrast with their breast-fed peers who treated them nearly equally (cf prediction 1).

3.3. Newborn Organisms Sould Recognize the Odor of their Familiar Amniotic Fluid and Milk

It is a prerequisite of the hypothesis of transnatal olfactory continuity that newborns should behave in a selective way in response to the odors of amniotic and lacteal fluids. Regarding the amniotic odor, infants from different species were shown to selectively orient to their own amniotic fluid when exposed to a choice test between it and amniotic fluid of an unfamiliar neonate. This was the case in murine (Hepper, 1987), ovine (Schaal *et al.*, 1995b), and human newborns (Marlier *et al.*, 1998a; Schaal *et al.*, 1998).

To our knowledge, a selective response to the odor of pure maternal milk was only reported in ovine and human newborns. Breast-fed infants aged 4 days evinced longer head orientation in response to the odor of their mother's milk when it was paired with milk collected from another woman who was in the same lactational stage (Marlier and Schaal, 1997). Previous studies using pads applied to the mother's breast, hence collecting the complex exocrine secretions of that region, with milk presumably being the main substrate, provided convergent results supporting a discriminatory preference for the genetic mother's milk (Macfarlane, 1975; Russell, 1976).

3.4. The Exposure to an Aroma *in utero* should Lead to a Selective Response to the Same Aroma *ex utero*

Transnatal continuity in odor cues may depend on species-specific or individual-specific factors pertaining to the fetus or the mother. As described above, early differentiation of familiar and nonfamiliar amniotic fluid or milk point to the impact of individual-specific influences due to the mother. Several experimental manipulations of the chemical composition of either amniotic or lacteal fluids suggest that neonatal discrimination and preferences can be based on simple compounds, which had been prominent during the prenatal interval. Rat pups born to females that had consumed either garlic or ethanol evinced a preference for that same flavor at 12 days of age (Hepper, 1988; Chotro and Molina, 1990). In the same line, newborn lambs displayed reduced aversion to cumin odor when they experienced it in utero through their mothers' diet (Schaal *et al.*, 1995b). Finally, human infants exposed to the flavor of anise evinced more appetitive oral responses and less facial responses indexing aversion as compared to control infants when both groups were tested with anise approximately 3 hours after birth. Interestingly, the retention of the positive response to anise in the anise-exposed group persisted for at least 4 days after birth (Schaal *et al.*, 2000), suggesting that fetal experience with odorants fosters memories upon which newborns can rely in the period when they face their first pre-ingestive and ingestive experiences.

3.5. The Exposure to an Aroma in the Amniotic Context should Lead to a Selective Response in the Milk Context

This prediction includes the expectation that olfactory perception by neonates should be able to discriminate sensory information that is common to distinct biological substrates, a concept we tested using rabbits. Two groups of rabbit's pups were obtained through differentially feeding their mothers during pregnancy. Some females received a cumin-added diet while the others were exposed to the aromas of the standard diet. In the day after birth, before any sucking experience, the pups were administered a test consisting in the presentation of three glass rods dipped in either milk from a cumin-fed female, milk from a standard female, or water. Both groups of pups were highly reactive to both conspecific milk, but they were additionally discriminatively responsive to them. Cumin-exposed pups orally seized the glass rod with milk from cumin-fed females more often than the rod with milk from standard-fed mothers, and the pattern was reversed for the standard pups (Coureaud et al., 1998). This result indicates that prenatal experience with a given odorant can alter the later response to that same odorant in spite of the distraction caused by the alteration of the biological matrix carrying the odorant.

4. BEHAVIORAL EFFECTS OF DISRUPTING THE TRANSNATAL CHEMOSENSORY CONTINUITY

If the neonatal perception of the chemosensory similarities of amniotic and lacteal fluids has any adaptive value, chemosensory perturbations should have immediate consequences at the behavioral or physiological levels. The effects of such violations have been assessed in newborn rats, rabbits and humans.

The rate of survival of rat pups delivered prematurely (on day 20 of gestation) and maintained in warm containers in isolation from the dam for the first postnatal hour has been shown to depend on the olfactory ambiance of the container (Smotherman et al., 1987). Four groups of animals were assigned to the following odor conditions: a) amniotic odor, b) odor of dimethyl disulfide, a compound that is present in the saliva of both mothers and pups, and which activates nipple seizing behavior in the pup, c) a novel odorant, mint, and d) no odor. After the isolation period, the survival rate was higher in condition a (90%), although it did not significantly differ from conditions b and d (80 and 75%, respectively). But the introduction of the unfamiliar odorant (condition c) was followed by a significantly higher level of pup mortality (50%), compared to the familiar amniotic odor. This differential effect of the 'continuous' and 'discontinuous' perinatal environments might be mediated by the behavioral activation of the pups: while the familiar amniotic odor activated, the novel odorant depressed their general motor activity. Alternatively, the novel odorant in this case could have initiated a pathological stress response.

In rabbits, the consequences of the disruption of perinatal odor continuity was assessed by cross-fostering pups born to females exposed to olfactorily-contrasted regimen during gestation (Coureaud et al., 1997). Immediately after (watched or induced) delivery, half of each litter was fostered to a female that had eaten the same fodder as their biological mother, the other half being fostered to a female that was exposed to a regimen that was olfactorily distinct. In this way, we obtained a group of pups exposed to perinatal continuity, called 'continuous' here, and a group of 'discontinuous' pups. These pups were followed for their sucking success and amount of milk consumed during the first three sucking opportunities (i.e., during the first 3 days). The 'continuous' pups evinced higher success at sucking than the 'discontinuous' pups at the first two sucking bouts. In addition, those who acceded to the nipples in the 'discontinuous' group ingested on average less milk than the successful pups in the 'continuous' group. Thus, the condition in which the olfactory properties of milk and/or of the mother's abdomen are 'continuous' with the olfactory experience acquired *in utero* can stimulate more efficient sucking performance in rabbit pups.

Finally, in the human case, we dispose of a cultural model of disruption of the transnatal olfactory continuity in the practice of feeding infants with formulas of heterospecific origin. To assess the potential impact of the feeding method, we considered breast-fed and bottle-fed infants to be exposed to olfactorily continuous and discontinuous perinatal environments, respectively. We compared the relative response to the odors of amniotic fluid and of milk, i.e., the postnatal substrate associated with satiation, in exclusively breast- and bottle-fed infants. Both groups of infants were simultaneously presented with the odors of their own amniotic fluid and those of their familiar milk. As already mentioned above, the relative responses of breast-fed infants to these two stimuli were not distinct at age 2 days, but changed to a clear discriminatory preference for the milk odor over the amniotic odor as they gained nursing experience (cf prediction 1). In a functionally similar test choice between their own amniotic odor with the odor of their formula milk, bottle-fed infants demonstrated a clear differential response in favor of the amniotic odor at age 2 days. Thus, from this first test, both groups of infants behaved differently. On the same test at age 4 days, bottle-fed infants continued to respond more insistently toward the odor of amniotic fluid presented along with formula milk (Marlier *et al.*, 1998b), whereas their breast-fed peers displayed a clear preference for the odor of their postnatal food over the prenatal odor. In other words, when introduced into a preference test contrasting two biologically-relevant odors experienced either prenatally or associated postnatally with satiation, breast-fed subjects develop a preference response for the latter in a short delay, while bottle-fed infants do not follow the same pattern in the same delay. The results for the bottle-fed babies are not in accordance with the general principle of a positive relationship between exposure duration and preference development.

The different pattern of preference acquisition for postnatal food odor between both categories of infants can receive several explanations: 1) an acute, as opposed to a chronic, olfactory transition may affect the differential engagement of learning mechanisms in the neonate organism; 2) human milk may contain olfactory factors that are especially attractive to newborns; 3) human milk may carry particular substances which increase the conditioning qualities of the odor cues it contains; 4) finally, human milk may have non-specific biological (endocrine, absorptive, digestive) effects that formula milks do not have, and which may differentially contribute to the faster conditioning to food odor in breast-feeding than in bottle-feeding infants.

5. CONCLUSION

The empirical examination of the above predictions supports the view that olfactory acquisitions in various mammalian fetuses can function as a mechanism that ensures attention cueing in the neonatal organism. These acquisitions may enforce the organism's attention to certain environmental features in one developmental niche and increase the efficiency of stimulus recognition and selective response in the subsequent niche. Fetal specification of information may thus increase efficiency and precision of search behavior and decision making in relation to nursing responses, and, accordingly, decrease exposure of neonates to mortality factors.

REFERENCES

Beauchamp, G. K., Yamazaki, K., Currant M., Bard, J., and Boyse, E. A., 1994, Fetal H-2 odortypes are evident in the urine of pregnant female mice, *Immunogenet.* **39**:109-113.

Beauchamp, G. K., Katahira, K., Yamazaki, K., Mennella, J. A., Bard, J., and Boyse, E. A., 1995, Evidence suggesting that odortypes of pregnant women are a compound of maternal and fetal odortypes, *Proc. Natl. Acad. Sci. USA* **92**:2617-2621.

Bilko, A., Altbäcker, V., and Hudson, R., 1994, Transmission of food preference in the rabbit: the means of information transfer, *Physiol. Behav.* **56**:907-912.

Chotro M. G., and Molina, J. C., 1990, Acute ethanol contamination of the amniotic fluid during gestational day 21: postnatal changes in alcohol responsiveness in rats, *Dev. Psychobiol.* **23**:535-547.

Coureaud, G., Schaal, B., Orgeur, P., Hudson, R., Lebas, F., and Coudert, P., 1997, Perinatal odour disruption impairs neonatal milk intake in the rabbit, *Adv. Ethol.* **32**:134.

Coureaud, G., Schaal, B., Orgeur, P. and Hudson, R., 1998, A test of transnatal olfactory continuity in the rabbit, *Devel. Psychobiol.* **33**:370.

Coureaud, G., Schaal, B., Langlois, D., LeQuéré, J. L., and Perrier G., 1999, The mammary pheromone of the rabbit: a milk odor fraction that equals odor cues from lactating females' abdomen and from milk, *16th Annual Meet. Intern. Soc. Chem. Ecol.*, Marseille, France.

Desage, M., Schaal, B., Orgeur, P., Soubeyran J., and Brazier, J. L., 1996, Gas chromatographic-mass spectrometric method to characterise the transfer of dietary odorous compounds into plasma and milk, *J. Chromatogr. Biomed. Appl.* **678**:205-211.

Galef, B. G., and Henderson, P. W., 1972, Mother's milk: A determinant of the feeding preferences of weaning rat pups, *J. Comp. Physiol. Psychol.* **78**:213-219.

Hepper, P. G., 1987, The amniotic fluid: an important priming role in kin recognition. *Anim. Behav.* **35**:1343-1346.

Hepper, P. G., 1988, Adaptive fetal learning: prenatal exposure to garlic affects postnatal preferences. *Anim. Behav.* **36**:935-936.

Keil W., von Stralendorff, F. and Hudson, R., 1990, A behavioral bioassay for analysis of rabbit nipple-search pheromone, *Physiol. Behav.* **47**:525-529.

Kodama, N., and Smotherman, W. P., 1996, Effects of amniotic fluid on head movement in cesarean delivered rat pups, *29th Annual Meeting of the International Society for Developmental Psychobiology*, Washington, DC.

Lambers, D. S., and Clark, K. E., 1996, The maternal and fetal physiologic effects of nicotine, *Sem. Perinatol.* **20**:115-126.

Macfarlane, A. 1975, Olfaction in the development of social preferences in the human neonate, *Ciba Found. Symp.* **33**:103-113.

Marlier, L., and Schaal, B., 1997, Familiarité et discrimination olfactive chez le nouveau-né: influence différentielle du mode d'alimentation? *Enfance* **1**:47-61.

Marlier, L., Schaal, B., and Soussignan, R., 1997, Orientation responses to biological odours in the human newborn. Initial pattern and postnatal plasticity, *C. R. Acad. Sci., Paris, Life Sci.* **320**:999-1005.

Marlier, L., Schaal, B., and Soussignan, R., 1998a, Neonatal responsiveness to the odor of amniotic and lacteal fluids: A test of perinatal chemosensory continuity, *Child Devel.* **69**:611-623.

Marlier, L., Schaal, B., and Soussignan, R., 1998b, Does human milk carry odour cues that are specially attractive to newborns: Breast-fed and bottle-fed infants compared, *Dev. Psychobiol.* **33**:377.

Mennella, J. A., Johnson, A., and Beauchamp, G. K., 1995. Garlic ingestion by pregnant women alters the odor of amniotic fluid, *Chem. Senses* **20**:207-209.

Parfet, K. A. R., and Gonyou, H. W., 1991, Attraction of newborn piglets to auditory, visual, olfactory and tactile stimuli, *J. Anim. Sci.* **69**:125-133.

Russell, M. J., 1976, Human olfactory communication, *Nature* **260**:520-522.

Schaal, B., and Orgeur, P., 1992, Olfaction in utero: can the rodent model be generalized? *Q. J. Exp. Psychol.* **44B**:245-278.

Schaal, B., Orgeur, P., and Porter, R. H., 1994, Short-term flavor preference induced by dietary flavours transferred into mother's milk, *Infant Behav. Devel.* (**ICIS Issue**): 927.

Schaal, B., Orgeur, P. and Arnould, C., 1995a, Olfactory preferences in newborn lambs: possible influence of prenatal experience, *Behaviour* **132**:351-365.

Schaal, B., Marlier, L., and Soussignan, R., 1995b, Responsiveness to the odour of amniotic fluid in the human neonate, *Biol. Neonate* **67**:397-406.

Schaal, B., Orgeur, P., and Rognon, C., 1995c, Odor sensing in the human fetus: anatomical, functional and chemo-ecological bases, in: *Prenatal Development. A Psychobiological Perspective* (J. P. Lecanuet, N. A. Krasnegor, W. A. Fifer, and W. Smotherman, eds.), Lawrence Erlbaum: Hillsdale, NJ, pp. 205-237.

Schaal, B., Orgeur, P., Desage M., and Brazier, J. L., 1995d, Transfer of the aromas of the pregnant and lactating mother's diet to the fetal and neonatal environments in the sheep. *Chem. Senses*, **20**:93-94.

Schaal, B., Marlier, L., and Soussignan, R., 1998, Olfactory function in the human fetus: evidence from selective neonatal responsiveness to the odor of amniotic fluid, *Behav. Neurosci.* **112**:1438-1449.

Schaal, B., Marlier, L., and Soussignan, R., 2000, Human foetuses learn odours from their pregnant mother's diet, *Chem. Senses* (in press).

Smotherman, W. P., and Robinson, S. R., 1987, Psychobiology of fetal experience in the rat, in: *Perinatal Development. A Psychobiological Perspective* (N. E. Krasnegor, E. M. Blass, M. A. Hofer, and W. P. Smotherman, eds.), Academic Press, Orlando, pp. 39-60.

Smotherman, W. P., and Robinson, S. R., 1995. Tracing developmental trajectories into the prenatal period, in: *Prenatal Development. A Psychobiological Perspective* (J. P. Lecanuet, N. A. Krasnegor, W. A. Fifer, and W. Smotherman, eds.), Lawrence Erlbaum: Hillsdale, NJ, pp. 15-32.

Smotherman, W. P., Robinson, S. R., La Vallée, P. A., and Henessy, M. D., 1987. Influence of early olfactory environment on the survival, behavior and pituitary-adrenal activity of cesarean delivered preterm rat pups, *Devel. Psychobiol.* **20**:415-423.

Teicher, M. H., and Blass E. M., 1977, First suckling response in the newborn albino rat: the roles of olfaction and amniotic fluid, *Science* **198**:635-636.

Yamazaki, K., Curran, M., and Beauchamp, G. K., 2000, Development and modification of behavioral responses to MHC-determined odor types. *Chemical Signals in Vertebrates IX,* 26-30 July 2000, Kraków, Poland, (abstract).

OLFACTION IN PREMATURE HUMAN NEWBORNS: DETECTION AND DISCRIMINATION ABILITIES TWO MONTHS BEFORE GESTATIONAL TERM

Luc Marlier[1], Benoist Schaal[1], Christophe Gaugler[2], and Jean Messer[2]

[1]Centre Européen des Sciences du Goût, CNRS
21000 Dijon, France
[2]Centre Hospitalier Universitaire
Service de Néonatalogie, Pédiatrie 2
67000 Strasbourg-Hautepierre, France

INTRODUCTION

It remains unclear at what stage during early ontogeny humans do have the ability to process olfactory information. Numerous studies have demonstrated olfactory detection, discrimination, preference and memory in full-term infants examined within hours or days after birth (see Schaal, 1988 for a review). Recent data have revealed that olfaction is functional before birth. Newborns evince selective responsiveness to complex or pure odorants, which they could only, or mainly contact in the prenatal environment (Marlier et al., 1998; Schaal et al., 1998, 2000).

Preterm birth provides the opportunity to further obtain informations about the early, prenatal, course of nasochemosensory function. To date, only few data are available on the reactivity of premature infants to olfactory stimuli. For example, Sarnat (1978) noted increasingly reliable behavioral responses to peppermint as a function of gestational age (in the range of 26-35 weeks) in prematures tested on the first postnatal days. He suggested that olfaction could be functional from this early age onwards. However, several caveats may be raised in this work, including the omission of the reference data with an odorless stimulus, and the exclusive use of peppermint, which introduced a trigeminal confound to the olfactory stimulus. Finally, the use of only one stimulus precluded any insight about the discriminative ability of premature infants. This latter point was assessed by Pihet et al. (1997) who exposed premature infants (gestational age: 31-37 weeks, mean: 34 weeks) to three stimuli (an olfactant: nonanoic acid, an irritant: eucalyptol, and a blank) within the first week of life. The recording of general and facial activities revealed detection (differential reaction between either stimulus and the blank) and discrimination (differential response between both odorants). The differential response between both odorants could not be clearly ascribed to their quality, intensity, or irritative potency, however. Thus, while effective chemoreceptive detection is fairly sure from gestational week 29 (Sarnat, 1978) or

31 (Pihet et al., 1997), olfactory discrimination and the age at which it may be functional remain to be ascertained.

The present study aimed at 1) verifying olfactory detection ability in preterm infants born from 28 gestational weeks, and 2) examining the ability to discriminate qualitatively distinct olfactory stimuli bearing weak or no trigeminal properties. We report preliminary results on the variations of respiratory responses to vapors of vanillin, butyric acid and water.

METHOD

Subjects

Twenty-four preterm infants (9 males, 15 females) were involved. They were born at 28-33 weeks of amenorrhea (mean gestational age ± SD: 31.2 ± 1.1 weeks) and tested during the second week after birth (mean age at test ± SD: 10.1 ± 4.1 days). Infants were delivered (12 spontaneous vaginal and 12 caesarean deliveries) without severe complications (Apgar score ≥ 4 at 1 min, and >9 at 10 min) and presented no diagnosed nor suspected anomalies. Birthweight ranged from 1030 to 2080 g (mean weight ± SD: 1548 ± 305 g; 3 subjects had low birthweight for age). All infants were clinically stable at testing. To prevent episodes of apnea and bradycardy, they were treated with cafeine citrate (5 mg/kg/day). They received enteral nutrition via an orogastric tube. Their mothers (mean age ± SD: 28.7 ± 4.5 years) provided written informed consent. The present study received the approval of the local Ethical Committee.

Stimulations

Two pure odor qualitaties were selected: vanillin and butyric acid. The intensities of both odorants (diluted in distilled water) were equalized (dilutions: 0.039% for vanillin and 0.0078% for butyric acid) and matched to the odor of amniotic fluid by a panel of adults (for the method of stimuli preparation, cf. Soussignan et al., 1997). Distilled water was used as control stimulus. The three stimuli were presented on sterile cotton swabs.

Procedure

All newborns were tested while lying supine in their incubator during episodes of active (n=14) or regular (n=10) sleep. The incubator (mean temperature ± SD: 31.2 ± 0.9°C) was located in an isolation room which was maintained as quiet as possible. The stimuli were positioned 1 cm before the infant's nostrils for a duration of 10s so that several respiratory cycles were covered. A minimal inter-stimulus interval of 2 min was respected. Respiratory responses were recorded using a Kontron monitor. The testing procedure was conducted by two experimenters using a double-blind design. Stimuli were presented in random order.

Recording and Analysis of the Respiratory Responses

Respiration was recorded by three electrodes placed on the infant's torso. Instantaneous values (in cycles per min) were directly obtained from the monitor. For each trial, the changes in respiratory rhythm were computed by subtracting the mean value of the 5s-pre-stimulus block (baseline condition) from the mean of two subsequent 5s-blocks in the stimulus period (S1 and S2) and of four 5s-blocks in the post-stimulus period (P1 to P4).

To assess the effect of the odor stimuli on the variation of respiratory rate, the data were submitted to bivariate analyses of variance (ANOVA) for repeated measures where the 'nature of the stimulus' and the 'period' were considered as within-subject factors. To localize the effects revealed by this analysis, post-hoc t-tests were conducted. Due to technical recording problems, statistical analyses could not be run on the whole sample (the number of subjects was nevertheless ≥ 21 in all comparisons).

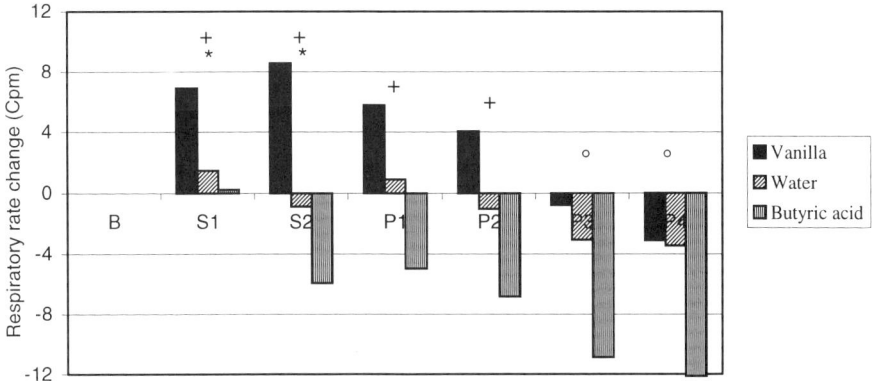

Figure 1. Changes in mean respiratory rate in premature newborns in response to the odors of vanillin, butyric acid and control (water). Changes are relative to the prestimulus period (B: baseline condition). S1-S2: stimulation periods; P1-P4: poststimulus periods. Symbols indicate significant differences between: *vanillin and control, °butyric acid an control, $^+$vanillin and butyric acid.

RESULTS

The statistical analyses yielded a significant main effect of the odor factor on the respiratory rate (RR) change when vanillin was compared to the control stimulus [$F(1, 20)=4.53$; $p<.05$]. The effect of butyric acid revealed to be only marginally significant [$F(1, 20)=3.37$, $p=.08$]. These comparisons further indicated significant main effects of the time period on the RR [for vanillin, $F(5, 100)=5.72$; $p<.001$; for butyric acid, $F(5, 100)=2.57$, $p=.03$]. Interaction between both factors reached not significance for one or the other stimulus. The newborns displayed accelerated breathing rate when they were exposed to vanillin as compared to the control stimulus (significant differences were noted in periods S1-S2: $t=3.14$ and 3.38, respectively; in both cases, $p<.01$, and marginally significant in period P1: $t=-1.78$, $p=.09$; Figure 1). In contrast, decelerative breathing rate followed the exposure to butyric acid as compared to the blank stimulus. In this case, differential response is marginally reliable in period P2 ($t=-1.76$, $p=.09$) and significant in periods P3 and P4 ($t=-2.18$ and -2.26, respectively; $p<.05$, in both cases ; Figure 1).

When the response to vanillin and butyric acid were directly compared to assess the infants' ability to differentiate between them, significant main effects of the 'stimulus' and 'period' factors on the RR were reached [$F(1, 21)=5.63$, $p<.03$ and $F(5, 105)=8.32$, $p<.001$, respectively] while their interaction was not significant. The changes in RR to both odorants were clearly distinct in periods S1, S2, P1 and P2 (t=2.87, 2.37, 2.51 and 2.0, in all cases p<.05) and marginally significant in period P3 (t=1.58, p=.08) (see Figure 1).

DISCUSSION

The present data provide evidence that premature infants born from 28 weeks of amenorrhea respond to low concentrations of pure odorants when tested around the 10th postnatal day. Vanillin and butyric acid induced significant respiratory rate changes as compared to the control, indicating reliable odor detection. These changes in the neonatal respiration index presumably reflect detection by the olfactory, rather than by the trigeminal system. Both stimuli were indeed assessed as non irritant by adult raters at the dilution used in the present study (see Soussignan et al., 1997). However, it cannot be excluded that the threshold for trigeminal stimulation may be lower in very young infants than in adults. Regardless of that possibility, these findings refine the results from earlier studies (Sarnat, 1978, Pihet et al., 1997) in two ways: first, they indicate that premature infants can detect odorants administered in an intensity range which mimics the one they may encounter in the amnion ; second, this keen detection ability is effective on average at 31 weeks of gestational age, which ranges in the earliest part of the age range examined in the studies mentioned in the introduction.

In addition, the chemoreception of premature newborns evinced the ability to differentiate vanillin and butyric acid matched for intensity and trigeminal features. The present data provide thus the strongest evidence for discriminative responsiveness by premature infants for odorants differing only along their qualitative 'dimension'.

It remains unclear whether an acceleration (as triggered by the odor of vanillin) or a decrease in respiratory rate (as triggered by the odor of butyric acid) indicate an ability to process the hedonic value of odorants. Respiratory measures should be complemented with behavioral measures, the interpretation of which is less ambiguous in terms of affective significance. Additional analyses are underway to characterize the ability of premature infants to express affectively differentiated behavioral responses to odorants that are hedonically contrasted to full-term infants, children and adults.

In sum, it seems reasonable to conclude from the present data that the main olfactory system is functionally ready to detect and discriminate olfactory qualities two months before gestational term.

ACKNOWLEDGMENTS

The mothers who accepted to let their infants participate are warmly acknowledged. We are also grateful to the nurses and midwifes for their continued cooperation. This work was supported by a grant from the Fyssen Fondation (to LM).

REFERENCES

Marlier, L., Schaal, B., and Soussignan, R., 1998, Neonatal responsiveness to the odor of amniotic and lacteal fluids: a test of perinatal chemosensory continuity, *Child Devel.* **69**:611-623.
Pihet, S., Mellier, D., Bullinger, A., and Schaal, B., 1997, Réponses comportementales aux odeurs chez le nouveau-né prématuré: étude préliminaire, *Enfance* **1997/1**:33-46.
Sarnat, H.B., 1978, Olfactory reflexes in the newborn infant, *J. Pediatrics* **92**:624-626.
Schaal, B., 1988, Olfaction in infants and children: developmental and functional perspectives, *Chem. Senses* **13**:145-190.

Schaal, B., Marlier, L., and Soussignan, R., 1998, Olfactory function in the human fetus: evidence from selective neonatal responsiveness to the odor of amniotic fluid, *Behav. Neurosci.* **112**:1-12.

Schaal, B., Marlier, L., and Soussignan, R., 2000, Human foetuses encode odors from their pregnant mother's diet, *Chem. Senses* (in press).

Soussignan, R., Schaal, B., Marlier, L., and Jiang, T., 1997, Facial and autonomic responses to biological and artificial olfactory stimuli in human neonates: re-examining early hedonic discrimination of odors. Physiol. Behav. **62**:745-758.

INTRAUTERINE POSITION EFFECTS ON RODENT URINARY CHEMOSIGNALS

Lee C. Drickamer

Department of Biological Sciences
Northern Arizona University
Flagstaff, AZ 86011-5640

INTRODUCTION

In litter bearing mammals the intrauterine position of a fetus with respect to other fetuses in each of the uterine horns has profound effects on its later morphology, physiology, and behavior. This effect was first reported in rats (Clemens et al., 1978). Similar work on mice has produced a considerable knowledge concerning intrauterine position effects (vom Saal and Bronson, 1978, 1980; reviews by vom Saal, 1989; vom Saal et al., 1992). Sexual differentiation of female mouse fetuses that are positioned between two males is affected by androgen from the neighboring male fetuses (vom Saal and Dhar, 1992). Females that are located in utero between two males are designated as from the 2M position, those located between two females are designated as from the 0M position (or these are sometimes referred to as 2F animals), and females positioned in utero between a male and a female are designated as 1M females. A similar terminology can be applied to males with respect to their in utero position in relation to other males.

Various studies have demonstrated the nature of intra-uterine position effects. In both rats (Clemens et al., 1978) and mice (vom Saal and Bronson, 1978) the anogenital distance (measured from the base of the genital papilla to the nearest point of the anus) is largest for 2M females, intermediate for 1M females, and shortest for 0M females. This same effect occurs for male mice (Drickamer, unpublished data). Serum concentrations of testosterone from fetuses of both sexes show this same variation (vom Saal et al., 1990) with the highest levels occurring in females or males from the 2M position. Both females and males from the 2M position are more aggressive in same sex encounters with mice from the 0M position (vom Saal et al., 1983). Also, 2M females show more aggression toward an intruder when they are in the postpartum condition (lactating) than 0M females (vom Saal and Bronson, 1978). Females from the 2M position have a greater tendency to be infanticidal than females from the 0M position (Yousif et al., 1989). Males from the 2M position show more parental care and are less infanticidal than males from the 0M position (vom Saal and Howard, 1982). These findings serve as the basis for some predictions about odor preferences in the next section of this paper.

In addition to the initial work on rats and the work on house mice, intrauterine position effects have been demonstrated in other rodents, most notably in gerbils (Clark and Galef, 1989; Clark et al., 1989, 1992). The gerbil findings include intrauterine position effects on rates of sexual development in females, variations in schedules of urination by pups, and effects on levels of circulating testosterone in male gerbils. In addition, Clark and Galef (1995) reported that the sex ratio of progeny produced by a female gerbil was influenced by the prior intrauterine position she occupied in utero during her own development. This latter report replicated a similar finding for house mice (Vandenbergh and Huggett, 1994). In both species, 2M females produce more sons in their litters and 0M females produced a higher proportion of daughters.

Vandenbergh and Huggett (1995) reported that the anogenital distance can be used as an index of intrauterine position. Measurements of the anogenital distance at weaning (21 days after birth) were corrected for body weight at weaning to create the anogenital distance index. Their findings indicate that the anogenital distance index is largest for females from the 2M position, intermediate for females from the 1M position, and significantly lower for females from the 0M position. Knowing that there is a predictable relationship between the intrauterine position and the anogenital distance in female mice permits us to take measurements on wild mice living under seminatural conditions where it would be otherwise impossible to know their exact intrauterine position.

My purposes in this short review of the relationship between intrauterine position and mouse urinary chemosignals are threefold. I will describe two findings relating intra-uterine position to urinary chemosignals. The first study concerns odor preferences of male and female mice under field conditions. We tested certain predictions about odor preferences based on the effects of intrauterine position on physiology and behavior. The second example relates intrauterine position to urine marking in male house mice. Lastly, I provide a brief synopsis of the information that is available on the relationships between intrauterine position effects on male and female mice and puberty-influencing urinary chemosignals.

INTRAUTERINE POSITION AND ODOR PREFERENCES

Based on the physiological and behavioral effects of prior intrauterine position, we can make several predictions concerning the potential mate preferences of male and female house mice. Females should prefer males from the 2M position. Males should prefer females from the 0M position. Laboratory investigations have shown that in fact male mice find 0M females more attractive than 2M females and tend to mount and inseminate them preferentially when given a choice (vom Saal and Bronson, 1978, 1980). Conversely, females, regardless of their own prior intrauterine position, prefer 2M males (vom Saal, 1984; Rines and vom Saal, 1984) as mates.

These laboratory findings were tested under field conditions using eight 0.1 ha enclosures (see Drickamer and Mikesic, 1990 for enclosure details). We used the odor-baited trap technique (Drickamer, 1995) and the known anogenital distances for each of the mice in the enclosures (Drickamer et al., 2001). As noted previously, the anogenital distance index (Vandenbergh and Huggett, 1995) is a reliable predictor of the prior intrauterine position of the mice. In our study there was no relationship between body mass and anogenital distance, so no correction was necessary.

We determined the median value for anogenital distance (AGD) for all females and divided the individuals into those above the median (high AGD) and those below the median (low AGD). A similar procedure was used to divide males into high and low AGD groups. Given that the effects of intrauterine position are not categorical, but generally provide a continuous distribution in terms of an outcome measure such as anogenital distance, the procedure we followed is quite conservative. Any significant effects should thus be quite robust.

By knowing the home ranges of each mouse we could determine the numbers of traps available odorized by males and females with high or low AGDs for each particular mouse. We could then analyze the captures of say females that were in the high AGD group that were caught in traps that were odorized by high or low AGD males as a proportion of those available. The same procedure was followed for low AGD females and for both types of males. In this way we could ask four questions: (1) Do high AGD females prefer traps odorized by low or high AGD males? (2) Do low AGD females prefer traps odorized by low or high AGD males? (3) Do low AGD males prefer traps odorized by low or high AGD females? and (4) Do high AGD males prefer traps odorized by low or high AGD females?

The results conform with both the original predictions and the findings from laboratory investigations. Females, regardless of their own AGD, prefer to associate with the odors from males with higher AGD values, those that would more likely come from the 2M position. Males, regardless of their own AGD, prefer to associate with odors associated with females having lower AGDs, those that would more likely come from the 0M position. This experiment then (Drickamer *et al.*, 2001) provides a nice confirmation of a phenomenon previously demonstrated in the laboratory. House mice can use odor cues to discriminate opposite-sexed conspecifics based on their prior intrauterine position under field conditions.

INTRAUTERINE POSITION AND URINE MARKING

Urine marking in house mice has been studied extensively in both the laboratory and under commensal conditions where the mice are living in poultry houses and similar farm buildings (Desjardins *et al.*, 1973; Bronson, 1976; Brown, 1985a, b; Hurst, 1987, 1993). Urine marking in house mice is influenced by novel environments and social stimuli (Maruniak *et al.*, 1974), water deprivation (Maruniak *et al.*, 1988), sex steroids (Maruniak *et al.*, 1975; Kimura and Hagiwara, 1985), and removal of the vomeronasal organ (Labov and Wysocki, 1989).

I tested the hypothesis that intrauterine position would effect male urinary marking behavior. I used the relationship between anogenital distance and intrauterine position established for females (Vandenbergh and Huggett, 1995) as a bioassay for intrauterine position and assumed that there was a similar relationship for males. I investigated two ideas. First, I tested the urinary marking behavior of males with different anogenital distances. Based on the knowledge that 2M males are more aggressive and have higher levels of circulating testosterone than 0M males, I predicted that males with larger anogenital distances would engage in more urinary marking behavior than males with smaller anogenital distances. Second, as an adjunct, I wanted to know whether there was any relationship between anogenital distance at 10 days of age, at weaning, and when the males were adults (75 days of age).

I used the same 40 males from 40 different litters to test both ideas. I measured the anogenital distance of each test male at 10, 24, and 60 days of age. Each male was tested for 24 hours at 75 days of age in a urine marking arena measuring 20 cm x 40 cm. The floor of the arena consisted of wire mesh and the floor was located 3 cm above a layer of filter paper placed to absorb the urine. The urine marks were visualized using ultraviolet light. The filter paper was placed on a light table marked off with a 1 cm square grid. As dependent measures, there were 684 possible interior squares and 116 perimeter squares that could contain urine marks.

The urine marking data were analyzed using two regressions, one for each dependent measure. I regressed the known anogenital distance at 60 days of age (x variate) against the numbers of squares marked with urine (y variates). For both tests there was a significant effect. With internal squares marked $r = 0.378$, $p = 0.0161$, and for the perimeter squares marked $r = 0.363$, $p = 0.0212$. For both dependent measures the effect was such that the larger the anogenital distance of the mouse, the more squares were marked with urine. This finding conforms with the hypothesis that males with larger anogenital distances, which are presumably more likely to come from 2M intrauterine positions, will have higher levels of circulating testosterone and this, in turn, relates to more urine marking.

The anogenital (AGD) distance data were analyzed using correlations. AGD at age 10 days was correlated with AGD at age 24 days with $r = 0.903$. AGD at age 10 days was correlated with AGD at age 60 days with $r = 0.898$. AGD at age 24 days was correlated with AGD at age 60 days with $r = 0.979$. All of these are significant at $p < 0.001$. Measurement of the anogenital distance as a bioassay for prior intrauterine position can occur at any age; the sensitivity of this assay is not apparently age dependent. This finding has significant implications for the use of the anogenital distance index with free living mice where it is not possible to obtain data on exact intrauterine positions of individuals.

INTRAUTERINE POSITION AND URINARY CHEMOSIGNAL EFFECTS ON PUBERTY

Urine from male mice accelerates puberty of young female mice (Vandenbergh, 1969). Urine from grouped females delays puberty in young females (Drickamer, 1974). Females from the 2M position are less responsive to both the puberty-delaying chemosignal from grouped females and the puberty-accelerating chemosignal from males (vom Saal *et al.*, 1981; Vandenbergh and Huggett, 1995). To date no one has explored the possible effects of prior intrauterine position on the production and release of these puberty-influencing urinary chemosignals. More work is needed to explore these phenomena. I would predict that males from the 2M intrauterine position would, as adults, be likely to produce and release more of the puberty-accelerating urinary chemosignal than males from the 0M position. In other words, males with larger anogenital distances will have a greater capacity to accelerate puberty in young female mice than males with shorter anogenital distances. Given that 0M females have shorter estrous cycles than 2M females (vom Saal and Bronson, 1978) but 0M females are more responsive to the puberty-delaying chemosignal from grouped females, I would be hesitant to predict the effects of intrauterine position of the production and release of the delay chemosignal by females from differing intrauterine positions.

SUMMARY

Intrauterine position affects several key phenomena that involve the communications value of urinary chemosignals in house mice. Based on laboratory findings, predictions can be made concerning the odor preferences of male and female mice for opposite sex partners from different intrauterine positions. These effects, tested initially in a laboratory setting, are, it turns out, also valid under field conditions in seminatural enclosures. Female mice prefer males from the 2M position and male mice prefer female mice from the 0M position. Urinary marking is used by mice in a variety of contexts. For male mice, the urinary marking behavior is related to their anogenital distance and thus to their prior intrauterine position. Males with larger AGDs have a tendency to mark more squares in a test arena than males with smaller AGDs. This is true regardless of whether we measure perimeter or internal squares with urine deposition. Lastly, additional work is needed on the relationships between intrauterine position and both the production and release of urinary chemosignals that influence puberty in female mice and the sensitivity of those females to urinary chemosignals.

ACKNOWLEDGMENTS

I especially thank Dr. John G. Vandenbergh for his mentoring when I was a post-doctoral student in his laboratory beginning 30 years ago. I thank Southern Illinois University at Carbondale and Northern Arizona University for the use of their facilities and for some financial support for these investigations. The research summarized here was supported by the National Institutes of Health (Grant No. HD-08585), and the National Science Foundation (Grant Nos. BNS 8796315, IBN 8616204, and IBN 9896250). I thank Ami Sessions Robinson, and Catherine A. Mossman for assistance in both the laboratory and field venues.

REFERENCES

Bronson, F. H., 1976, Urine marking in mice: Causes and effects, in: *Mammalian Olfaction, Reproductive Processes, and Behavior*, (R. L. Doty, ed.), Academic Press, New York, pp. 119-141.

Brown, R. E., 1985a, The rodents I. Effects of odours on reproductive physiology, in: *Social Odours in Mammals*, Vol. 1 (R. E. Brown, and D. W. Macdonald, eds.), Oxford University Press, Oxford, pp. 245-344.

Brown, R. E., 1985b, The rodents II. Suborder Myomorpha, in: *Social Odours in Mammals*, Vol. 1 (R. E. Brown, and D. W. Macdonald, eds.), Oxford University Press, Oxford, pp. 345-457.

Clark, M. M., and Galef, B. G. Jr., 1989, Measuring rates of sexual development in female Mongolian gerbils, *Dev. Psychobiol.* **22**:173-182.

Clark, M. M., and Galef, B. G. Jr., 1995, A gerbil's dam's fetal intrauterine position affects the sex ratios of litters she gestates, *Physiol. Behav.* **57**:297-299.

Clark, M. M., Bone, S., and Galef, B. G. Jr., 1989, Uterine positions and schedules of urination: Correlates of differential anogenital stimulation, *Dev. Psychobiol.* **22**:389-400.

Clark, M. M., vom Saal, F. S., and Galef, B. G. Jr., 1992, Intrauterine positions and testosterone levels of adult male gerbils, *Physiol. Behav.* **51**:957-960.

Clemens, L., Gladue, B., and Coniglio, L., 1978, Prenatal and endogenous androgenic influences on masculine sexual behavior and genital morphology in male and female rats, *Horm. Behav.* **10**:40-53.

Desjardins, C., Maruniak, J. A., and Bronson, F. H., 1973, Social rank in house mice: Differentiation by ultraviolet visualization of urinary marking patterns, *Science* **182**:239-241.

Drickamer, L. C., 1974, Sexual maturation of female house mice: Social inhibition, *Dev. Psychobiol.* **7**:257-265.

Drickamer, L. C., 1995, Odors in traps: Does most recent occupant influence capture rates for house mice? *J. Chem. Ecol.* **21**:541-555.

Drickamer, L. C., and Mikesic, D. M., 1990, Urinary chemosignals, reproduction, and population size for house mice (*Mus domesticus*) living in field enclosures, *J. Chem. Ecol.* **16**:2955-2968.

Drickamer, L. S., Robinson, A. S., and Mossman, C. A., 2001, Differential responses to same and opposite sex odors by adult house mice associated with anogenital distance, *Ethology* (in press).

Hurst, J. L., 1987, The functions of urine marking in a freeliving population of house mice, *Mus domesticus*, Rutty, *Anim. Behav.* **35**:1433-1442.

Hurst, J. L., 1993, The priming effects of urine substrate marks on interactions between male house mice, *Mus musculus domesticus* Schwarz and Schwarz, *Anim. Behav.* **45**:55-81.

Kimura, T., and Hagiwara, Y., 1985, Regulation of urine marking in male and female mice: Effects of steroids, *Horm. Behav.* **19**:64-70.

Labov, J., and Wysocki, C. J., 1989, Vomeronasal organ and social factors affect urine marking by male mice, *Physiol. Behav.* **45**:443-447.

Maruniak, J. A., Owen, J., Bronson, F. H., and Desjardins, C., 1974, Urinary marking in male house mice: Responses to novel environmental and social stimuli, *Physiol. Behav.* **12**:1035-1039.

Maruniak, J. A., Owen, J., Bronson, F. H., and Desjardins, C., 1975, Urinary marking in female house mice: Effects of ovarian steroids, sex experience, and type of stimulus, *Behav. Biol.* **13**:211-217.

Maruniak, J. A., Taylor, J. A., and Perrigo, G., 1988, Effects of water deprivation on urine marking and aggression in male house mice, *Physiol. Behav.* **42**:47-51.

Rines, J., and vom Saal, F. S., 1984, Fetal effects on sexual behavior and aggression in young and old female mice treated with estrogen and testosterone, *Horm. Behav.* **18**:17-26.

Vandenbergh, J. G., 1969, Male odor accelerates female sexual maturation in mice, *Endocrinology* **84**:658-660.

Vandenbergh, J. G., and Huggett, C. L., 1994, Mother's prior intrauterine position affects the sex ration of her offspring in house mice, *Proc. Nat. Acad. Sci. USA* **91**:11055-11059.

Vandenbergh, J. G., and Huggett, C. L., 1995, The anogenital distance index, a predictor of the intrauterine position effects on reproduction in female house mice, *Lab. Anim. Sci.* **45**:567-573.

vom Saal, F. S., 1984, the intrauterine position phenomenon: Effects on physiology, aggressive behavior and population dynamics in house mice, in: *Biological Perspectives on Aggression*, (K. Flannelly, B. Blanchard, and D. Blanchard eds.), Alan R. Liss, New York, pp. 135-179.

vom Saal, F. S., 1989, Sexual differentiation in litter-bearing mammals: influence of sex of adjacent fetuses in utero, *J. Anim. Sci.* **67**:1824-1840.

vom Saal, F. S., and Bronson, F. H., 1978, In utero proximity of female mouse fetuses to males: effect on reproductive performance during later life, *Biol. Reprod.* **19**:842-845.

vom Saal, F. S., and Bronson, F. H., 1980, Sexual characteristics of adult female mice are correlated with their blood testosterone levels during prenatal development, *Science* **208**:597-599.

vom Saal, F. S., and Dhar, M., 1992, Blood flow in the uterine loop artery and loop vein is bidirectional in the mouse: implications for transport of steroids between fetuses, *Physiol. Behav.* **52**:163-171.

vom Saal, F. S., and Howard, L., 1982, The regulation of infanticide and parental behavior: Implications for reproductive success in male mice, *Science* **220**:1306-1308.

vom Saal, F. S., Pryor, S., and Bronson, F. H., 1981, Effects of prior intrauterine position and housing on oestrous cycle length in adolescent mice, *J. Reprod. Fert.* **62**:33-37.

vom Saal, F. S., Grant, W., McMullen, C., and Laves, K., 1983, High fetal estrogen titers correlate with enhanced adult sexual performance and decreased aggression in male mice, *Science* **220**:1306-1308.

vom Saal, F. S., Quadagno, D., Even, M., Keisler, L., Keisler, D., and Khan, S., 1990, Paradoxical effects of maternal stress on fetal steroids and postnatal reproductive traits in female mice from different intrauterine positions. *Biol. Reprod.* **43**:751-761.

vom Saal, F. S., Montano, M. S., and Wang, M. H., 1992, Sexual differentiation in mammals, in: *Chemically Induced Alterations in Sexual and Functional Development: The Wildlife-Human Connection*, (T. Colburn, and C. Clement, eds.), Princeton Scientific, Princeton, New Jersey.

Yousif, Y. Y., Brian, P. F., Parmagiani, S., and Mainardi, M., 1989, Effects of genotype and intrauterine location on the propensity for infanticide by primiparous female mice, *Ethol. Ecol. Evol.* **1**:283-290.

SOCIAL STATUS, ODOUR COMMUNICATION AND MATE CHOICE IN WILD HOUSE MICE

Nicholas Malone,[1] Caroline E. Payne,[1] Robert J. Beynon,[2] and Jane L. Hurst[1]

[1]Animal Behaviour Group and [2]Protein Function Group
Faculty of Veterinary Science
University of Liverpool
Leahurst, Neston, CH64 7TE, UK

1. INTRODUCTION

Due to the differential costs of sexual reproduction (Bateman, 1948; Orians, 1969), females are assumed be the more selective sex when it comes to mate choice. Females can maximise their reproductive investments by choosing high quality mates. A number of factors may influence a male's quality as a mate, in particular competitive ability, disease resistance and relatedness to the female. Competition between males results in the most competitive males accruing the most resources and highest social status. Females may identify males of high competitive ability by using signals of these competitive gains (Zahavi, 1987). Due to the detrimental effects of inbreeding and disease, females also need to detect relatedness and infection status from male signals. Here we review evidence showing how female house mice may use urinary scents to detect a male's competitive ability, his disease resistance and relatedness, and thus his desirability as a mate.

2. MALE SOCIAL STRATEGIES

Male house mice (*Mus domesticus*) are fiercely territorial and will attempt to exclude other males from their territories. Good quality territories are limited in number as space containing suitable resources is finite. Consequently, males either opt to compete by attempting to defend a territory, emigrate or adopt a non-competitive subordinate strategy, each of which have different costs and benefits for the male.

Between them, competitive territory owners account for virtually all of the matings within a mouse population (Wolff, 1985; Hurst, 1987; Potts *et al.*, 1994). However, competitive males have the costs of gaining and defending territories in which they can mate, including the costs of signalling their status as territory owners. Males that lack the competitive ability to secure their own territory have two options: leave the population and set up a territory away from more competitive males, or live within another male's territory. Emigration is perhaps the more risky option. House mice usually live commensally where food is super-abundant but patchily distributed. Emigrating away from the original population may remove the cost of competition but may also mean abandoning critical resources. Starvation and predation are major risks for emigrants, and the lack of mating opportunities a major cost. Gerlach (1990) found that males in a semi-natural enclosure were reluctant to leave their group, only doing so once aggression from other males became excessive.

Staying within another male's territory allows access to food and shelter but is not without costs. Dominant territory owners are relatively tolerant of familiar subordinates that within their territory and do not compete although the risk of attack is not removed completely (Hurst, 1987). However, the main cost of being a subordinate male, from an evolutionary perspective, is the lack of mating opportunities. Any attempts to mate with resident females are not tolerated by dominant males (Huck,1982), and females show a strong mate preference for dominant territory owners (Wolff, 1985; Hurst 1987; Potts et al., 1994). However, when the dominant male's competitive ability starts to wane with age or disease, a subordinate should be ideally located to monitor his decline and take over the territory at the opportune moment.

2.1 Modelling Male Social Strategies

One reason that mice are often used in mate choice research may be the apparent ease with which dominant-subordinate relationships can be established in the laboratory. Many studies have used the paradigm of matched dominant-subordinate pairs, where two males interact and dominance is assigned to the male displaying the most aggression (e.g. Desjardins et al., 1973; Harvey et al., 1989; Novotny et al., 1990; Drickamer, 1992; Gosling et al., 1996). However, methods of establishing such pairs can vary considerably. For example, Desjardins et al. (1973) housed their male pairs together for 10-hours and used the relative number of bite wounds received by each male to decide dominance while Gosling et al. (1996) housed their male pairs together for 4-weeks and used fleeing behaviour during a single 15-minute encounter to assign social status. Harvey et al. (1989) and Drickamer (1992) compared the number of attacks and chases between males to assign social status but only brought their male pairs together during 10-min interaction sessions, returning each male to a separate home cage afterwards. While Harvey et al. (1989) used a series of daily 10-min interactions to decide dominance, Drickamer (1992) used only a single interaction. These separately housed 'dominant' and 'subordinate' males were thus like the 'winners' and 'losers' used by Sawyer (1978) who used individually housed males paired together for brief and infrequent interactions in a neutral arena.

Such differences in the nature and extent of social contact between males are likely to be crucial to the social strategies they adopt. Jones and Nowell (1989) demonstrated the importance of close, continuous olfactory contact in maintaining dominant-subordinate male pairs. The urine of dominant or isolate males contains a substance aversive to other males, whereas the urine of subordinate males does not (Jones and Nowell, 1973). However, if males that are submissive during interactions are housed individually between brief interaction sessions, they retain the aversive factor in their urine (Sawyer, 1978; Jones and Nowell, 1989). Such procedures, rather than establishing dominant-subordinate pairs that live in the same territory, may model interactions between animals from two separate territories where the submissive behaviour of one male reflects the response of a less competitive intruder rather than that of a resident subordinate. Appropriate space to establish territories, as well as olfactory contact and regular interaction, may also be required for realistic modelling of male social strategies. Bishop and Chevins (1987) found that housing males in a large free-range room resulted in the urine of the dominant animal becoming aversive to other males, whereas housing males in large cages produced dominants whose urine lacked this aversive quality.

It is thus essential that laboratory studies of social status use appropriate models that reflect the different strategies that males adopt under natural conditions and that such differences in social experience and housing conditions are taken into account when interpreting the behavioural and ecological significance of results.

3. SOCIAL STATUS AND FEMALE CHOICE

3.1 Scent Mark Quality

Early work showed that mouse urine contains chemicals capable of eliciting changes in the reproductive behaviour and physiology of recipient animals. The pheromones within male urine have effects on pregnancy block (Huck, 1982), and age of maturation in females (Lombardi and Vandenbergh, 1977), oestrus synchronization sexual attraction (Caroom and Bronson, 1971), male investigatory behaviour (Jones and Nowell, 1973; Novotny et al., 1990), and inter-male aggression (Jones and Nowell, 1973). Several of these effects are androgen-dependent, for example, onset of oestrus (Bronson and Whitten, 1968) and inter-male aggression (Jones and Nowell, 1973). Harvey et al. (1989) found a number of androgen-dependent volatile components in male urine, including four compounds that were not observed in immature males and were strongly depressed by castration: α- and β-farnesenes (farnesenes), 2-(sec-Butyl)-4,5-dihydrothiazole (thiazole) and dehydro-exo-brevicomin (brevicomin). These compounds have a number of effects on female attraction and reproductive priming, and on male competitive behaviour. Females prefer urine samples with high levels of farnesenes over samples with low levels or no farnesenes at all (Jemiolo et al., 1991) while males, especially subordinates, are more reluctant to scent mark areas containing either natural or artificial farnesenes (Jemiolo et al., 1992). Brevicomin and thiazole are attractive to females (Jemiolo et al., 1985) and increase aggressive behaviour in males (Novotny et al., 1985) though, interestingly, only act synergistically. Each of these chemicals has also been shown to accelerate sexual maturation in females (Novotny et al., 1999).

Social subordination suppresses androgen levels in mice (Bronson, 1973) and Harvey et al. (1989) demonstrated that dominant males have more androgen-dependent farnesenes and thiazole in their urine compared to subordinates (housed individually). Androgens have been shown to influence the social status of males in the laboratory and their competitive ability in the field (Zielinski and Vandenbergh, 1993). It thus appears that females use the levels of androgen-dependent semiochemicals in male urine to assess individual social status and competitive ability.

Thiazole and brevicomin are highly volatile (Harvey et al., 1989; Novotny et al., 1990), which should reduce their longevity and, therefore, their usefulness as olfactory signals. However, these volatiles are associated with lipocalins, proteins commonly found in the urine, saliva and scent-gland secretions of mammals. The major urinary proteins (MUPs) in male mouse urine bind thiazole and brevicomin in a central cavity (Robertson et al., 1993) and provide a slow release of the volatiles from deposited scent marks, thus prolonging the longevity of a scent mark as a signalling medium (Hurst et al., 1998). Expression of the MUPs themselves is regulated in part by testosterone, which also makes them candidates for signalling androgen-dependent social status. As the loss of these proteins is costly (see Beynon et al., this volume), the prolific scent marking patterns of competitive males might represent a Zahavian handicap (Zahavi, 1975). If only competitive males able to defend high quality food resources could support the metabolic turnover required to deposit so many costly scent marks, females might use this to assess male competitive ability. However, there is no difference in the amount of urine produced by dominant and subordinate males (Desjardins et al., 1973; Daumae and Kimura, 1988) and recent work from our laboratory has found no quantitative difference in the protein output between dominant and subordinate males (unpublished data). Thus MUP production *per se* does not appear to signal male social status (see Hurst et al., this volume).

3.2 Scent Mark Distribution

Dominant and subordinate male mice show a dramatic difference in their scent marking behaviour. While dominant or isolated males deposit urine marks at a very high rate in small spots and streaks throughout their territory, subordinate males mark at a much lower rate and deposit more large spots and pools (e.g. Desjardins et al., 1973; Jemiolo et al., 1992). Dominant but not subordinate males also increase their rate of scent marking to counter-mark scent marks of other males (Hurst, 1990; Nevison et al., 2000). The widely scattered pattern of scent marks and counter-marks deposited by competitive males appears to play an important role in communicating their ability to defend their territory (Hurst et al., this volume). However, it is not yet known whether individual differences in the rate and pattern of scent mark deposits affects a male's attractiveness as a mate or whether females use status-related differences in scent marking to discriminate between dominant and subordinate males as well as differences in the quality of their scents. Results of experiments of our laboratory suggest that females are attracted to dominant male urine from a distance, regardless of the size or pattern of scent marks, but only discriminate against scent from a subordinate male and fail to approach this if the subordinate's scent mark is large (>5 µl) (unpublished data). The status related difference in scent mark size may thus have an important signalling function.

4. HEALTH STATUS AND FEMALE CHOICE

As well as providing females with cues about competitive ability, male scent marks may give females information about the author's health status. Avoidance of males infected by disease can preserve female fitness in two ways, by avoiding fitness costs of illness (e.g. reduced foraging efficiency, reduced fertility, increased mortality risk), and by avoiding matings that might pass on genetic susceptibility to their offspring. Two possible mechanisms for this avoidance are readily suggested. First, females may be able to detect odours of parasite by-products directly from male urine. Second, infection may have a suppressive effect on androgen levels (Larralde et al., 1995) causing down-regulation of the androgen-dependent semiochemicals in urine and their deposition in scent marks. Thus infections may prevent males from signalling high competitive ability and reduce their attractiveness to females.

Female mice show an aversive response to the odour of males infected with the nematode *Heligmosomoides polygurus* (Kaveliers and Colwell, 1995). Females also spend less time investigating urine from males infected with influenza, though the urine is not aversive (Penn et al., 1998). Such a reduction in the attractiveness of male urine might be caused by a decrease in serum androgen concentration as a result of the infection. Morales et al. (1996) demonstrated that male sexual behaviour decreased as a result of the oestrogenation and deandrogenation associated with tapeworm (*Taenia crassiceps*) infection, and that these behavioural consequences were fully reversible if testosterone was administered.

As yet, little is known about the effects of disease upon scent marking and counter-marking behaviour. The costs associated with disease, such as reduced testosterone levels and reduced motility, are likely to affect the rate and spatial deposition of scent signals. Androgen levels are known to affect marking patterns as well as urine composition in mice (Wolff and Powell, 1984; Harvey et al., 1989; Daumae and Kimura, 1988; Kimura and Hagiwara, 1985), and certain pathogens are known to disrupt androgen levels (Larralde et al., 1995).

As androgens facilitate both the signalling of, and intensity of, competitive ability but also have a suppressive effect on the immune system, there may be a testosterone trade-off between maximising signals of competitive ability and minimising immuno-suppression. This would be an example of an immunocompetence handicap (after Folstad and Karter, 1992); only males with 'good genes' for immunity could afford the immuno-suppressive effect of the high testosterone levels associated with intensely competitive behaviour and signalling. As yet, there is little evidence to either support or refute the Immunocompetence Handicap Hypothesis. However, in the case of house mice, the lack of a simple relationship between exposure to parasites and testosterone levels, which are affected by a wide range of other environmental variables, calls into question whether the immuno-suppressive effects of testosterone could provide the basis of a reliable signal of innate disease resistance or competitive ability. Instead, health status appears to be one of a number of factors that affects male competitive ability and it seems more likely that females use scent signals to detect the combined outcome of these factors on male social status.

5. MHC AND RELATEDNESS

The immune system is also known to influence mate choice via the major histocompatibility complex. The MHC is a large, hyper-variable chromosomal region containing several closely linked genes critical for immunological recognition (Klein, 1986). Mice can discriminate between one another by using urinary odours that are influenced by the MHC (reviewed by Penn and Potts, 1998a). Mice can distinguish between individuals from inbred strains that are genetically identical except in the MHC region (Yamazaki et al., 1979) and at single MHC loci (Yamazaki et al., 1983; Penn and Potts, 1998b). How MHC genes influence odour is not understood, but MHC types are associated with differences in the relative concentrations of volatile acids held by MHC peptides or MUPs in mouse urine (Singer et al., 1997). Although it was suggested that these differences might derive from commensal microorganisms determined by MHC type (Singh et al., 1990), this does not appear to be the case in mice (Yamazaki et al., 1990). Inbred females prefer to mate with males whose MHC odour-types are dissimilar from the female's familial MHC odour-type, a preference that is maintained by cross-fostered females which will select males of their own odour-type over those that share the odour-type of their foster mother (Egid and Brown, 1989; Eklund, 1997; Penn and Potts, 1998c). However, Eklund (1999) failed to find such preferences in wild-derived mice, which preferred one specific MHC type rather than non-familial types against a standard genetic background. Laboratory studies also suggest that MHC recognition is disrupted by other variations in genetic background (Eggert et al., 1996) or by environmental changes (Hurst et al., this volume) that would normally occur in wild populations.

If further evidence confirms that wild females prefer MHC-dissimilar mates, the adaptive significance may be because this produces MHC heterogeneity in the offspring and thus greater disease resistance (Potts and Wakeland, 1993). Alternatively, females might use scents derived from the highly polymorphic MHC as a marker of genetic relatedness more generally to avoid inbreeding (Penn and Potts, 1998b). It is clear, though, that social status overrides MHC-type as a mate choice criterion, as females invariably select dominant males over subordinates (Wolff, 1985; Hurst, 1987) regardless of the relative MHC similarities (Potts *et al.*, 1994). In populations of wild mice crossed with inbred strains of known MHC-type, female preferences among dominant territorial males resulted in a bias against those of their familial MHC-type, though not complete avoidance, and the bias was apparent in only half the replicate populations studied (Potts *et al.*, 1991; Penn and Potts, 1998c). Since differences in the competitive ability of territory owners are also known to affect female choice (Wolff, 1985; Rich and Hurst, 1998, 1999; Hurst *et al.*, this volume), it will be important to establish how preferences based on MHC type interact with preferences based on individual competitive ability.

6. CONCLUSION

The competitive social status of a male as a dominant territory owner or a subordinate is a very important factor in female mate choice in semi-natural mouse populations. Female discrimination appears to be based on qualitative differences in a number of androgen-dependent volatiles in male urine, resulting in an almost exclusive mating preference for dominant territory owners. It is not yet known whether females also use status-related differences in scent deposition patterns to discriminate between males. However, variations in the laboratory models used to examine the behavioural and chemical basis of male social strategies call into question how well such studies reflect normal dominant-subordinate relationships between males in natural populations. Further studies are thus required to confirm the chemical basis of status signalling under natural or semi-natural conditions, while long-term studies of lifetime reproductive success would also allow estimation of the fitness value of different male social strategies. Androgen-dependent semiochemicals and scent marking may allow females to select mates that are healthy and resistant to diseases but, as yet, little is known about the chemical mechanisms used to detect infection status. Social status appears to be more important than MHC odour-types in determining female mate choice, while evidence for the importance of MHC odour-type in determining mate choice among wild mice is still equivocal. Further studies are needed under conditions reflecting the normal genetic and environmental variation found in wild mouse populations and individual variation in competitive ability and attractiveness of territorial males.

ACKNOWLEDGMENTS

We would like to thank Liverpool University, the Rhoda Le Marchant Bankes Travel Fund and the BBSRC for financial support. We also wish to thank members of the Animal Behaviour and Protein Function Groups at Liverpool for comments, suggestions and encouragement.

REFERENCES

Bateman, A. J. P., 1948, Intra-sexual selection in *Drosophila*, *Heredity*, 2:349-368.
Bishop, M. J., and Chevins, P. F. D., 1987, Urine odours and marking patterns in territorial laboratory mice (*Mus musculus*), *Behav. Process.* 15:233-248.
Bronson, F. H., 1973, Establishment of social rank among grouped male mice: relative effects on circulating FSH, LH, and corticosterone, *Physiol. Behav.* 10:947-951.
Bronson, F. H., and Whitten W. K., 1968, Oestrus-accelerating pheromone of mice: assay, androgen dependency and presence in bladder urine, *J. Reprod. Fertil.* 15:131-134.

Caroom, D., and Bronson, F. H., 1971, Responsiveness of female mice to preputial attractant: effects of sexual experience and ovarian hormones, *Physiol. Behav.* **7**:659-662.

Daumae, M., and Kimura, T., 1988, Factors regulating urination patterns in male and female mice (*Mus musculus*), *Zool. Sci.* **5**:855-861.

Desjardins, C., Maruniak, J. A., and Bronson, F. H., 1973, Social rank in house mice: differentiation revealedby ultraviolet visualization of urinary marking patterns, *Science* **40**:939-941.

Drickamer, L. C., 1992, Oestrus female house mice discriminate dominant from subordinate males and sons of dominant from sons of subordinate males by odour cues, *Anim. Behav.* **43**:868-870.

Eggert, F., Höller, C., Luszyk, D., Müller-Ruchholtz, W., and Ferstl, R., 1996, MHC-assortive and MHC-independent urinary chemosignals in mice, *Physiol. Behav.* **59**:57-62.

Egid, K., and Brown , J., 1989, The major histocompatibility complex and female mating preferences in mice, *Anim. Behav.* **38**:548-550.

Eklund, A., 1997, The effect of early experience on MHC-based mate preferences in two B10.W strains of mice (Mus domesticus), *Behav. Genet.* **27**:223-229.

Eklund, A., 1999, Use of the MHC for mate choice in wild house mice (*Mus domesticus*), *Genetica* **104**:245-248.

Folstad, I., and Karter, A., 1992, Parasites, bright males and the immunocompetence handicap, *Am. Nat.* **139**:603-622.

Gerlach, G., 1990, Dispersal mechanisms in a captive wild house mouse population (*Mus domesticus* Rutty), *Biol. J. Linn. Soc.* **41**:271-277.

Gosling, L. M., Atkinson, N. W., Dunn, S., and Collins, S. A., 1996, The response of subordinate male mice to scent marks varies in relation to their own competitive ability, *Anim. Behav.* **52**:1185-1191.

Harvey, S., Jemiolo, B., and Novotny, M., 1989, Pattern of volatile compounds in dominant and subordinate male mouse urine, *J. Chem. Ecol.* **15**:2061-2071.

Huck, U. W., 1982, Pregnancy block in laboratory mice as a function of male social status, *J. Reprod. Fertil.* **66**:181-184.

Hurst, J. L., 1987, Behavioural variation in wild house mice (*Mus domesticus* Rutty): a quantitative assessment of female social organisation, *Anim. Behav.* **35**:1846-1857.

Hurst, J. L., 1990, Urine marking in populations of wild house mice *Mus domesticus* Rutty 1. Communication between males, *Anim. Behav.* **40**:209-222.

Hurst, J. L., Robertson, D. H. L., Tolladay, U., and Beynon, R. J., 1998, Proteins in urine scent marks of male house mice extend the longevity of olfactory signals, *Anim. Behav.* **55**:1289-1297.

Jemiolo, B., Alberts, J., Sochinski-Wiggins, S., Harvey, S., and Novotny, M., 1985, Behavioral and endocrine responses of female mice to synthetic analogs of volatile compounds in male mice, *Anim. Behav.* **33**:1114-1118.

Jemiolo, B., Xie, T., and Novotny, M., 1991, Socio-sexual olfactory preferences in female mice: attractiveness of synthetic chemosignals, *Physiol. Behav.* **50**:1119-1122.

Jemiolo, B., Xie, T., and Novotny, M., 1992, Urine marking in male mice: responses to natural and synthetic chemosignals, *Physiol. Behav.* **52**:521-526.

Jones, R. B., and Nowell, N. W., 1973, Aversive and aggression-promoting properties of urine from dominant and subordinate male mice, *Anim. Learn. Behav.* **1**:207-210.

Jones, R. B., and Nowell, N. W., 1989, Aversive potency of urine from dominant and subordinate male laboratory mice (*Mus musculus*): resolution of a conflict, *Aggr. Behav.* **15**:291-296.

Kaveliers, M., and Colwell, D. D., 1995, Odours of parasitized males induce aversive responses in female mice, *Anim. Behav.* **50**:1161-1169.

Kimura, T., and Hagiwara, Y., 1985, Regulation of urine marking in male and female mice: effects of sex steroids, *Hormon. Behav.* **19**:64-70.

Klein, J., 1986, *Natural history of the histocompatibility complex*, Wiley, New York.

Larralde, C. J., Morales, J., Terrazas, I., Govezensky, T., and Romano, M. C., 1995, Sex hormone changes induced by the parasite lead to feminization of the male host in murine *Taenia crassiceps* cysticercosis, *J. Steroid Biochem. Molec. Biol.* **52**:575-580.

Lombardi, J. R., and Vandenbergh, J. G., 1977, Pheromonally induced sexual maturation in females: regulation by the social environment of the male, *Science* **196**:545.

Morales, J., Larralde, C., Arteaga, M., Govezensky, T., Romano, M. C., and Morali, G., 1996, Inhibition of sexual behavior in male mice infected with *Taenia crassiceps* cysterci, *J. Parasitol.* **82**:689-693.

Nevison, C. M., Barnard, C. J., Beynon, R. J., and Hurst, J. L., 2000, The consequences of inbreeding for recognising competitors, *Proc. R. Soc. B* **267**:687-694.

Novotny, M., Jemiolo, B., and Alberts, J., 1985, Synthetic pheromones that produce inter-male aggression in mice, *Proc. Natl. Acad. Sci. USA* **82**:2059-2061.

Novotny, M., Harvey, S., and Jemiolo, B., 1990, Chemistry of male dominance in the house mouse *Mus domesticus*, *Experientia* **46**:109-113.

Novotny, M., Ma, W. D., Wiesler, D., and Zidek, L., 1999, Positive identification of the puberty-accelerating pheromone of the house mouse: the volatile ligands associating with the major urinary protein, *Proc. R. Soc. Lond. B* **266**:2017-2022.
Orians, G. H., 1969, On the evolution of mating systems in birds and mammals, *Am. Nat.* **103**:589-603.
Penn, D., and Potts, W. K., 1998a, The evolution of mating preferences and major histocompatibility complex genes, *Am. Nat.* **153**:145-164.
Penn, D., and Potts, W. K., 1998b, Untrained mice discriminate MHC-determined odours, *Physiol. Behav.* **63**:235-243.
Penn, D., and Potts, W. K., 1998c, MHC-dissortive mating preferences reversed by cross-fostering, *Proc. R. Soc. Lond. B* **265**:1299-1306.
Penn, D., Schneider, G., White, K., Slev, P., and Potts, W., 1998, Influenza infection neutralizes the attractiveness of male odour to female mice (*Mus musculus*), *Ethology* **104**:685-694.
Potts, W. K., and Wakeland, E. K., 1993, Evolution of MHC genetic diversity: a tale of incest, pestilence and sexual preference, *Trends Genet.* **9**:408-412.
Potts, W. K., Manning, C. J., and Wakeland, E. K., 1991, Mating patterns in seminatural populations of mice influenced by mhc genotype, *Nature* **352**:619-621.
Potts, W. K., Manning, C. J., and Wakeland, E. K., 1994, The role of infectious disease, inbreeding, and mating preferences in maintaining MHC diversity: an experimental test, *Phil. Trans. R. Soc. Lond. B* **346**:369-378.
Rich, T. J., and Hurst, J. L., 1998, Scent marks as reliable signals of the competitive ability of mates, *Anim.Behav.* **56**:727-735.
Rich, T. J., and Hurst, J. L., 1999, The competing countermarks hypothesis: reliable assessment of competitive ability by potential mates, *Anim. Behav.* **58**:1027-1037.
Robertson, D. H. L., Beynon, R. J., and Evershed, R. P., 1993, Extraction, characterization, and binding analysis of two pheromonally active ligands associated with major urinary proteins of the house mouse (*Mus musculus*), *J. Chem. Ecol.* **19**:1405-1416.
Sawyer, T. F., 1978, Aversive odours of male mice: experimental and castration effects and the predictability of the outcomes of agonistic encounters, *Aggr. Behav.* **15**:291-296.
Singh, P. B., Herbert, J., Roser, B., Arnott, J., Tucker, D., and Brown, R., 1990, Rearing rats in a germ-free environment eliminates their odours of individuality, *J. Chem. Ecol.* **16**:1667-1682.
Singer, A. G., Beauchamp, G. K., and Yamazaki, K., 1997, Volatile signals of the major histocompatibility complex in male mouse urine, *Proc. Natl. Acad. Sci. USA* **9**:2210-2214.
Wolff, R. J., 1985, Mating behavior and female choice: their relation to social structure in wild caught house mice *Mus musculus* housed in a semi-natural environment, *J. Zoology* **207**:43-51.
Wolff, P. R., and Powell, A. J., 1984, Urine Patterns in mice:an analysis of male/female counter-marking, *Anim. Behav.* **32**:1185-1191.
Yamazaki, K., Yamaguchi, M., Baranoski, L., Bard, L. D., Boyse, E. A., and Thomas, L., 1979, Recognition among mice: evidence from the use of a Y-maze differentially scented by congenic mice of different major histocompatibility types, *J. Exp. Med.* **150**:755-760.
Yamazaki, K., Beauchamp, G. K., Kupniewski, D., Bard, J., Thomas, L., and Boyse, E. A., 1983, Sensory distinction between $H-2^b$ and $H-2^{bm1}$ mutant mice, *Proc. Natl. Acad. Sci. USA* **80**:5685-5688.
Yamazaki, K., Beauchamp, G. K., Imai, Y., Bard, J., Phelan, S. P., Thomas, L., and Boyse, E. A., 1990, Odor types determined by the major histocompatibility complex in mice, *Proc. Natl. Acad. Sci. USA* **21**:8431-8436.
Zahavi, A., 1975, Mate selection – a selection for a handicap, *J. Theor. Biol.* **53**:205-214.
Zahavi, A., 1987, The theory of signal selection and some of its implications, in: *International Symposium of Biological Evolution* (V. P. Delfino, ed.), Adriatica Editrice, Bari, Italy.
Zielinski, W. J., and Vandenbergh, J. G., 1993, Testosterone and competitive ability in house mice, *Mus domesticus*: laboratory and field studies, *Anim. Behav.* **45**:873-891.

EFFECTS OF INBREEDING AND SOCIAL STATUS ON INDIVIDUAL RECOGNITION IN MICE

Charlotte M. Nevison,[1] Christopher J. Barnard,[3] Robert J. Beynon,[2] and Jane L. Hurst[1]

[1]Animal Behaviour Group and [2]Protein Function Group
Faculty of Veterinary Science
University of Liverpool
Leahurst, Neston CH64 7TE, UK
[3]School of Biology
University of Nottingham
Nottingham NG7 2RD, UK

1. INTRODUCTION

Individual recognition modulates social behaviour between conspecifics, enabling an animal to assess its familiarity and kinship with other individuals (Barnard et al., 1991; Barnard and Aldhous, 1991). Wild mice identify each other through individually unique urinary odour cues that are determined, at least in part, by genetic differences (Eggert et al., 1996). By depositing these unique odour cues as scent marks, mice provide signals of their presence and social status (see Hurst et. al., this volume). Dominant males deposit scent marks at high frequency as a sign of their competitive quality and current territorial ownership, and increase their rate of scent marking where they encounter competing scent marks from other males in their territory. Countermarking of scent marks from other males by dominant males thus provides a specific test that the mice have recognised scent marks as being derived from another individual.

Although countermarking is costly in terms of the physical energy expended and protein loss, the system has benefits. For instance, females use these marks to assess and choose high quality mates (Rich and Hurst, 1999). Males also match deposited urinary odours with the body odours of males they encounter to identify territory owners or subordinate group members (Gosling and McKay, 1990; Hurst, 1993; Hurst et al., 1993), allowing them to modulate their competitive behaviour appropriately towards specific individuals upon encounter. Dominant males tolerate individuals they recognise as familiar and non-competitive (e.g. Hurst et al., 1993), but escalate their aggression towards unfamiliar and competitive individuals.

In captivity, laboratory mice are usually caged in single strain and sex groups. Under these confined conditions, mice modulate their aggression towards those with whom they are housed. However, most strains are inbred, thus individuals of the same sex within these strains will share the same genetically determined 'individuality' odour cues. This suggests that they may be unable to discriminate between different individuals of their own sex and strain, or their urinary odour cues. Individuals of highly inbred strains fail to discriminate each other's volatile odours when they are housed under identical conditions (Yamaguchi *et al.*, 1981). However, environmental factors such as food type, bacterial gut flora and social pressure induce changes in the volatile odours emitted by animals (Apps *et al.*, 1988; Jones and Nowell, 1989; Schellinck *et al.*, 1992). It is possible that such environmentally induced differences might be used as a basis for individual recognition against an identical genetic background. In this paper, we review some of our recent studies (Nevison, data in preparation) that test this and examine whether mice can discriminate between familiar and unfamiliar individuals of the same inbred strain.

2. SOCIAL STATUS AND INDIVIDUAL RECOGNITION

To assess whether environmentally-induced status differences in odours play a part in individual recognition in mice, we compared the ability of inbred and outbred male laboratory mice to discriminate between the odours of males from different social classes together with their ability to discriminate between odours of males from their own or another strain (Nevison *et al.*, 2000). Since outbred males are genetically heterozygous, we predicted that outbred males would recognise the scent marks of any other males as different from their own through genetically-determined odours. Like wild mice (Desjardins *et al.*, 1973; Hurst 1990), dominant males should increase their rate of counter-marking to urine from another male of their own strain, whether from another dominant or a subordinate male. They should also counter-mark urine from another male whether from their own or another strain. By contrast, inbred dominant males should only counter-mark scent marks from males of another strain that possess individuality odours distinct from their own. If status-induced odour differences (Apps *et al.*, 1988; Harvey *et al.*, 1989; Jones and Nowell, 1989; Novotny *et al.*, 1990) are used in individual recognition, inbred dominant males should counter-mark scent marks from subordinate males of their own strain that differ in social status.

These predictions were tested by presenting mice with different types of urine or water in a clean arena, alongside a control water mark, to measure both investigatory and counter-marking (Hurst, 1990; Humphries *et al.*, 1999) responses. Only dominant individuals showed significant counter-marking in response to scent marks; as expected, subordinate males did not deposit many urine marks (Desjardins *et al.*, 1973).

Dominant outbred males (ICR and TO strains) responded to a scent mark from an unfamiliar male of their own strain by counter-marking strongly (Figure 1). The status of the odour donor had no effect on their countermarking responses. Despite a difference in their readiness to counter-mark, dominant and subordinate outbred males spent similar amounts of time investigating urine scents from males of their own strain. Both spent longer investigating subordinate male urine than that from a dominant male (Figure 2).

BALB/c males were stimulated to investigate urine from males of their own strain and could discriminate a difference in the donor's dominance status (Figure 2). However, despite this investigatory behaviour, dominant BALB/c males did not counter-mark urine from another male of their own strain (Nevison *et al.*, 2000). This is consistent with a failure to recognise that the mark came from a potential competitor that was genetically identical to themselves. By contrast, they counter-marked urine from a male of another strain, genetically distinct from themselves.

The investigatory responses showed that the inbred males discriminated a difference in the odours of dominant and subordinate males, but this is not a specific test of individual recognition. Counter-marking is a more specific test – any dominant male should counter-mark scent marks from other males regardless of their status or familiarity. Although the urinary scents of other males from the same inbred strain came from animals that were apparently very close relatives and their scent would have been highly familiar, males should still counter-mark the scents of any other males to advertise their own competitive ability (see Hurst *et al.*, this volume). Wild house mice readily counter-mark the familiar scents of neighbours and close relatives (Hurst, 1990). We thus interpret these data as showing that chemical information signalling social status is additive to information conveying genetically-determined individual identity. Further, males do not use such environmentally-determined cues to discriminate between individuals against a common genetic background.

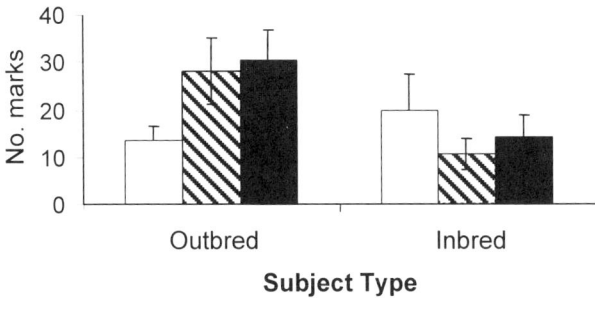

Figure 1. Countermarking responses of male mice (outbred ICR, TO and inbred BALB/c) to control stimuli (water) and urine from males of their own strain.

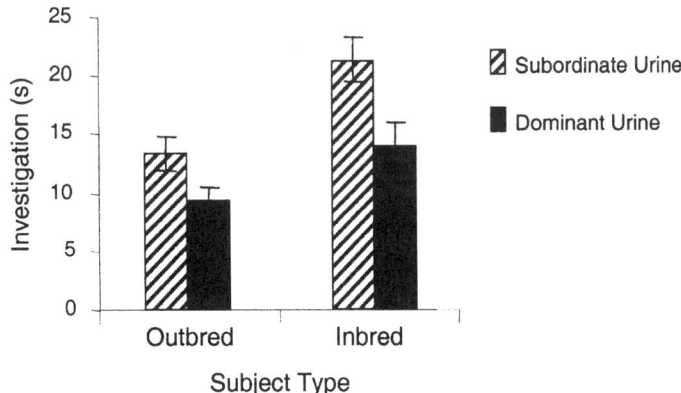

Figure 2. Investigation of own strain urinary odours from males of different social status by outbred (ICR, TO) and inbred (BALB/c) mice.

3. RECOGNITION OF FAMILIAR INDIVIDUALS

Our previous study examined whether mice recognised the urine scent marks of other males as different from their own. In a subsequent study, we have examined whether mice can discriminate between familiar or unfamiliar individuals, either from their urinary scent marks or when they meet each other (Nevison, data in preparation). Discrimination between scent marks from different individuals relies solely on chemical information in the scent. When animals meet each other, they receive additional information, such as individual differences in behaviour, acoustic or visual characteristics or body odour scents, which might also be used in individual recognition. If any of these characteristics are influenced by experience, status or the development of the animal, as well as by genetic make-up, animals may be able to make more sensitive discriminations between individuals during interactions.

We thus examined whether inbred (BALB/c) and outbred (ICR) males could discriminate between familiar cage companions and unfamiliar mice of their own strain, or between unfamiliar mice from their own or another strain. Firstly, we presented mice in their home cage with a choice between the odours (urine marks and soiled cage substrate) of mice of their own strain with whom they had been recently housed ('familiar') and odours from mice of their own strain that they not interacted with ('unfamiliar'). Odours were urine marks and soiled cage substrate from pairs of males, thus should represent both dominant and subordinate male scents. As expected, the outbred males investigated odours from unfamiliar individuals more than those from recent cagemates. By contrast, males from the genetically homozygous strain spent the same amount of time investigating odours whether from familiar recent cagemates or unfamiliar mice of the same strain, suggesting that they failed to recognise that some of the scents were unfamiliar. However, when animals were given a choice between scents from unfamiliar males of their own or the other strain, inbred males spent much more time investigating scents from the unfamiliar strain. Results were thus consistent with our previous study, suggesting that males only recognised that scents were from unfamiliar individuals when donors differed genetically from familiar individuals. Interestingly, outbred males spent as much time investigating scents from unfamiliar males of their own strain as they spent investigating BALB/c strain scents, suggesting that the scent of other individuals of their own strain was as novel as that of another strain.

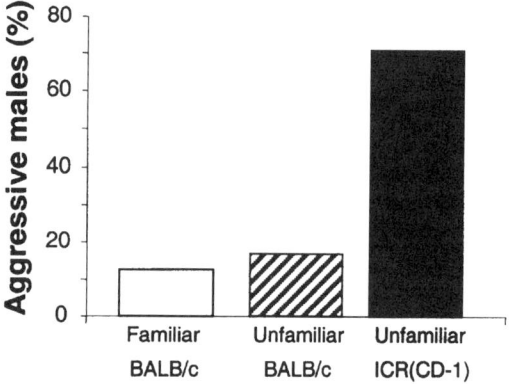

Figure 3. Aggressive responses of male inbred BALB/C mice to opponents of different familiarity or strain.

Secondly, we tested whether animals were more aggressive towards unfamiliar animals introduced into their home cage than towards familiar companions from whom they had been separated for 24h but maintained in olfactory contact with shared soiled cage substrate (Hurst et al., 1993). While outbred males were more aggressive towards unfamiliar individuals of their own strain as expected, inbred males failed to discriminate between familiar and unfamiliar individuals of their own strain either in terms of social investigation or aggression. Very few BALB/c males showed any aggression when animals of their own strain were introduced into their home cage. By contrast, most BALB/c males attacked males of the ICR strain when they met (Figure 3), even though these males were considerably larger than themselves. This response did not appear to be induced by aggressive behaviour from the ICR males since there was a negative correlation between the aggression shown by BALB/c and ICR opponents. They thus responded aggressively to unfamiliar males only when these were genetically distinct from themselves and their familiar cage companions.

4. CONCLUSIONS

Both counter-marking and aggressive responses were consistent with the hypothesis that mice use genetically-determined odours to discriminate between different individuals. Individual identity cues need to be stable and resistant to environmental changes that might cause an animal to suddenly change its identity. Differences in social status, hormone levels, pathogens and nutrients may all contribute to odour-based discriminations between individuals (e.g. see Brown, 1995). However, individual identity cues need to be stable. They also need to vary sufficiently between individuals to allow different animals to be identified even when they are quite closely related, or where there is relatively little genetic variation within a population. Status differences are unlikely to provide this degree of individual variability and will vary according to the recent social experiences of the individual concerned, unlike genetically-determined odours. Individual identity is thus best coded by genetically-determined differences in odours that are resistant to change by environmental influences. Animals in the two laboratory outbred strains ICR and TO were able to discriminate easily between the odours of different individuals from the same strain, even though they came from restricted laboratory strains in which individuals must share a high degree of similarity compared to wild populations (see Payne et al., this volume). The ability to discriminate between individuals on the basis of limited genetic differences is consistent with other studies (reviewed by Boyse et al., 1991; Brown, 1995), though most studies of individual identity cues have used assays based on investigatory responses such as habituation-dishabituation tests or an olfactory discrimination learning paradigm. As we have shown here, scents may stimulate differences in investigation but this does not imply that animals interpret these scents as coming from different individuals. Animals are likely to be interested in many different types of information in scents that stimulate an investigatory response. It remains to be determined whether other environmentally induced differences in odours, such as changes in hormone levels, pathogens or nutrients which are likely to occur in animals living under natural conditions, are used in identifying different individuals.

Although the fact that individuals from the same inbred strain share the same genetically determined odour template has been well recognised and used as a tool for olfactory communication research, the implications of this for the welfare of inbred strains in the laboratory and for husbandry practices have received little attention. Aggression between male laboratory mice can be a significant problem for maintenance in the laboratory, particularly if unfamiliar adult animals are mixed (Jennings et al., 1998;

Koolhaas, 1999). Such mixing is unlikely to present a problem in inbred strains as all individuals will be perceived as familiar, as long as only subordinate animals are moved between groups. However, since scent marks normally function to allow dominant and subordinate animals to identity each other, hierarchies are likely to be unstable in aggressive inbred strains and apparently spontaneous outbreaks of injurious aggression are often reported. Lastly, it would be interesting to investigate whether the restricted olfactory environment that inbred animals experience constrains the neuronal development of their olfactory system (e.g. Benson et al., 1984).

ACKNOWLEDGMENTS

This work was funded by a BBSRC research studentship to CMN and research grants to JLH, CJB and RJB.

REFERENCES

Apps, P. J., Rasa, A., and Viljoen, H. W., 1988, Quantitative chromatographic profiling of odours associated with dominance in male laboratory mice. *Aggr. Behav.* **4**:451-461.

Barnard, C. J., and Aldhous, P., 1991, Kinship, kin discrimination and mate choice, in: *Kin recognition*, (P. G. Hepper, ed.), Cambridge University Press, Cambridge.

Barnard, C. J., Hurst, J. L., and Aldhous, P., 1991, Of mice and kin: the functional significance of kin bias in social behaviour. *Biol. Rev.* **66**:379-430.

Benson, T. E., Ryugo, D. K., and Hinds, J. W., 1984, Effects of sensory deprivation on the developing mouse olfactory system: a light and electron microscopic, morphometric analysis. *J. Neurosci.* **4**:638-653.

Boyse, E. A., Beauchamp, G. K., Yamazaki, K., and Bard, J., 1991, Genetic components of kin recognition in mammals. In: *Kin Recognition* (P. G. Hepper, ed.), pp. 148-161. Cambridge University Press.

Brown, R. E., 1995, What is the role of the immune-system in determining individually distinct body odors? *Int. J. Immunopharm.*. **17**:655-661.

Desjardins, C., Maruniak, J. A., and Bronson, F. H., 1973, Social rank in the house mouse: differentiation revealed by ultra-violet visualisation of urinary marking patterns. *Science* **182**:939-941.

Eggert, F., Höller, C., Luszyk, D., Müller-Ruchholtz, W., and Ferstl, R., 1996, MHC-associated and MHC-independent urinary chemosignals in mice. *Physiol. Behav.* **59**:57-62.

Gosling, L. M., and McKay, H. V., 1990, Competitor assessment by scent-matching: an experimental test. *Behav. Ecol. Sociobiol.* **26**:415-420.

Harvey, S., Jemiolo, B., and Novotny, M., 1989, Pattern of volatile components in dominant and subordinate male mouse urine. *J. Chem. Ecol*. **15**:2061-2072.

Humphries, R. E., Robertson, D. H. L., Beynon, R. J., and Hurst, J. L., 1999, Unravelling the chemical basis of competitive scent marking. *Anim. Behav.* **37**:705-725.

Hurst, J. L., 1990, Urine marking in populations of wild house mice *Mus domesticus* Rutty I. Communication between males. *Anim. Behav.* **45**:55-81.

Hurst, J. L., 1993, The priming effects of urine substrate marks on interactions between male house mice, *Mus musculus domesticus* Schwarz and Schwarz. *Anim. Behav.* **45**:55-81.

Hurst, J. L., Fang, J. M., and Barnard, C. J., 1993, The role of substrate odours in maintaining social tolerance between male house mice, *Mus musculus domesticus*. *Anim. Behav.* **45**:997-1006.

Jennings, M., Batchelor, G. R., Brain, P. F., Dick, A., Elliott, H., Francis, R. J., Hubrecht, R. C., Hurst, J. L., Morton, D. B., Peters, A. G., Raymond, R., Sales, G. D., Sherwin, C. M., West, C., 1998, Refinements in rodent husbandry: the mouse. *Lab. Anim.* **32**:233-259.

Jones, R.B., and Nowell, N.W., 1989, Aversive potency of urine from dominant and subordinate male laboratory mice (*Mus musculus*): resolution of a conflict. *Aggr. Behav.* **15**:291-296.

Koolhaas, J. M., 1999, The laboratory rat. In: *The UFAW Handbook on the Care and Management of Laboratory Animals* (T. B. Poole, ed.), Oxford: Blackwell Science, pp.313-330.

Nevison, C. M., Barnard, C. J., Beynon, R. J., and Hurst, J. L., 2000, The consequences of inbreeding for recognizing competitors. *Proc. R.. Soc. Lond. B.* **267**:687-694.

Novotny, M., Harvey, S., and Jemiolo, B., 1990, Chemistry of male dominance in the house mouse *Mus domesticus*. *Experientia*, **46**:109-113.

Rich, T. J., and Hurst, J. L., 1999, The competing countermarks hypothesis: reliable assessment of competitive ability by potential mates. *Anim. Behav.* **88**:1027-1037.

Schellinck, H. M., West, A. M., and Brown, R. E., 1992, Rats can discriminate between the urine odors of genetically identical mice maintained on different diets. *Physiol. Behav.* **51**:1079-1082.

Yamaguchi, M., Yamazaki, K., Beauchamp, G. K., Bard, J., Thomas, L., and Boyse, E. A., 1981, Distinctive urinary odours governed by the major histocompatability locus of the mouse. *Proc. Natl. Acad. Sci. USA* **78**:5817-5820.

HETEROGENEITY OF MAJOR URINARY PROTEINS IN HOUSE MICE: POPULATION AND SEX DIFFERENCES

Caroline E. Payne,[1] Nick Malone,[1] Rick Humphries,[1] Carl Bradbrook,[1] Christina Veggerby,[2] Robert J. Beynon,[2] and Jane L. Hurst[1]

[1]Animal Behaviour Group and [2]Protein Function Group
Faculty of Veterinary Science
University of Liverpool
Leahurst, Neston CH64 7TE, UK

1. INTRODUCTION

House mice (*Mus domesticus*) and rats (*Rattus norvegicus*) secrete large quantities of protein into their urine as the normal condition (Finlayson and Baumann, 1958). These proteins are major urinary proteins (MUP) and alpha$_{2u}$-globulins respectively. Both are small proteins that have been assigned to the lipocalin protein family on the basis of sequence data (reviewed by Flower, 1996) and X-ray crystallography (Bocskei *et al.*, 1992). As such, both display a typical lipocalin beta-barrel structure enclosing an internal cavity in which volatile ligands may bind.

Over 99% of the protein that mice produce in their urine is MUP (Humphries *et al.*, 1999) and the remainder includes peptides of the major histocompatibility complex (MHC) (Singer *et al.*, 1993). MHC cell-surface glycoproteins are involved in cell recognition by the immune system and show extreme polymorphism. As such MHC peptides have been implicated in signalling individuality through odours (reviewed, Yamazaki *et al.*, 1992). Although rodents use volatile odours to identify MHC types and MHC peptides are not themselves volatile, two mechanisms of detection have been suggested that could account for this apparent contradiction. Small fragments of the MHC could themselves be volatile and hence provide a direct method of detection. Alternatively, MHC peptides could associate with a mixture of volatile metabolites unique to that individual (Singh *et al.*, 1987). However, these cues are susceptible to disruption by environmental effects such as diet (Schellink *et al.*, 1997) and bacteria gut flora (reviewed by Brown, 1995) and as such prove unreliable.

MUPs bind a number of volatile ligands and most of the research into chemical signalling in house mice has focused on the volatile component of the signal. Six volatile molecules that have possible signalling functions have so far been found in male mouse urine. In particular, 2-*sec*-butyl-4,5 dihydrothiazole (thiazole) and 3,4-dehydro-*exo*-

brevicomin (brevicomin), which stimulate puberty and synchronise oestrus (Novotny *et al.,* 1999a; Jemiolo *et al.,* 1986) are bound to MUP (Bacchini *et al.,* 1992; Robertson *et al.,*1993) and a key function of MUPs is to delay the release of volatile ligands from deposited scent marks (Hurst *et al.,* 1998).

Although volatile components of the male signal have been targeted as the most likely candidates for stimulating the behavioural and physiological responses to male mouse urine (see Leinders-Zufall *et al.,* 2000 for full list), the protein-ligand complexes have also been implicated. In particular, the protein fraction of the urine stimulates countermarking (Humphries *et al.,* 1999), causes pregnancy block (Marchlewska-Koj, 1981) and accelerates puberty (Vandenbergh *et al.,* 1975). The protein fraction of the urine stripped of most or all of its volatile ligands is more effective than thiazole and brevicomin at causing pregnancy block (Brennan *et al.,* 1999). In addition, an N-terminal MUP-like hexapeptide has been reported as being effective in accelerating puberty (Mucignat *et al.,* 1995), although this has not been confirmed in a second study (Novotny *et al.,* 1999b). Nonetheless, the potential importance of MUPs to the mouse communication system is clear.

Much research into the production of MUPs has focused on laboratory strains of mice and more specifically on males, probably because of the easy availability of laboratory mice and because male mice are reported to produce 20-30x more protein in their urine than females (Wicks, 1941; Szoka and Paigen, 1978). The MUP family is highly heterogeneous, with at least 13 different MUPs observed in two strains of inbred laboratory mouse (Robertson *et al.,* 1996). Each mouse (male or female) produces a profile of MUPs in its urine. All males and females of the same inbred strain have identical profiles, which is expected since the mice are genetically identical. The female pattern of expression in laboratory mice is simpler than the male pattern, although treatment with testosterone (Szoka and Paigen, 1978) or growth hormone (Al-Shawi *et al.,* 1992) not only raises the output of protein to the male level, but also produces a profile characteristic of the male of that strain. Testosterone and growth hormone in combination with thyroxine are responsible for regulating MUP expression (Clissold *et al.,* 1983; Al-Shawi, 1992; Knopf *et al.,* 1983).

The limited data that are available on the production and heterogeneity of MUPs in wild mice has raised some interesting questions. Like laboratory mice, wild mice (Robertson *et al.,* 1997; Pes *et al.,* 1999) produce a substantial amount of protein in their urine and, furthermore, produce MUPs not previously observed in laboratory strains (Robertson *et al.,* 1997; Pes *et al.,* 1999; Veggerby *et al.,* this volume). Due to the complex molecular characterisation used in these studies (high resolution liquid chromatography combined with electrospray ionisation and MALDI-TOF mass spectrometry), only a small number of individuals have been studied from different sites in the UK. However, this study still raised the intriguing possibility that, wild mice possess individually different MUP profiles. To further test this hypothesis we investigated population and sex differences in wild mice sampled from four geographically separate populations in the UK. We also bred F1 litters from wild caught mice to investigate individual development of protein output and MUP profile.

2. METHODS

2.1. Subjects

2.1.1. Wild caught subjects. Wild house mice were caught from four geographically separate sites in the UK. Feral mice (males only) were captured from the Isle of May (site A) a small island 1.6 km long and 0.5km wide situated in the Firth of Forth, UK. The island is uninhabited except for wardens and research staff present during the summer months, and the mouse population is not dependent on people for its survival. The population has been isolated for over a century except for the introduction (1982) of 77 mice from Eday, an Orkney island by Berry et al. (1991), and it is estimated that the island can support up to 3000 mice (Triggs, 1991).

Commensal mice were captured from pig farm infestations in Berkshire (site B), Oxfordshire (site C) and Cheshire (site D). The Berkshire farm (site B) housed a small infestation of approximately 60 mice in a single farm building (25m x 10m). The Oxfordshire farm housed a large infestation of approximately 300-500 mice living in two separate buildings (60m x 20m and 30m x 15m), which were approximately 300 metres apart. The Cheshire farm also housed a large infestation of approximately 200 – 300 mice spread over three buildings sited a few metres apart. Population estimates were based on daylight sightings, the number of fresh droppings and trapping success.

Mice from site A and site D were fed on laboratory mouse pellets (TRM9607, Harlan Teklad). Those from sites B and C were maintained on their pre-capture pig diet (maize-based pig diet, Growell Feeds, Melksham, UK. and wheat-based pig diet, Livestock Modelling System, Cheltenham, UK. respectively). Food and water were provided ad libitum. Urine evacuated when mice were held by the scruff of the neck, was collected into plastic Eppendorf tubes and stored at $-18^{\circ}C$.

2.1.2. Laboratory bred subjects. Outbred F1 offspring were descended from one wild-caught parent from site 4 and one parent caught from additional sites to those described here. Offspring remained with their mother until one month after birth at which time males were weaned and housed in male only sibling groups whilst female offspring remained with their mother. Urine samples were collected as above.

2.2. Sample Preparation

Whole urine samples were diluted prior to electrophoresis. Adult male samples were defrosted, diluted to 1:10 for isoelectric focusing (IEF) and refrozen ready for use. Female and juvenile samples were assayed for their protein concentration prior to electrophoresis and the dilution for IEF was calculated relative to the protein concentration, generally between 1:2 and 1:5 dilution.

2.3. Isoelectric Focusing

Proteins were separated on the basis of charge on pre-cast, narrow range (pH 4.2 - 4.9) immobilised pH gradient gels (Pharmacia, UK) rehydrated in 15% glycerol and 2.5% ampholine (pH 3.5-9.5), using the Multiphor flatbed electrophoresis system and MultiTemp III thermostatic circulator (Pharmacia UK). Diluted urine samples (5 microlitres) were applied to the gel using paper sample application pieces (Pharmacia UK). Samples were 'run in' for 200Vh, and then sample application pieces were removed and the gel focused for 14.8KVh. All gels were run at $10^{\circ}C$ and double distilled water used as electrolyte solution. Gels were fixed for 1 hour in 20% trichloroacetic acid (w/v), washed in destain solution (ddH$_2$O (60% v/v) methanol (30% (v/v)) acetic acid (10% (v/v)) for 20 min, stained in 0.02% (w/v) Coomassie blue with 0.1% (w/v) CuSO$_4$ for 15 min, destained using the above solution for 15 hours (two washes), and preserved in glycerol (12.5% (v/v)) for 1 hour. All development was carried out automatically in the Hoefer automatic stainer (Pharmacia, UK).

2.4. Protein and Creatinine Assays

Protein concentrations were calculated using bovine serum albumin (BSA) as standard using the „Coomassie Plus Assay" (Pierce Chemicals, Chester, UK.). Creatinine, a by-product of muscle metabolism, is excreted at a constant rate and hence was used as a metabolic standard. Protein creatinine ratios were calculated by dividing urinary protein (mg/ml) by urinary creatinine (mg/ml) to account for differences in the dilution of the urine. Creatinine concentrations were determined by the alkaline picrate assay (Sigma Chemicals, Poole, Dorset, UK.).

3. RESULTS AND DISCUSSION

Although isoelectric focusing is a high-resolution technique it has some limitations. Levels of expression of the same protein (band intensity) may differ between individuals, a single band in an individual's profile may represent more than one protein, or comparable bands between individuals may not be the same protein. Hence, the complexity and heterogeneity that we observe in wild mouse populations is likely to be a conservative underestimate.

3.1. MUP Profile Complexity

Isoelectric focusing was used to determine the MUP profiles of 35 males from site A; 11 males and 10 females from site B; 11 males and 12 females from site C and 30 males from site D. The MUP profiles from individual animals showed considerable complexity (Figure 1), with the number of bands observed in a single profile ranging from four to 15. A small number of bands were consistently present in both sexes and in all populations sampled. 'Signature patterns' present in all or most individuals caught from the same population but unique to that population were also observed.

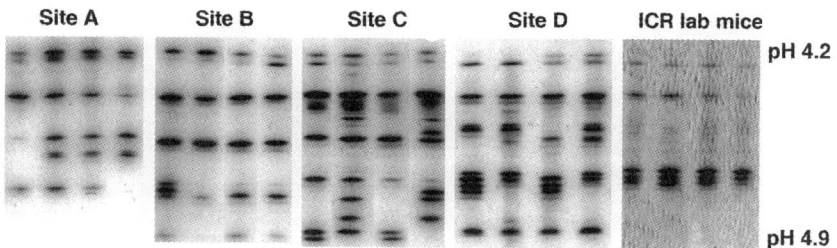

Figure 1. The complexity of male MUP profiles from the sampled sites and a typical laboratory mouse example for comparison. Profiles were separated on the basis of charge on narrow range (pH4.2 – 4.9) immobilised gradient gels (Pharmacia, UK).

From initial observation of the MUP profiles it appeared that males and females displayed similar levels of profile complexity (Figure 2). The total number of bands for each individual profile was counted by eye for all males and females from sites B and C. The effects of sex and population on the number of MUP bands in an individual's profile from these two sites were examined using a parametric analysis of variance. There was no significant difference in profile complexity between males and females (ANOVA: F1,40 = 0.209, NS) and no significant interaction between sex and population (ANOVA: F1,40 = 0.160, NS). However, there was a significant difference in profile complexity between individuals from site B and site C (ANOVA: F1,40 = 25.503, p<0.001) since mice from site C (a large, two building infestation) expressed more MUP bands than those from site B (a small single building infestation).

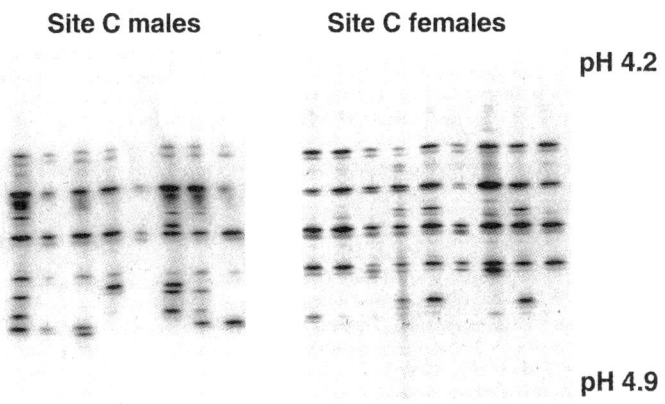

Figure 2. Comparison of MUP profiles produced by males and females from site C suggests that there is no sex difference in profile complexity in wild mice.

On average the profile for a wild mouse contained twice as many bands as for an inbred laboratory strain. This implies a high level of heterozygosity at each MUP locus and confirms the rich source of genetic diversity embedded in MUP alleles. The profile complexity did not differ between males and females. This is contrary to findings in laboratory mice where the pattern of MUP bands was less complex in females than in males. In laboratory mice, males produce more complex profiles than females because expression of MUPs is testosterone and growth hormone dependent and female laboratory mice only produce the full set of proteins observed in male mice when treated with either of these hormones. The lack of sex difference in wild mice suggests that there is a difference in either regulation or hormone levels between laboratory and wild mice. Hormone levels might be affected by the social environment of female wild mice. Unlike laboratory mice (which are often isolated or housed in single sex groups) female wild mice are reproductively active, exposed to males and involved in territory and nest defence, all of which could influence their hormone levels (Wolff, 1985; Hurst, 1987). An alternative explanation is that the inbreeding of laboratory mice has resulted in aberrant regulation of MUP gene expression.

3.2. MUP Profile Sharing

Individual profiles were compared within each sex and population to identify individuals that shared the same profiles. Two profiles were considered to be identical if all bands matched. For this analysis, band intensity was not taken into account, however it was noted that profiles that were identical on the basis of presence/absence of bands often differed in the level of expression (intensity) of certain bands. This intensity difference might indicate additional proteins with identical pI values or the inheritance of the gene for one particular MUP from both parents. Therefore, this analysis is likely to be a conservative overestimate of profile sharing within a population.

Although there was a difference in the amount of profile sharing between males from the four sites sampled (ANOVA: $F_{3,83} = 18.198$, $p<0.001$), this difference was caused by the high degree of sharing between males from the feral Isle of May population. There was no difference in the amount of profile sharing seen between males from the three mainland commensal populations (ANOVA: $F_{2,51} = 3.170$, NS).

Females shared profiles more than males, and females from site B (a small single building infestation) showed a higher degree of profile sharing than females from site C (a large two building infestation) (Mann-Whitney U test: $U = 21.50$, $N_1=10$, $N_2 = 12$, $p<0.01$).

Interestingly, complexity and profile sharing seemed to relate to population size. Large populations had more bands than small populations, and small populations showed more female profile sharing than large, which is likely to relate to restriction of the gene pool. There is more profile complexity in large populations than small ones because there are more animals and hence more alleles, and there is more profile sharing in small populations than large because there is a limit on the available mates and hence individuals are likely to be more closely related. Although there is a large number of animals on the Isle of May, they have been genetically isolated since 1982. There has been significant opportunity for a founder effect and as such it is the extent of differences between profiles, rather than the similarities that is surprising. The degree of profile sharing is higher in female mice than in males, which was not related to profile complexity (see above). It may reflect their behavioural ecology and a male biased dispersal model, where males tend to leave the population and females tend to remain and breed in closely related groups.

3.3. Development of MUP Expression

Serial urine samples were collected from males and females from the ten F1 litters bred in the laboratory from wild-caught mice. Urine samples were analysed for their protein and creatinine concentrations and the protein: creatinine ratios were calculated. Female protein: creatinine ratios remained around 10 from 20-100 days of age. Male ratios remained around 10 until the males reached 30-40 days of age, at which time most males showed a sharp increase in the levels of protein in their urine reaching a ratio of around 30. The timing of this increase in protein production is consistent with the increase in body weight after 20 days of age and testosterone after 30 days of age, witnessed in laboratory raised descendants of wild house mice (*Mus musculus*) (McKinney and Desjardins, 1973).

Serial urine samples taken from 3 males between the ages of 21 and 52 days were run on isoelectric focusing gels to monitor the development of the MUP profiles during the observed increase in protein production. Some bands were not present in the profiles of the males between 21-27 days of age, but were evident in profiles taken when the males were older (Figure 3). This suggests that during development, the expression of a complete MUP profile is not just gradually increased, but that the activation of certain MUPs is staged.

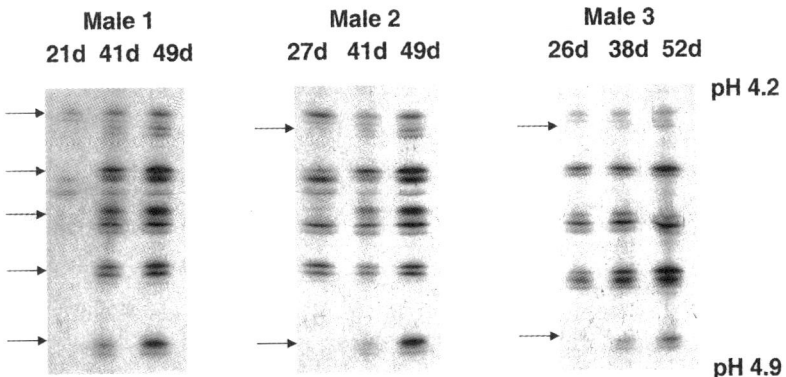

Figure 3. Comparison of MUP profiles from individual male mice taken at different ages suggests that the activation of some MUPs may be staged. MUP profiles were separated on the basis of charge on narrow range (pH4.2 – 4.9) immobilised gradient gels (Pharmacia, UK). Arrows indicate areas where there is evidence of stage-specific expression.

4. CONCLUSIONS

Wild mice differed from laboratory mice in that the mean profile complexity of each individual was twice as high, that there was considerable variation in the MUP proteins expressed by different individuals and that they did not exhibit a sex-specific expression pattern. The functional significance of MUP heterogeneity still remains to be determined, but the large number of polymorphic variants suggests a more complex role than that of a pheromone binding protein. We, and others, have suggested that MUPs may provide individuality information in scent marks (see Brennan *et al.*, 1999, Beynon *et al.*, and Hurst *et al.*, this volume) and the heterogeneity in MUP profiles between individuals and the complexity in both males and females would support such a role.

5. ACKNOWLEDGMENTS

This work was funded by a BBSRC studentship and Departmental Rhoda Le Marchant Bankes travel scholarship to CEP and BBSRC research grants to JLH and RJB

REFERENCES

Al-Shawi, R., Wallace, H., Harrison, S., Jones, C., Johnson, D., and Bishop, J. O., 1992, Sexual dimorphism and growth hormone regulation of a hybrid gene in transgenic mice, *Mol. Endocrinol.* **6**:181-190.

Bacchini, A., Gaetani, E., and Cavaggioni, A., 1992, Pheromone binding proteins of the mouse, *Mus musculus*, *Experientia* **48**:419-421.

Bocskei, Z., Grooms, C. R., Flower, D. R., Wright, C. E., Phillips, S. E.V., Cavaggioni, A., Findlay, J. B. C., and North, A. C. T., 1992, Pheromone binding to two rodent urinary proteins revealed by X-ray crystallography, *Nature* **360**:186-188.

Brennan, P. A., Schellinck, H. M., and Keverne, E. B., 1999, Patterns of expression of the immediate-early gene egr-1 in the accessory olfactory bulb of female mice exposed to pheromonal constituents of male urine, *Neurosci.* **90**:1463-1470.

Brown, R. E., 1995, What is the role of the immune system in determining individually distinct body odors, *Int. J. Immunopharmacol.* **17**:655-661.

Clissold, P. M., Hainey, S., and Bishop, J. O., 1983, Messenger RNAs coding for mouse major urinary proteins are differentially induced by testosterone, *Biochem. Genet.* **22**:379-387.

Finayson, J. S., and Baumann, C. A., 1958, Mouse proteinuria, *Am. J. Physiol.* **192**:69-72.

Flower, D. R., 1996, The lipocalin protein family: structure and function, *Biochem. J.* **318**:1-14.

Humphries, R. E., Robertson, D. H. L., Beynon, R. J., and Hurst, J. L., 1999, Unravelling the chemical basis of competitive scent marking in house mice, *Anim. Behav.* **58**:1177-1190.

Hurst, J. L., 1987, Behavioural variation in wild house mice (*Mus domesticus*, Rutty): a quantitative assessment of female social organisation, *Anim. Behav.* **35**:1846-1857.

Hurst, J. L., Robertson, D. H. L., Tolladay, U., and Beynon, R. J., 1998, Proteins in urine scent marks of male house mice extend the longevity of olfactory signals, *Anim. Behav.* **55**:1289-1297.

Jemiolo, B., Harvey, S., and Novotny, M., 1986, Promotion of the Whitten effect in female mice by synthetic analogs of male urinary constituents, *Proc. Natl. Acad. Sci. USA.* **83**:4576-4579.

Knopf, J. L., Gallagher, J. F., and Held, W. A., 1983, Differential multi-hormone regulation of the mouse urinary protein gene family in the liver, *Mol. Cell Biol.* **3**:2232-2240.

Leinders-Zufall, T., Lane, A. P., Puche, A. C., Ma, W., Novotny, M., Shipley, M. T., and Zufall, F., 2000, Ultrasensitive pheromone detection by mammalian vomeronasal neurones, *Nature* **405**:792-796.

Marchlewska-Koj, A., 1981, Pregnancy block elicted by male urinary peptides in mice, *J. Reprod. Fert.* **61**:221-224.

McKinney, T. D., and Desjardins, C., 1973, Intermale stimuli and testicular function in adult and immature house mice, *Biol. Reprod.* **9**:370-378.

Mucignat, C. C., Caretta, A., and Cavaggioni, A., 1995, Accelaration of puberty onset in female mice by male urinary proteins, *J. Physiol.* **486**:517-522.

Novotny, M. V., Ma, W., Zidek, L., and Daev, E., 1999a, Recent biochemical insights into puberty acceleration, estrus induction and puberty delay in the house mouse, in: *Advances in Chemical Communication,* (R. E. Johnston, D. Müller-Schwarze, and P. Sorenson, eds.), Plenum Press, New York, pp. 99-116.

Novotny, M. V., Ma, W. D., Wiesler, D., and Zidek, L., 1999b, Positive identification of the puberty-accelerating pheromone of the house mouse: the volatile ligands associating with the major urinary protein, *Proc. Roy. Soc., Lond. Series B* **266**:2017-2022.

Pes, D., Robertson, D. H. L., Hurst, J. L., Gaskell, S. J., and Beynon, R. J., 1999, How many major urinary proteins are produced by the house mouse *Mus domesticus*?, in: *Advances in Chemical Communication,* (R. E. Johnston, D. Müller-Schwarze, and P. Sorenson, eds.), Plenum Press, New York, pp. 149-161.

Robertson, D. H. L., Beynon, R. J., and Evershed, R. P., 1993, Extraction, Characterization, and Binding Analysis of 2 Pheromonally Active Ligands Associated With Major Urinary Protein of House Mouse (Mus-musculus), *J. Chem. Ecol.* **19**:1405-1416.

Robertson, D. H. L., Cox, K. A., Gaskell, S. J., Evershed, R. P., and Beynon, R. J., 1996, Molecular heterogeneity in the major urinary proteins of the house mouse Mus musculus, *Biochem. J.* **316**:265-272.

Robertson, D. H. L., Hurst, J. L., Bolgar, M. S., Gaskell, S. J., and Beynon, R. J., 1997, Molecular heterogeneity of urinary proteins in wild house mouse populations, *Rapid Commun. Mass Spectrom.* **11**:786-790.

Schellinck, H. M., Slotnick, B. M., and Brown, R. E., 1997, Odors of individuality originating from the major histocompatibility complex are masked by diet cues in the urine of rats, *Anim. Learn. and Behav.* **25**:193-199.

Singer, A. G., Tsuchiya, H., Wellington, J. L., Beauchamp, G. K., and Yamazaki, K., 1993, Chemistry of odortypes in mice: fractionation and bioassay, *J. Chem. Ecol.* **19**:569-579.

Singh, P. B., Brown, R. E., and Bruce, R., 1987, MHC antigens in urine as olfactory recognition cues, *Nature* **327**:161-164.

Szoka, P., and Paigen, K., 1978, Regulation of mouse major urinary protein produced by the *MUP-a-gene*, *Genetics* **90**:597-612.

Triggs, G. S., 1991, The population ecology of house mice (Mus domesticus) on the Isle of May, Scotland, *J. Zool.* **206**:286-291.

Vandenbergh, J. G., Whisett, J. M., and Lombardi, J. R., 1975, Partial isolation of a pheromone accelerating puberty in female mice, *J. Reprod. Fert.* **43**:515-523.

Wicks, L. F., 1941, Sex and proteinuria of mice, *Proc. Soc. Exp. Biol. Med.* **48**:395-400.

Wolff, R. J., 1985, Mating behaviour and female choice: their relation to social structure in wild caught house mice *Mus musculus* in a seminatural enclosure, *J. Zool.* **207**:43-51.

Yamazaki, K., Beauchamp, G. K., Imai, Y., Bard, J., Thomas, L., and Boyse, E. A., 1992, MHC control of odortypes in the mouse, in: *Chemical Signals in Vertebrates VI* (R. L. Doty, and D. Müller-Schwarze, eds.), Plenum Press, New York, pp. 189-196.

GENETIC DIFFERENCES IN ODOR DISCRIMINATION BY NEWBORN MICE AS EXPRESSED BY ULTRASONIC CALLS

Joanna Kapusta and Hajnalka Szentgyörgyi

Department of Mammalian Reproduction
Institute of Environmental Sciences, Jagiellonian University
Ingardena 6, 30-060 Kraków, Poland

1. INTRODUCTION

Infant mice, like many other small mammals, are poikilothermic. They depend on the warmth and care provided by the mother. Communication between mother and offspring is one of the most important factors determining the survival of pups during the first days of life. Interaction between newborns and the mother is based on olfactory and vocal communication. In newborn rodents, olfaction is the only well-developed system serving social communication. Late-term fetuses of mice (Coppola and Millar, 1997) and rats (Smotherman, 1982) demonstrate the ability to detect chemical stimuli before birth. It is clear that in murine rodents the olfactory system or part of it functions already from the end of the gestation period. Newborn animals also demonstrate a preference for some odors, which evoke different kinds of behavior, such as ultrasonic calling. Neonates of rats (Noirot, 1968), mice (Okon, 1970), and bank voles (Kapusta *et al.*, 1995) removed from the nest elaborate ultrasounds. These calls influence maternal behavior, as shown for rats (Jams and Leon, 1983) and mice (Hennessy *et al.*, 1980). Bank voles recognize the odor of home bedding, producing fewer and shorter calls than pups lacking of any source of scent (Kapusta *et al.*, 1995). Infant rats increase ultrasound production in the absence of familiar nest odor (Hofer and Shair, 1987). The ultrasonic response of mice pups is also modified by various kinds of bedding (D'Amato and Cabib, 1987; Marchlewska-Koj *et al.*, 1999).

Sensitivity to chemosignals can be determined genetically, or learned starting at birth as well as during fetal life (Schaal and Orgeur, 1992). Ultrasonic responses to different odors can also be genetically controlled or acquired.

Our hypothesis is that mice can discriminate between their own and an unrelated genotype's odor and that genetically different strains produce different ultrasounds. To test this we compare the ultrasonic vocalizations of two strains of mice, CBA and C57BL, in the presence of four types of odor: clean bedding, bedding from their home nest, and bedding from nests of a related or unrelated genotype.

2. MATERIALS AND METHOD

2.1. Animals

The subjects of this experiment were CBA and C57BL strains of mice. Nulliparous females and males from the colony at the Polish-American Children's Hospital (Collegium Medicum, Jagiellonian University), were purchased and bred in the laboratory of the Department of Mammalian Reproduction, Jagiellonian University. All animals were housed in polyethylene cages (37x20x15 cm) at 18-21 °C with a 14 h photoperiod (lights on at 7:00 a.m.) and 20-60% relative humidity. Standard pelleted diet and water were given *ad libitum*. Wood shavings, used as bedding material, were changed weekly. One month after arrival the animals were paired, one male with one female. Pregnant females were separated from the male and kept one per cage. The day of the neonates' birth was defined as day 0 of life. The number of pups in the nest was reduced to 6 animals. They stayed with the mother until testing.

2.2. Experimental Procedure

All experiments were carried out on 3-4-day-old mice. The animals in their home cages were transferred from the colony room to the laboratory a day before the test was performed. In all experimental groups each pup was used only once; and 3-4 pups per nest were tested.

The vocalizations of 43 CBA pups and 41 C57BL pups were examined in the presence of different odors: clean bedding, home bedding, and bedding of related (from the same strain) or unrelated (from the other tested strain) nests. All bedding material was taken from cages with pups of the same age as the tested animals.

An individual pup was removed from it's home cage, placed on a plastic dish and kept at room temperature. After 10 min. the bedding material (approximately 2 cc) was placed on the dish next to the pup. Then the tested animals were transferred to an acoustically isolated chamber with a microphone suspended 5 cm above the pup. The microphone was connected to a QMC ultrasound detector, type S30, coupled to a cassette recorder (Sony, HX PRO). The vocalizations of the individual pup were monitored for 1 min. The number of calls was estimated for each sample. For comparison of duration and fundamental frequency, 5 pulses of each 1 min record (or less if not enough calls were produced) were randomly selected and analyzed using the Canary Bioacoustic Workstation Version 1.2 (see Figure1).

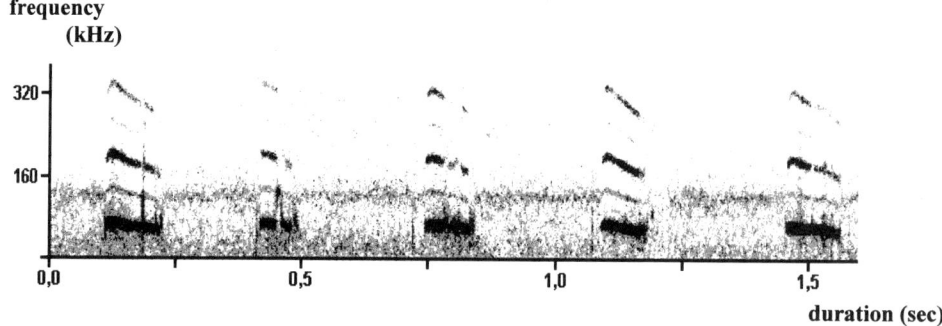

Figure 1. Sonogram of ultrasonic calls emitted by 4-day-old *Mus musculus*.

2.3. Statistical Analysis

The influence of genotype and the presence of bedding on the number and duration of ultrasounds were compared using two-way analysis of variance (ANOVA). Differences between CBA and C57BL pups were assessed using one-way ANOVA. The post-hoc comparison Tukey's test was used for analysis of the number and duration of calls within a strain. The frequencies of calls were compared with the Kruskal-Wallis test for comparisons across groups, and the Mann-Whitney U-test for paired comparisons (Sokal and Rohlf, 1981).

3. RESULTS

Pups of CBA and C57BL mice differed in ultrasonic response to bedding odor. Statistical analysis (two-way ANOVA) of the number of ultrasounds emitted by CBA and C57BL pups indicated a significant effect of genetic background ($F_{1,76}=13.22$, $p<0.001$) and olfactory stimulant ($F_{3,76}=3.09$, $p<0.05$). There was no interaction between factors. The duration of calls produced by newborns of CBA and C57BL was also influenced by both bedding odor and genotype. Interaction between experimental factors was statistically significant (two-way ANOVA).

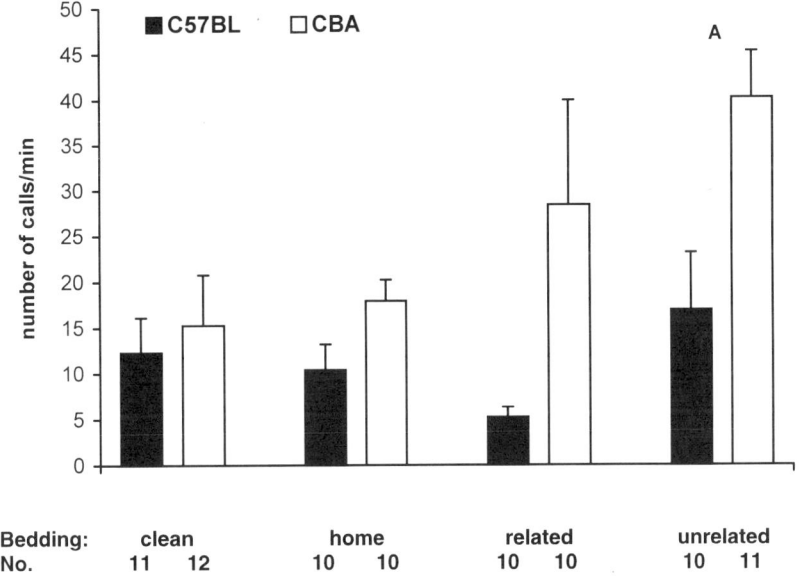

Figure 2. Number of ultrasonic calls emitted by 3-4-day old C57BL and CBA mice in the presence of different odor cues. Mean ± SE; A=p<0.01.

As shown in Figure 2, CBA and C57BL produced similar numbers of calls in the presence of clean bedding ($F_{1,21}=0.2$, $p>0.05$). The number of pulses produced by CBA pups in the presence of C57BL bedding was significantly higher than for C57BL mice exposed to CBA bedding ($F_{1,19}=8.28$, $p<0.01$). C57BL pups produced fewer calls than CBA did in the presence of home bedding and related odor, although the differences were not significant ($F_{1,18}=4.17$, p=0.05; $F_{1,18}=4.41$, p=0.05 respectively). The duration of ultrasounds (Figure 3) produced by CBA newborn mice was significantly longer than the duration of calls emitted by C57BL mice in similar conditions (clean bedding: $F_{1,72}=17.7$, related bedding: $F_{1,54}=44.8$, unrelated bedding $F_{1,88}=210.5$, all p<0.001; home bedding: $F_{1,83}=4.3$, p<0.05).

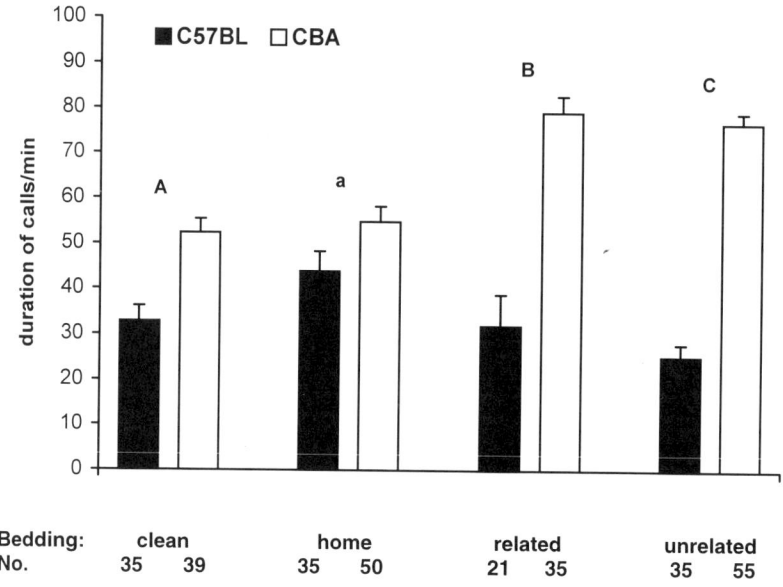

Figure 3. Duration of ultrasonic calls emitted by 3-4-day old C57BL and CBA mice in the presence of different odor cues. Mean ± SE; **A, B, C**=p<0.001; **a**=p<0.05.

Statistical analysis of the pups' reaction to different odors within a strain was also performed. C57BL newborns emitted similar numbers of ultrasounds during 1-min. tests regardless of bedding type (C57BL: $F_{3,37}=1.7$, p>0.05). Pups of CBA tended to increase the number of calls in the presence of a foreign odor, but only the differences between unrelated and clean bedding were significant ($F_{3,39}=2.88$, p<0.05). The source of odor affected the duration of calls produced by newborn mice. CBA pups emitted significantly longer pulses when exposed to unrelated nest material from the same and different genotype in comparison with the response to clean or home bedding ($F_{3,175}=20.9$, p<0.001). There were no differences in the duration of calls between clean and home bedding tests, or between genetically related and unrelated nest material. C57BL mice were less sensitive in the response to different odors. The only significant difference was found in the duration of calls produced by pups exposed to home and unrelated odor ($F_{3,122}=3.83$, p<0.05). The presence of home bedding almost doubled the duration.

Calls of CBA pups were emitted at a higher fundamental frequency than those of C57BL mice (Figure 4a, 4b). The differences between strains were highly significant in all experimental groups (clean: U=325.5; home: U=110.0; related: U=98.0; unrelated bedding: U=247.0; all p<0.001).

The bedding conditions modified the fundamental frequency of calls emitted by CBA newborn mice ($H_{3,179}=21.4$, p<0.001), as presented in Figure 4b. The pulses produced by pups in the presence of unrelated bedding were significantly higher in frequency than those emitted by animals placed in contact with clean (U=598.0, p<0.001) and own (U=953.0, p<0.001), and related (U=525.5, p<0.001) odors. The frequency of ultrasounds produced by animals in the presence of own bedding was also higher than that of animals on clean bedding. There were no differences in the frequencies of sounds emitted by animals exposed to own bedding and to a related odor. The fundamental frequency of sounds produced by C57BL mice (Figure 4a) in the presence of an unrelated odor were higher ($H_{3,126}=5.8$, p>0.05) than in the rest of the experimental groups similarly to the reaction of CBA newborns in the same condition.

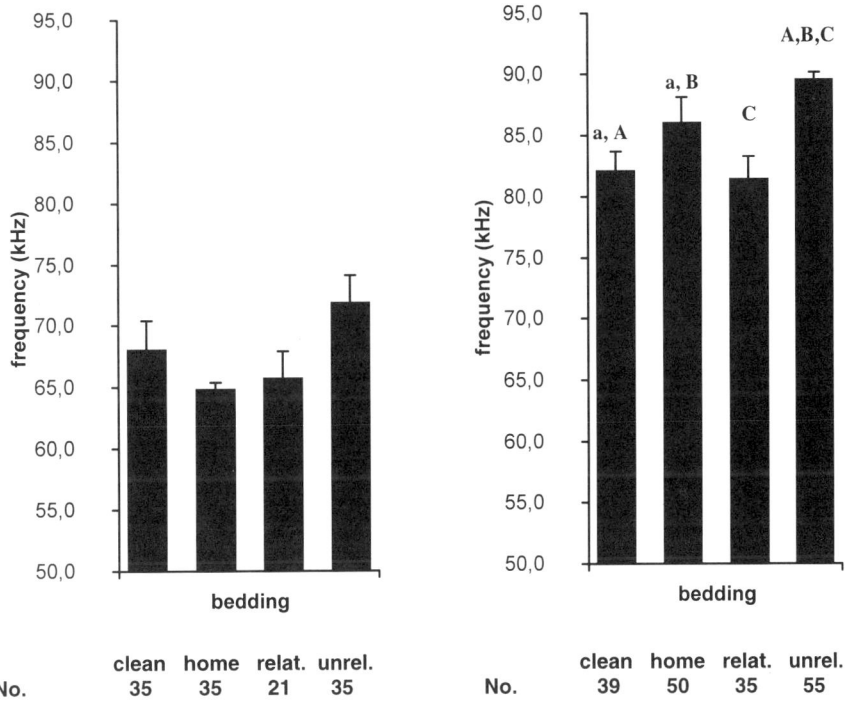

Figure 4/a. Frequency of calls emitted by 3-4-day-old C57BL mice in the presence of different odor cues. Mean ± SE.

Figure 4/b. Frequency of calls emitted by 3-4-day-old CBA mice in the presence of different odor cues. Mean ± SE; A,B,C=p<0,01; a= p<0.05

4. DISCUSSION

The vocal response of newborn rodents to isolation is clearly affected by the genotype of the tested animal. Studies conducted on rats showed that Wistar rat pups generally vocalize more and louder than Lister pups do in similar conditions (Sales, 1979). Comparative analysis of infant murids revealed that wild type mice, C_3H, EN and C_3HxEN pups emit ultrasounds differing in their duration and/or frequency depending on the strain (Sales and Smith, 1978). Nitschke and coworkers (1972) showed that pups of C_3H and BALB mice produce longer and louder calls than C57BL at the same age.

The results reported here show that the ultrasonic response of pups to similar olfactory cues depends on the genotype of the animals. CBA pups vocalized more than C57BL pups, and produced longer calls at a higher fundamental frequency. Analysis of between-factor interaction (two-way ANOVA) indicates that the duration of ultrasounds emitted by pups in response to different chemical stimuli is influenced by genetic background. The lower fundamental frequency of C57BL calls, regardless of bedding, suggests that this parameter also could be genetically controlled.

The sensitivity of mice to olfactory signals is modified by external factors. CBA pups nursed by their own mother and exposed to CBA bedding vocalize at a higher fundamental frequency then CBA neonates fostered by C57BL females and tested in similar conditions (Marchlewska-Koj et al., 1999). Olfactory cues presented in the close environment of the pups can also change their ultrasonic calls. Bank voles (Kapusta et al., 1995) recognize the home odor, vocalizing less in the presence of home bedding than in its absence. Laboratory rats (Lyons and Banks, 1982) are capable of discriminating between the home cage odor and that of an unrelated nursing mother. Albino mouse pups exposed to clean bedding emit a considerable number of ultrasonic calls, even if they are not separated from their littermates and are not exposed to low temperature. The familiar odor of the nest significantly reduces the number of calls emitted by animals (D'Amato and Cabib, 1987).

CBA as well as C57BL pups can identify chemical signals. They were capable of discriminating between the odor of their own genotype and that of an unrelated stimulus. The calls of CBA mice emitted in the presence of home bedding tended to be shorter in duration and lower in frequency than those produced in contact with alien nest odors. While the reaction of C57BL pups to the absence of familiar olfactory cues was expressed clearly only by the duration of the emitted ultrasounds.

Our results confirm the hypothesis that the genetically different strains, CBA and C57BL produce different ultrasounds. Newborns of these strains are able to discriminate between their own and an unknown odor, and these chemosignals modify their ultrasonic calls. Analysis of ultrasonic calls of infant mice can serve as a test for identification of chemosignals.

ACKNOWLEDGMENTS

This work was supported by grant no. BW/V/INoŚ/13/99. We thank Dr. P. Olejniczak for help and advice in preparing statistical analysis of the presented data.

REFERENCES

D'Amato, F. R., and Cabib, S., 1987, Chronic exposure to a novel odor increases pups vocalizations, maternal care, and alters dopaminergic functioning in developing mice, *Behav. Neur. Biol.* **48**:197-205.

Coppola, D. M., and Millar, L. C., 1997, Olfaction in utero: Behavioral studies of the mouse fetus, *Behav. Proc.* **39**:53-68.

Hennessy, M. B., Li, Y., Lowe, E. L., and Levina, S., 1980, Maternal behavior, pup vocalization and pup temperature changes following handling in mice of two inbred strains, *Dev. Psychobiol.* **13**:573-586.

Hofer, M. A., and Shair, H. N., 1987, Isolation distress in 2-week-old rats: Influence of home cage, social companions, and prior experience with litter mates, *Dev. Psychobiol.* **20**:465-576.

Jams, J. E., and Leon, M., 1983, Determinants of mother - young contact in Norway rats, *Physiol. Behav.* **30**:919-935.

Kapusta, J., Marchlewska-Koj, A., and Sales, G. D., 1995, Home bedding modifies ultrasonic vocalization of infant bank voles, *J. Chem. Ecol.* **21**:577-583.

Lyons, D. M., and Banks, E. M., 1982, Ultrasounds in neonatal rats: novel, predator and conspecific odor cues, *Dev.Psychobiol.* **15**(5):455-460.

Marchlewska-Koj, A., Kapusta, J., and Olejniczak, P., 1999, Ultrasonic response of CBA newborn mice to the odor of bedding, *Behaviour* **136**:269-278.

Nitschke, W., Bell, R. W., and Zachman, T., 1972, Distress vocalizations of young in three inbred strains of mice, *Dev. Psychobiol.* **5**: 363-370.

Noirot, E., 1968, Ultrasounds in young rodents. II. Changes with age in albino rats, *Anim. Behav.* **16**:129-134.

Okon, E. E., 1970, The effect of environmental temperature on the production of ultrasounds by isolated non-handled albino mouse pups, *J. Zool. Lond.* **162**:71-81.

Sales, G. D., 1979, Strain differences in the ultrasonic behavior of rats (Rattus norvegicus), *Amer. Zool.* **19**:513-527.

Sales, G. D., and Smith, J. C., 1978, Comperative studies of the ultrasonic calls of infant murid rodents, *Dev. Psychobiol.* **11**(6):595-619.

Schaal, B., and Orgeur, P., 1992, Olfaction in utero: can the rodent model be generalized?, *Quart. J. Exp. Psychol.* **44**B:245-278.

Smotherman, W. P., 1982, Odor aversion learning by the rat fetus. *Physiol. Behav.* **28**:927-931.

Sokal, R. R., and Rohlf, F. J., 1981, *Biometry: The Principles and Practice of Statistics in Biological Research,* Freeman, San Francisco.

ENHANCED IMMUNE FUNCTION DECREASES ODOR ATTRACTION OF MALE LABORATORY MICE

Katherine Litvinova,[2] Irene Kolosova,[1] Viktoria Mak,[1] and Mikhail Moshkin[1]

[1]Institute of Systematics and Ecology of Animals
Siberian Branch of RAS
Frunze str., 11, 630091 Novosibirsk, Russia
[2]Kemerovo State University
Krasnay str., 6650043 Kemerovo, Russia

INTRODUCTION

It is possible that parasite-dependent modification of the chemical signals used in mammal communication might be perceived by healthy conspecifics. For example, since antiquity, changes in human body odors have been used as diagnostic indicators for many diseases, including infections (Keith *et al.,* 1970; Rudiger, 1970; O'Down and Bourne, 1994; Penn and Potts, 1998). Despite the reliable relationship between infectious diseases and body odor, there is a lack of experimental tests of the hypothesis that sexual attraction odors are altered in parasitized mammals.

In experiments on house mice, Kavaliers and co-workers (1993, 1995a, b; 1997) showed that the odor of males infected enterically with the protozoan parasite *Eimeria vermiformi*, or with the nematode *H. polygyrus*, is less attractive to females than is the odor of control males. The decrease of odor attractiveness was detected even at the subclinical stage of infection. Penn *et al.* (1998) also reported a change of odor attractiveness of male mice infected with influenza virus, i.e., females preferred the urine of control males than the urine of infected males or water.

Pathogen-dependent changes of odor attractiveness could result from the direct addition of parasite metabolites to volatile components of the host urine and/or the modification of host metabolism due to immuno-neuroendocrine responses to parasite-derived antigens (Penn and Potts, 1998; Lochmiller and Moshkin, 1999). Artificial antigenic challenge seems to be an appropriate investigative approach for studies of immuno-neuroendocrine interactions that might produce parasite-dependent changes in chemical signals. Since injection with sheep red blood cells (SRBC) has no directly toxic effects but does stimulate the immune and endocrine reactions that are typical of parasitic infections (Besedovsky and Sorkin, 1975; Korneva and Shkinek, 1988), this method has become popular in ecobehavioral studies (Williams et al., 1999; Saino et al., 2000). Here we report that enhanced immune function decreased the attractiveness to females from sexual naive mice, whether or not those males had received exposure to female odors.

METHODS

House mice of the outbred ICR strain, of age 10-12 weeks, were kept in plastic cages with sawdust as nesting material and ad libitum feeding, at room temperature ($20 \pm 2°C$) and under an imposed light-dark cycle of 14L : 10D, with „lights on" at 06:00 a.m. Forty females were maintained in groups of four in plastic cages in a separate room.

Two weeks before, and during treatment, males were caged singly in two separate rooms, with 20 males in each room. Males from one room were exposed to soiled bedding from female cages during 15 or 30 days; the bedding was presented to each male in small net boxes before the light was turned off. Males from the other room were exposed to clean sawdust on the same terms. Five days before the end of exposure all males were injected intraperitoneally with 0.5 ml of SRBC (2×10^8 cells).

The cages of the males were cleaned once every 5 days. Samples of soiled bedding from each male's cage were collected during the 5-day intervals before and after SRBC treatment. The samples were frozen (-20°C) for 2 or 7 days until it was presented to females in olfactory tests.

Females were separated into individual cages two hours before olfactory tests. Estrus was determined by vaginal smear to classify each subject as an estrous or non-estrous subject. Two boxes with male-soiled bedding were placed for 2 min in different locations in the female's cage. Soiled bedding in one box was collected before SRBC-treatment and the bedding in the other box was collected from the same male after SRBC-treatment; the duration that each subject's nose was in close proximity (less than one cm) to each odor sample was recorded. An index of odor attractiveness of male after SRBC treatment (IOA_{im}) was calculated (*sensu* Kavaliers and Colwell, 1995a, b) by dividing the amount of time the female spent in close proximity to the post-immunization stimulus box by the total duration of time she spent in proximity to both stimulus boxes. The index of attraction pre-immunization (IOA_c) was calculated by subtracting the post-immunization index from one. If both durations were zero, data were not included in the analyses.

Five days after injections with SRBC, aggressive behaviors of the male odor donors were observed in 15-min dyadic encounters. Social conflict was provoked by placing two males from one group in a standard mouse cage with fresh sawdust, covered by a transparent top. The following types of event for each individual were recorded: aggressive lunges, bites of contestant, tail rattles, defensive uprights, and vocalization (Moshkin et al., 1993). Males classified as dominant engaged in aggressive lunges, tail rattles, and bites, whereas subordinate males demonstrated vocalization and defensive uprights. Immediately after these dyadic encounters the mice were killed by chloroform overdose, and the numbers of splenic plague-forming cells (PFC) were estimated as by Cunningham (1965), standardized by dividing by body mass.

STATISTICAL ANALYSIS

Paired Student's t-tests were used for comparing the pre- and post-immunization indices of attraction. Statistical significance of behavioral differences between dominant and subordinate males were assessed by Student's t-test. Pearson's coefficients of correlation were calculated to determine interrelations between measures.

RESULTS

Activation of the immune system led to a decline in the attractiveness of odors from males that had not been exposed to female odors (Table 1). Only the estrous females discriminated between the odor of soiled bedding collected before and after SRBC-treatment.

Table 1. Effect of SRBC-treatment on the male odor attractiveness (IOA_{im}) to female mice. Male mice were maintained with or without female odor.

(Values are means ± s.e. Numbers of olfactory tests are in parentheses)

Stage of ovarian cycle	The male housing conditions	
	Without female odor	With female odor[1]
Estrus	0,354 ± 0,049 (21) P<0,001[2]	0,491 ± 0,08 (18) NS[3]
Non-estrus	0,441 ± 0,052 (10) NS	0,488 ± 0,091 (18) NS

[1] - Time of exposure to female odor (15- and 30-days) have not significant effect on IOA (P>0.1), therefore this data were combined.
[2] - P<0,001 versus IOA_c
[3] - NS- non significant.

The decreased attractiveness to estrous females of odors from unexperienced, immuno-stimulated males raises a question about individual variability in immune responses. A statistically significant, positive correlation between the post-immunization index of attraction and the number of PFCs is noted in Figure 1. It will be noted that a significant decrease of odor attractiveness in response to antigenic stimulation is associated with males, which had lower numbers of PFCs. There were no significant correlations between IAOim and humoral immune response in males that were regularly exposed to female odor (Figure 1).

It is known that odor attractiveness of male mice depends on their social status (Gerlinskaya et al., 1995). According to our data, estrous females spent different lengths of time in close proximity to soiled bedding collected from dominant and subordinate males before administration of SRBC. Females preferred the odor of dominant males, without any effect from the conditions of the male's maintenance. Contrary to this clear difference, estrous females could not distinguish between odor signals of dominants and subordinates, when soiled bedding was collected after SRBC treatment. These data demonstrated that development of the humoral immune response modified the odor, lowering the attractiveness of odors from high-ranking males (Figure 2).

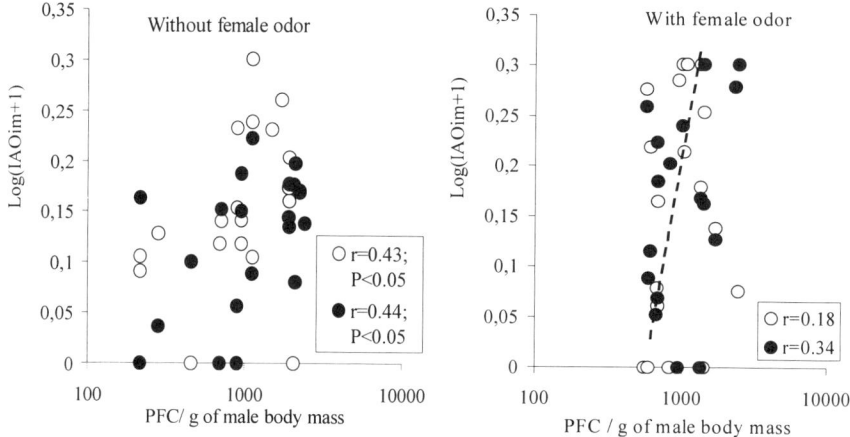

Figure 1. Correlation of odor attractiveness of SRBC-treated males with their humoral immune response. Male mice were maintained with or without female odor.
The value of IOA_{im} ws added to one and then these data were log-transformed to obtain linearity. The solid lines represent IOA_{im} equal 0,5. Circles that disposed under solid line testified to decline odor attractiveness after injection with SRBC ($IOA_{im}<0,5$).
Closed circles – estrous female, open circles – non-estrous female.

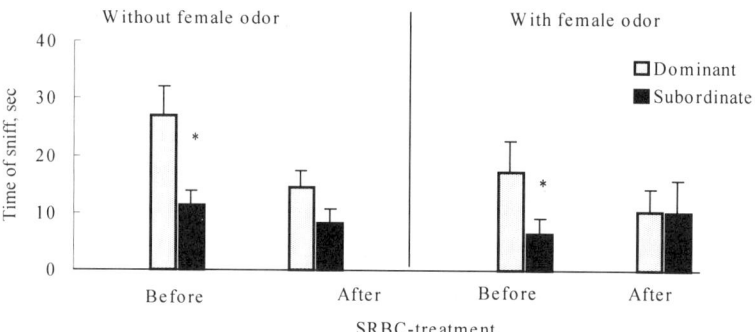

Figure 2. Effect of SRBC-treatment on the odor attractiveness of dominant and subordinate males to females. *-$P<0,05$ dominants versus subordinates;

CONCLUSION

Our study of ICR house mice demonstrated that estrous females spent less time inspecting the soiled bedding from female-odor-unexperienced males after these males were injected with SRBC. This decreased odor attractiveness was especially evident in males with low immunoresponsiveness. Along with the decrease of male attractiveness in general, antigenic stimulation cancelled the differences in odor attractiveness between dominants and subordinates. So, the immune response to antigenic challenge as an obligatory consequence of an infection contributes to decreased odor attractiveness of infected animals.

ACKNOWLEDGMENTS

We thank an anonymous referee for his help in improving the paper. We also thank Mr. V. Yudushkin. His observation on himself propelled this study. This study was supported by the Russian Foundation for Basal Research (grant of 99-04-49927).

REFERENCES

Besedovsky, H., and Sorkin, E., 1975, Changes in blood hormone levels during the immune response, *Proc. Soc. Exp. Biol. Med.* **150**:466-470.

Cunningham, A. J., 1965, A method of increased sensitivity for detecting single antibody-forming cells, *Nature* **207**:1106-1107.

Gerlinskaya, L. A., Rogova, O. A., Yakushko, O. F., and Evsikov, V. I., 1995, Female olfactory choice and its influence on pregnancy in mice, *Adv. Biosci.* **93**:297-302.

Kavaliers, M., and Colwell, D. D., 1993, Aversive responses of female mice to the odors of parasitized males: neuromodulatory mechanisms and implications for mate choice, *Ethology* **95**:206-212.

Kavaliers, M., and Colwell, D. D., 1995a, Discrimination by female mice between the odours of parasitized and non-parasitized males, *Proc. R. Soc. Lond. B.* **261**:31-35.

Kavaliers, M., and Colwell, D. D., 1995b, Odours of parasitized males induce aversive response in female mice, *Anim. Behav.* **50**:1161-1169.

Kavaliers, M., Colwell, D. D., Ossenkopp, K-P., and Perrot-Sinal, T. S., 1997, Altered responses to female odors in parasitized male mice: neuromodulatory mechanisms and relations to female choice, *Behav. Ecol. Sociobiol.* **40**:373-384.

Keith, L., Bush, I. M., Dranieks, A., and Krotoszynski, B. K., 1970, Changes of vaginal odors of 6 patients under nitrofurazone treatment. A study in applied olfactorics, *J. Reprod. Med.* **4**:69-76.

Korneva, E. A., and Shkinek, A. A., 1988, *Hormones and Immune System*, Nauka, Leningrad, (in Russian)

Lochmiller, R. L., and Moshkin, M. P., 1999, The adaptive significance or the variability of immunocompetence in populations of small mammals, *Siberian J. Ecol.* **6**:37-58 (in Russian).

Moshkin, M. P., Potapov, M. A., Frolova, O. F., and Evsikov, V. I., 1993, Changes in aggressive behavior, thermoregulation, and endocrine responses in BALB/cLac and C57Bl/6j mice under cold exposure, *Physiol. Behav.* **53**:535-538.

O'Down, T. C., and Bourne, N., 1994, Inventing a new diagnostic test for vaginal infection, *BMJ.* **309**:40-42.

Penn, D. J., and Potts, W. K., 1998, Chemical signals and parasite-mediated sexual selection, *Trens. Ecol. Evol.* **13**:391-396.

Penn, D. J., Schneider, G., White, K., Slev, P., and Potts, W. K., 1998, Influenza infection neutralizes the attractiveness of male odor to female mice *(Mus musculus), Ethology* **104**:685-694.

Rudiger, W., 1970, Differential diagnostic significance of smell in diseases of the upper respiratory tract, *Med. Welt.* **9**:340-345.

Saino, N., Ninni, P., Calza, S., Martinelli, R., De Bernardi, F., and Moller, A. P., 2000, Better red than dead: carotenoid-based mouth coloration reveals infection in barn swallow nestlings, *Proc. Royal Society Lond. B.* **267**:57-61.

Williams, T. D., Christians, J. K., Aiken, J. J., and Evanson, M., 1999, Enhanced immune function does not depress reproductive output, *Proc. R. Soc. Lond. B.* **266**:753-757.

THE OLFACTORY SEXUAL PREFERENCES OF GOLDEN HAMSTER (*MESOCRICETUS AURATUS*): THE EFFECTS OF EARLY SOCIAL AND SEXUAL EXPERIENCE

Alexei V. Surov, Antonina V. Solovieva, and Alexander N. Minaev

Institute of Animal Ecology and Evolution
Russian Academy of Sciences
Leninsky pr., 33 Moscow 117071, Russia

INTRODUCTION

Sex discrimination is one of the most important attributes of natural social behavior promoting mate search and therefore breeding success as well as maintenance of spatial and social structure of population. Many studies suggest that in mammals young animals tend to develop social preferences for stimuli associated with the species that provides parental care (Cairns and Johnson, 1965; Denenberg et al., 1964; Fox, 1969; Mason and Kenney, 1974). Animals reared under an alien species afterwards interact with them or their signals more often, than animals reared in the usual way (Cairns, 1966; Hudgens et al., 1968; Fox, 1969; Gilbert, 1974).

Interspecific cross-fostering in mammals results in a reorientation of sexual behavior as well. For example, Lagerspetz and Heino (1970) found that mice raised by rat mothers demonstrated more active sexual behavior towards small, prematurely estrous rat than did mouse-reared animals. Sexual behavior directed toward mice decreased in rat-reared animals. Similarly, male *Mus musculus* that had been cross-fostered to *Baiomys taylori* dams subsequently mounted *Baiomys* (Quadagno and Banks 1970). So, cross-fostering of one young of strain or species to another shortly after birth with subsequent exposure to chemical signals in adulthood allow us to analyze the development of olfactory preferences.

Another factor influencing sexual preferences is the copulatory experience (Carr et al. 1965; Lydell and Doty 1972; Hayashi and Kimura 1974). Sexually experienced male rats prefer receptive over non-receptive females (Stern, 1970). Experienced males of the golden hamster directed significantly more mounts to anaesthetized females than did sexually inexperienced males (Landauer et al., 1978).

In the present paper we addressed the questions of how the olfactory image of the sexual partner in mammals is development in ontogeny and what basic factors exert influence on formation of sexual preference of olfactory signals of conspecifics in adulthood.

METHODS

Animals. The subject chosen for this study was the golden hamster *(Mesocricetus auratus)*. Animals were reared by their natural parents (group 1: 16 males and 21 females) or cross-fostered by laboratory rats (group 2: 16 males, 21 females). Each animal in group 2 (at age 4-12 days) was removed from its home cage with a minimum of disturbance and placed in the nest area with foster rat dam and its 2-4 own pups (with at least one male). The rest of the rat's litter was moved to another lactating female. The entire procedure required less than 2 min. All hamsters were weaned at 52-57 days of age and were housed subsequently in individual plastic cages measuring 25 x 18 x 13 cm with wood shavings serving as the bedding material. All animals were maintained on inverted light: dark cycle (14L: 10D) with lights off at 10 a.m. Food and water were available ad libitum.

Apparatus. Hamsters' olfactory preferences were tested in a Plexiglas four-choice chamber (700 x 700 x 500 mm) with four stimulus boxes (250 x 210 x 150 mm), each containing some test stimulus. The entrance of each stimulus box was a Plexiglas door with a circular hole 3 mm diameter, through which the subject could presumably smell stimulus. The wall of the passage to each stimulus box contained an infra-red detector system, permitting automatic registration of the number of approaches to stimulus box, and the duration of time spent near each of them. Recording and data processing were conducted with a special computer program.

Stimuli collecting. For each series of tests with choice among urine stimuli approximately equal volumes of urine were collected and pooled from three individuals of each of the following classes of adult animals: rat and hamster females (estrous and anestrous) and males of both species. None of the donors had any contact with the experimental subjects. All donors were maintained on the same diet as tested animals. To minimize disturbance of donors, urine was collected by placing a clean sheet of polyethylene under the donor's wire-bottomed home cage. Every hour urine samples (if appeared) were collected with a syringe, the pool was prepared and frozen. Prior to the tests, samples were defrosted and warmed to room temperature.

The same method of collecting was used for experiments with choice among four bedding stimuli. Sample pools were prepared from 7-days old soiled bedding of three individuals of *Mesocricetus auratus* or *Rattus norvegicus*. Collection was proceeded before the tests and approximately the same volume of soiled bedding from each individuals was collected.

Procedure. Beginning at two months of age, each subject was given a series of 2 four-choice tests in olfactometer with the intervals by 3 to 5 days, conducted in random order. In the two four-choice tests males were presented the following stimuli: 1) urine of estrous female rat vs. urine of anestrous female rat vs. urine of estrous female hamster vs. urine of anestrous female hamster; 2). soiled bedding of female rat vs. soiled bedding of male rat vs. soiled bedding of female hamster vs. soiled bedding of male hamster. Two four-choice tests for estrous and diestrous females recipients employed the same stimuli as in test #2 for males.

During a week prior to testing, each subject was given a two 10-min habituation trials in the olfactometer with clean stimulus boxes. At the beginning of each test, subjects were placed into the center of the chamber and time recording started immediately after the first approach to any one of the stimulus boxes. Tests lasted for 10 min. Before starting the test with each next animal camera and entry in stimulus boxes were washed by 50% ethanol and were carefully aerated. During the series position of stimulus was changed several times randomly: to exclude an influence of casual attracting factors.

Within a week upon completion males and females of both groups received sexual experience. After that experiments that are described above were repeated.

Statistical analysis. We conducted simple t-test for matched-paired to determined whether a significant preference was established for one of the exposed stimulus comparative of each of the rest in particular test. We did not compare results in different tests intentionally to avoid the interference of stimuli and calculated percent of time spent by recipient investigating the stimulus. For 100% we took the total time of investigation of all four stimuli.

RESULTS

Experiment 1 was conducted to analyze the response of cross-fostered or normally reared males to the stimuli from estrous and anestrous females of the foster species.

At the first stage sexually naive cross-fostered males demonstrated strong preference for the odor of estrous female rat compared to all other stimuli ($P<0,001$). Besides that, in this test they have shown preference of odor from estrous female hamster to the anestrous female hamster ($P<0,01$) (Figure 1a). Normal males (group 1) also showed preference of estrous stimuli comparative to anestrous regardless of whether the donor was rat ($P<0,05$) or hamster ($P<0,001$), but they have spent longer time near the stimuli from estrous female hamster (Figure 1b). After receiving of sexual experience cross-fostered males still discriminated physiological status of rats ($P<0,05$), however, hamster's stimuli became more attractive for them than at the first stage (Figure 1a). Sexually experienced normally fostered males kept showing preference of olfactory signal of estrous and anestrous female hamster significantly preferring them relative to rat's stimuli ($P<0,001$). Preference of estrous rat's stimuli to anestrous ones was not demonstrated in this test (Figure 1b).

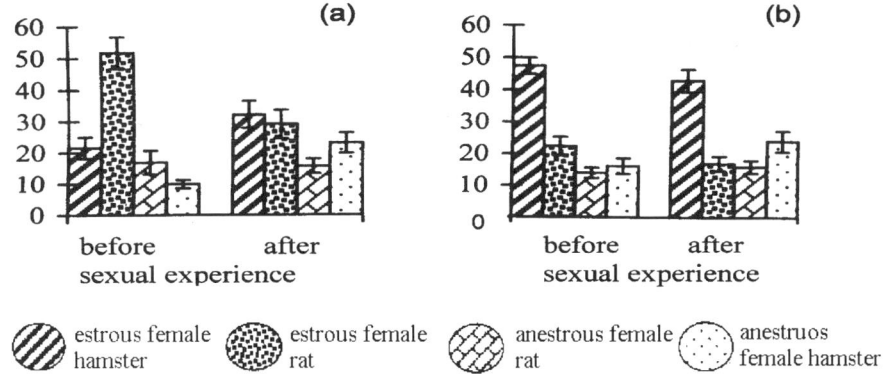

Figure 1. Mean (%)(±) time spent investigating urine samples of female rats and hamsters in estrus and anestrous a) by cross-fosters; b) by non-crossfostered male hamsters.

Experiment 2 was carried out to determine the influence of cross-fostering and sexual experience upon discrimination of sex odors.

Sexually naive cross-fostered males in situation of choice of bedding material from males and females showed difference between reactions to males and females of rats only (P<0,01). The rest of stimuli did not differ much for them (Figure 2a). The control male group, on the contrary, preferred females of hamsters to males of both species (P<0,001) and to female rats (P<0,05). They also preferred bedding material from females of rats to the bedding of male rats (P<0,05) (Figure 2b). After receiving of sexual experience the cross-fostered males investigated bedding from female hamster longer than bedding of females (P<0,05) and male (P<0,01) rats (Figure 2a). Normal males after receiving of sexual experience reduced time of investigation of bedding from females of rats and retained the preference of stimulus from females of hamster (Figure 2b).

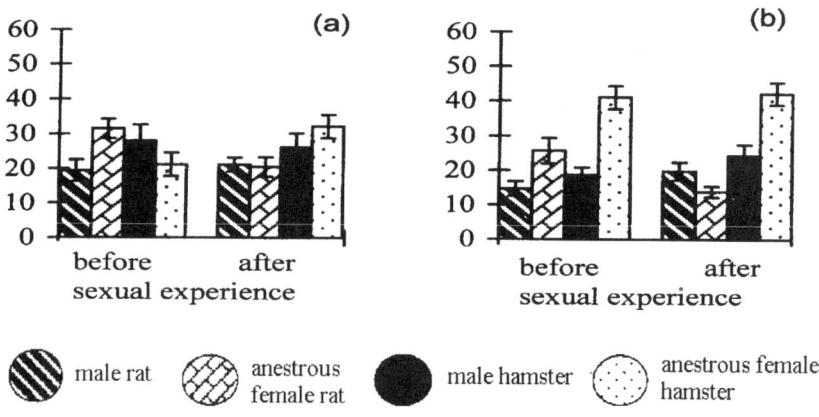

Figure 2. Mean (%)(±) time spent investigating bedding samples of female and male rats a) by cross-fosters; b) by non-crossfostered male hamsters.

Sexually naive cross-fostered females in anestrous significantly preferred stimuli of female rats (P<0,01). No other stimuli in this test were discriminated significantly (Figure 3a). Control inexperienced females spent significantly longer time investigating stimuli from male hamsters (P<0,05) as did sexually experienced females of this group (P<0,05) (Figure 3b). With receiving of sexual experience cross-fostered females changed preferences when tested in anestrous as they began to demonstrate discrimination of the stimuli from female and male rats (P<0,05) and from female and male hamsters (P<0,05) in favor of males in either case (Figure 3a).

Sexually naive cross-fostered females in estrus reacted equally to soiled bedding of male and female rats in preferring them to both hamster's stimuli (P<0,01) (Figure 4a). After receiving of sexual experience the stimulus of male hamster was significantly preferred to any of the other (P<0,01) (Figure 4a). Irrespective of sexual experience control estrous females demonstrated strong preference for the odor of soiled bedding from conspecific males (P<0,001) (Figure 4b).

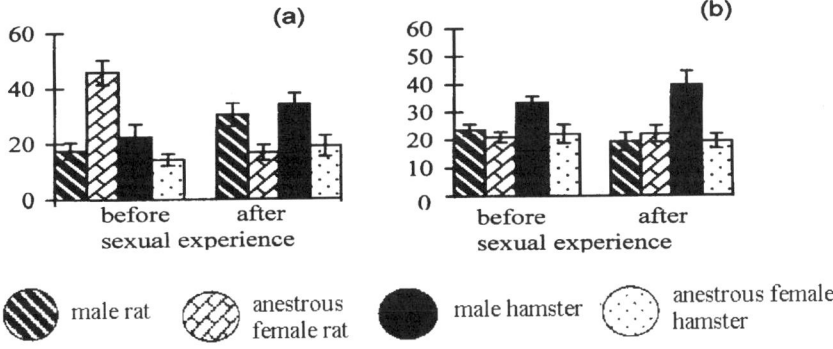

Figure 3. Mean (%)(±) time spent investigating bedding samples of female and male rats and hamsters a) by cross-fosters; b) by non-crossfostered anestrous female hamsters.

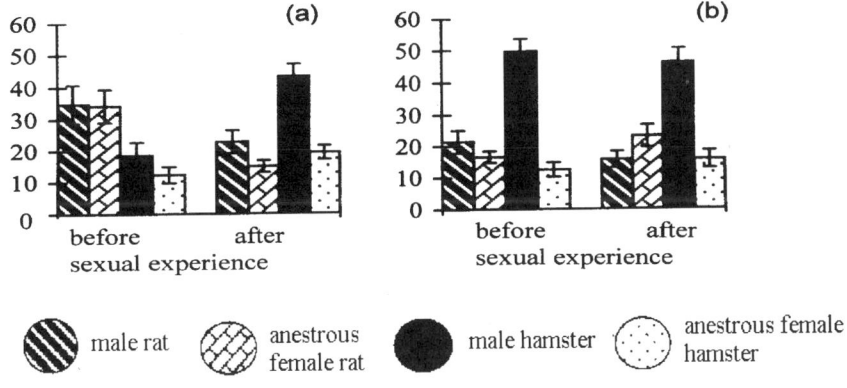

Figure 4. Mean (%)(±) time spent investigating bedding samples of female and male rats and hamsters a) by cross-fosters; b) by non-crossfostered estrus female hamsters.

DISCUSSION

Cross-fostering is one of the most wide-spread method to analyze the role of hereditary factors in species preferences in mammals. Porter *et al.* (1977) showed that 2-days-old spiny mice fostered to *Mus musculus* preferred bedding soiled by laboratory mice over bedding soiled by spiny mice. Denenberg *et al.* (1964) and Hudgens *et al.* (1968) reared mice with rats and found that such mice as adults preferred to spend time with rats. Quadagno and Banks (1970) reciprocally cross-fostered pigmy mice *(Baiomys taylori)* and laboratory mice *(Mus musculus)* and found that cross-fostered Baiomys female and both cross-fostered males and females of mice spent less time near conspecifics than controls, when given a choice between Baiomys and mice. McCarty and Southwick (1977) cross-fostered grasshopper mice *(Onyhomys torridus)* with *Peromyscus leucopus*. Cross-fostered

Peromyscus males preferred soiled bedding of Onyhomys. Thus, postnatal experience with non-conspecific parents may influence the choice of chemical signals of mature animals and these reactions depend on particular species and sex of recipients. In most of studies researchers deal with species-specific signals and practically ignore the sexual component. In fact, these two parameters in many cases could be just parts of one common cue of a sexual partner. According to one concept mammal pheromones comprise a number of compound, which are perceived as a unique „olfactory image" (Sokolov *et al.*, 1978). In other words, having contacted with a source of an olfactory cue the animal receives the whole information about the subject which is not subdivided into discrete „units of meaning" and response of the animal depends on the context and its individual features (Surov, 1986; Kotenkova, 1988). However, there are certain data concerning discrimination of sex and physiological state of nonconspecific animals even on normally reared animals. For example, sexually experienced males of brown and collared lemmings preferred estrous female to anestrous independent of the species to which they belong (Huck and Banks, 1984).

We hypothesized that in a number of species there are some common substances coding sexual and physiological information. If it is true the animals have to prefer the opposite sex even in heterospecific animals. We have obtained some evidence for it in our experiments where rats and dogs were trained to discriminate between the odors of heterospecific males and females. Then they were tested under extinction for the ability to discriminate between the male and female odor samples of novel species. In most cases rats and dogs showed the capability to recognize the sex of exposed species and performance accuracy was as high as during training trials or close to that (Surov *et al.*, 1999a, b).

The role of species-specific components should increase during the life history (early social and sexual experience). Our study has shown that cross-fostered males of golden hamster demonstrated strong preference of estrous female rat odors relative to other exposed stimuli even conspecific ones. After these animals had received sexual experience their capability to discriminate between estrous and anestrous female rats was retained but stimuli from estrous female hamsters became more attractive compared to naive stage. Yet more important result is that control naive males, as well, have shown some preference of estrous rat's stimuli compared to anestrous ones, although estrous hamster stimuli were most preferable both before and after the sexual experience. This phenomenon leads us to conclusion that attractiveness of estrous status depends on early olfactory learning as well as on receiving of sexual experience in adulthood. We could not say that it was exclusively olfactory experience because experimental animals were admitted to any social relationship with dam rat and cagemate young rats although the role of olfactory signals in this process could be critical.

In our study cross-fostered naive females spent more time investigating stimuli of rats. For anestrous females it was a stimulus from female rat and for estrous female – both male and female rat. This result is hard to interpret but, probably, this phenomenon is due to learning from foster mothers or step siblings. After receiving of sexual experience females in anestrous began to prefer the opposite sex of both species but for those in estrus only conspecific males were the most attractive.

Our results correspond partly to hypothesis of Huck and Banks (1980) that in case the preference for conspecific odor is learned in the maternal environment, cross-fostered animals should show increased preference for the foster species odor. According to our data it seems likely that process of formation of the „image of sexual partner" in the course of ontogeny species-specific components depend much on early experience and can be changed in adulthood. The sexual and physiological components depend on early experience to a less extent. In our experiments the animals reared by alien species respond to species-specific, sexual and receptivity signals from foster species which are added and/or replaced by conspecific ones in the course of sexual interactions. We suggest the simple model of this process which is shown in Figure 5.

The figures in the center and to the left demonstrate a schematic representation of the process of formation of olfactory images of a sexual partner in rats and hamsters. The basal level of each cone corresponds to the components associated with sex and receptivity which could be similar in many mammalian species. The upper levels (species information) are formed under the influence of early social and sexual experience. Rat-reared hamsters present an intermediate case which can be changed into conspecific direction after sexual experience is obtained (to the right).

So, we suppose that olfactory image of a sexual partner is gradually formed on the basis of the inherited capacity of mammals for discrimination of sex and receptivity. Since many mammalian species share common components conveying sex-related information the enhancement of species-specific component of the olfactory image is necessary for effective communication. In our opinion the process of its development in ontogenesis is affected by a number of factors among which early social and sexual experience are the most important. In its course common 'crude' signals of sex and receptivity are supplemented with species-specific characteristics which are responsible for the formation of an image of a conspecific sexual partner.

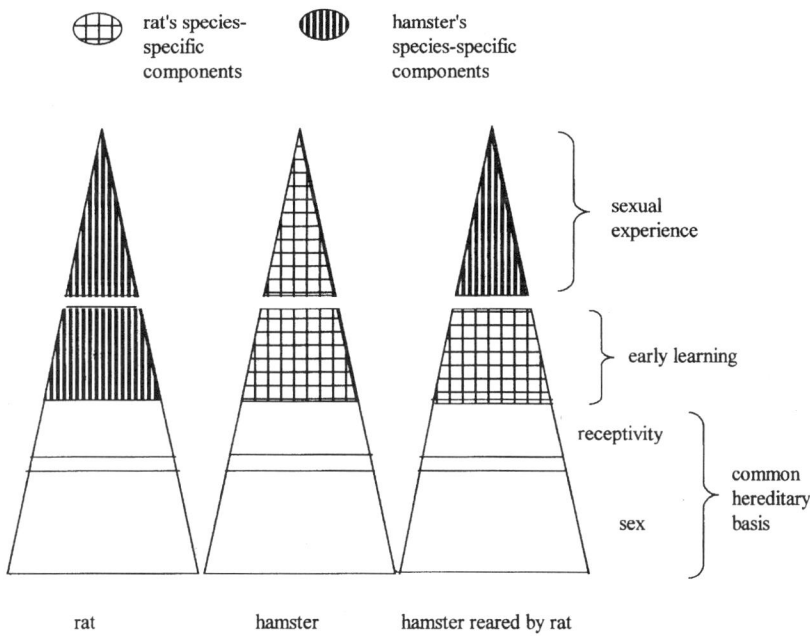

Figure 5. Scheme showing the process of the formation of the olfactory image of a conspecific sexual partner.

So, considering both studied factors – early learning and sexual experience we may conclude that preference of olfactory signals of the opposite sex and receptivity is more or less an hereditary factor, but perception of species-specific component could be change with age and experience.

ACKNOWLEDGMENTS

This study was supported by the Russian Foundation for Basic Research N 99-04-48192. We kindly thankful to Natalya Bodyak for design of experiments, to Vladimir Lebedev for critical notes and English corrections and technician of Moscow Zoo Elena Nenarokova for providing us number of pregnant rats for fostering experiments.

REFERENCES

Cairns, R. B., 1966, Development, maintenance, and extinction of social attachment behavior in sheep, *J. Comp. Physiol.* **62**:298-306.
Cairns, R. B., and Johnson, D. L., 1965, The development of interspecies social attachments, *Psychol. Sci.* **2**:337-338.
Carr, W. J., Loeb, L. S., and Dissinger, M. L., 1965, Responses of rats to sex odors. *J. Comp. Physiol. Psychol.* **59**:370-377.
Denenberg, V. H., Hudgens, G. A., and Zarrow, M. X., 1964, Mice reared with rats: Modification of behavior by early experience with another species. *Science* **143**: 380-381.
Fox, M. W., 1969, Behavioural effects of rearing dogs with cats during the "critical period of socialisation", *Behaviour* **35**:273-280.
Gilbert, B. K., 1974, The influence of foster rearing on adult social behavior in fallow deer (Dama dama), in: *The Behaviour of Ungulates and its Relation to Management., (1)* (V. Geist, and F. Valter, eds.), Gresham Press, Surrey, UK, pp. 247-273.
Hayashi, S., and Kimura, T., 1974, Sex-attractant emitted by female mice, *Physiol. Behav.* **13**:563-567.
Huck, W., and Banks, E., 1980, The effects of cross-fostering on the behaviour of two species of north american lemmings, Dicrostonyx groenlandicus and Lemmus trimucronatus: olfactory preferences, *Anim. Behav.* **28**:1046-1052.
Huck, W., and Banks, E., 1984, Social olfaction in male brown lemmings (Lemmus sibiricus = trimucronatus). 1. Discrimination of species, sex, and estrous condition, *J. Comp. Physiol. Psychol.* **98**:54-59.
Hudgens, G. A., Denenberg, V. H., and Zarrow, M. X., 1968, Mice reared with rats: effects of preweaning and postweaning social interactions upon adult behavior, *Behaviour* **30**:259-274.
Kotenkova, E. V., 1988, Methodological approaches to the study of chemical communication of mammals. Olfactory signals as mechanisms of ethological isolation between species. in.: *Itogi nauki i techniki.* VINITI, ser. *Zoologiya pozvonochnikh,* **15**: 92-151. (In Russian).
Lagerspetz, K., and Heino, T., 1970, Changes in social reaction resulting from early experience with other species, *Psychol. Reprod.* **27**:255-262.
Landauer, M. R., Banks, E. M., and Carter, C. S., 1978, Sexual and olfactory preferences of naive and experienced male hamsters, *Anim. Behav.* **26**:611-621.
Lydell, K., and Doty, R. L., 1972, Male rat odor preferences for female urine as a function of sexual experience, urine age and urine source, *Horm. Behav.* **3**:205-212.
McCarty, and Southwick, 1977, Cross-species fostering: effects on the olfactory preference of Onychomys torridus and Peromyscus leucopus, *Behav. Biol.* **19**:255-260.
Mason, W. A., and Kenney, M. D., 1974, Redirection of filial attachment in rhesus monkeys: dogs as mother surrogates, *Science* **183**:1209-1211.
Porter, R. H., Deni, R., and Doane, H. M., 1977, Responses of Acomys cahirinus pups to chemical cues produced by a foster species, *Behav. Biol.* **20**:244-251.
Quadagno, D. M., and Banks, E. M., 1970, The effect of reciprocal cross fostering on the behaviour two species of rodents, Mus musculus and Baiomys tailori ater, *Anim. Behav.* **18**:379-390.
Sokolov, V. E., Lyapunova, K. L., Surov, A. V., and Zinkevich, E. P., 1978, Species-specific pheromone of Norway rat – „olfactory image", in: *II International Theriological Congress*, Brno, pp. 363, (abstract)
Stern, J. J., 1970, Responces of male rats to sex odors, *Physiol. Behav.* **5**:519-524.
Surov, A. V., 1986, Influence of Olfactory Signals on the Behavior of Golden Hamster (Mesocricetus auratus Waterhouse, 1839). (Cand. Sci. (Biol.) Dissertation, Moscow). 206 pp. (In Russian)
Surov, A. V., Bodyak, N. D., and Solovieva, A. V., 1999a, Do Common Sex Markers Exist in the Olfactory Signals of Mammals? *Dokl. Biol.Nauk* **383**:499-501.
Surov, A. V., Solovieva, A. V., Krutova, V. I., and Bodyak, N. D., 1999b, Olfactory signals of sex: is interspecific discrimination possible?, in: *3-rd European Congress of Mammology*, pp. 220, (abstract).

MATE RECOGNITION VIA WATERBORNE CHEMICAL CUES IN THE VIVIPAROUS CAECILIAN *TYPHLONECTES NATANS* (AMPHIBIA: GYMNOPHIONA)

Andrea Warbeck and Jakob Parzefall

Zoological Institute and Zoological Museum
University of Hamburg,
Martin-Luther-King-Platz 3
20146 Hamburg, Germany

INTRODUCTION

Caecilians, members of the order *Gymnophiona*, are elongate, limbless and annulated animals with reduced eyes (Wake, 1985) and a unique organ at the head, the paired tentacle, which is connected with the vomeronasal organ via the tentacular duct (Schmidt and Wake, 1990). These morphological characters and behavioral studies suggest that the paired tentacle is involved in chemoreception (Himstedt and Simon, 1995).

Caecilians are nocturnal and mostly burrowing or litter-dwelling amphibians, found in the tropics. The members of the *Typhlonectinae* are secondarily aquatic or semi-aquatic. The majority of these animals inhabits slow-moving lowland rivers, side channels and swamps in South America. *Typhlonectes natans* (Fischer, 1879) is fully aquatic and is restricted to northwestern Venezuela and waters of the Rio Cauca/Rio Magdalena drainage in Colombia (Wilkinson, 1996; 1999).

Since caecilians are rarely collected and seldom observed in the field, very little is known about their natural history, ecology, and behavior. Although typhlonectids have been maintained and bred in aquaria, information about mating behavior is limited to only few reports (e.g. Murphy *et al.*, 1977; Billo *et al.*, 1985). Caecilians practice internal fertilization by an intromittent organ, the phallodeum, formed by the posterior part of the male cloaca (Taylor, 1968). *T. natans* is viviparous providing maternal nutrition to the intra-oviductally developing young (Wake, 1992). At best, females give birth every second year to 3-9 young after 6-9 month of pregnancy (Warbeck, unpublished data).

Pheromonal communication is used by vertebrates in many contexts and can provide information on species, gender, kin, and reproductive state. In many amphibians, chemical communication plays a major role in sex recognition and courtship behavior. Courtship pheromones have been reported in urodeles (e.g. Belvedere *et al.*, 1988; Kikujama *et al.*, 1997; Rollmann, *et al.*, 1999) and anurans (Wabnitz *et al.*, 1999). However, very little is known about communication in caecilians although morphological (tentacle organ) and ecological adaptations (nocturnal and fossorial life) suggest a well developed chemical sense.

In *T. natans*, females, males, and juveniles spend extensive time periods curled together in shelters. Previous behavioral studies have shown the importance of chemical communication in social behavior and conspecific recognition. *T. natans* not only mark their communal shelters but also release waterborne chemical cues that attract conspecifics (Warbeck *et al.*, 1996). Y-maze experiments gave indication of sex-specific odors in adults (Warbeck and Parzefall, 1997), however, only a limited number of animals were available for experiments. Juveniles preferred female-scented water to male-scented water (Warbeck and Parzefall, 1997) but no preference for their own mother versus other maternal females could be detected (Warbeck *et al.*, 1996).

In the present study we present additional data on sex discrimination in adults and on the role of waterborne chemical cues in mate recognition. The following questions were addressed: First, do females produce chemical cues that give information to males about their receptivity? Second, are females and males able to discriminate between unrelated and related potential mates? If mate choice in adult *T. natans* is influenced by kin recognition, mating preferences for non-relatives would be a way to avoid inbreeding. If so, we hypothesize that females should be choosier than males because of their much higher reproductive costs.

MATERIAL AND METHODS

Study Animals

Experimental animals came from stock populations kept in our institute since 1990. The founder animals of this colony were obtained from pet stores and probably originated from Colombia. Specimens were examined according to Nussbaum and Wilkinson (1989) and Wilkinson (1996) and have been positively identified as *T. natans*. Animals were kept in groups of 4 to 7 individuals in 200 – 360 l aquaria (water level 10 - 25 cm) under controlled temperature (26 – 30°C) and light conditions (12 hours light and dark cycle). Plants, terracotta tiles, plexiglas tubes, and stones provided shelter. Animals were fed once a week ad libitum, using a variety of foods, beef heart, beef liver, fish, dried pellet fish, turtle food, dried or frozen shrimps, tubifex, earthworms, chironomidae larvae. Rain seasons were simulated once a year for 4 to 8 weeks by lowering water temperature (20 – 26 °C) and increasing water level. Females and males of each litter were separated at an age of 1 - 1 1/2 years when the male cloacal disc was clearly developed. All tested caecilians and their maternal relationships were individually known. For behavioral tests, only adult specimens, age 2 1/2 years or more (sizes ranging from 35 to 55 cm), were used.

Test-Equipment and Preparation of Stimulus Water

Experiments were conducted in a plexiglas choice-chamber based on a modified Y-maze, as described by Parzefall (1976). A main chamber called the neutral compartment (45 x 30 x 30 cm) was connected via a perforated sliding door to two dark inlet arms called choice compartments (16 cm length x 4.5 cm diameter). Test waters were supplied from two identical stimulus tanks via PVC tubes (inner diameter 9 mm) into the two choice compartments at a rate of 150-200 ml per minute. Preliminary tests using colored water revealed that the two opposing stimuli mixed only in the neutral compartment. The water level in the choice-chamber was held constant (10 cm) by an overflow.

Three days before an experiment was started, single scent-donor animals were transferred into the two stimulus tanks. Each tank (100 l aquaria filled with approximately 40 – 50 l of tap water) was aerated at room temperature (22 - 24 °C) and equipped with a clean terracotta shelter. Molted skin and feces were removed from the tanks on a daily basis to prevent unwanted odor build up. The animals were removed from the tanks 1 h prior to the experiment to exclude any other stimulus except odor cues. To prevent food odor biases, all caecilians were fed the same food several weeks prior to the experiments. No feeding took place during experiments. To minimize biases of odor concentration due to size difference, stimulus and tested animals differed by not more than 5 cm in length. Stimulus individuals were unknown and unrelated to the respective test animal to prevent biases of familiarity or kin. To test the influence of mate choice by kin discrimination, we used brother or son / sister or mother as related stimulus donors that had been isolated from the test animal for more than a year. Each pair of stimulus animals was used only once in each experiment.

EXPERIMENTAL PROCEDURE

Two experiments (I and II) were performed outside of the reproductive period in order to test the animals' ability to discriminate sex of odor donors. In experiment I, non-reproductive females were given the choice between female and male odor cues; in experiment II, we repeated these tests with males. Experiments III-VI, were conducted during four mating periods (1996-1999). In experiment III, we tested whether receptive females change their behavior compared to experiment I. In experiment IV, reproductive females were given the choice between an unrelated and a related male (brother or son) to test the influence of kin on mate choice. Experiment V, tested whether males are able to recognize receptive females. And in experiment VI, we tested the ability of males to discriminate between related (sister or mother) and unrelated females (both receptive).

All tests were conducted in semi-darkness (<1 lux). Test animals were caught with a hand held net and transferred into the choice chamber. An acclimatization period of 5 to 10 minutes was allowed before the experiment was started. During this time, stimulus water was flowing, however, a perforated door closed both choice compartments. Opening the door started the experiment. As an indicator of preference, the total amount of time spent in each choice compartment was measured during the ensuing 30 minute time period. However, if a test animal had not entered any choice compartment after 15 minutes, the test was terminated and these data were excluded from analyses. Stimulus water inflow to choice compartments was connected by random and identified afterwards, thus the observer was not aware which choice compartment carried which signal. After each experiment, the choice-chamber, tubing, and stimulus tanks were thoroughly cleaned and rinsed with tap water.

Data were analyzed using Wilcoxon matched-paired signed-ranks test (Siegel, 1956). Calculations were done using InStat 2.01 for Macintosh (InStat, 1993). All p values are two tailed.

RESULTS

The results of all tests are shown in Table 1. Experiment I, showed that non-reproductive females of *T. natans* spent significantly more time in the choice-compartment with water scented by females rather than scented by males (Wilcoxon matched-paired signed-ranks test: n=16, p=0.0003). Correspondingly, non-reproductive males chose male-scented water over female-scented water (Exp. II: n=15, p=0.0043). During the mating period, receptive females changed their preference and chose male-scented water over female-scented water (Exp. III: n=15, *p*=0.0003). Given the choice between odors from unrelated and related males, they significantly preferred the unrelated male (n=14, *p*=0.0023). Reproductive *T. natans* males showed a greater attraction to odors of receptive females than of non-receptive females (Exp. V: n=15, p=0.0003), but did not significantly prefer the unrelated or the related receptive female in experiment VI (n=9, p=0.3594).

Table 1. Results of choice tests : Amount of time (s) spent in each stimulus compartment.

Experiment	Sample size (n)	Test animals	stimulus water produced by	Mean (s)	Median (s)	Range (s)	T*	*p**
I	16	females	females	422	367	45-881	2	0.0003
			males	170	171	65-299		
II	15	males	females	131	98	32-284	12	0.0043
			males	294	291	62-559		
III	15	receptive females	females	188	189	45-348	3	0.0003
			males	459	485	137-734		
IV	14	receptive females	related males	159	117	65-360	7	0.0023
			unrelat. males	282	255	43-521		
V	15	reproductive males	receptive fem.	397	371	12-854	3	0.0003
			non-recept. fem.	130	103	15-279		
VI	9	reproductive males	related females	421	352	34-821	31	0.3594
			unrelated fem.	562	612	82-901		

* Wilcoxon matched paired signed ranks test, two-tailed

DISCUSSION

Our results indicate that *T. natans* produces waterborne chemical signals that communicate about gender, reproductive state, and kin.

During the non-reproductive period, females and males preferred odors from conspecifics of the same sex, which provides evidence for sex discrimination and confirms our previous study (Warbeck and Parzefall, 1997). During mating periods, females changed their preference and chose male-scented water to female-scented water, which suggests an active search for mates. On the other hand, reproductive males continued to prefer males-scented water when given the choice between a male and a non-receptive female (Warbeck and Parzefall, 1997). However, males possess the ability to discriminate between receptive and unreceptive females, suggesting the production of waterborne chemical cues by females that give information about their reproductive state. Furthermore, *T. natans* males show a specific head shaking behavior followed by mating attempts after sniffing at a receptive female. We interpret this behavior as an indication for a contact pheromone that elicits courtship behavior in males.

The benefit for females to attract males remains somehow unclear. Males often form large aggregations around a receptive female for several weeks. Male-male competition is high during mating period and aggressive encounters between males can lead to serious injuries, and females often suffer bites. Copulations in *T. natans* lasts up to four hours and only old and heavy males have been observed to copulate successfully. However, most mating attempts and copulations were interrupted by the female (not by other males), which give indication of female choice rather than male-male competition.

Although we could not find evidence for kin (mother) recognition in juvenile *T. natans* (Warbeck *et al.*, 1996), our data presented here strongly suggest that adult females discriminate kin from non-kin, using waterborne olfactory cues and thus could avoid inbreeding. Receptive females chose to spend time near the odors of unrelated males over those from related males, even after more than one year away from the related males.

Consequently, mate choice in females is most likely influenced by kin recognition, as suggested. The mechanisms and the adaptive value of kin recognition in most species that have been investigated to date is not fully understood (for discussion see e.g. Grafen, 1990; Blaustein *et al.*, 1991). However, animals that avoid mating with close relatives benefit from the ability to discriminate kin. In contrast, we could not find any significant preference for unrelated females in reproductive *T. natans* males, and thus males seem to be less choosey than females, as hypothesized. However, we cannot exclude the possibility that they are lacking the ability to discriminate kin at all, or were not motivated to discriminate under these experimental conditions.

The biochemical nature and source of the olfactory cues released by *T. natans* are not yet known. Peptides and proteins are reported to have important functions as chemical signals in urodeles (e.g. Kikujama *et al.*, 1995; Rollmann *et al.*, 1999). Preliminary attempts to isolate chemical cues in *T. natans* from male-scented water indicate that the biologically active molecule(s) that attract conspecifics are polar and extremely water-soluble (Jansen, pers. communication).

ACKNOWLEDGMENTS

We thank Markus A. Wetzel for valuable comments on the manuscript. This research was supported by grants from the Deutsche Forschungsgemeinschaft (DFG-Pa 148/15-1) and the PhD Scholarship program of the University of Hamburg given to A. Warbeck.

REFERENCES

Belvedere, P., Colombo, L., Giacoma, C., Malacarne, G., and Andreoletti, G. E., 1988, Comparative ethological and biochemical aspects of courtship pheromones in european newts, *Monitore Zool. Ital. (N.S.)* **22**:397-403.

Billo, R. R., Straub, J. O., and Senn, D. G., 1985, Vivipare Apoda (Amphibia: Gymnophiona) *Typhlonectes compressicaudus* (Dumeril and Bibronm 1841): Kopulation, Tragzeit und Geburt, *Amphibia-Reptilia* **6**:1-9.

Blaustein, A. R., Bekoff, M., Byers, J. A., and Daniels, T. J., 1991, Kin recognition in vertebrates: what do we really know about adaptive value?, *Anim. Behav.* **41**:1079-1083.

Grafen, A., 1990, Do animals really recognize kin?, *Anim. Behav.* **39**:42-54.

Himstedt, W., and Simon, D., 1995, Sensory basis of foraging behavior in caecilians (Amphibia, Gymnophiona), *Herpetol. J.* **5**:266-270.

InStat, 1993, GraphPad Software, San Diego, USA.

Kikuyama, S., Toyoda, F., Ohmiya, Y., Matsuda, K., Tanaka, S., and Hayashi, H., 1995, Sodefrin: a female-attracting peptide pheromone in newt cloacal glands, *Science* **267**:1643-1645.

Kikuyama, S., Toyoda, F., Yamamoto, K., Tanaka, S., and Hayashi, H., 1997, Female-attracting pheromone in newt cloacal glands, *Brain Res. Bull.* **44**:415-422.

Murphy, J. B., Quinn, H., and Campbell, J. A., 1977, Observations on the breeding habits of the aquatic caecilian *Typhlonectes compressicaudus*, *Copeia* **1977**:66-69.

Nussbaum, R. A., and Wilkinson, M., 1989, On the classification and phylogeny of caecilians (Amphibia: Gymnophiona), a critical review, *Herpetol. Monogr.* **3**:1-42.

Parzefall, J., 1976, Die Rolle der chemischen Kommunikation im Verhalten des Grottenolms Proteus anguinus Laur. (Proteidae, Urodela), *Z. Tierpsychol.* **42**:29-49.

Rollmann, S. M., Houck, L. D., and Feldhoff, R. C., 1999, Proteinaceous pheromone affecting female receptivity in a terrestrial salamander, *Science* **285**:1907-1909.

Schmidt, A., and Wake, M. H., 1990, Olfactory and vomeronasal systems of caecilians (Amphibia: Gymnophiona), *J. Morphol.* **205**:255-268.

Siegel, S., 1956, *Nonparametric Statistics,* Mc Graw-Hill Inc., New York.

Taylor, E. H., 1968, *The Caecilians of the World*, University of Kansas Press, Lawrence.

Wabnitz, P. A., Bowie, J. H., Tyler, M. J., Wallace, J. C., and Smith, B. P., 1999, Aquatic sex pheromone from a male tree frog, *Nature* **401**:444-445.

Wake, M. H., 1985, The comperative morphology and evolution of the eyes of caecilians (Amphibia, Gymnophiona), *Zoomorphology* **105**:277-95.

Wake, M. H., 1992, Reproduction in caecilians, in: *Reproductive Biology of South American Vertebrates*, (W.C. Hamlett, ed.), Springer, New York.

Warbeck, A., and Parzefall, J., 1997, Sex-specific pheromones in the caecilian *Typhlonectes natans* (Amphibia: Gymnophiona), in: *Advances in Ethology 32*, (M. Taborsky and B. Taborsky, eds.), Blackwell, Berlin, Vienna, pp.137.

Warbeck, A., Breiter, I., and Parzefall, J., 1996, Evidence for chemical communication in the aquatic caecilian *Typhlonectes natans* (Typhlonectidae, Gymnophiona), *Mém. Biospéol.* **23**:37-41.

Wilkinson, M., 1996, The taxonomic status of *Typhlonectes venezulensis* Fuhrmann (Amphibia: Gymnophiona: Typhlonectidae), *Herpetol. J.* **6**:30-31.

Wilkinson, M., 1999, Evolutionary relationships of the lungless caecilian *Atretochoana eiselti* (Amphibia: Gymnophiona: Typhlonectidae), *Zool. J. Linn. Soc.* **126**:191-223.

MATERNAL ANOGENITAL LICKING IN RATS : EXPLORING THE DAM'S DIFFERENTIAL SEXUAL TREATMENT OF PUPS

Isabelle Brouette-Lahlou[1], Evelyne Vernet-Maury[1], Francine Chastrette[2], and Jacques Chanel[1]

[1]Emotion et Vigilance, Capteurs Biomédicaux
INSA Lyon - bâtiment 401
Léonard de Vinci, RdC
F-69621 Villeurbanne cedex – France
[2]Laboratoire de Chimie Organique Physique
CNRS – Université Claude Bernard/Lyon
F-6922 Villeurbanne cedex – France

INTRODUCTION

Licking the neonate's anogenital region is a maternal behavior seen in most rodents. This behavior, termed maternal anogenital licking (MAGL), was described for domestic Norway "laboratory" rats by Rosenblatt and Lehrman (1963) and was also analyzed by Charton et al., (1971). In laboratory rats, MAGL lasts up to weaning; it stimulates reflexive urination and facilitates urine production in the neonates. Its behavioral role in recycling water and ions is part of the complex nutritional interaction between mothers and neonates (Gubernick and Alberts, 1983). In a previous work (Brouette-Lahlou et al., 1991b), we found that chemicals in the pup's preputial gland secretions apparently provide a target for MAGL behavior and contribute to its homeostatic regulation. We found that after pup's preputial glands had been ablated, abbreviated as PPX, MAGL behavior was not only present but its duration was enhanced. We propose that the increased duration of MAGL after pup PPX is due to the mothers' unmet appetite to detect/ingest their pup's preputial – gland secretions. Moreover, Moore and Samonte (1986) demonstrated that a pup's preputial gland secretions are used by dams to discriminate that pup's sex.

In previous work, combined gas chromatograhy and mass spectroscopy of a lipid extract from pup preputial glands allowed Brouette-Lahlou et al. (1991a) to identify these constituents: dodecyl propionate (DP), iso-propylmyristate (traces), and di (2-ethyl-hexyl) adipate. Among these, DP appeared to be the best candidate for having pheromone-like qualities related to MAGL regulation.

We have reported that ultrasonic vocalizations from pups were emitted just prior to the dam's initiation of MAGL (Brouette-Lahlou et al., 1992). We observed that dams placed in test cages with one 5-day old pup (male or female) systematically responded to pups' vocalization by moving the pup to a corner of the cage and initiating MAGL

Moore (1979, 1981, 1982) has reported in several studies that dams are able to discriminate between male and female pups and that the dams direct more licking to males than to female pups. However, the sex but not the individual identity of the pups was recorded in those studies, so it's possible disproportionate MAGL to "favorite" male pups might have skewed those results. All of those observations took place for pups aged 2 to18 days and during a 3-hr interval of the dark phase of the light/dark cycle. Richmond and Sachs (1984) conducted similar tests except that each pup was individually identified; observations took place only on days 4, 7 and 10 after delivery.

The present report addresses whether or not a male-female difference in the production of a preputial gland chemical secretion, eg., DP, could explain the greater amount of licking directed toward male pup's anogenital region throughout lactation period. We sought to answer this question by measuring AGL durations and frequencies in intact pups and PPX pups. The discussion ensuing was conducted to evaluate what stimulus from pups (DP or US calls) may be helpful to solve the question of differential MAGL behavior throughout lactation licking cycle.

METHODS

Subjects were primiparous Wistar IOPS rat dams weighing 200 to 250g reared in the IFFA-CREDO laboratory colony in France and mated with males from the same strain. All animals were housed with food and water available ad libitum and were in natural light/dark cycle, with "lights on" from 0080 to 0016hrs.

At least two days before parturition, each animal was housed singly in a plexiglas cage (39x19x18cm) provided with pine shavings. The animal room was maintained approximately at 20-22°C. Day 1 was the day following the day of birth. Each litter was culled to 4 males and 4 females and each pup was marked with potassium permanganate stain on different parts of the body (head, back, tail, right foreleg, left foreleg, right hind leg, left hind leg, small of the back). Four litters were observed intermittently from 0080 to 0014hrs for 30 days. Behavioral observations were made one hour per day in permutation over six consecutive days for three litters, the fourth litter was observed daily for 30 days.

Maternal contact is high during the light phase and depressed during the dark phase of the light/dark cycle, therefore observations of maternal behavior were carried out during the middle of the light cycle (0080-0014hrs) corresponding to the peak of maternal behavior, according to Ader and Grota (1969). MAGL activity alone was observed independently of the other maternal activities. MAGL duration and frequency were recorded on stopwatches in relation to pup sex. Pup sex was identified at birth on the basis of anogenital distance, of genital prominence in males and nipple location in female pups.

Preputial-glandectomy was designed to determine whether the lack of chemical cues could modify AGL durations and frequencies according to pup sex. Five litters were observed on different days as explained above: 2 served as control and pups from 3 litters underwent PPX at birth. As described by Brouette-Lahlou et al. (1991b), pups were anesthetized with ether, an incision was performed by cutting a pane of skin about $2mm^2$ on three sides, 0.5cm above the perineal area so that the dam's wound-cleaning movements could be easily distinguished from MAGL. As we did in this previous work, we carried out PPX at parturition to eliminate learning during a pre-exposure to preputial gland odor of pups before surgery. The control pups received a sham surgical procedure which was identical with PPX pups except that the glands were not removed. The frequency and duration of MAGL bouts were recorded on days 1-4-7-10-14-16, and 19 postpartum in order to compare data from SHAM and PPX litters. As in the first experiment, direct observations were carried out throughout the six hours of the day corresponding to the peak of maternal behavior (Ader and Grota, 1969).

Data were analyzed using: i) the Mann-Whitney U test to compare data from two populations: male pups vs.female pups and PPX.vs.control pups , ii) the Kruskal-Wallis one-way analysis of variance was used to assess group differences in maternal behavior, iii) As far as frequency data are concerned, differences between control and PPX pups were analyzed using Chi-square tests. Pups were the units of analysis. Frequency and iiii) AGL duration values of each population are express as medians (semi-semi interquartile ranges).

RESULTS

MAGLD Delivery According to Pup Age and Sex

MAGL bout durations. Homogeneity of results made it possible to pool data from the four experimental series of four litters. There was no significant effect of litter on MAGL durations, concerning male pups (Kruskal-Wallis test: H=3.84, df=3, NS) and female pups (H=3.97, df=3, NS). The results are presented in Figure 1, each point represents MAGL durations cumulated over the 6 hours per day observation.

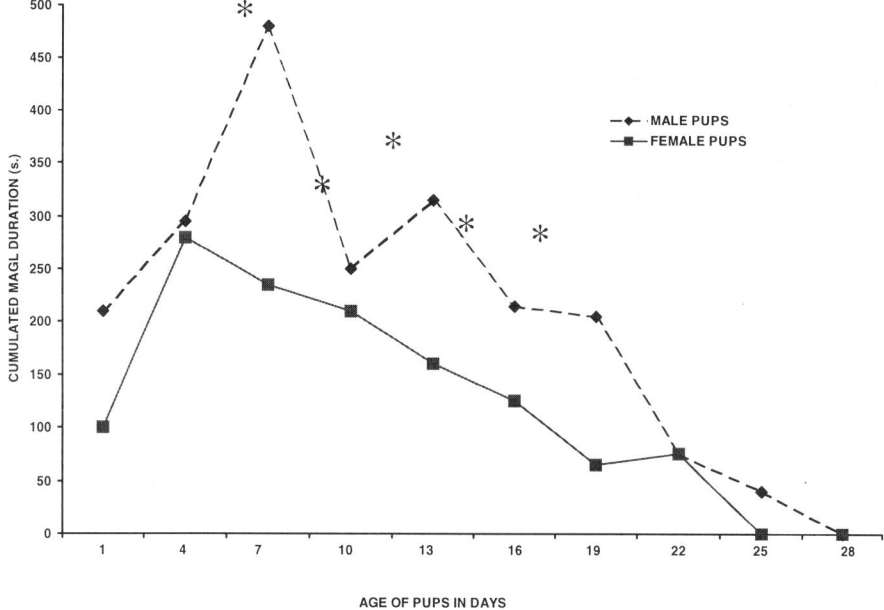

Figure 1. Duration of maternal anogenital licking (MAGL) of pups throughout lactation. Each point represents the cumulated data for four litters directly observed continuously for 6 hrs/day until day 30 post-partum.

As seen in Figure 1, the duration of MAGL directed to males follows a pattern different to that directed to female pups. In male pups, MAGL duration increases until day

7 when it has the longest duration, decreases on day 9, and increases again on 13th day, followed by a steady decrease until the 28th day. In female pups, the highest MAGL duration is observed on the 4th day, and then steadily decreases until the 25th day. No further MAGL was observed towards female pups after day 22 and on day 25 one male pup was still licked. Thus dams spent significantly more time licking their male than their female pups. Subsequent comparisons of bout duration day by day using a Mann-Whitney U test confirmed this difference: day 7, z =-2.38, p=0.0125; day 10, z = -1.87, p=0.0307;day 13, z =-1.54, p = 0.057; day 16, z = -2.19, p=0.0143; day 19, z=-1.80, p=0.035. These differences remained when MAGL durations were analyzed by a Mann-Whitney U test across litters and days of observations (z=-3.91, p>0.0003), medians (semi-interquartile range) being respectively, 13 (4.5) s. for male pups and 10 (2.5) s. for females.

MAGL frequency. No significant effect of litter was observed on MAGL frequencies concerning male pups (Kruskal-Wallis test: H=7.70, df=3, NS) and concerning female pups (H=1.52,df=3.2, NS), thus we pool the data from the four litters observed. MAGL frequency values from females are always lower than those for males. The difference is especially obvious on day 7 (z=-2.13, p<0.0166).

The data analyzed by a Mann-Whitney U test confirmed this difference throughout lactation period until 28th postpartum day (U=10,n_1=n_2=9, p<0.001).

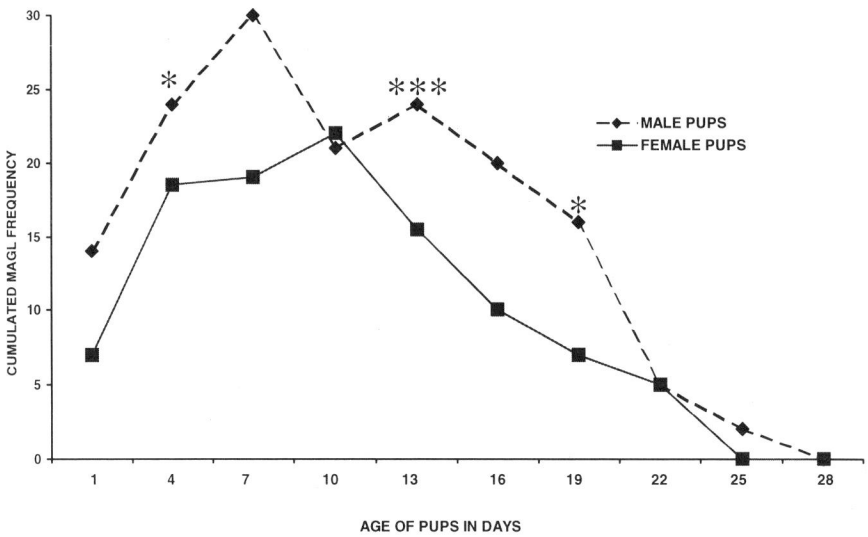

Figure 2. Frequencies of maternal licking of pups (MAGL) throughout lactation. (Conditions as in Figure1).

MAGL Durations and Frequencies after Preputialectomy

There was no litter effect on MAGL durations in PPX pups (MAGL durations to male pups: Kruskal-Wallis test: H=4.16, df=2,NS), and to female pups (H=4.35, df=2,NS); or on MAGL frequencies in PPX pups (MAGL frequencies to male pups (H=3.92, df=2,NS) and to female pups (H=4.82, df=2,NS). Likewise there is no litter effect on MAGL duration in control pups (MAGL duration to male pups, n_1=n_2=7; U=28, NS, and to female pups, n_1=n_2=7; U=14,NS) or on MAGL frequencies to control pups (MAGL frequencies to male pups, χ^2=0.42, df=1, NS and to female pups, χ^2=0.1, df=1, NS).Thus, we can pool the data from the three PPX litters and the two control litters: data from pups were summed over six hours of observations throughout the seven days.

Results appear in Figure 3. Several findings emerged from the analysis of this figure. First, in the control pups, male pups were licked more frequently than were female pups (124 versus 69 bouts respectively, p<0.001 using a χ^2 one sample-test). Males were also licked for longer periods than were females [11(4) s. versus 9 (3) s., p< 0.001]. Second, comparing data from control pups versus PPX pups, MAGL duration was longer for PPX males [11(4) s. versus 7 (4.5) s., respectively, z=-6.71, p<0.001]. Third, also comparing data from control pups versus PPX pups, MAGL duration was longer for PPX female pups [9(3) s. versus 15(3) s., respectively, z=-6.66, p<0.001]. The PPX-related increase also reached statistical significance for MAGL frequencies in male PPX pups (172 in PPX pups versus 124 in control pups, χ^2 =3.59, df=1.7, p<0.005 for one-tailed test) and in female pups (118 in PPX pups versus 69 in control females, χ^2=6, df=1 0.002>p>0.001).

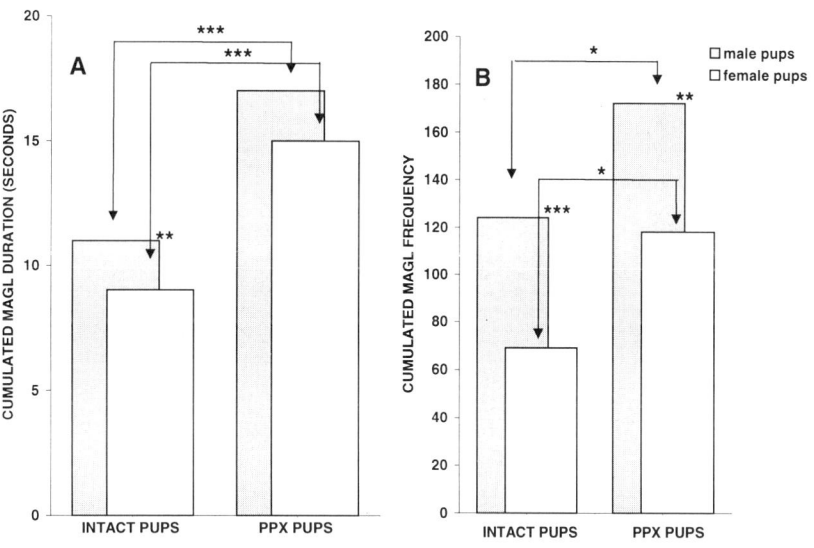

Figure 3. Medians of MAGL duration (A) and frequency (B) directed respectively to male pups and to female pups from day 1 to day 20 postpartum when pups were preputialectomized (PPX) at birth.**p < 0.01; ***p < 0.001; without * = NS.

An important finding was the lack of statistical difference in MAGL duration between male and female pups: 17(4.5) s. versus 15(3) s., z=0.57, NS. It is noteworthy that there was a statistically significant difference between male and female PPX pups in MAGL frequency need (172 versus 118 licks, respectively, and χ^2=10, p<0.001 using a χ^2 one-sample test). Thus, dams' discrimination between male and female pups remains only for MAGL frequencies whereas it disappears for MAGL durations after PPX.

DISCUSSION

The differential treatment of male and female offspring by rat dams, originally reported by Moore and Morelli (1979), supported by the studies of Richmond and Sachs (1984), was further confirmed by the present study. Our study provided new evidence that pups are no longer licked in the anogenital area when females are 22, and males 28 days of age, respectively.

At weaning on day 25, pups are mature enough to urinate and defecate without assistance by MAGL; in 1983 Gubernick and Alberts described the water and electrolyte flow from the mother rat to the pups in the form of milk and their returned to the dam via pup urine. Thus, MAGL plays a major role in the homeostatic maintenance of the nursing dams' body fluids The decline in MAGL after day 15 postpartum corresponds developmentally to the decrease in water transfer. At weaning, pups start to consume water and solid foods while reducing the amount of nursing, so the energetic demands of lactation are reduced, and the mother's body fluid balance returns to its pre-nursing mechanisms of homeostatic balance.

The developmental sequence present above does not explain why dams stopped licking female pups on day 22 but males pups received MAGL until age 28 days. We propose that males produce more DP, and that this preputial pheromone is a strong stimulus for MAGL. It is now well established that the onset of maternal behavior depends on oestrogens secreted during the period around parturition. In the postpartum period, stimuli from pups promote maternal responsiveness. Among these stimuli, DP from the preputials gland play an important role by regulating and targeting MAGL behavior, as we noted when we first used the term of "cyberomone" to emphasize this (Vernet-Maury and Brouette-Lahlou, 1991). DP's persistence in preputial gland secretions of maturing male pups may help the dam to continue licking them and this has important consequences for males' sexual behavior later in life: male rats that receive less MAGL copulated less robustly than other males (Moore, 1984). On day 15 after parturition, di (2-ethyl-hexyl) adipate, another preputial gland secretion product, appeared precisely at the moment when MAGL declined, so it might serve to inhibit MAGL responses to the maturing pups (Vernet-Maury *et al.,* 1987; Brouette-Lahlou, 1992).

Sex bias was maintained in PPX pups when frequency of licking was used as the measure; this indicates a discriminable cue other than preputial glands secretions. Naito and Tonoue (1987) established that male pups produce more ultrasonic vocalizations than do female pups when they coexist in the same litter. We propose that male pups might be licked more often than females as a result of a higher amount of ultrasonic emissions.

In previous work (Moore, 1981), chemical cues in the urine of male and female pups were identified by Moore as one source for maternal discrimination of sex. Male pups secrete testosterone perinatally; Moore (1982) suggested that testosterone or one of its metabolites is responsible for producing the more effective odor of males; she demonstrated that female pups treated with dihydrotestosterone receive the same amount of MAGL as males. Preputial gland openings and urine ducts are not separated by more than 2 mm in pups; therefore, urine may contain preputial gland secretions such as DP. Moreover, preputial glands are known as the source of androgen-dependent olfactory stimuli (Mugford and Nowell, 1971). The amount of DP may also be under hormonal stimulation, but it remains for future research to demonstrate that DP secretion may be androgen dependent.

Finally, we propose two main stimuli to explain the greater amount of MAGL directed to male pups: i) From birth to the 22ndday, the dam responds to pup calls and to DP, thus performing MAGL more frequently on male pups who emitted more calls and more DP, ii) DP is a regulating factor which helps the dam to target MAGL and determine MAGL duration. After the 22nd day, pup's emissions decrease drastically (Duveau *et al.,* 1981;

Vernet-Maury et al. 1982) dams only lick pups when DP is still present in preputial secretion i.e., in male pups. The reproductive function being vital to species conservation, it is not surprising that two stimuli act synergically to ensure an optimal potentiality to males.

REFERENCES

Ader, R., and Grota, J., 1969, Continuous recording of maternal behavior in *Rattus norvegicus*, *Anim. Behav.* **17**:722-729.

Brouette-Lalhou, I., 1992, Le Concept De Cyberomone, Thèse de doctorat d'état es sciences, Université Hassan II, Casablanca.

Brouette-Lahlou, I., Amouroux, R., Chastrette, F., Cosnier, J., Stoffelsma, J., and Vernet-Maury, E., 1991a, Dodecyl propionate, attractant from rat pup preputial gland: characterization and identification, *J. Chem. Ecol.* **17**(7):1343-1354.

Brouette-Lahlou, I., Vernet-Maury, E., and Chanel, J., 1991b, Is rat dam licking behavior regulated by pups' preputial gland secretion? *Anim. Learn. Behav.* **19**(2):177-184.

Brouette-Lahlou, I., Vernet-Maury, E., and Vigouroux, M., 1992, Role of pup's ultrasonic calls in a particular maternal behavior in Wistar rat: pups anogenital licking, *Behav. Brain Res.* **50**:147-154.

Charton, D., Adrien, J., and. Cosnier, J., 1971, Déclencheurs chimiques du comportement de léchage des petits par la ratte parturiente, *Revue du Comportement Animal* **5**:89-94.

Duveau, A., Vernet-Maury, E., and Chanel, J., 1981, Ontogenèse des communications perceptuelles: vocalisations ultrasonores à point de départ olfactif chez le rat, *J. Physiol.* **77**(9):53A.

Gubernick, D. J., and Alberts, J. R., 1983, Maternal licking of young: resource exchange and proximate controls, *Physiol. Behav.* **31**:593-601.

Moore, C. L., 1981, An olfactory basis for maternal discrimination of sex of offspring in rats (Rattus norvegicus), *Anim. Behav.* **29**: 83-386.

Moore, C. L., 1982, Maternal behavior of rats is affected by hormonal condition of pups, *J. Comp. Physiol. Psychol.* **96**:123-129.

Moore, C. L., 1984, Maternal contributions to the development of masculine sexual behavior in laboratory rats, *Dev. Psychol.* **17**:347-356.

Moore, C. L., and Morelli, G. A., 1979, Mother rats interact differently with male and female offspring, *J. Comp. Physiol. Psychol.* **93**:677-684.

Moore, C. L., and Samonte, B., 1986, preputial glands of infant rats (Rattus norvegicus) provide chemosignals for maternal discrimination of sex, *J. Comp. Physiol. Psychol.* **100**(1):76-80.

Mugford, R. A., and Nowell, N. W., 1971, The preputial glands as a source of aggression-promoting odors in mice, *Physiol. Behav.* **6**: 247-249.

Naito, H., and Tonoue, T., 1987, Sex difference in ultrasound distress call by rat pups, *Behav. Brain Res.* **25**:13-21.

Richmond, G., and Sachs, B. D., 1984, Maternal discrimination of pup sex in rats. *Dev. Psychol.* **17**:98-99.

Rosenblatt, J. S., and Lehrman, D. S., 1963, Maternal behavior of the laboratory rat, in: *Maternal behavior in mammals*, (Wiley N.Y., ed.) pp. 8-57.

Vernet-Maury, E., and Brouette-Lahlou, I., 1991, The regulating factor from pup's preputial gland secretion as a cyberomone. *The VIII Meeting of the International Society of Chemical Ecology*, Dijon (abstract).

Vernet-Maury, E., Brouette-Lahlou, I., Chanel, J., 1987, Ontogenetic analysis of the rat pup pheromone implicated in perigenital licking, *Chem. Senses.* **12**:186.

Vernet-Maury, E., Duveau, A., and Chanel, J., 1982, Ontogeny of ultrasonic vocalizations in the rat and their alteration by fox odor, *Fifth ECRO Congress*, Regensburg, p.97 (abstract).

PREDATOR DIET CUES AND THE ASSESSMENT OF PREDATION RISK BY AQUATIC VERTEBRATES: A REVIEW AND PROSPECTUS

Douglas P. Chivers and Reehan S. Mirza

Department of Biology, University of Saskatchewan
112 Science Place, Saskatoon, SK S7N 5E2, Canada

1. INTRODUCTION

Chemosensory assessment of predation risk is widespread in aquatic systems (Chivers and Smith, 1998; Kats and Dill, 1998). Studies completed primarily in the last decade suggest that the level of this assessment is probably much more sophisticated than was previously thought. For example, many prey animals appear to be able to distinguish between predators that are fed different diets. Being able to differentiate that predators are fed different diets means that prey animals can potentially use this information to mediate the intensity of their responses to predators.

The aim of this paper is to provide a comprehensive review of studies examining the importance of predator diet cues in chemosensory assessment of predation risk by primary aquatic vertebrates (fishes and amphibians). We highlight several papers that have demonstrated the importance of diet in chemically mediated responses. These papers should serve to illustrate why it is critical for researchers to consider predator diet in the design of future studies.

2. LITERATURE SURVEY OF CHEMOSENSORY ASSESSMENT OF PREDATION RISK BY AQUATIC ANIMALS

We have identified 229 papers published between 1956 and 2000 that have documented that aquatic animals respond to chemical cues of predators. Of these 229 papers, 79 papers examine the responses of vertebrate prey. The focus of this review will be to examine predator-diet effects in primary aquatic vertebrates (fishes and amphibians). However, it is important to be aware that predator-diet effects are also seen in terrestrial vertebrate systems (e.g. Madison *et al.*, 1999) as well as in aquatic invertebrate systems. According to our tally, 11 of the 150 invertebrate papers have examined the importance of predator diet in the response to predators (e.g. Chivers *et al.*, 1996).

Of the 79 papers examining responses of fishes and amphibians to chemical cues of predators, only 52% report the diet of the predator. Even fewer studies have manipulated the diet of the predator. The vast majority of studies that have manipulated diet have been completed in the past decade. Table 1 provides a summary of the results of the studies that have tested whether anti-predator responses are influenced by predator diet. It is important to note that diet dependent responses have not been documented in all systems.

3. EXAMPLES OF DIET-DEPENDENT ANTI-PREDATOR RESPONSES

We have chosen to highlight three predator-prey systems that exemplify that predator diet influences morphological, behavioral and life historical defenses of prey animals. The examples we highlight reflect the sophistication of diet dependent responses.

3.1. Adaptive Changes in Body Morphology of Crucian Carp

Brönmark and Miner (1992) conducted a series of experiments examining predator-induced morphological defenses in crucian carp. They showed that 12 weeks after introducing pike into one side of a divided pond, carp from the two sides diverged in body shape, with carp from the pike side being deeper bodied. The results of additional laboratory experiments showed that the observed divergence in body shape was a predator-induced morphological change. Increased body depth results in carp being less vulnerable to gape limited predators such as pike; however, this change in morphology incurs a cost through an increase in hydrodynamic drag when the carp are swimming (Brönmark and Miner, 1992).

Two studies have closely examined the importance of predator-diet cues in the morphological change in carp. Both of these studies have found that diet is an important variable to consider. Brönmark and Pettersson (1994) showed that carp exhibit an adaptive change in morphology in response to perch fed carp but not to perch fed chironomids. Similarly, Stabell and Lwin (1997) documented that carp exhibited a change in body morphology in response to cues of pike fed carp but not pike fed Arctic charr, *Salvelinus alpinus*. Carp use cues of conspecifics in the diet of the predator to recognize that the predator represents a substantial threat. Exposure to a predator that is piscivorous (i.e. eats fish), but not feeding on carp, is not sufficient to induce the change in morphology.

3.2. Behavioral and Life History Responses of Red-legged Frogs to Predator Diet Cues

Red-legged frogs and rough skinned newts can be found together in many of the same breeding habitats where their ranges overlap in western North America. Rough skinned newts are effective predators of red-legged frog larvae, although, the presence of alternative prey may lower predation risk for red-legged frogs.

Wilson and Lefcort (1993) examined the importance of predator-diet cues in interactions between tadpoles and adult newts. Tadpoles showed substantial reductions in movement in response to chemical cues of newts that had been fed tadpoles. However, they failed to respond to chemical cues of newts fed insects. In a subsequent study, Kiesecker *et al.* (unpub. data) showed that these behavioral effects might translate into differences in life history traits. Larval red-legged frogs that were repeatedly exposed to cues of newts feeding on tadpoles initiated metamorphosis more than six days earlier and were 20% smaller than their counterparts exposed to newts fed insects. By metamorphosing earlier and leaving the pond, red-legged frogs would likely decrease their predation risk to adult newts. However, metamorphosing at a smaller size may be costly in terms of lower survival in the terrestrial environment.

Table 1. Summary of experiments that have tested whether the responses of aquatic vertebrates are influenced by predator diet.

Prey	RESPONSE
brook trout *(Salvelinus fontinalis)*	trout avoid water conditioned with Atlantic salmon *(Salmo salar)* fed goldfish *(Carassius auratus)* but not salmon fed earthworms (Keefe, 1992)
	trout trained to recognize pickerel *(Esox americanus)* fed goldfish do not distinguish pickerel fed goldfish from pickerel fed mealworms *(Tenebrio spp.)* (**no diet effect observed**) (Keefe et al., 1992)
	trout emerge from gravel faster when exposed to cues of slimy sculpins *(Cottus cognatus)* fed trout eggs versus sculpins fed brine shrimp (Mirza et al., unpub.)
fathead minnows *(Pimephales promelas)*	naive minnows decrease activity, seek shelter and alter space use in response to pike *(Esox lucius)* fed non-breeding minnows but not pike fed breeding minnows or other fish; minnows acquire predator recognition through cues detected in diet of predators (Mathis and Smith, 1993a, b; Brown et al., 1995a)
	minnows avoid feces of pike fed minnows or brook stickleback *(Culaea inconstans)* but not pike fed other fish (Brown et al., 1995b)
	male minnows show reduced territorial behavior in response pike fed minnows compared to pike fed stickleback (Jones and Paszkowski, 1997)
bleak *(Alburnus alburnus)*	bleak reduce foraging and have slower growth when exposed to pike feeding on bleak than pike that were unfed (Jachner, 1997)
Finescale dace *(Chrosomus neogaeus)*	dace increase shoaling and alter predator inspection behavior to perch *(Perca flavescens)* fed dace but not perch fed other fish (Brown and Cowan, 2000)
golden shiner *(Notemigonus chrysoleucas)*	shiners show more dashing behavior to cues of northern water snakes *(Nerodia sipedon)* fed shiners than snakes fed other fish (Goddard et al., 1998)
crucian carp *(Carassius carassius)*	carp show adaptive change in body morphology to perch *(Perca fluviatilis)* fed carp but not perch fed chironomids (Brönmark and Petterson, 1994)
	carp show adaptive change in morphology to pike fed carp but not pike fed other fish (Stabell and Lwin, 1997)
goldfish *(Carassius auratus)*	goldfish show adaptive change in body morphology to pike fed goldfish but not pike fed other fish (Mirza, 1998)
glowlight tetra *(Hemigrammus erythrozonus)*	tetras show altered predator inspection behavior to cues of cichlids *(Cichlasoma octofaciatum)* fed tetras but not starved cichlids or cichlids fed other fish (Brown and Godin, 1999; Brown et al., 2000)
brook stickleback *(Culaea inconstans)*	stickleback decrease activity in response to pike fed stickleback but not pike fed other fishes (Gelowitz et al., 1993)
slimy sculpin *(Cottus cognatus)*	sculpins decrease activity to brook trout *(Salvelinus fontinalis)* fed sculpins, brine shrimp and commercial pellets (**no diet effect observed**) (Bryer et al., unpub.)

Table 1. Continuation

Prey	RESPONSE
red-legged frog *(Rana aurora)* tadpoles	tadpoles decrease movement and alter metamorphic characteristics in response to newts *(Taricha granulosa)* fed tadpoles but not newts fed insects (Wilson and Lefcort, 1993; Kiesecker *et al.*, unpub.)
wood frog *(Rana sylvatica)* tadpoles	tadpoles decrease activity and reduce foraging in response to dragonflies *(Anax junius)* fed tadpoles over starved dragonflies (**no diet effect observed**) (Petranka and Hayes, 1998)
wood frog *(Rana sylvatica)* tadpoles	tadpoles decrease activity more in response to perch *(Perca flavescens)* or dragonflies *(Anax spp.)* fed tadpoles than perch or dragonflies fed invertebrates (Chivers and Mirza, in press)
common frog *(Rana temporaria)* tadpoles	tadpoles show greater reductions in activity and spatial avoidance of dragonflies *(Aeshna juncea)* fed tadpoles than dragonflies fed insects (Laurila *et al.*, 1997)
Columbia spotted frog *(Rana luteiventris)*	tadpoles increase refuge use more in response to sunfish *(Lepomis macrochirus)* fed tadpoles than sunfish fed invertebrates (Lefcort *et al.*, 1999)
American toad *(Bufo americanus)* tadpoles	tadpoles decrease activity and reduce foraging in response to both dragonflies *(Anax junius)* fed tadpoles and dragonflies that were starved (**no diet effect observed**) (Petranka and Hayes, 1998)
southern toad *(Bufo terrestris)*	tadpoles exhibited avoidance behavior (increased movement) to cues of amphiuma *(Amphiuma means)* fed toad tadpoles but not amphiuma fed leopard frog *(Rana utricularia)* tadpoles (Lefcort, 1998)
	tadpoles exhibited avoidance behavior (increased movement) to cues of sirens *(Siren intermedia)* fed toad tadpoles but not sirens fed leopard frog *(Rana utricularia)* tadpoles (Lefcort, 1998)
western toad *(Bufo boreas)* tadpoles	juvenile toads avoid cues of garter snakes *(Thamnophis sirtalis)* fed juvenile toads but not larval toads (Belden *et al.*, 2000)
long-toed salamanders *(Ambystoma macrodactylum)* larvae	salamanders reduced activity to cues of intraspecific predators fed conspecifics and Tubifex (**no diet effect observed**) (Chivers *et al.*, 1997)
	salamanders metamorphose faster and at a smaller size in response to intraspecific predators fed conspecifics versus Tubifex (Wildy *et al.*, 1999)

We need additional experiments to determine the specificity of the diet response. Would red-legged frogs show a behavioral or life history response if exposed to cues of newts feeding on any tadpole or would the response occur only if the newts were fed red-legged frog tadpoles?

3.3. Fathead Minnows Acquire Recognition of Predators Based on Predator Diet Cues

Fathead minnows that are experienced with predatory pike show strong anti-predator responses to pike chemical cues, whereas naive minnows do not respond to pike cues (Mathis *et al.*, 1993). Laboratory experiments have shown that fathead minnows can learn to recognize predators and that predator diet cues are critical to this learning.

Mathis and Smith (1993a) demonstrated that pike-naive minnows exhibited an anti-predator response to cues of pike fed fathead minnows but not to cues of pike fed swordtails *(Xiphophorous helleri)*. When minnows that were initially exposed to pike fed minnows were subsequently tested for a response to cues of pike fed a different diet, they exhibited a strong response. This demonstrates learned recognition of the predator based on detecting conspecific cues in the diet of the predator.

In a follow-up experiment, Mathis and Smith (1993b) elucidated the nature of the cues responsible for the diet effect. They exposed pike-naive fathead minnows to cues of pike fed breeding male fathead minnows, non-breeding fathead minnows or swordtails. Breeding male fathead minnows temporarily lose their alarm pheromone cells during the breeding season. Test minnows showed an anti-predator response only to chemical stimuli from pike that had been fed non-breeding minnows. These results demonstrate that the alarm pheromone contained in the skin of minnows is the diet cue to which the minnows respond.

4. DISCUSSION

Research conducted over the past decade has revealed that prey animals are relatively sophisticated when it comes to assessing chemical cues released by predators. We have provided a few examples of behavioral, morphological and life history experiments that have exemplified this sophistication. There are several important issues related to predator diets effects that deserve consideration. For example, what is the nature of the chemical cues that elicit the diet dependent responses? What measures may predators use to counteract being labeled by a diet cue? Why are predator diet effects seen in some but not all predator/prey systems? We briefly address each of these questions in the discussion that follows.

4.1. What is the Nature of Chemicals that Elicit Diet-dependent Responses?

Notwithstanding the experiments by Mathis and Smith (1993a, b) we know very little about the chemical nature of the cues that are responsible for eliciting the diet-dependent responses that we observe. However, the chemical cues are probably highly variable given that there is obvious gradation in the specificity of diet dependent responses. Keefe (1992) observed a general diet response. She found that brook trout responded to cues of salmon that were fed goldfish but not to cues of salmon that were fed invertebrates. In this example the predator does not need to feed on conspecifics of the prey, but instead just needs to be piscivorous. This contrasts with other studies in which the prey will respond to the predator only if the predator is fed conspecifics of the prey. For example, stickleback respond to pike fed stickleback but not to the same predator fed a different fish diet (Gelowitz *et al.*, 1993). Crucian carp respond to cues of predators fed carp but not to cues of the same predators fed another fish (Brönmark and Petterson, 1994; Stabell and Lwin, 1997). Belden *et al.* (2000) showed a highly specific diet dependent response. In that study, juvenile toads would respond to cues of snakes fed juvenile toads but not to cues of the same predators fed toad tadpoles. Here prey only respond to the predator if it is feeding on conspecifics in the same life history stage.

Further experiments are needed to elucidate the specific chemicals to which the prey animals are responding. Having knowledge of the chemicals that elicit responses should allow us to improve our ability to design future experiments.

4.2. Predator counter-responses to dietary labeling

Predator-prey interactions can often be described as an evolutionary arms race whereby responses of prey animals to avoid being captured are counteracted by predators. It is clear that prey animals benefit by being able to differentiate cues of predators fed different diets. Consequently, it seems likely that predators would benefit by being able to counteract these defenses.

One means by which predators could avoid being labeled by their diet would be to break down or deactivate the cues that prey animals use to recognize them. Researchers seem not to have considered this possibility. We suggest that comparative studies examining the responses of many prey animals to different classes of predators would be needed in order to address this issue. Moreover, we suggest that identifying the nature of the chemicals is essential to make any substantial advances in this area.

Brown *et al.* (1995a, 1996) have taken a different approach to test whether predators can minimize being labeled by diet cues. They studied interactions between pike and minnows. The cue that fathead minnows respond to in the pike's diet is minnow alarm pheromone (Mathis and Smith, 1993a, b). When pike are fed minnows, the pheromone is released in the feces of pike and minnows actively avoid areas where they detect feces of pike that have been fed minnows (Brown *et al.*, 1995a, b). Brown *et al.* (1995a, 1996) examined whether pike would defecate away from their foraging area in order to avoid labeling their foraging area as dangerous to minnows. In trials where pike were fed fathead minnows, pike defecated away from their foraging area. In contrast, pike defecated in their foraging area when they were fed other diets. By localizing their defecation away from their foraging area, pike feeding on fathead minnows would avoid labeling their foraging area as dangerous to their prey.

4.3. Why are diet effects seen in some but not all predator-prey systems?

An examination of Table 1 makes it clear that diet effects are not seen in all predator-prey systems. It is critical for future researchers to address this disparity. It seems likely that we do not observe diet effects in some systems because predators are able to breakdown the cues to which the prey respond. However, several other possibilities exist.

Bryer *et al.* (unpub. data) documented that slimy sculpins exhibit anti-predator behaviour to chemical stimuli from brook trout regardless of the diet of the trout. They argue that brook trout are always a threat to slimy sculpins from this population and hence knowledge of the last meal eaten by the trout may provide little valuable information to the sculpin. The same logic may apply to other systems that have failed to find predator-diet effects.

We speculate that the absence of a diet effect in some predator-prey systems might reflect the fact that there is no variation in the intensity of predation over time. If a predator specializes on other prey types during certain times of the year, then it makes little sense for the prey to respond to the predator if it is not actively foraging on conspecifics. Belden *et al.* (2000) showed that it makes little sense for juvenile toads to respond to cues of garter snake that are concentrating their foraging effort on tadpoles.

Knowledge of the history of interactions between a particular predator and prey is an important factor to consider when explaining why a predator diet effect may be absent from a particular system. The fathead minnow/northern pike system provides a great example. Fathead minnows collected from locations with pike show strong anti-predator responses to pike (Mathis *et al.*, 1993). They respond to the pike regardless of the pike's recent diet. However, minnows from areas where there are no pike, only respond to the pike if the pike is fed minnows (Mathis and Smith, 1993a). These examples clearly show that the population source and experience of the prey can influence whether a diet effect is observed. Researchers need to consider the population source and experience of the prey they test.

5. CONCLUSION

The sophistication of chemosensory assessment of predation risk seems remarkable. As chemical ecologists we have likely just begun to understand the importance of predator diet in this assessment. We implore researchers to carefully consider predator diet in the design and interpretation of their future studies. This means that predator diets always need to be reported in publications. Likewise, we need to consider the experience of our prey animals with different predators in order to understand responses. We feel that it is critical to explore why we see predator diet effects in some but not all predator-prey systems. Advancements in identifying the chemical nature of predator diet responses would undoubtedly take us to new levels as we endeavor to understand chemosensory assessment by prey animals.

REFERENCES

Belden, L. K., Wildy, E. L., Hatch, A. C., and Blaustein, A. R., 2000, Juvenile western toads, *Bufo boreas*, avoid chemical cues of snakes fed juvenile, but not larval, conspecifics, *Anim. Behav.* **59**:871-875.

Brönmark, C., and Miner, J. G., 1992, Predator-induced phenotypical change in body morphology in crucian carp, *Science* **258**:1348-1350.

Brönmark, C., and Petterson, L. B., 1994, Chemical cues from piscivores induce a change in morphology in crucian carp, *Oikos* **70**:396-402.

Brown, G. E., and Cowan, J., 2000, Foraging trade-offs and predator inspection in an ostariophysan fish: switching from chemical to visual cues, *Behaviour* **137**:181-196.

Brown, G. E., and Godin, J.-G. J., 1999, Who dares learns: chemical inspection behaviour and acquired predator recognition in a characin fish. *Anim. Behav.* **57**: 475- 481.

Brown, G. E., Chivers, D. P., and Smith, R. J. F., 1995a, Localized defecation by pike: a response to labelling by cyprinid alarm pheromone? *Behav. Ecol. Sociobiol.* **36**: 105-110.

Brown, G. E., Chivers, D. P., and Smith, R. J. F., 1995b, Fathead minnows avoid conspecific and heterospecific alarm pheromones in the faeces of northern pike, *J. Fish Biol.* **47**: 387-393.

Brown, G. E., Chivers, D. P., and Smith, R. J. F., 1996, The effects of diet on localized defecation by northern pike, *Esox lucius*, *J. Chem. Ecol.* **22**: 467-475.

Brown, G. E., Paige, J. A., and Godin, J.-G. J., 2000, Chemically-mediated predator inspection behaviour in the absence of predator visual cues by a characin fish, *Anim. Behav.* **60**:315-321.

Chivers, D. P., and Mirza, R. S., The importance of predator diet cues in the responses of larval wood frogs to fish and invertebrate predators, *J. Chem. Ecol.* (in press).

Chivers, D. P., and Smith, R. J. F., 1998, Chemical alarm signalling in aquatic predator/prey interactions: a review and prospectus, *Écoscience* **5**:338-352.

Chivers, D. P., Wisenden, B. D., and Smith, R. J. F., 1996, Damselfly larvae learn to recognize predators from chemical cues in the predator's diet, *Anim. Behav.* **52**:315-320.

Chivers, D. P., Wildy, E. L., and Blaustein, A. R., 1997, Eastern long-toed salamander *(Ambystoma macrodactylum columbianum)* larvae recognize cannibalistic conspecifics, *Ethology*, **103**: 187-197.

Gelowitz, C. M., Mathis, A., and Smith, R. J. F., 1993, Chemosensory recognition of northern pike (*Esox lucius*) by brook stickleback (*Culaea inconstans*): population differences and the influence of predator diet. *Behaviour* **127**: 105-118.

Goddard, R. E., Bowers, B. B., and Wannamaker, C., 1998, Responses of golden shiner minnows to chemical cues of snake predators, *Behaviour* **135**:1213-1228.

Jachner, A., 1997, The response of bleak to predator odour of unfed and recently fed pike, *J. Fish. Biol.* **50**:878-886.

Jones, H. M., and Paszkowski, C. A., 1997, Effects of exposure to predatory cues on territorial behaviour of male fathead minnows, *Env. Biol. Fish.* **49**:97-109.

Kats, L. B., and Dill, L. M., 1998, The scent of death: chemosensory assessment of predation risk by prey animals, *Écoscience* **5**:361-394.

Keefe, M., 1992, Chemically mediated avoidance behaviour in wild brook trout, *Salvelinus fontinalis*: the response to familiar and unfamiliar predaceous fishes and the influence of fish diet, *Can. J. Zool.* **70**:288-292.

Keefe, M., Whitesel, T. A., and Winn, H. E., 1992, Learned predator avoidance behavior and a two-level system for chemosensory recognition of predatory fishes in juvenile brook trout, in: *Chemical Signals in Vertebrates* 6 (R. L. Doty and D. Müller-Schwarze, eds.), Plenum Publ., New York, pp. 375-381.

Laurila, A., Kujasalo, J., and Ranta, E., 1997, Different antipredator behaviour in two anuran tadpoles: effects of predator diet, *Behav. Ecol. Sociobiol.* **40**:329-336.

Lefcort, H., 1998, Chemically mediated fright response in southern toad (*Bufo terrestris*) tadpoles, *Copeia* **1998**:445-450.

Lefcort, H., Thomson, S. M., Cowles, E. E., Harowicz, H. L., Livaudais, B. M., Roberts, W. E., and Ettinger, W. F., 1999, Ramifications of predator avoidance: predator and heavy metal-mediated competition between tadpoles and snails. *Ecol. Appl.* **9**:1477-1489.

Madison, D. M., Maerz, J. C., and McDarby, J. M., 1999, Optimization of predator avoidance by salamanders using chemical cues: diet and diel effects, *Ethology* **105**:1073-1086.

Mathis, A., and Smith, R. J. F., 1993a, Fathead minnows, *Pimephales promelas*, learn to recognize northern pike, *Esox lucius*, as predators on the basis of chemical stimuli from minnows in the pike's diet, *Anim. Behav.* **46**:645-656.

Mathis, A., and Smith, R. J. F., 1993b, Chemical labelling of northern pike (*Esox lucius*) by the alarm pheromone of fathead minnows (*Pimephales promelas*), *J. Chem. Ecol.* **19**:1967-1979.

Mathis, A., Chivers, D. P., and Smith, R. J. F., 1993, Population differences in responses of fathead minnows (*Pimephales promelas*) to visual and chemical stimuli from predators, *Ethology* **93**: 31-40.

Mirza, R. S., 1998, Induced morphological change in fishes mediated by chemical stimuli associated with predation, M.Sc. thesis. Univ. Saskatchewan, Saskatoon, Canada.

Petranka, J. W., and Hayes, L., 1998, Chemically mediated avoidance of a predatory odonate (*Anax junius*) by American toad (*Bufo americanus*) and wood frog (*Rana sylvatica*) tadpoles, *Behav. Ecol. Sociobiol.* **42**:263-271.

Stabell, O. B., and Lwin, M. S., 1997, Predator-induced phenotypic changes in crucian carp are caused by chemical signals from conspecifics, *Env. Biol. Fish.* **49**:145-149.

Wildy, E. L., Chivers, D. P., and Blaustein, A. R., 1999, Shifts in life history traits as a response to cannibalism in long-toed salamanders *(Ambystoma macrodactylum)*, *J. Chem. Ecol.* **25**: 2337-2346.

Wilson, D. J., and Lefcort, H., 1993, The effect of predator diet on the alarm response of red-legged frog, *Rana aurora*, tadpoles, *Anim. Behav.* **46**:1017-1019.

FIELD OBSERVATIONS CONFIRM LABORATORY REPORTS OF DEFENSE RESPONSES BY PREY SNAKES TO THE ODORS OF PREDATORY SNAKES

William H.N. Gutzke

Department of Biology
The University of Memphis
Memphis, TN, 38152

INTRODUCTION

Organisms rely on their senses to obtain information about the biotic and abiotic environment. Snakes are dependent upon their chemical and visual senses in obtaining information about their surroundings (Halpern and Frumin, 1979). The structure largely responsible for gathering much of this information in snakes is the vomeronasal organ, which detects environmental chemicals. Many studies have demonstrated the role of this organ in prey detection/prey trailing for all the major snake groups, however, studies examining the prey's ability to detect and respond to a potential predator have been confined to a few laboratory experiments and isolated field observations. I report here preliminary experiments on responses of prey snakes, cottonmouth moccasins, to odors of one of their predators, kingsnakes.

One of the problems with studying chemical communication in the laboratory is to make meaningful ecological extensions of lab findings by replicating the observations in the field. I addressed this concern by observing defense responses of free-living cottonmouth moccasin snakes in the field to odors of kingsnakes, and discussed my observations in an ecological context.

An organism's ability to recognize and respond effectively to potential predators increases its probability of survival and thus its reproductive potential. North American pit vipers (*Reptilia: Serpentes: Viperidae: Crotalinae*) when disturbed usually display a threat posture consisting of coiling the body and raising the head above the coils. The approach of a large, non-prey animal is generally accompanied by an audible warning produced by shaking of the cartilaginous "rattle" on the tip of the tail. Snakes in the crotaline genus *Agkistrodon* the "prey" species in this report, lack the rattle, but will often vibrate their tails on the substrate which produces a sound remarkably similar to that of rattlesnakes

Kingsnakes of the genus *Lampropeltis* attack and consume other snakes including venomous crotalines and are insensitive to the crotaline venom (Cowles, 1938; Marchisin, 1980). Previous laboratory studies have demonstrated that the venomous crotaline snakes are capable of recognizing and defensively responding to the ophiophagic (snake eating) kingsnake, *Lampropeltis getula* (Carpenter and Gillingham, 1975; Weldon and Burghardt, 1979; Gutzke et al., 1993). These studies also report that when confronted by a kingsnake, crotalines will not initiate or cease tail rattling and will demonstrate several unique predator-avoidance behaviors (Weldon and Burghardt, 1979; Miller and Gutzke, 1999), which usually do not include biting. The lack of biting is not surprising in that most of ophiophagic snakes would actually benefit by being provided the location of the potential prey's head (see Miller and Gutzke, 1999). The recognition of kingsnakes by crotalines has been demonstrated to be mediated via the vomeronasal organ (Gutzke et al., 1993; Miller and Gutzke, 1999) and apparently requires none of the other senses (Klauber, 1936; Bogert, 1941; Miller and Gutzke, 1999). It is interesting to note that the defense responses of crotaline snakes are seen even in populations where *Lampropeltis* does not presently occur.

In order to address the questions posed in the first paragraph of this paper, I conducted a field study using a population of cottonmouth (water) moccasins *(Agkistrodon piscivorus)*, which had been the source of animals used in laboratory experiments previously conducted to determined the types and frequencies of ophiophagic defense responses (Miller and Gutzke, 1999). The experiment reported herein was designed to test the following hypotheses: (1) that cottonmouths in the field would respond to the presence of kingsnake semiochemicals in the same manner/intensity as those in the lab; and (2) the proportion of cottonmouth responders in various size/age groups would increase with size until the snakes reach a length that predation by kingsnakes would cease (generally around a meter; Gutzke et al., 1993).

MATERIALS AND METHODS

A population of cottonmouths located in and near T.O. Fuller State Park, Memphis, Tennessee was used in this study. Animals were tested in the wild from June through July 2000. This population is ideal to conduct such studies in that: (a) the population of cottonmouths is large and thus many individuals can be tested; (b) the site also has a relatively large population of speckled kingsnakes (an ophidian ophiophage); and (c) the population had been the source of animals for a laboratory study on the role of the vomeronasal organ in predator detection (Miller and Gutzke, 1999). The previous study had determined that approximately 50% of both neonates, and juvenile and sub-adults less than 1.0 m in total length show a defense response when exposed to semi-volatile chemicals from the skin of kingsnakes (Gutzke et al., 1993). These percentages did not differ from those observed when intact kingsnakes were used

Upon locating a cottonmouth, care was taken not to disturb it; this was usually accomplished by maintaining a two-meter distance between the snake and the observer. Any animal appeared to be disturbed by my presence was given the opportunity to calm down after I backed away from it. Animals that moved away, continued to take note of my presence, or mouth gaped, were not tested.

Testing of animals was accomplished by using long tongs to slowly present two or three (see below) small cotton pads to the cottonmouth. The first pad to be presented was a "blank". For a subset of animals a pad with "POLO" cologne was then presented to determine subjects' response to a novel scent. After this, a pad, which had been rubbed on a live kingsnake, was presented and the behavior of the cottonmouth noted. Finally the animal was captured and assigned to one of four categories: neonate (young of the year), juveniles (one to two year old with "yellow tails"), sub-adults/small adults (no yellow on tail and less than one meter total length), and large adults (> one meter). Each animal was marked with a waterproof yellow magic marker on its back to avoid inclusion of the same individual more than once in the study.

Data collected for each size/age category were calculated as a percentage of individuals from each category that responded positively to the presence of the kingsnake semiochemicals. Size classes were tested using a Chi-squared analysis of proportions to determine if significant difference among the classes were present. Significance was set at the Alpha ≤ 0.05 level.

RESULTS

Data on a total of 62 individuals were obtained from the field. The findings demonstrate that the frequency of responders in natural settings is approximately the same as those, which responded in the laboratory: approximately 52% field responders (Table 1) vs. 48% in the laboratory (Miller and Gutzke, 1999). The fraction of neonates that responded were also approximately the same in both studies 7 of 15 in this study vs. 6 of 11 in the laboratory study. If separated into age/size categories (excluding individuals > 1 m), there is a significant increase in the frequency of response with size (Table 1) ranging from 46.7% for neonates to 66.7% for juveniles (12/18) to 81.3% (13/16) for sub-adults. Adults over one meter in length responded only seldomly (1/13) probably reflecting their size-based resistance to kingsnake predation. In addition, the field work revealed that another ophiophagic defense response was noted in some neonates: head hiding (5 of the 7 neonate responders; see Miller and Gutzke 1999 for a description of this behavior). None of the animals tested with the cologne demonstrated a defensive response.

Chi-square analysis of proportions indicates that these proportions are significantly different at the 0.05 level and that neonates respond less frequently than subadults/small adults and neonates/juveniles/subadults all respond more frequently than large adults (Table 1). The inability to obtain statistically significant differences in the proportion of responders between neonates and juveniles (46.6% vs. 66,7% respectively) is most likely due to lack of power associated with the small sample size in each category (13 and 18 individuals).

Table 1. Proportions of four size/age groups of field tested cottonmouths demonstrating the ophiophagic defense response when subjected to kingsnake semiochemicals. Values with different letter are significantly different, $p \leq 0.05$.

Size Category	Responders	Total	% Responding
Neonate	7	15	46.6^a
Juvenile	12	18	$66.7^{a,b}$
Subadults/Adults	13	16	81.3^b
Large Adults	1	13	7.7^c
Totals	32	62	51.6

DISCUSSION

The findings presented herein extend to field settings our laboratory observations of the vomeronasal-organ mediated ophiophagic defense responses of venomous snakes. The present study also supports the assumption that the ophiophagic defense response is ecologically relevant in this population of cottonmouths. Thus a fruitful discussion of the specific factors, which result in these findings, can be addressed. In addition, the lack of response to a novel scent indicates that the cottonmouths were responding to specific chemicals from the kingsnake rather than a generalized response to any novel scent in the selected cologne.

Data reported herein can be logically interpreted in two ways that assume kingsnakes are potentially a significant predator of cottonmouths in this population. The first possible explanation is that the defense response by cottonmouths is at least somewhat successful in warding off attacks by kingsnakes. Thus any individual who does not display this behavior is at added risk of being consumed by a kingsnake. This would account for the increase with size in the percentage of responders, until a size is reached at which the snakes were too large for kingsnake predation (see Table 1).

The second possible explanation is that cottonmouths learn the defense response as they mature. This hypothesis requires than an innate defense response seen in some, but not all, neonates, becomes more readily evoked as the snakes mature. We know from other laboratory studies that defense responses decline or habituate when predator odors are repeatedly presented without any physical consequences, *i.e.* predation attempts (Weldon and Burghardt, 1979; Marchisin, 1980). Following the odor presentation with an actual attack by a kingsnake or with pursuit by the researcher's tongs maintains defense responsiveness (these "attacks" are halted prior to physical injury to the subjects; Miller and Gutzke, 1999). Thus, if individuals can learn to "not respond", they should be able to learn to respond to the appropriate stimulus. Actual predation attempts by kingsnakes that fail to result in a "kill" can increase the proportion of responders. Both hypotheses lead to the results reported in this study. Unfortunately the present study was not designed to determine which of these competing hypotheses is correct. Thus this interesting question awaits further study.

ACKNOWLEDGMENTS

The author wishes to thank Craig Wilmhoff, John Farrell, and Daniel French for their assistance in the field. In addition, earlier drafts of this manuscript were improved by suggestions of Drs. Mike Ferkin and Jerry Wolff, and Mr. Craig Wilmhoff. The work was sponsored in part by the Biology Department of The University of Memphis.

REFERENCES

Bogert, C., 1941, Sensory cure used by rattlesnakes in their recognition of ophidian enemies, *Ann. New York. Acad. Sci.* **41**:329-343.

Carpenter, C., and Gillingham, J., 1975, Postural responses to kingsnakes by crotaline snakes, *Herpetologica* **31**:293-302.

Cowles, R., 1938, Unusual defense postures assumed by rattlesnakes. *Copeia* **1938**:13-16.

Gutzke, W. H. N., Tucker, C., and Mason, R., 1993, Chemical recognition of kingsnakes by crotalines: effects of size on the ophiophage defensive response, *Brain, Behav. Evol.* **41**:234-238.

Halpern, M., and Fruman, N., 1979, Roles of the vomeronasal and olfactory systems in prey attack and feeding in adult garter snakes, *Physiol. Behav.* **22**:1183-1189.

Klauber, L., 1936, The California kingsnake, a case of pattern dimorphism. *Herpetologica* **1**:18-29.

Marchisin, A., 1980, *Predator-prey Interactions between Snake-eating Snakes*, Ph.D. thesis, Rutgers University, Newark, New Jersey.

Miller, L. R., and Gutzke, W. H. N., 1999, The role of the vomeronasal organ of crotalines (Reptilia: Serpentes: Viperidae) in predator detection, *Anim. Behav.* **58**:53-57.

Weldon, P., and Burghardt, G., 1979, The optiophage defensive response in crotaline snakes: extension to new taxa, *J Chem. Ecol.* **5**:141-151.

SPATIAL RESPONSES OF FIELD (*MICROTUS AGRESTIS*) AND BANK (*CLETHRIONOMYS GLAREOLUS*) VOLES TO WEASEL (*MUSTELA NIVALIS*) ODOUR IN NATURAL HABITAT

Zbigniew Borowski[1] and Edyta Owadowska[2]

[1] Section of Wildlife Management, Forest Research Institute
Bitwy Warszawskiej 1920r. 3, 00-973 Warsaw, Poland
[2] Kampinos National Park
Tetmajera 38, 05-090 Izabelin, Poland

INTRODUCTION

The odors of predatory mammals, i.e. mustelids, might be recognised by potential prey, i.e., small rodents, as a signal of predation risk. Using the sense of smell, rodents could detect the presence of predators and respond accordingly by avoiding areas where they have encountered the predator's scent (Jędrzejewski *et al.*, 1993). Early recognition of predation risk by detection of predator odors might thus increase the chances for survival for small rodents (Borowski, unpublished data). For example, experiments in both laboratory and field conditions demonstrated that exposure to weasel odors caused voles to exhibit antipredatory reactions (Jędrzejewski *et al.*, 1993). Voles that are in danger of predation by mustelids apparently reduce the risk of predation by moving to a microhabitat that lacks evidence of predators (Korpimäki *et al.*, 1996). Voles under predation risk might also limit or change their feeding preferences (Sullivan *et al.*, 1988; Borowski, 1998a) and reduce their locomotor activity to a smaller home range area (Gorman, 1984; Jędrzejewski *et al.*, 1993; Borowski, 1998b). Seasonal differences in responses to weasel odors in field populations of root voles (*Microtus oeconomus*) have been demonstrated: during the breeding season, voles avoided the weasel odour to a greater extent than during the non-breeding season (Borowski, 1998b).

The radiotracking study described here reports our efforts to determine and compare the spatial reactions to weasel odors in field populations of two vole species associated with forest habitat, namely, field voles, *Microtus agrestis*, and bank voles, *Clethrionomys glareolus*. We also compared responses during the breeding season with those of the non-breeding season. To pursue these objectives, we assessed the following parameters of the vole populations: home range size, home range drift, and the mean distance covered by the rodents between the successive radiolocations.

MATERIALS AND METHODS

The study was conducted in the Sudety Mountains, approximately 3 km south of Szklarska Poręba in south-eastern Poland. The study was performed on a 1.2 ha trapping grid in the area covered by spruce forests and young forest plantations with birch (*Betula verrucosa*), Norway spruce (*Picea exelsa*) and rowan (*Sorbus aucuparia*).

Live-trapping of rodents was conducted for 7 consecutive days before, and for 3-4 days after, two sessions of radiotracking conducted in August and in October, 1999. Traps were baited with oats and distributed at a distance of 10 m between each other. The traps were checked twice daily and all the voles were individually marked with toe clipping. Species, individual identification number, trap station, body mass, sex and reproductive conditions were recorded for all captures. During this study, the population density of field voles declined from 75 individuals per ha during August 1999 to 39 individuals per ha during October 1999; among bank voles, the density declined from 21 to 9 individuals per ha in that same interval (Borowski unpublished data).

For the radiotracking sessions, selected voles were taken to the laboratory to be fitted with radiocollars and then released at the capture site 2-5 hours later. Field voles, of body mass greater than 30 g, were fitted with SS-2 radiocollars (2 g mass; Biotrack, Dorset, England). Bank voles, of body mass greater than 17 g, were fitted with LTI radiocollars (0.8 g mass; Titley Electronics, Balinea, Australia). Radiotracked voles were located at least twice a day, morning and evening, during the peak times of locomotor activity. A grid of 10 m^2 was used to record the telemetric positions of the voles. Positions were determined during two intervals: during the first 5 days, when there was no exposure to the predator or its odour, and during the subsequent 5 days, when some of the radiotracked voles were exposed to two female weasels placed in the home range centre (Borowski, 1998b). Exposure to the predator occurred for 20 min in the morning and 20 min in the evening of each day. The location of each radiotracked was determined about 6 times a day during the exposure intervals. Overall, the location of each vole was recorded ~30 times before the start of the period of predator exposure and 30 times after it.

During the breeding season (August 1999), the group exposed to weasel odors consisted of 4 field and 2 bank voles, whereas 3 field and 2 bank voles were in the control (no odour) group. During the non-breeding season (October 1999), the group exposed to weasel odors consisted of 3 field and 3 bank voles; there were 5 field voles in the control group.

Home range analyses were performed using the computer package Tracker vision 1.1. Home ranges were computed as 90 % Probability Convex Polygons centred on the recalculated arithmetic mean. To assess the impact of weasel odour on home range stability, we compared home range centre drift before and during the weasel exposure interval (Agrell, 1995). All of the statistical analyses were performed using procedures available in Statistica 5.1.

RESULTS

Spatial responses of the two vole species to weasel odour showed interspecific variations. For field voles (Figure 1a), there were no significant differences in home range sizes before and after the intervals with or without exposure to weasel odors. Furthermore, there were no significant differences in mean home range sizes in the breeding versus the non-breeding seasons. The home ranges were similar between the control and weasel odour periods both in the breeding (Wilcoxon's signed rank test, $Z = 1.826$. $p = 0.068$) and non-breeding seasons (Wilcoxon's signed rank test, $Z = 0.0$. $p = 1.0$). In contrast, bank voles

(Figure 1b) decreased their home range sizes in August when they were exposed to weasels (Mann Whitney U-test, U = 6.0, Z = 2.309, p = 0.0209), as compared with the control group of these animals (Mann Whitney U-test, U = 3.0, Z = 0.654, p = 0523) (Figure 1b). However, the home range sizes of bank voles exposed to weasel odors in October did not change (Wilcoxon's signed rank test, Z = 1.069, p 0.285).

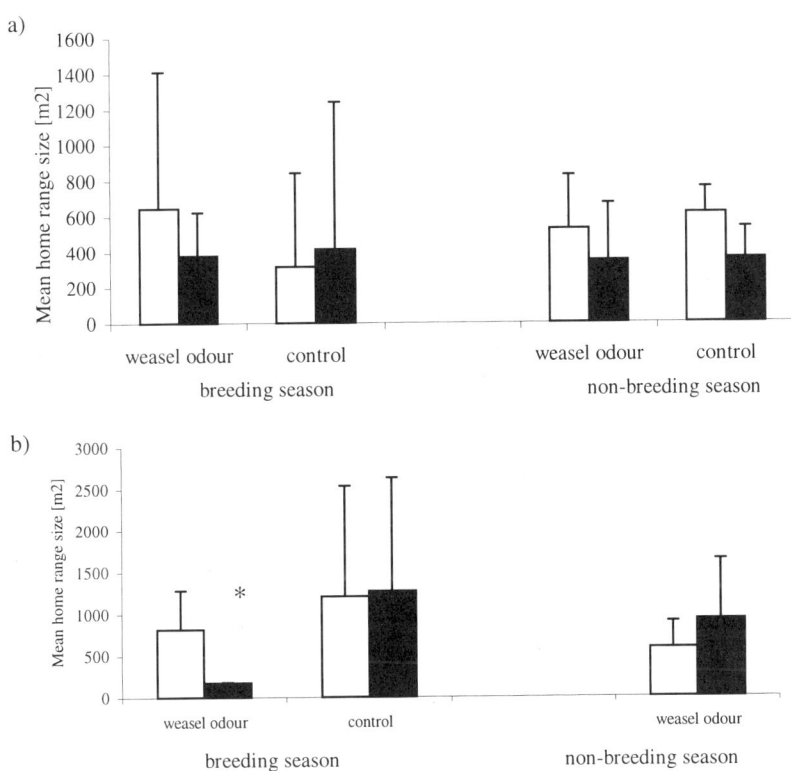

Figure 1. Comparisons of home range sizes of field (a) and bank (b) voles before (open bars) and after (filled bars) exposure to weasel odors or to control odors in breeding (August) and non-breeding (October) seasons.

For field voles (Figure 2a), distances travelled between sequential radiolocations were shorter when the voles were exposed to predator cues during August (n = 68, t = 1.52, p = 0.02, paired t test), but were statistically equivalent during October (n = 84, t = 1.32, p = 0.189, paired t test). Bank voles did not show a significant effect of the weasel odors on the mean distance travelled in August (n = 50, t = 0.39, p > 0.5, paired t test) or in October (n = 69, t = 1.29, p = 0.5, paired t test).

For both field and bank voles, the stability of the home ranges, as estimated by statistical analysis of home range center shifts, showed no significant differences between odor exposures or between seasons (data not shown).

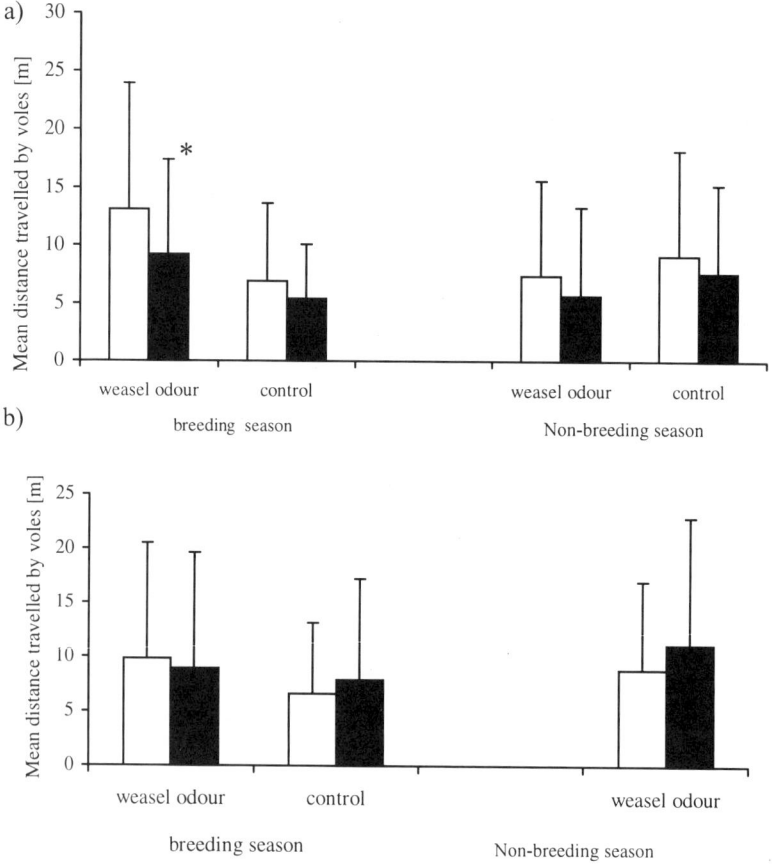

Figure 2. Comparisons of distance travelled by field (a) and bank (b) voles before (open bars) and after (filled bars) exposure to weasel odors or to control odors in breeding (August) and non-breeding (October) seasons.

DISCUSSION

Our radiotracking studies of free-ranging populations of field voles and bank voles in their natural, multi-predator, habitat show that these animals did not exhibit overt escaping behaviors from sites of exposure to weasel cues, i.e., sites with an assumed high risk of predation. In particular, the home range center drift between the control and weasel exposure intervals was similar in the breeding (August) and non-breeding (October) seasons. These findings call into question earlier laboratory findings that voles might attempt to escape from sites intensively penetrated by small mustelids (Jędrzejewski and Jędrzejewska, 1990). The present results are further strengthened by similar observations that movements of root voles did not change substantially after predator exposure (Borowski, 1998a).

An interesting aspect of this work is the seasonal difference were observed in the spatial responses of field voles to weasel presence/odors (Figure 2a). Even though the weasel's presence did not influence the distance travelled during the breeding and non-breeding seasons by bank voles (Figure 2b), for field voles there was a small but statistically significant decrease in travel distances when the weasel was present during the breeding season (Figure 2a). The weasel odors, simulating high predation risk, thus influenced the field voles' locomotor movements only during the breeding season (August). For bank voles exposed to weasels during the breeding season, home range sizes was significantly smaller (Figure 1b); a similar tendency was observed for field voles in August but the differences were not significant (Figure 1a). In the non-breeding season, both the bank and field voles did not alter their home range sizes when exposed to the weasel cues.

We expected bank voles and field voles to show the same seasonal differences to predator cues that we earlier reported for root voles, i.e., in decreased home range sizes in the non-breeding rather than in the breeding season (Borowski, 1999). We propose that seasonal differences in antipredatory behaviour between the bank, field and root voles are due to biological differences resulting from differences in habitat structure. Bank voles and field voles in Poland are closely connected with forested areas (Pucek, 1984), contrary to root voles, which thrive in flood-prone grasslands (Pucek, 1984; Raczyński et. al., 1984). Predation pressure from raptors is greater in grasslands than in forested areas (Cushing and Cawthorn, 1996; Borowski, 1998b)., and in many open areas predation risk by avian predators is the major factor influencing the spatial behaviour of small mammals (Boonstra, 1977; Sonerud, 1986). In contrast, small mammals that live in forested areas are under less selective pressures from predatory raptors. We therefore conclude that habitat structures, through changing the type of predation risk, can modify the antipredatory spatial behaviour during the breeding and non-breeding seasons.

REFERENCES

Agrell, J., 1995, A shift in female social organization independent of relatedness: an experimental study on the field vole (*Microtus agrestis*), *Behav. Ecol.* **2**:182-191.

Borowski, Z., 1998a, Influence of predator odour to feeding behaviour of the root vole (*Microtus oeconomus* Pallas, 1776). *Can. J. Zool.* **76**:1791-1794.

Borowski, Z., 1998b, Influence of weasel (*Mustela nivalis* Linneaus, 1776) odour on spatial behaviour of root voles (*Microtus oeconomus* Pallas, 1776), *Can. J. Zool.* **76**:1799-1804.

Borowski, Z., 1999, *Influence of predator odour on spatial behaviour of the root voles (Microtus oeconomus) in free-living population, Poland*, PD thesis, Forest Research Institute, Warsaw. (in Polish).

Boonstra, R., 1977, Predation on *Microtus townsendii* populations: impact and vulnerability, *Can. J. Zool.* **55**: 1631-1643.

Cushing, B. S., and Cawthorn, J. M., 1996, Species differences in activity patterns during oestrus, *Can. J. Zool.* **74**:473-479.

Gorman, M. L., 1984, The response of prey to stoat (*Mustela erminea*) scent, *J. Zool.* **202**: 419-423.

Jędrzejewski, W., and Jędrzejewska, B., 1990, Effect of predator's visit on the spatial distribution of bank voles: experiments with weasels, *Can. J. Zool.* **68**:660-666.

Jędrzejewski, W., Rychlik, L., and Jędrzejewska, B., 1993, Responses of bank voles to odours of seven species of predators: experimental data and their relevance to natural predator-vole relationships, *Oikos* **68**:251-257.

Korpimäki, E., Koivunen, K., and Hakkarinen, H., 1996, Microhabitat use and behaviour of voles under weasel and raptor predation risk: predator facilitation? *Behav. Ecol.* **7**:30-34.

Pucek, Z. (ed.), 1984, *Key to identification of Polish mammals]. PWN-Polish Scientific Publishers, Warszawa: 1-387.* (in Polish).

Raczyński, J., Fedyk, S., Gębczyńska, Z., and Pucek, M., 1984, Distribution of Micromamalia against natural differentation of the Biebrza Valley habitats. *Pol. Ecol. Stud.* **10**:425-445.

Sonerud, G. A., 1986, Effect of snow cover on seasonal changes in diet, habitat and regional distribution of raptors that prey on small mammals in boreal zones of Fennoscandia, *Holarctic Ecol.* **9**:3-45.

Sullivan, T. P., Crump, D. R., and Sullivan, D. S., 1988, The use of predator odour repellents to reduce feeding damage by herbivores. III Montane (*Microtus montanus*) and meadow (*M. pennsylvanicus*) voles, *J. Chem. Ecol.* **14**:363-377.

DO NEWTS AVOID CONSPECIFIC ALARM SUBSTANCES: THE PREDATION HYPOTHESIS REVISITED

Jason R. Rohr and Dale M. Madison

Department of Biological Sciences
State University of New York at Binghamton
Binghamton, NY 13902-6000

INTRODUCTION

Predator detection and avoidance are important for prey survival. Upon noticing a predator, most prey will decrease activity, move into refugia, or flee (Lima, 1998). Many prey rely on visual cues to reveal threats, but dense vegetation and turbidity often make vision unreliable in aquatic environments (Kats and Dill, 1998). Therefore, aquatic prey may benefit more from using chemical cues than visual cues to detect predators (Dodson *et al.*, 1994). Numerous studies have demonstrated that aquatic animals use their chemical senses to identify predators and release antipredatory behaviors (Kats and Dill, 1998). In fact, Stauffer and Semlitsch (1993) demonstrated that predators' chemical cues amplified antipredatory behaviors in *Rana lessonae* and *Rana esculenta* tadpoles, but visual cues did not.

In addition to predator odors, aquatic prey utilize chemical cues from injured conspecifics to detect threats (e.g. sea anemones: Howe and Sheikh, 1975; sea urchins: Snyder and Snyder, 1970; fish: Pfeiffer, 1974). Since predators are often the cause of prey tissue damage, chemical substances from injured conspecifics may alert others to avoid the vicinity of foraging predators. These conspecific alarm pheromones induce antipredatory behaviors in many aquatic vertebrates, including amphibians (Pfeiffer, 1974; Hews and Blaustein, 1985; Hews, 1988; Petranka, 1989; Lutterschmidt *et al.*, 1994; Chivers *et al.*, 1996).

Recently, red-spotted newts, *Notophthalmus viridescens*, have been shown to avoid skin extracts from conspecifics in the laboratory (Marvin and Hutchinson, 1995) and field (Woody and Mathis, 1997). This lead to the hypothesis that conspecific alarm signals are released and that these substances induce antipredatory avoidance behavior. Adult red-spotted newts are cryptically colored (Bishop, 1941), suggesting a significant predation threat. However, red-spotted newts possess a very potent and effective neurotoxin (Brodie, 1968; Hurlbert, 1970) and are unpalatable to most predators (Formanowicz and Brodie,

1982). In fact, the only reports of consistent predation on adult red-spotted newts have occurred in the laboratory where predators usually preferred alternative prey (Hurlbert, 1970). Additionally, predators that consumed adult newts in the laboratory have not done so in the field, based on gut analyses (*Chrysemys picta*: Raney and Lachner, 1942; Gibbons, 1967; Lindeman, 1996, *Chelydra serpentina*: Lagler, 1943; Bush, 1959, *Rana catesbeiana*: Frost, 1935; Korschgen and Moyle, 1955; Bush, 1959; Korschgen and Baskett, 1963, *Thamnophis sirtalis*: Hamilton, 1951).

Since questions surround predation and anti-predatory behaviors in red-spotted newts, the purpose of this study was to re-examine avoidance behavior in response to conspecific tissue extracts in both the laboratory and field, and to collect and evaluate field evidence for predation on newts. We predicted that newts would avoid conspecific extract in the laboratory and field, and that avoidance would be facultative depending on the degree of predation threat. The latter prediction was based on previous findings of threat-sensitive antipredatory behaviors in amphibians (e.g. Madison *et al.*, 1999; Puttlitz *et al.*, 1999; Belden *et al.*, 2000).

EXPERIMENT I: DO NEWTS AVOID CONSPECIFIC EXTRACTS IN THE LABORATORY?

Methods

Adult eastern red-spotted newts were collected from a private pond in Chenango County, New York in early fall of 1998. We separated the newts by sex and placed a maximum of 40 same-sex newts in cattle troughs. Each trough contained 567 liters of aged tap water and simulated vegetation (multiple 75-cm segments of partially unraveled green nylon twine). The tanks were maintained in a controlled environment chamber at 18°C on a 14:10 h light:dark cycle. The newts were fed chopped earthworms and freeze-dried blood worms weekly.

We constructed 6 test troughs with white, U-shaped gutter (11.4 cm wide x 6.4 cm high). Each trough was 101-cm long and sealed at each end with plastic caps. We marked the insides of the troughs with cross lines creating four equal subdivisions.

Rectangular release cages (21 x 7.6 x 6.4 cm) were made from clear plastic mesh (2 mm square holes). Grey tape covered the inside top and two longest sides of the cage to discourage climbing by the newts. The cages were centered in the troughs with their longest sides parallel to the trough. We filled each trough with 3.5 l of aged tap water at 20°C. Clear plastic was placed over each trough on either side of the cage to negate possible water currents induced by the room ventilation system.

Experiments began two days after the newts were collected by randomly selecting six reproductive males and placing one under each release cage. Newts were allowed to habituate in the release cages for 75 min before test substances were added to the trough ends.

Test substances included macerated newt extract (MNE) and control water. We prepared and froze male newt extract according to Woody and Mathis (1997), except that the extract was collected by macerating entire male newts rather than only newt skin. Three males were macerated in dechlorinated water and filtered through glass wool. The filtrate was divided into 3 ml aliquots, placed in a freezer, and thawed just before use. Control water was also kept frozen. Six ml of dechlorinated water was added to each aliquot after thawing. Test solutions were assigned to the trough ends (right or left) by stratified randomization, and an equivalent amount of water (9 ml) was assigned to the opposite end to control for possible disturbance effects. Two treatment comparisons were performed: water vs water as a control and water vs MNE. We slowly pipetted the extract or water at the appropriate ends of the troughs through a small hole in the plastic cover.

Trials began 5 min after injecting the test solutions by slowly removing the release cages. The observer remained as motionless as possible while recording the quadrant that each newt occupied at the beginning of each minute for 30 min. The six troughs were run concurrently for 10 consecutive days producing 30 replicates for each treatment comparison. The gutters were cleaned and rinsed thoroughly after each trial. We used a Mann-Whitney U test (U) to compare the average quadrant locations relative to the test solutions.

Results

Eighteen newts initially moved away from MNE and seven toward MNE (5 did not move) (χ^2=4.84, P<0.05). Of the latter seven, six remained predominantly immobile on the MNE side for the first 25 min, producing variance heterogeneity and a slightly bimodal frequency distribution for MNE treatments. This required the use of nonparamateric analyses, which revealed significant avoidance of alarm pheromone over the thirty minutes (units from treatment: mean ± se=1.75 ± 0.14) when compared to control treatments (mean ± se=1.47 ± 0.09, U=298.0, N=30, P<0.025).

EXPERIMENT II: DO NEWTS AVOID CONSPECIFIC EXTRACTS IN THE FIELD? 1999 TRIALS

Methods

Extract and control treatments were prepared as before, except that 5 ml aliquots were placed on cellulose sponges (1 x 2.5 x 4 cm) before freezing. The sponges were placed in Ziploc bags and kept frozen until use in the field.

We used Woody and Mathis' (1997) field methods. Treatment sponges were transported on ice to minnow traps in a pond, and then the number of newts captured per trap was counted after two hours. We deviated slightly from Woody and Mathis by placing 24 traps a minimum of 12.5 m apart (rather than 1 m apart) along various shores of Binghamton University's Nature Preserve pond complex in Broome County, NY. The larger trap spacing reduced possible treatment effects between traps. Thirteen traps were placed along the shore of the main pond, and 11 in a side pool. The two sites, only separated by a beaver dam, were considered a single newt population. Trapping occurred daily between 11:00 and 15:00 h during April 1999, otherwise the traps were left open.

Field treatments were assigned using a stratified randomization process. The 24 traps were divided into four groups of six. We randomly selected traps from each group of six to receive a control and MNE sponge. This procedure was repeated on subsequent days except that the same trap was not used more than once for the same treatment. On every third day we set all traps without any sponges and checked them after 2 h. Traps without sponges will be referred to as „non-treatment controls", while traps containing a sponge with just water were considered „treatment controls". Every trap received each treatment once over nine days (six treatment days and three non-treatment days). This entire nine-day squence was repeated after one non-trapping day. We hypothesized avoidance of MNE at the two field sites, and used a Wilcoxon matched-pairs signed-ranks test (T) to analyze for treatment effects.

Results

Two traps in the main pond did not catch any newts, so we excluded them from the analyses. No difference in newt capture rates occurred between MNE (mean ± SE=0.70 ± 0.16) and control traps (mean ± SE=0.91 ± 0.21; T=126.0, N=44, P=0.493). However, the main pond showed significantly fewer newts captured per trap (mean ± SE=0.45 ± 0.08) than the side pool (mean ± SE=1.58 ± 0.17; U=2810, N=99, P<0.001), and so we decided to analyze the results for the two sites separately. The main pond showed a near significant difference between control and MNE treatments for total newts captured (T=3.0, N=22, P=0.06; Figure 1A) and significant avoidance by male newts (T=0.0, N=22, P<0.05). No statistical tests were run on female capture rates in the main pond because only one female was captured in traps containing treatments.

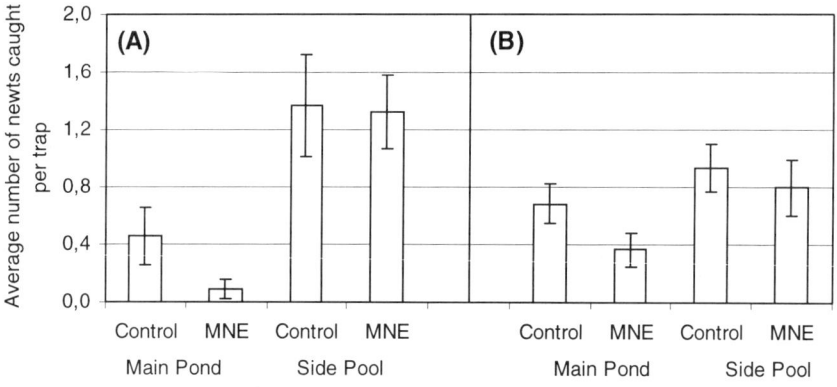

Figure 1. Newt response to macerated newt extract (MNE) in the main pond and side pool in 1999 (A) and 2000 (B).

In the side pool, neither males (T=41.0, N=22, P=0.47), total newts (T=73.5, N=22, P=0.89; Figure 1A), nor females demonstrated avoidance of MNE. Eight females were captured in five MNE traps, and one female was captured in a control trap (χ^2=3.13, P=0.08).

EXPERIMENT III: DO NEWTS AVOID CONSPECIFIC EXTRACTS IN THE FIELD? 2000 TRIALS

Methods

To determine whether the site difference in 1999 was reproducible and therefore a possible site-specific difference in predation, we repeated the study during March and April of 2000 with minor changes. First, 22 minnow traps were placed at each site (44 total traps) and the minimum inter-trap distance was decreased to 7 m to make space for the additional traps. The minimum inter-trap distance was still larger than the estimated home range size of the eastern red-spotted newt (Bellis, 1968) and much larger than the 1 m used by Woody and Mathis (1997). Second, the number of fish captured per trap was recorded in 2000 to determine whether differences in fish captures occurred between sites. To decrease the number of newts that had to be sacrificed to generate MNE, we also reduced the MNE concentration to 70% of the 1999 level.

Results

Once again, fewer newts were captured in the main pond than the side pool, and the combination of the 1999 and 2000 control data demonstrated significantly lower capture rates in the main pond (mean ± SE=0.52 ± 0.07) than the side pool (mean ± SE=1.38 ± 0.13; U=6771, N=143, P<0.001).

The 2000 treatment effects were consistent with the 1999 results. In the main pond, newts significantly avoided MNE. Significantly fewer males (T=35.0, N=44, P<0.05) and total newts (T=35.0, N=44, P<0.05; Figure 1B) were captured in MNE traps than control traps, and therefore, fewer newts were caught in MNE traps than control traps for both years combined (T=58.0, N=66, P<0.01). In the side pool, there was no difference between the number of males (T=164.5, N=44, P=0.38) or total newts (T=163.0, N=44, P=0.36; Figure 1B) captured per control and MNE traps, and thus, no difference between MNE and control traps for the two years combined (T=463.5, N=66, P=0.54).

Too few females were captured in 2000 to justify statistical analyses. However, female capture rates did not differ between the two treatments for 1999 and 2000 combined. Nine females were captured in six MNE traps and three females were captured in three control traps (χ^2=1.125, P=0.29).

If a female or male newt in a trap attracted other newts, the attraction could artificially amplify differences between treatment groups. So, we first reanalyzed the data from both sites excluding traps where females were captured. The analysis revealed that no trends changed and no probability values crossed the alpha significance threshold, indicating that female captures could not account for the results. Second, to assess the effect of conspecific attraction, we used a chi-square test to compare the number of traps with MNE and control treatment that captured newts (1999+2000), as opposed to the number of newts captured per trap. In the main pond, significantly more control traps contained newts (26 of 66, range 0-3 newts/trap) than MNE traps (13 of 66, range 0-2 newts/trap) (χ^2=6.15, P<0.02). However, in the side pool, the proportion of control traps that captured newts (38 of 66, range 0-5 newts/trap) did not differ from the proportion of MNE traps that caught newts (32 of 66, range 0-5 newts/trap) (χ^2=1.09, P=0.29). Thus, conspecific attraction also failed to sufficiently explain the differences between treatment groups at the two sites.

EXPERIMENT IV: IS THERE FIELD EVIDENCE TO SUPPORT THE PREDATOR AVOIDANCE HYPOTHESIS?

Methods

We predicted that closer examination of the two sites would show evidence of reduced predation threat in the side pool, and we tested this prediction in three ways. First, we assessed possible habitat differences between the sites that might be associated with vulnerability to predation. Since a gradual sloping littoral zone usually provides greater foraging area for potential predators, such as wading birds, we measured the slope of the shoreline at each trap location by recording the distance from the top of a submerged minnow trap (22.5 cm) to the shoreline. Greater distances indicated more gradual slopes and more extensive littoral zones. We hypothesized a more gradual sloping littoral zone in the main pond and used a Mann-Whitney U test to test for site effects.

We also counted the number of open perches at the two sites. Since many birds found near aquatic environments use open perches to gain better foraging vantage points, we thought the number of open perches per unit area might reflect the quality of the foraging habitat. We defined open perches as leafless, upright or partially downed trees and tall stumps. We used global positioning system (GPS) to estimate the area of the two sites so we could compare the densities of open perches. The main pond was 8524 m^2 with a perimeter of 383 m and the side pool was 5623 m^2 with a perimeter of 589 m, making the main pond approximately 1.5 times as large as the side pool. We used a chi-square test with an expected frequency for the main pond 1.5 times that of the side pool to test for differences in open perch densities. We hypothesized that the main pond would have a higher open perch density than the side pool.

Second, to assess the likelihood of a site disparity in potential newt predators, we spent 30 min at each site on six different afternoons in June and July 2000 and recorded all potential bird, reptile, and rodent predators observed (no animal was recorded more than once). We hypothesized a lower potential predator density at the side pool, and used a *t* test (t) to evaluate our predictions.

Third, we evaluated the influence of fish on newts, since fish are also potential newt predators. Our concern was that newt extract may attract fish and the fish, rather than the extract, may be causing MNE avoidance. To ascertain whether fish were attracted to MNE, we counted the number of fish caught in MNE and control traps in 2000. To determine whether newts avoided traps with fish, we used minnow traps in cattle troughs to test whether newts would avoid traps with bluegill sunfish, *Lepomis macrochilus*, the most frequently caught fish in our field experiments. Thirty male newts were placed in the trough with a minnow trap at both ends of the trough. During the end of June 2000, we randomly placed two sunfish (6-8 cm total length and recently captured from the same location as the newts) in one of the two minnow traps and subsequently counted the number of newts captured in each trap after two hours. We ran trials daily, one at 10:00 h and the other at 13:30 h, and the entire experiment ran five days. The minnow traps were left open in the newt tank when we were not running trials. No sunfish was used more than twice. A *t* test was used to examine for treatment effects. We hypothesized fish and newt indifference to one another, because of newt toxicity.

Results

The results demonstrate significant habitat differences between the two sites. The average littoral zone slope in the side pool in 2000 (mean ± SE=27.85° ± 3.79) was significantly greater than that in the main pond (mean ± SE=8.66°± 0.92; U=43.5, N1=26, N2=22, P<0.001). Thus, the side pool had steeper banks and, most likely, a smaller littoral zone. Additionally, the side pool had 29 open perches, significantly fewer than the 249 in the main pond ($\chi2$=101.27, P<0.001).

There were also significantly more numbers (t=13.07, df=5, P<0.001; Table 1) and species (t=12.12, df=5, P<0.001; Table 1) of potential newt predators at the main pond than at the side pool. Significantly more birds (t=12.95, df=5, P<0.001), reptiles (t=9.50, df=5, P<0.001), and rodents (t=5.00, df=5, P<0.005; Table 1) were observed at the main pond, and the side pool had significantly fewer species of all three potential predators (birds: t=11.26, df=5, P<0.001; reptiles: t=2.86, df=5, P<0.05; rodents: t=5.00, df=5, P<0.005; Table 1).

Finally, there was no effect of treatment on fish captures in the main pond (U=963.5, N=44, P=0.97) or side pool (U=946.5, N=44, P=0.86), and significantly more fish were captured per trap in the main pond (mean ± SE=1.25 ± 0.36) than side pool (mean ±

SE=0.15 ± 0.05; U=3102.0, N=88, P<0.025). There was also a positive correlation between traps that captured newts and those that captured fish (r=0.17, N=176, P<0.025) and vice versa (r=0.20, N=176, P<0.01). These correlations could be explained by newt attraction to fish, fish attraction to newts, or cohabitation in similar microhabitats. A 2000 field study at the main pond and side pool (unpublished) showed no correlation between newt and fish captures when zero to four newts were placed in traps from the onset of the experiment (r=0.01, N=266, P=0.86), and our 2000 laboratory study showed no difference in newt capture rates between traps with fish (mean ± SE=1.90 ± 0.43) and without fish (mean ± SE=1.90 ± 0.46; t=0.00, df=9, P=1.00). Therefore, newts and small bluegills seem to be indifferent to one another at our sites, and the correlation in occurrence was most likely due to both species being attracted to traps in microenvironments with abundant resources, such as vegetative cover or food.

Table 1. Average number and species of potential and/or perceived predators of *Notophthalmus viridescens* observed at the main pond and side pool during six 30 min observation periods. A T-test was used to calculate P values for site effects.

	Main Pond	Side Pool	P
	Mean (± SE) number of animals observed per km^2		
Birds	2.48 ± 0.33	0.65 ± 0.39	<0.001
Reptiles	1.27 ± 0.15	0.03 ± 0.03	<0.001
Rodents	0.10 ± 0.02	0.00 ± 0.00	<0.005
Total	3.85 ± 0.37*	0.68 ± 0.38**	<0.001
	Mean (± SE) number of species observed per km^2		
Birds	0.94 ± 0.11	0.33 ± 0.12	<0.001
Reptiles	0.16 ± 0.03	0.03 ± 0.03	<0.05
Rodents	0.10 ± 0.02	0.00 ± 0.00	<0.005
Total	1.19 ± 0.12*	0.36 ± 0.11**	<0.001

* 21 tree swallows, 20 red-winged blackbirds, 20 canada geese (CG), 16 song sparrows (SS), 16 american goldfinches, 8 eastern pheobes, 8 common grackles (Gr), 5 belted kingfishers, 3 common yellowthroats, 3 eastern kingbirds, 2 yellow warblers, 2 brown thrashers, 1 great blue, heron, 1 baltimore oriole (BO), 1 mallard, 62 midland painted turtles, 3 northern brown water snakes, 5 muskrats.
** 10 CG, 4 SS, 4 BO, 2 american robins, 1 green-backed heron, 1 garter snake

DISCUSSION

Our findings that male newts avoided MNE in the laboratory and main pond were consistent with the results of Marvin and Hutchinson (1995) and Woody and Mathis (1997). The more surprising result was the absence of male avoidance to MNE in the side pool. Conspecific attraction and female captures failed to explain our results, and anecdotal evidence suggests that conspecific aversion does not account for MNE avoidance (Woody and Mathis 1997). Our results, therefore, suggest that the MNE avoidance discrepancy between the main pond and side pool could have resulted from facultative responses to local predation pressure.

Magnitude of predatory threat can be affected by predator densities, prey densities, or both. Assuming that significantly higher capture rates of newts in the side pool reflected a higher newt density, the probability of an individual newt being attacked would have been reduced in the side pool, and this „dilution effect" (Krebs and Davies, 1993) could have caused males to relax their predator and MNE avoidance.

The fewer potential predators captured and observed in the side pool could have also produced a lower predation threat. The side pool had significantly fewer numbers and species of birds than the main pond. Although most birds find newts noxious (Hurlbert, 1970; Howard and Brodie, 1973), birds often have to learn to avoid unpalatable amphibians (e.g. Brodie and Howard, 1973; Howard and Brodie, 1973; Hensel and Brodie, 1976; Brandon et al., 1979), and thus, young birds may be attempting to prey on newts. In addition, adult birds may know newts are toxic but mistake them for palatable prey. We found four dead newts (three females and one of unknown sex) along the shore during our study that had obvious wounds but no signs of consumption. Lamoureux (2000) observed two great blue herons and a kingfisher forcefully manipulate and then expel red-spotted newts from their beaks in Binghamton University's Nature Preserve.

Unlike the main pond, the side pool would be relatively unattractive to birds that may accidentally prey on newts. The significantly steeper banks and smaller littoral zone would discourage wading birds, such as herons, and the scarcity of open perches would be less attractive to kingfishers. Additionally, kingfishers and herons forage on fish, and therefore would not prefer the lower fish density in the side pool. We are aware that most of the birds observed at the sites probably do not prey on newts. However, newts may not possess the ability to discriminate between harmful and innocuous birds, and therefore may perceive most birds as predators. Birds learning that newts are noxious, or mistaking them for palatable prey, may have been sufficient to drive the evolution of cryptic coloration in adult newts and facultative avoidance of injured conspecifics.

In addition to fewer birds, the side pool had fewer reptiles. Painted turtles, *Chrysemys picta*, and Snapping turtles, *Chelydra serpentina*, have both consumed either part or entire newts in the laboratory (Hurlbert, 1970), and certain populations of garter snakes, *Thamnophis sirtalis*, prey upon the highly toxic rough-skinned newt, *Taricha granulosa* (Brodie and Brodie, 1990; Brodie and Brodie, 1991). A female newt was left in a minnow trap in the main pond over night and by morning her limbs had been eaten, consistent with the turtle predation described by Hurlbert (1970).

Finally, adult newts did not avoid sunfish in our laboratory study or in a previous field experiment (Kesler and Munns, 1991). This result was not surprising since adult newts are toxic and unpalatable to fish (Pope, 1924; Webster, 1960; Brodie, 1968; Hurlbert, 1970).

Although direct fish predation on adult newts seems doubtful, we believe fish have two important nonlethal, indirect effects on adult newts that may have accounted for our MNE avoidance discrepancy. First, the higher fish density in the main pond would have attracted more predators, such as kingfishers and herons, that mistakenly prey upon adult newts. Second, fish seem to directly prey on newt eggs and larvae, which would have indirectly decreased adult newt densities more in the main pond (due to the higher fish densities), diminishing any dilution effect. Fish predation on newt eggs and larvae accounted for the significant declines of the California newt, *Taricha torosa* (Gamradt and Kats, 1996), and the most common fish genus in Binghamton's Nature Preserve, *Lepomis*, has been observed consuming larval newts (Kesler and Munns, 1991). Researchers have also reported newt preference for ponds with lower fish densities (Beebee, 1979; Miaud et al., 1993), significant declines after fish introductions (Gamradt and Kats, 1996; Beebee, 1997), and data suggesting that newts migrate from fish to fishless ponds (Miaud et al., 1993). Thus, reduced indirect fish effects on adult newts in the side pool may have produced the higher newt density, which could have enhanced the dilution effect and allowed for relaxation of MNE avoidance.

The results of our experiments support the hypothesis that newt avoidance of conspecific tissue extracts is a threat-sensitive antipredatory behavior and that learning and mistakes by predators may pose a significant threat to adult newts. Future studies should systematically control predation pressure while testing newt response to conspecific tissue extracts.

REFERENCES

Beebee, T. J., 1979, Habitats of the British amphibians (2): suburban parks and gardens, *Biol. Conserv.* **15**:241-257.
Beebee, T. J., 1997, Changes in dewpond numbers and amphibian diversity over 20 years on Chalk Downland in Sussex, England, *Biol. Conserv.* **81**:215-219.
Belden, L. K., Wildy, E. L., Hatch, A. C., and Blaustein, A. R., 2000, Juvenile western toads, *Bufo boreas*, avoid chemical cues of snakes fed juveniles, but not larval, conspecifics, *Anim. Behav.* **59**:871-875.
Bellis, E. D., 1968, Summer movement of red-spotted newts in a small pond, *J. Herpetol.* **1**:86-91.
Bishop, S. C., 1941, The salamanders of New York, *Bull. N.Y. State Mus.* **324**:365.
Brandon, R. A., Labanick, G. M., and Huheey, J. E., 1979, Learned avoidance of brown efts, *Notophthalmus viridescens louisianensis* (Amphibia, Urodela, Salamandridae), by chickens, *J. Herpetol.* **13**:171-176.
Brodie, E. D., Jr., 1968, Investigations on the skin toxin of the red-spotted newt, *Notophthalmus viridescens*, *Am. Midl. Nat.* **80**:276-280.
Brodie, E. D., III, and Brodie, E. D., Jr., 1990, Tetrodotoxin resistance in garter snakes: an evolutionary response of predators to dangerous prey, *Evolution* **44**:651-659.
Brodie, E. D., III, and Brodie, E. D., Jr., 1991, Evolutionary response of predators to dangerous prey: reduction of toxicity of newts and resistance of garter snakes in island populations, *Evolution* **45**:221-224.
Brodie, E. D., Jr., and Howard, R. R., 1973, Experimental study of Batesian mimicry in the salamander *Plethodon jordani* and *Desmognathus ochrophaeus*, *Am. Midl. Nat.* **90**:38-46.
Bush, F. M., 1959, Foods of some Kentucky herpetiles, *Herpetologica* **15**:73-77.
Chivers, D. P., Kiesecker, J. M., Anderson, M. T., Wildy, E. L., and Blaustein, A. R., 1996, Avoidance response of a terrestrial salamander *(Ambystoma macrodactylum)* to chemical alarm cues, *J. Chem. Ecol.* **22**:1709-1716.
Dodson, S. I., Crowl, T. A., Peckarsky, D. L., Kats, L. B., Covich, A. P., and Culp, J. M., 1994, Nonvisual communication in freshwater benthos: an overview, *J.N. Am.Benthol. Soc.* **13**:268-282.
Frost, S. W., 1935, The food of *Rana catesbeiana*, *Copeia* **1**:15-18.
Gamradt, S. C., and Kats, L. B., 1996, Effect of introduced crayfish and mosquitofish on California newts, *Conserv. Biol.* **10**:1155-1162.
Gibbons, J. W., 1967, Variation in growth rates in three populations of the painted turtle, *Chrysemys picta*. *Herpetologica* **23**:296-303.
Hamilton, W. J., Jr., 1951, The food and feeding behavior of the garter snake in New York State, *Am. Midl. Nat.* **46**:385-390.
Hensel, J. L., Jr., and Brodie, E. D., Jr., 1976, Experimental study of aposematic coloration in the salamander *Plethodon jordani*, *Copeia* **1976**:59-65.
Hews, D. K., 1988, Alarm response in larval western toads, Bufo boreas: release of larval chemicals by a natural predator and its effect on predator capture efficiency. *Anim. Behav.* **36**:125-133.
Hews, D. K., and Blaustein, A. R., 1985, An investigation of the alarm response in *Bufo boreas* and *Rana cascadae* tadpoles, *Behav. Neural Biol.* **43**:47-57.
Howard, R. R., and Brodie, E. D., 1973, A Batesian mimetic complex in salamanders: responses of avian predators, *Herpetologica* **29**:33-41.
Howe, N. R., and Sheikh, Y. M., 1975, Anthopleurine, a sea anemone alarm pheromone, *Science* **189**:386-388.
Hurlbert, S. H., 1970, Predator responses to the vermillion-spotted newt *(Notophthalmus viridescens)*, *J. Herpetol.* **4**:47-55.
Fomanowicz, D. R., Jr., and Brodie, E. D., Jr., 1982, Relative palatabilities of members of a larval amphibian community, *Copeia* **1982**:91-97.
Kats, L. B., and Dill, L. M., 1998, The scent of death: chemosensory assessment of predation risk by prey animals, *Ecoscience* **5**:361-394.

Kesler, D. H., and Munns, W. R., Jr., 1991, Diel feeding by adult red-spotted newts in the presence and absence of sunfish, *J. Freshwater Ecol.* **6**:267-273.

Korschgen, L. J., and Baskett, T. S., 1963, Food of impoundment- and stream-dwelling bullfrogs in Missouri, *Herpetologica* **19**:89-99.

Korschgen, L. J., and Moyle, D. L., 1955, Food habits of the bullfrog in central Missouri farm ponds, *Am. Midl. Nat.* **54**:333-341.

Krebs, J. R., and Davies, N. B., 1993, *An Introduction to Behavioral Ecology*, 3rd edn., Blackwell Scientific Publications, London.

Lagler, K. F., 1943, Food habits and economic relations of the turtles of Michigan with special reference to fish management, *Am. Midl. Nat.* **29**:257-312.

Lamoureux, V. S., 2000, Ecology and seasonal behavior of the green frog, *Rana clamitans*, Ph.D. Thesis, Binghamton University, Binghamton, NY.

Lima, S. L., 1998, Nonlethal effects in the ecology of predator-prey interactions, *BioScience* **48**:25-34.

Lindeman, P. V., 1996, Comparative life history of painted turtles *(Chrysemys picta)* in two habitats in the inland Pacific northwest, *Copeia* **1996**:114-130.

Lutterschmidt, W. I., Marvin, G. A., and Hutchinson, V. H., 1994, Alarm response by a plethodontid salamander *(Desmognathus ochrophaeus)*: conspecific and heterospecific "schreckstoff", *J. Chem. Ecol.* **20**:2751-2759.

Madison, D. M., Maerz, J. C., and McDarby, J. H., 1999, Optimization of predator avoidance by salamanders using chemical cues: diet and diel effects, *Ethology* **105**:1073-1086.

Marvin, G. A., and Hutchinson, V. H., 1995, Avoidance response by adult newts *(Cynops pyrrhogaster* and *Notophthalmus viridescens)* to chemical alarm cues, *Behaviour* **132**:95-105.

Miaud, C., Joly, P., and Castanet, J., 1993, Variation in age structures in a subdivided population of *Triturus cristatus*, *Can. J. Zool.* **71**:1874-1879.

Petranka, J. W., 1989, Response of toad tadpoles to conflicting chemical stimuli: predator avoidance versus "optimal" foraging, *Herpetologica* **45**:283-292.

Pfeiffer, W., 1974, Pheromones in fish and amphibia, in: *Pheromones: Frontiers of Biology*, Vol. 32, (M. C. Birch ed.), North-Holland, Amsterdam.

Pope, P. H., 1924, The life history of the common water newt *(Notophthalmus viridescens)*, together with observations on the sense of smell, *Ann. Carnegie Mus.* **15**:305-368.

Puttlitz, M. H, Chivers, D. P., Kiesecker, J. M, and Blaustein, A. R., 1999, Threat-sensitive predator avoidance by larval Pacific treefrogs (Amphibia, Hylidae), *Ethology* **105**:449-456.

Raney, E. C., and Lachner, E. A., 1942, Summer food of *Chrysemys picta marginata*, in Chatauqua Lake, New York, *Copeia* **1942**:83-85.

Snyder, N. F. R., and Snyder, H. A., 1970, Alarm response of *Diamdema antillarum*, *Science* **168**:276-278.

Stauffer, H., and Semlitsch, R. D., 1993, Effects of visual, chemical, and tactile cues of fish on the behavioral responses of tadpoles, *Anim. Behav.* **46**:355-364.

Webster, D. A., 1960, Toxicity of the spotted newt, Notophthalmus viridescens, to trout, *Copeia* **1960**:74-75.

Woody, D. R., and Mathis, A., 1997, Avoidance of areas labeled with chemical stimuli from damaged conspecifics by adult newts, *Notophthalmus viridescens*, in a natural habitat, *J. Herpetol.* **31**:316-318.

RESPONSES TO NITROGEN-OXIDES BY CHARACIFORME FISHES SUGGEST EVOLUTIONARY CONSERVATION IN OSTARIOPHYSAN ALARM PHEROMONES

Grant E. Brown,[1] James C. Adrian, Jr.,[2] Ilyssa H. Kaufman,[1] Jody L. Erickson,[1] and Devon Gershaneck[1]

[1]Department of Biological Sciences
[2]Department of Chemistry, Union College
Schenectady, New York 12308

INTRODUCTION

Many fishes of the superorder Ostariophysi (and members of their prey guild) possess chemical alarm signals or alarm pheromones (Smith, 1992; Mathis and Smith, 1993; Brown and Godin, 1997; Brown and Smith, 1997; Chivers and Smith, 1998; Brown and Brennan, 2000), which are sequestered in specialized epidermal club cells (Smith, 1992). Although controversial (Magurran *et al.*, 1996; Smith, 1997; Brown and Godin, 1999a), there exists considerable evidence for the anti-predator function of alarm pheromones for both signal senders and receivers in Ostariophysan fishes (Smith, 1992; Chivers and Smith, 1998; Brown and Godin, 1999b). When detected by nearby conspecifics and sympatric heterospecifics, the alarm pheromone(s) elicits responses characterized by an increase in a variety of species-specific, anti-predator behaviors such as refuging, shoaling and immobility (Chivers and Smith, 1998).

Pfeiffer *et al.* (1985) tentatively identified the Ostariophysan alarm pheromone as hypoxanthine-3-N-oxide (H3NO, Table 1). Hypoxanthine-3-N-oxide is a relatively small molecule comprised of a purine skeleton with a nitrogen-oxide functional group at the three position. Brown *et al.*, (2000) have demonstrated, however, that H3NO is likely only one of several active compounds within the Ostariophysan alarm pheromone system. Brown *et al.* ,(2000) compared the behavioural responses of two cyprinid species, fathead minnows (Pimephales promelas) and finescale dace (Phoxinus neogaeus), to conspecific alarm pheromones, H3NO and a suite of structurally and functionally similar molecules. Both minnows and dace exhibited significant increases in anti-predator behaviour when exposed to conspecific skin extract, H3NO and pyridine-N-oxide, but not to other aromatic compounds lacking a nitrogen-oxide functional group. These data suggest that the nitrogen-oxide functional group is the molecular trigger that elicits the downstream cascade of behavioural patterns, and that any aromatic compound with a nitrogen-oxide functional group will serve as an alarm pheromone.

The superorder Ostariophysi is comprised of five extant families (Nelson, 1994), Gonorychiformes, Cypriniformes, Charicaformes, Gymnotiformes, and Siluriformes. All orders, with the exception of the Gymnotiformes, contain species that utilize alarm pheromones and possess specialized epidermal club cells (Pfeiffer, 1977; Smith, 1992). Characins (Characidae; Characiformes), a diverse group of primarily neotropical species, have been demonstrated to possess an alarm pheromone (Brown and Godin, 1999b). Characins are closely related to the Cyprinids and probably diverged sometime during the Miocene period (Fink and Fink, 1981; Orti and Meyer, 1997). Glowlight tetras (*Hemigrammus erythrozonus*) and neon tetras (*Paracheirodon innesi*) are common prey species native to South America (Axelrod and Vonderwinkler, 1983; Robins *et al.*, 1991). Both species are typically found in slow moving streams, characterized by high levels of biotic and abiotic structure (Axelrod and Vonderwinkler, 1983). As such, they occupy a similar ecological niche, as do fathead minnows and finescale dace in North America.

In order to address the hypothesis that the nitrogen-oxide moiety is conserved within the alarm pheromones used in the superorder Ostariophysi, we examined anti-predator behaviors in glowlight and neon tetras exposed to conspecific skin extract, swordtail skin extract, and a suite of seven synthetic compounds tested previously (Brown *et al.*, 2000). As described below, the tetras showed alarm responses only to compounds containing a nitrogen-oxide functional group, suggesting that it is conserved within the superorder.

METHODS

Test Fish

Glowlight tetras, neon tetras and swordtails were obtained from a commercial supplier and held in 60 L aquaria, filled with continuously filtered dechlorinated tap water (pH 7.8, 24° C) on a 12:12 L:D cycle. All fish were fed ad libitum, twice daily, with commercial flake food. Mean (± S.E.) standard body length at time of testing was 2.52 ± 0.09 cm and 2.63 ± 0.08 cm (glowlight and neon tetras respectively).

Experimental Stimuli

We generated natural skin extracts from both test species and swordtails. Swordtails lack the Ostariophysan alarm pheromone (Pfeiffer, 1977) and were used as a control for the response to any injured fish. To generate skin extracts, donor fish were killed with a blow to the head (in accordance with Union College Institutional Animal Care and Use Protocol # 2-27-98). The skin was removed from either side of the donor fish, and immediately placed in 50 ml of chilled, glass-distilled water. Skin samples were then homogenized, filtered through glass wool and the final volumes adjusted by adding glass-distilled water. Total skin areas of 10.49 and 10.53 cm2 in 125 ml of distilled water were collected for glowlight and neon tetras respectively. A total skin area of 22.54 cm2 in 225 ml of distilled water was collected for swordtails. These concentrations were similar to those employed in Brown *et al.* (2000). All skin extracts were frozen in 15 ml aliquots at –20° C until needed.

In addition to the H3NO, synthesized in the laboratory (as previously described, Brown et al., 2000), six other chemical to be used as experimental stimuli were obtained from commercial suppliers (Table 1). Hypoxanthine, xanthine, and guanine are structurally similar purine compounds. The purine ring system contains a fused six member pyrimidine ring and a five membered imidazole ring, which share a common side. The six membered

4(3H)-pyrimidone resembles, nearly identically, the pyrimidine moiety of hypoxanthine-3-N-oxide. The final two compounds, pyridine and pyridine N-oxide, are both six membered heterocycles incorporating a single nitrogen. We chose these compounds because of their structural or functional similarities to hypoxanthine-3-N-oxide. It is important to note that hypoxanthine-3-N-oxide and pyridine N-oxide are related only by the nitrogen-oxide functional group, which is absent in the remaining compounds. We generated our experimental stimuli by dissolving 0.002 g of our compound of interest in 200 ml of glass distilled water and stirring for 15 minutes. The resulting solutions were frozen in 15 ml aliquots at $-20°C$ until required. These concentrations are similar to those used by Pfeiffer et al., (1985) and Brown et al., (2000). In addition, it has been demonstrated that this concentration is near the minimum stimulus response threshold (Brown, Adrian, and Shih, submitted).

Experimental Protocol

Test tanks consisted of 37-L aquaria, filled with dechlorinated tap water (pH 7.8) and containing a gravel substrate and a single airstone, mounted along the back wall of the tank. An additional length of tubing was attached to the airstone, which allowed us to introduce a chemical stimulus from behind a black viewing curtain, at a distance of about 3 m, without disturbing the test fish. Test tanks were illuminated, on a 12:12 L:D cycle, with 40-W incandescent lamps. Shoals of four glowlight or neon tetras (matched for size) were placed in the test tanks and allowed a 24-hr acclimation period (N = 10 shoals for each treatment condition). Control and experimental trials consisted of a 10-min pre-stimulus and a 10 min post-stimulus observation period. Prior to the pre-stimulus period, 60 ml of tank water was withdrawn from the stimulus injection tube and discarded (to remove any stagnant water). An additional 60 ml was withdrawn and retained. At the beginning of the post-stimulus period, 5 ml of distilled water (control trials) or 5 ml of conspecific skin extract (CSE), swordtail skin extract (SWT) or one of the seven synthetic stimuli (Table 1) was injected and slowly flushed into the tank using the retained tank water. Dye tests demonstrated that this technique results in the distribution of the stimuli throughout the tank in about 16 sec. Control trials were conducted between 08:00 and 11:00 hr daily, and experimental trials were conducted between 13:00 and 16:00 hr. Control trials were conducted before experimental trials because any response to the experimental stimuli may have masked any response to the control stimuli (Lawrence and Smith, 1989; Hazlett, 1997).

Table 1: Chemical structures of synthetic stimuli.

Hypoxanthine-3-N-oxide	Pyridine-N-oxide	Guanine	Hypoxanthine
Pyridine	4(3H)-Pyrimidone	Xanthine	

During both the pre- and post-stimulus observation periods, we observed the position of each fish, and recorded 'area use' and 'shoaling index' every 15 sec. Area use was defined as the position of each fish (1 = bottom third of the test tank, 3 = top third of the test tank) giving scores ranging from 4 (all fish on the bottom of the tank) to 12 (all fish near the surface). Shoaling index ranged from 1 (no fish within one body length of each other) to 4 (all fish within one body length of each other). A decrease in area use and an increase in shoaling index are indicative of an anti-predator response under laboratory conditions (Brown and Godin, 1999b; Brown et al., 1999). We also recorded the frequency of occurrence of three additional anti-predator behaviours. Dashing was defined as a sudden burst of, apparently, disoriented swimming. Freezing was defined as the cessation of movement by at least one tetra, with the fish settling to the substrate and remaining motionless for at least 30 sec. Fin flicking was defined as the very rapid and simultaneous movement of the dorsal and caudal fins by a tetra whose body position remained unchanged. All three behaviour patterns are reliable indicators of anti-predator behaviour in characins (Brown and Godin, 1999b; Brown et al., 1999).

For area use and shoaling index scores, we calculated the difference between the pre- and post-stimulus observation periods for paired control and experimental trials and compared them using the Mann-Whitney U test. Frequency of occurrence of dashing, freezing and fin flicking were compared between control and experimental trials using the Fisher's Exact Probability test.

RESULTS

We observed significant differences in area use scores (i.e. individuals remained closer to the substrate) in both glowlight and neon tetras when exposed to conspecific skin extract or H3NO (compared to distilled water controls; Figure 1).

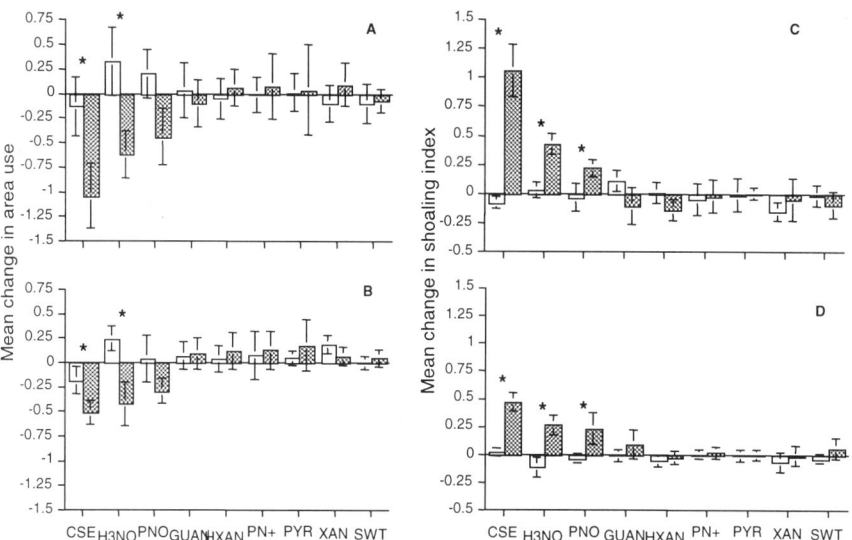

Figure 1: Mean (± S.E.) change in anti-predator responses for glowlight and neon tetras exposed to distilled water controls (open bars) and experimental stimuli (solid bars). A = Glowlight tetra area use; B = Neon tetra area use; C = Glowlight tetra shoaling index; D = Neon tetra shoaling index. CSE = conspecific skin extract; H3NO = hypoxanthine-3-N-oxide, PNO = pyridine-N-oxide, GUAN = guanine, HXAN = hypoxanthine, PN+ = pyridine, PYR = 4(3H)-pyrimidone, XAN = xanthine; SWT = swordtail skin extract.

In addition, glowlight tetras exhibited a significant decrease in area use when exposed to PNO (Figure 1). No significant differences in area use were observed for the remaining experimental treatments (Table 2). We found a similar trend in the shoaling index data as well. Both glowlight and neon tetras significantly increased their shoal cohesiveness when exposed to conspecific skin extract and H3NO (Figure 1, Table 2). Neon tetras also exhibited a significant increase in shoaling index when exposed to PNO (Figure 1, Table 2). As with the area use scores, no differences were found in shoaling index scores for the remaining experimental treatments (Table 2). In addition, both glowlight and neon tetras increased their rates of dashing, freezing and fin flicking when exposed to conspecific skin extract, H3NO and PNO (Table 3).

Table 2: Individual comparisons for glowlight and neon tetras' response to each experimental stimulus versus distilled water control based on Mann-Whitney U tests. Probabilities are corrected for multiple comparisons using the modified Bonferonni test. Stimulus abbreviations as in Figure 1.

	Area Use				Shoaling Index			
	Glowlight		Neon		Glowlight		Neon	
	Z	P	Z	P	Z	P	Z	P
CSE	-2.37	<0.05	-2.34	<0.05	-3.46	<0.05	-3.44	<0.05
H3NO	-2.08	<0.05	-2.20	<0.05	-2.54	<0.05	-2.80	<0.05
PNO	-1.40	NS	-0.91	NS	-2.57	<0.05	-2.09	<0.05
GUAN	-0.45	NS	-0.45	NS	-1.40	NS	-0.16	NS
HXAN	-0.38	NS	-0.83	NS	-0.95	NS	-0.27	NS
PN+	-0.04	NS	-0.11	NS	-0.45	NS	-0.11	NS
PYR	-0.53	NS	-0.23	NS	-0.05	NS	-0.53	NS
XAN	-0.38	NS	-0.70	NS	-0.34	NS	-0.27	NS
SWT	-0.11	NS	-0.08	NS	-0.57	NS	-0.83	NS

Table 3: Frequency of occurrence of dashing, freezing and fin flicking behaviour of glowlight and neon tetras exposed to each of the chemical stimuli vs. distilled water controls (C = dw controls, E = experimental stimuli). Numbers denote trials (N = 10, except for glowlight tetras exposed to CSE, where N = 12) in which behaviour was observed.
P (based on Fisher's Exact Probability): * < 0.05, ** < 0.01, # < 0.001. Stimulus abbreviations as in Table 2.

	Dashing						Freezing						Fin Flicking					
	Glowlight			Neon			Glowlight			Neon			Glowlight			Neon		
	C	E	P	C	E	P	C	E	P	C	E	P	C	E	P	C	E	P
CSE	3	9	*	2	8	*	2	10	**	2	9	#	0	12	#	0	9	#
H3NO	1	6	*	2	8	*	1	7	**	3	9	**	3	8	NS	1	5	NS
PNO	0	5	*	1	7	**	3	7	NS	2	8	*	1	7	**	1	5	NS
GUAN	0	1	NS	3	2	NS	3	4	NS	3	3	NS	0	2	NS	1	2	NS
HXAN	2	3	NS	3	2	NS	2	4	NS	2	2	NS	1	1	NS	1	1	NS
PN+	0	0	NS	1	3	NS	1	4	NS	3	2	NS	0	2	NS	0	0	NS
PYR	0	0	NS	1	2	NS	3	2	NS	3	3	NS	0	1	NS	0	1	NS
XAN	0	4	NS	1	1	NS	2	4	NS	4	3	NS	1	2	NS	1	1	NS
SWT	3	3	NS	1	3	NS	3	4	NS	4	4	NS	1	0	NS	0	1	NS

DISCUSSION

These data demonstrate that the nitrogen oxide functional group is a significant component of the chemical alarm signalling system in at least two species of Characin fishes. We observed significant increases in anti-predator behaviour to conspecific skin extract and hypoxanthine-3-N-oxide and a less intense, but still significant response to pyridine-N-oxide. We found no significant difference in anti-predator responses in tetras exposed to the remaining test compounds. Fathead minnows and finescale dace exhibited similar behavioural responses to conspecific skin extract, hypoxanthine-3-N-oxide and pyridine-N-oxide (Brown et al., 2000). The similarity of response between characins and cyprinids strongly suggests that the nitrogen-oxide molecular trigger is evolutionarily conserved within the superorder Ostariophysi.

In addition, the data presented here provide further support for our hypothesis that the nitrogen-oxide functional group is the chief molecular trigger. Significant increases in anti-predator behaviour were observed when tetras (and previously tested cyprinids; Brown et al., 2000) were exposed to compounds containing a nitrogen-oxide functional group. No response was seen in tetras (and cyprinids) exposed to structurally similar compounds lacking this functional group. This suggests that the purine skeleton has little or no role as a signalling agent. It is likely that any purine compound, which contains a nitrogen-oxide functional group, will act as an alarm signal (Brown et al., 2000).

Additional sources support our hypothesis. Fathead minnows exposed to natural alarm pheromone or hypoxanthine-3-N-oxide under weakly acidic conditions (pH 6.0) did not exhibit a significant anti-predator response, but did respond to these chemicals when re-tested four days later at pH 8.0 (Brown, Adrian, Lewis and Tower, unpublished data). When hypoxanthine-3-N-oxide is acidified, the nitrogen oxide functional group is lost, due to a non-reversible covalent change (Kawashima and Kumashiro, 1969; Wölcke and Brown, 1969). Thus, these data suggest that as a result of the weakly acidic conditions, the nitrogen-oxide molecular trigger is lost and the alarm pheromone is rendered inactive (Brown et al., 2000).

Preliminary data suggest that replacement of the double bonded oxygen at the 6' position of H3NO with physically larger functional groups, e.g. 6-Methoxy- and 6-Ethoxy-purine-N-oxide, does not reduce the effectiveness of the chemical alarm signal (i.e. responses were not significantly different from H3NO, Brown, Adrian and Tower, unpublished data). This further supports our hypothesis that the nitrogen-oxide is the chief molecular trigger in this chemical alarm signalling system.

Why Nitrogen-oxides?

In order to maximize potential benefits associated with aquatic chemical alarm signalling in fishes, such signals should: 1) be energetically inexpensive to produce, 2) be sufficiently simple to allow for reliable cross-species responses within prey guilds, and 3) be stable under normal environmental conditions.

There are at least three discrete metabolic costs associated with Ostariophysan alarm pheromone production: producing the pheromone, sequestering the pheromone into specialized epidermal club cells, and storing the pheromone in club cells. Purine-N-oxides likely evolved as alarm signalling compounds, as they should minimize at least two of these costs. The purine starting materials are intermediate metabolic byproducts associated with the purine degradation pathway (Stryer, 1981). As such, beyond the oxidation of the three position, there is likely little energetic input required for the alarm pheromone production. Wisenden and Smith (1997) have demonstrated that there is an energetic cost to the production of the specialized club cells. Thus, while the biosynthesis of the alarm pheromone is not costly in terms of energy input, there is a demonstrated energetic cost associated with sequestering it within the club cells. Purine-N-oxides are strongly polarized molecules, which are readily water-soluble. Hence, once in the club cell, they are unlikely to passively cross the lipophilic club cell membrane. Therefore, Ostariophysan alarm pheromones can only be released following mechanical damage to the skin (Smith, 1992), thus not requiring energetically costly cellular transport mechanisms to allow for the voluntary release of the pheromone.

Cross species responses have been demonstrated for a variety of Ostariophysans and their prey guild members (Smith, 1992; Chivers and Smith, 1998). Individuals can increase their likelihood of detecting a predator if they respond to the alarm pheromones of prey guild members. Our data suggest that any compound containing a nitrogen-oxide functional group may act as a chemical alarm pheromone for these fishes. It is possible that species-specific pheromones may simply be produced by differences in the relative proportions of various purine-N-oxides. If this is indeed the case, then the detection and response to a single functional group (nitrogen-oxide) could easily explain the observed prey guild responses. Lawrence and Smith (1989) demonstrated an increased latency to respond in fathead minnows exposed to decreased concentrations of natural conspecific alarm pheromones. Brown, Adrian, and Shih (submitted) did not find a similar latency-concentration interaction by fathead minnows when exposed to purified hypoxanthine-3-N-oxide, further supporting our hypothesis of multiple active compounds within the Ostariophysan alarm pheromone.

Finally, purine-N-oxides are stable under neutral to basic pH, as would be found in most natural waterways. Given that they are water soluble, they would be easily transmitted through an aquatic environment. Upon release, the alarm pheromone would remain active long enough for conspecifics to detect and assess local predation risk. Alarm signals, however, should not remain active for too long, as this would provide inaccurate information to signal receivers (Kats and Dill, 1998). We have shown that Ostariophysan alarm pheromones have a relatively high behavioural stimulus response threshold (Brown, Adrian, and Shih, submitted). The functional concentration of the pheromone decreases as a function of the cube of the distance from the signal sender. This would limit the detection of the signal ('active space', Lawrence and Smith, 1989) to those conspecifics and heterospecifics within the immediate area, thus providing useful information regarding predation risk to signal receivers.

ACKNOWLEDGMENTS

The authors wish to thank Leo Fleishman for comments on an earlier version of this manuscript. Financial support was provided by Union College and the National Science Foundation (DUE 97-174).

REFERENCES

Axelrod, H. R., and Vonderwinkler, W., 1983, *Encyclopedia of Tropical Fishes*. T.F.H. Publications: Neptune City, N.J.

Brown, G. E., and Brennan, S., 2000, Chemical alarm signals in juvenile green sunfish (*Lepomis cyanellus*, Centrarchidae). *Copeia* **2000**:1079-1082.

Brown, G. E., and Godin, J.-G.J., 1997, Anti-predator response to conspecific and heterospecific skin extracts by threespine sticklebacks: Alarm pheromones revisited, *Behaviour* **134**:1123-1134.

Brown, G. E., and Godin, J.-G. J., 1999a, Chemical alarm signals in wild Trinidadian guppies (*Poecilia reticulata*), *Can. J. Zool.* 77:562-570.

Brown, G. E., and Godin, J.-G. J., 1999b, Who dares, learns: Chemical inspection behaviour and acquired predator recognition in a characin fish, *Anim. Behav.* **57**:475-481.

Brown, G. E., and Smith, R. J. F., 1997, Conspecific skin extract elicits antipredator behaviour in juvenile rainbow trout (*Oncorhynchus mykiss*), *Can. J. Zool.* **75**:1916-1922.

Brown, G. E., Adrian, J. C. Jr., Smyth, E., Leet, H., and Brennan, S., 2000, Ostariophysan alarm pheromones: Laboratory and field tests of the functional significance of nitrogen oxides. *J. Chem. Ecol.* **26**:139-154.

Brown, G. E., Godin, J.-G. J., and Pedersen, J., 1999, Fin-flicking behaviour: a visual antipredator alarm signal in a characin fish, *Hemigrammus erythrozonus Anim. Behav.* **58**:469-475.

Chivers, D. P., and Smith, R. J. F., 1998, Chemical alarm signalling in aquatic predator-prey systems: A review and prospectus, *Écoscience* **5**:338-352.

Fink, S. V., and Fink, W. L., 1981, Interrelationships of the Ostariophysan fishes (Teleosti*) Zool. J. Linn. Soc.* **72**:297-353.

Hazlett, B. A., 1997, The organisation of behaviour in hermit crabs: Response to variation in stimulus strength, *Behaviour* **134**:59-70.

Kats, L. B., and Dill, L. M., 1998, The scent of death: Chemosensory assessment of predation risk by prey animals. Écoscience **5**:361-394.

Kawashima, H., and Kumashrio, I., 1969, Studies of purine *N*-oxides. III. The synthesis of purine-*N*-oxides. *Bull. Chem. Soc. Jpn.* **42**:750-755.

Lawrence, B., and Smith, R. J. F., 1989, The behavioral response of solitary fathead minnows, *Pimephales promelas*, to alarm substance, *J. Chem. Ecol.* **15**:209-219.

Magurran, A. E., Irving, P. W., and Henderson, P. A., 1996, Is there a fish alarm pheromone? A wild study and critique, *Proc. R. Soc. Lond. Ser. B.* **263**:1551-1556.

Mathis, A., and Smith, R. J. F., 1993, Intraspecific and cross-superorder responses to chemical alarm signals by brook sticklebacks (*Culaea inconstans*), *Ecology* **74**:2395-2404.

Nelson, J. S., 1994, *Fishes of the World* (3rd ed.) John Wiley and Sons: New York.

Orti, G., and Meyer, A., 1997, The radiation of characiform fishes and the limits of resolution of mitochondrial ribosomal DNA sequences, *Syst. Biol.* **46**:75-100.

Pfeiffer, W., 1977, The distribution of fright reaction and alarm substance cells in fishes *Copeia* **1977**: 653-665.

Pfeiffer, W., Riegelbauer, G., Meier, G., and Scheibler, B., 1985, Effect of hypoxanthine-3-*N*-oxide and hypoxanthine-1-*N*-oxide on the central nervous excitation of the black tetra, *Gymnocorymbus ternetzi* (Characidae, Ostariophysi, Pisces) indicated by dorsal light response, *J. Chem. Ecol.* **11**:507-523.

Robins, C. R., Bailey, R. M., Bond, C. E., Brooker, J. R., Lachner, E. A., Leau, R. N., and Scott, W.B., 1991, World fishes important to North Americans, exclusive of species from the continental waters of the United States and Canada, *Am. Fish. Soc. Spec. Publ.* **21**:243.

Smith, R. J. F., 1992, Alarm signals in fishes, *Rev. Fish Biol. Fish.* **2**:33-63.

Smith, R. J. F., 1997, Does one result trump all others? A response to Magurran, Irving and Henderson, *Proc. R. Soc. Lond. Ser. B.* **264**:445-450.

Stryer, L., 1981, *Biochemistry*. W. H. Freeman: San Fransisco.

Wisenden, B. D., and Smith, R. J. F., 1997, The effect of physical condition and shoalmate familiarity on proliferation of alarm substance cells in the epidermis of fathead minnows, *J. Fish Biol.* **50**:799-808.

Wölcke, U., and Brown, G. B., 1969, Purine *N*-oxides. XXII. On the structure of 3-hydroxyxanthine and Guanine 3-*N*-oxide, *J. Org. Chem.* **34**:978-981.

ANAL SCENT GLAND SECRETION OF THE EUROPEAN OTTER (*LUTRA LUTRA*)

Adeline Bradshaw,[1] Manfred Beckmann,[2] Rebecca Stevens,[2] and Fred Slater[1]

[1]Llysdinam Field Centre
Cardiff School of Biosciences
Newbridge-on-Wye
Llandrindod Wells, LD1 6NB, UK
[2]Cardiff School of Biosciences
Cardiff, CF10 3TL, UK

INTRODUCTION

European otters (*Lutra lutra*) typically lead a solitary lifestyle, except for the times when females are rearing cubs. Territorial ranges of individuals vary between 4 and 80 km (Kruuk, 1995) and communication between neighboring individuals is achieved mostly through olfactory signals (Trowbridge, 1983).

One source of olfactory communication signals used by otters appears to be the pair of vesicular sacs laying either side of the rectum (Gorman *et al*., 1978). The ducts of these glands lead directly to the alimentary canal and the contents are smeared onto the passing faeces to form an excretion called the „spraint".

It has been suggested that fresh otter spraint provides otters with information of individual identity, age, sex, and reproductive condition. Furthermore, otters might be able to gauge the time of deposition of this faecal material, as its more volatile components might dissipate more rapidly than those of higher molecular mass (Chanin, 1985).

Erlinge (1968) showed that intense spraintiing occurred in the „meeting zones" of otters from different areas, indicating that spraintiing is a function of territorial behaviour. Radio-tracking using isotopic markers has shown that an individual's spraintiing behaviour increases in the presence of other otters. In females, spraintiing was found to be more frequent at activity centres rather than at territorial boundaries (Green *et al*., 1984).

Although the overall communication impact of spraint remains unclear, chemical analyses combined with behavioural observations have provided some evidence that it contains specific messages (Gorman *et al*., 1978). Chemical analysis of anal sac secretions of a number of mustelids revealed that a species-specific chemical composition exists, providing a basis for the exchange of information between coexisting mustelids (Brinck *et al*., 1983). Gas-liquid chromatography and acrylamide gel electrophoresis revealed individual differences in the composition of secretions. The stability of protein bands was explored for its potential use in following the free-ranging movements of individual otters. In addition, a pair of captive otters exhibited a periodicity in the pattern of spraint deposition that was similar to that of the oestrus cycle (Gorman et al., 1978). Technological advances in the development of GC-MS (gas chromatography-mass spectrometry) made since the early work carried out by Gorman means that our ability to identify the profile of an otter is more accurate.

The potential for the use of animal odours in the management of vertebrate populations was reviewed by Müller-Schwarze (1990). Priming pheromones, for example, might be used to manipulate reproductive activity. Attractants have been tested for their efficacy as lures to traps, and repellents have been investigated for their potential deployment in programs to prevent damage caused by wildlife or predation. While these applications require detailed knowledge of the scents involved in animal behaviour, a variety of potential applications may be possible for otter populations: the identification of attractive aspects of the scent may assist in luring otters away from busy roads; a repellent scent may be applied to prevent fish predation by otters at fish farms.

Here we report our chemical analysis of volatile organic substances in the aqueous extract of anal scent gland secretions, as determined by coupled purge and trap-gas chromatography-mass spectrometry (PT-GC-MS). Our principal aims were to identify the main volatile organic chemical constituents of the scent secretion, to investigate any differences between the scent profiles obtained from the two glands from an individual otter, and to investigate the differences in scent profile between the sexes and age classes.

METHODS

Gland Removal and Storage

Anal scent glands were removed from 24 otters (see Table 1 for age and sex information) during the routine necropsy. Following dissection the glands were stored intact under argon in a sterile glass vial covered in parafilm^{TM}. The vials were kept frozen at $-20°C$ until analysis could be arranged. Individual otters were classified according to age as immature (<18 months) or mature (>18 months).

Sample preparation

The secretion of anal scent glands were obtained by squeezing the contents of defrosted otter glands directly into a screw capped glass vial with PTFE coated rubber seals, which were stored under 99.999% argon in a freezer at -20°C until required. Prior to analysis, 200 µl of each defrosted sample was transferred with a pasteur pipette into a glass vial containing 5 ml of purified water (Elgastat UHP), pre-purged with helium, and capped. The mixture was agitated for 1 minute (Wirley-mixer), centrifuged for 2 min at 1000 g in a centrifuge at room temperature (22°C), and 4 ml of the aqueous phase was sampled with a gas-tight syringe and immediately injected into the sparger of the purge and trap system.

Coupled Purge and Trap-Gas Chromatography-Mass Spectrometry

Compounds were tentatively identified by coupled purge and trap - gas chromatography - mass spectrometry (GC-MS, GC 8000 series - MD 800, Finnigan). Ionization was by electron impact at 70 eV, 220 °C. Volatile substances of the aqueous gland extract were purged with helium at 22°C for 11 min and trapped on a Tenax/Silica Gel/Charcoal-phase in the purge and trap system (Teckmar 3000). Substances were desorbed onto a fused silica column (DB-5MS, 30 m, 0.32 mm ID, 0.25 µm film) of the gas-chromatography system with helium as carrier gas. Column temperature was held for 5 min at 30°C and increased at a program rate of 10°/min to 260°C. The identities of the compounds 2-methylpropanal, 3-methylbutanal, 2-methylbutanal, 2-ethylfuran, 2-pentylfuran and benzaldehyde, were confirmed in the liquid samples by standards obtained from commercial sources. All the other substances were tentatively identified from GC-MS data alone (NIST-library, MassLab V.1.4, Finnigan).

Subsequent to an analytical run, the sparger of the purge and trap-system was rinsed with purified water, and 5.5 ml purified water was analyzed to „blank" the PT-GC-MS-system; 5 ml of the pre-purged water was used for sample preparation. As controls, 200 µl purified water samples were processed identically as a freshly obtained gland extract.

Data Analysis

Peak analysis of each chromatogram was carried out using Masslab software. The more common substances were compared with pure standards in order to confirm their identity. The relative concentration (area under the peak) was calculated for each substance that had been positively identified. A bar chart was produced for each secretion showing the main chemicals identified and their relative concentration.

Kruskal-Wallis analyses were performed to examine any differences between mean of the main chemicals for the sexes and age classes (Sokal, 1981). Chemical data were log-transformed, following which a Principal Components Analysis (PCA) was applied to the covariance matrix (Fry, 1993).

RESULTS

Table 1 provides a general description of the sample. Table 2 summarises the main substances identified.

Table 1. Description of otters used in this study

	Males	Females	Total
Immature otters	4	1	5
Mature otters	15	5	20
Total	19	6	25

A total of 47 volatile organic substances were tentatively identified in the aqueous extract of scent using GC-MS data alone, the main ones identified are shown in Table 2.

In general, these were classified as alkylated furans, alcohols, aldehydes, up to C-6 acid alkyl esters and organic sulphur compounds. A typical GC-MS chromatogram after purge and trap of the aqueous extract of anal scent gland secretion is shown in Figure 1.

Chromatograms were visually examined for differences between a pair of glands for one individual. Each pair revealed identical profiles in terms of the substances present and the relative quantities observed. The only significant difference between the glands was the volume obtained (between 200 and 1200 µl). Table 3 lists the substances that were confirmed using standards from commercial sources. Benzaldehyde was by far the most abundant substance identified.

Table 2. The main substances identified and their code numbers (used in Figure 2).

SUBSTANCE CODE NUMBER	SUBSTANCE NAME
6	2-methyl-propanal
12	3-methyl-butanal
13	2-methyl-butanal
16	2-ethyl-furan
21	Dimethyl-disulphide
22	1-pentanol
29	Hexanal
34	2-methyl-butanoic acid ethyl-ester
35	3-methyl-butanoic acid ethyl-ester
39	5-methyl-hexanal
41	Benzaldehyde
42	Dimethyl-trisulphide
43	2-pentyl-furan

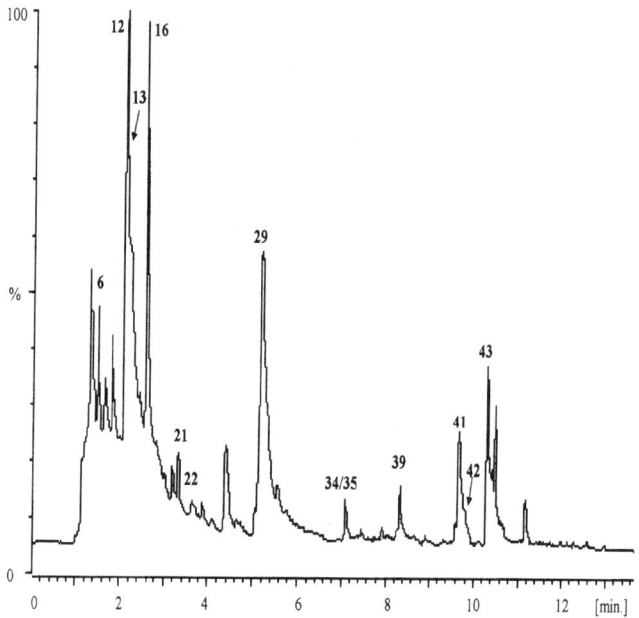

Figure 1: A typical GC-MS-chromatogram after purge and trap of the aqueous extract of anal scent gland secretion. Substance numbers refer to those used in Table 2.

There was a significant difference between the age classes in chemical 35 (3-methyl-butanoic acid-ethyl-ester), with immature animals having higher levels than mature otters (Kruskal-Wallis H=7.4, df = 2, p<0.05). No other significant differences were established between age classes or sexes for any other chemicals.

Chemical data for the main 13 compounds were $\log^{10}$ transformed for all 25 individuals. This matrix was then used for a "sex-out" Principal Components Analysis. The scores were plotted and individuals labelled male or female. The resulting patterns consistently showed female anal gland volatile profiles clustered within a broader male distribution. (Figure 2).

Table 3. Amount of substances in 4 ml processed samples (= 160 µl gland secretion).

SUBSTANCE	AMOUNT (AVERAGE)	MAXIMUM:
2-methylpropanal	1550 ng	7700 ng
3-methylbutanal	260 ng	1180 ng
2-methylbutanal	160 ng	960 ng
2-ethylfuran	440 ng	960 ng
2-pentylfuran	360 ng	950 ng
benzaldehyde	10,2 µg	18,1 µg

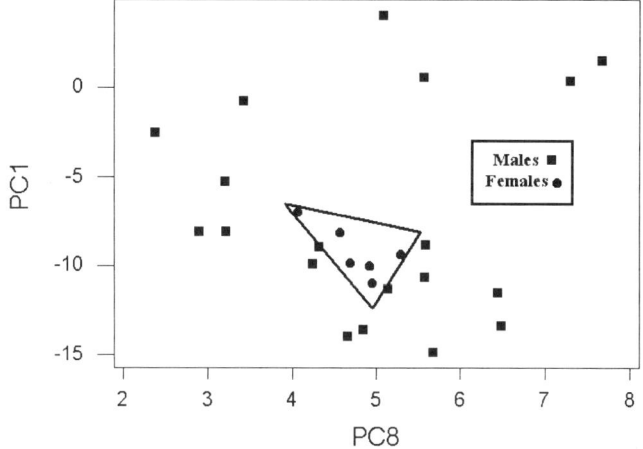

Figure 2. Results of Principal Components Analysis – plot of PC1 and PC8.

DISCUSSION

Olfaction often constitutes a substantial part of the communication systems used by solitary, wide-ranging mammals such as otters. An early study carried out by Trowbridge (1983) combined behavioural observation with gas chromatography to conclude that not only do otters produce a characteristic odour, they are also able to discriminate between other individual otters using only the spraint odour. Communication related to the maintenance of the dominance hierarchy among males and the maintenance of widely overlapping home ranges in both sexes are thought to be the principal functions assisted by spraint odour recognition. Furthermore, it was concluded that the source of the odour was not the faecal material itself by the contents of the anal scent glands.

In the present study, the chemical profile of anal gland contents was further elucidated to identify the main volatile components of otter scent and to establish trends between gland extracts from a single individual, between the sexes and between age classes. These data support conclusions by Trowbridge (1983) and Gorman *et al.*, (1978), who reported that individual differences allow the use of chemical „fingerprints" in otter scent. This has also been identified in other mustelids such as mink (*Mustela vison*), (Brinck *et al.*, 1983) and stoats (*Mustela erminea*), (Erlinge *et al.*, 1982).

It appears that a pair of scent glands from one individual tends to produce identical profiles and differ only in the volume present. It is possible that one gland is more efficient at secretion production, or that at the time of death, one gland had recently been voided.

Few scent glands from immature otters were available for this study, which may have hampered statistical analysis. The comparison of adult and subadult samples by the Kruskal-Wallis resulted in statistically equivalent median concentrations of the main 13 chemicals identified with the exception of methyl-butanoic acid-ethyl-esters, which were found at higher levels in immature animals. It is possible that this is evidence of under-developed gland activity.

Due to the complexity of the chemical profile produced for each individual otter, a multivariate statistical test was applied as a hypothesis-forming examination. The „sex-out" analysis refers to the inclusion of only chemical data in the test in an attempt to display any inherent patterns. The Principal Components Analysis plots suggested that the female odour profiles cluster within a broader range of male profiles. This was a recurring pattern observed over many combinations of plots. There are several possible explanations for the presence of a characteristic female odour. Females become increasingly secretive approaching the birth of young. It has been suggested that the use of secretive behaviour and a natal holt some distance from other otter activity sites is due to the high risk or infanticide from male otters, a phenomenon common in other carnivore species (Kruuk, 1995). It is possible that due to the length of cub dependence the female otter would repel male attention using anal gland odour.

These preliminary data have provided a protocol that we will apply in future studies. It is likely that adult or dominant males have a distinctive odour, but this and other aspects have yet to be identified. There are three main areas we will pursue. First, the study of volatile organic compounds will continue using animals of known sex and age, expanding the sample to include more females and younger animals. Second, we will investigate semi-volatile compounds present in otter scent such as fatty acids. Third, we will conduct a series of behavioural trials in a captive setting to examine the response of otter behaviour to certain groups of compounds identified in otter scent with a view to identifying its' attractant properties.

ACKNOWLEDGMENTS

This work was generously supported by the Esmée Fairbairn Charitable Trust and Cardiff School of Biosciences.

REFERENCES

Brinck, C., Erlinge, S., and Sandell, M., 1983, Anal sac secretion in mustelids: a comparison, *J. Chem. Ecol.* **9**:727.

Chanin, P., 1985, *The Natural History of Otters*, Croom Helm, Beckenham, Kent, UK.

Erlinge, S., 1968, Territoriality of the otter *Lutra lutra* L, *Oikos* **19**:81.

Erlinge, S., Sandell, M., and Brink, C., 1982, Scent –marking and its territorial significance in stoats, *Mustela erminea, Anim. Behav.* **30**:811.

Fry, J. C. (ed.), 1993, *Biological Data Analysis: a Practical Approach*, IRL Press at Oxford University, Oxford, UK.

Gorman, M.L., Jenkins, D., and Harper, R.J., 1978, The anal scent sacs of the otter (*Lutra lutra), J. Zool., Lond.* **186**:463.

Green, J., Green, R., and Jefferies, D.J.,1984, A radio tracking survey of otters (*Lutra lutra*) in a Perthshire river system, *Lutra* **27**:85.

Kruuk, H., 1995, *Wild otters: predation and populations.* Oxford University Press, Oxford, UK.

Müller-Schwarze, D., 1990, Leading them by their noses: animal and plant odours for managing vertebrates, in: *Chemical Signals in Vertebrates 5* (D. W. Macdonald, D. Müller-Schwarze and S. E. Natynczuk, eds.), Oxford University Press, Oxford, UK.

Sokal, R. R., 1981, *Biometry,* WH Freeman and Co, USA.

Trowbridge, B. J.,1983, *Olfactory communication on the European otter Lutra lutra L*. Ph.D. Dissertation, University of Aberdeen, UK.

SCENT-MARKING BEHAVIOUR OF THE EUROPEAN BADGER (*MELES MELES*): RESOURCE DEFENCE OR INDIVIDUAL ADVERTISEMENT?

Christina D. Buesching and David W. Macdonald

Wildlife Conservation Research Unit
Department of Zoology, University of Oxford
Oxford, OX2 9DD, UK

INTRODUCTION

European badgers (*Meles meles*) are nocturnal mustelids. Unlike most other members of this family, which are mainly solitary or pair-living (Powell, 1979; Johnson *et al.*, 2000), badgers live in closely related groups of up to 30 individuals (Rogers *et al.*, 1997). However, the adaptive significance of group-living in this species remains unclear (e.g. Woodroffe and Macdonald, 1993). Incidences of truly co-operative behaviour have not as yet been reported, and although there is some evidence of allo-parental behaviour (Woodroffe, 1993), the benefit of apparent "helpers" for cub-survival appears to be negligible (Woodroffe and Macdonald, 2000). Other behaviours, such as allo-grooming, which are often considered as primitive forms of co-operation, appear to follow a "tit for tat" pattern in badgers (Macdonald *et al.*, 2000). Thus, sociality in Meles meles is often explained instead on the basis of ecological factors, such as food distribution (Kruuk, 1978; Macdonald, 1983). It has only recently become possible to study in detail the behavioural repertoire underlying social group dynamics in badgers, as advances in remote infra-red video surveillance (Stewart *et al.*, 1997b) have overcome the traditional problems of working with nocturnal species. Observations using this technique have already provided insights into sett-maintenance (Stewart *et al.*, 1999) and dominance (Murphy *et al.*, in press). It now offers great potential regarding scent-marking behaviour. In this paper we review the scent-marking behaviour of badgers along territory boundaries, at the sett, and in social context, and we discuss the significance of olfactory signals in the context of both resource defence and individual advertisement.

Badgers have poor eyesight, and although their hearing is acute, they rely mostly on their sense of smell and have a well developed vomeronasal organ. In the context of intraspecific communication badgers make extensive use of faeces, urine, and secretions of specialised glands, such as anal-, and subcaudal glands as olfactory signals. They scent-mark their environment (sett, paths, territory boundaries), resources (e.g. bedding material, food patches), and conspecifics (Kruuk, 1989; Delahay et al., in press). A wealth of information appears to be encoded in these different odour sources.

SOURCES OF BADGER SCENT-MARKS

(i) The subcaudal gland of the European badger is unique amongst the Carnivora (Brown and Macdonald, 1985). It consists of several layers of apocrine and sebaceous cells, which secrete a margarine-like paste into a common lumen, the subcaudal pouch (Stubbe, 1971). The pouch opens at a 20-80 mm wide slit situated between the anus and the base of the tail. Both sexes possess this gland and use the secretion for scent-marking by pressing this slit onto the substrate (Östborn, 1976).

Cubs first start to produce traces of subcaudal gland secretion when they are approximately 4 months old (Buesching and Macdonald, unpublished data). Although secretion volume exhibits seasonal variation in both sexes, males always store significantly larger amounts in their subcaudal pouches than do females (Buesching et al., submitted). Secretion colour is related to sex, age, body condition and reproductive status of individuals and is dependent on the season when the secretion was collected (Buesching et al., submitted). Gas chromatographic analyses have shown that the secretion contains mainly long-chained carboxylic acids, water and protein (Gorman et al., 1984). The chemical composition of subcaudal gland secretions is highly individual-specific and also encodes information about group-membership, sex, age, body condition, and reproductive status of the donor (Östborn, 1976; Gorman et al., 1984; Buesching et al., submitted). Two types of subcaudal scent-marking behaviour can be distinguished (Östborn, 1976; Kruuk, 1978): allo-marking (i.e. scent-marking of conspecifics; Buesching et al., submitted), and object-marking (i.e. scent-marking of the ground/objects in the environment; Buesching and Macdonald, submitted).

(ii) The anal glands of the badger comprise paired sac-like glandular structures, which open into the rectum just internally of the anal sphincter (Stubbe, 1971). The secretion is an orange coloured jelly of low volatility and contains mainly long-chained fatty acids and water (Davies et al., 1988). Chemical composition of anal gland secretions varies little between individuals and sexes, but is likely to encode group-membership (Davies et al., 1988). Anal gland secretion is deposited at latrine sites, usually together with faeces and is thought to act as a long-term territorial defence signal (Roper et al., 1986; Davies et al., 1988).

(iii) Badger faeces vary considerably in appearance and consistency according to diet (Kruuk, 1989). In many species it is possible to determine sex and reproductive status of the marking individual by quantifying metabolites of sex-steroids excreted in faeces and/or urine (Heistermann et al., 1994), and members of several carnivore species use urine and/or faeces as reliable advertisement signals in the context of reproduction (Brown and Macdonald, 1985). In badgers, faeces have long been recognised as an important olfactory signal in intra-specific communication (e.g. Neal, 1977; Kruuk, 1978).

SCENT-MARKING BEHAVIOUR

Scent-marking at Latrine Sites

European badgers have developed a system of shared defecation sites (i.e. latrines), which are generally understood to play an important role in territorial scent-marking (Kruuk, 1978; Roper et al., 1986; Stewart et al., 1997a). One latrine site may comprise as many as 70 individual dung pits, which may contain faeces from several individuals (Kruuk, 1978). Faeces are usually coated in anal gland secretion Roper *et al.*, 1986), and badgers also frequently urinate and/or scent-mark with subcaudal gland secretion at latrine sites.

Two types of latrines can be distinguished: a) hinterland latrines, which are distributed throughout the territory and are used exclusively by members of the same social group, and b) boundary latrines, which are located along the periphery of the territory, and are shared by members of neighbouring groups (Kruuk, 1978; Cheeseman *et al.*, 1981). Bait-marking studies (for method see Delahay *et al.*, in press) showed that boundary latrines usually occur equidistant between neighbouring setts, and are relatively evenly distributed along the border (Stewart *et al.*, 1997a). Their location varies little between years (Delahay *et al.*, 2000). Neighbouring groups tend to deposit similar amounts of faeces in border latrines, an effect which appears to be stronger the closer the respective latrine is located to the sett (Stewart *et al.*, submitted). Therefore boundary latrines are often considered to be active territorial defence signals.

However, radio-tracking studies have shown that individuals from both groups may occasionally transgress this border and feed in close proximity to each other along this line of peripheral latrines (Roper *et al.*, 1986; Woodroffe, 1993; Christian, 1993). Thus, peripheral latrines appear to indicate a general boundary area between the home ranges of neighbouring groups rather than a definitive territory border. In support of this hypothesis is the observation that although the latrine system appears to be strongly affected by the removal of one social group (e.g. in the context of bovine tuberculosis control operations; Tuyttens and Macdonald, 2000), recolonisation of unpopulated territories has been shown to be very slow (Cheeseman *et al.*, 1993). Also, latrine activity is not correlated to food shortage, rather it is highest in spring (high abundance of earthworms) and early autumn (high abundance of cereal), and lowest in summer when food is scarcest (Roper *et al.*, 1986) and is therefore unlikely to function purely as a food-based territorial defence signal as often assumed. Moreover, peaks in latrine use coincide with the main mating and cub-rearing season in spring and the second mating season in early autumn (Roper *et al.*, 1986) with males and females differing in their pattern of latrine use. Whereas males defecate predominantly in boundary latrines, females often use hinterland latrines (Stewart, 1997). Thus, Roper *et al.* (1986) propose that latrine activity is implicated in reproductive advertisement rather than resource defence.

Faeces volume and consistency are likely to be honest and unfakable signals indicating the type and richness of the food resource being exploited. Behavioural observations at latrine sites using remote video cameras have established that individuals spend over 40% of their total time at the latrine sniffing (Stewart, 1997). Hence, badgers visiting a latrine should be able to decipher reliable information about number, group-membership, and possibly individual-specific parameters, such as sex and reproductive status of other individuals using the same latrine.

An alternative explanation for the location of boundary latrines relates to resource depletion shaped by foraging patterns of individuals (Stewart *et al.*, 1997a). According to this model, boundary latrines are located on the food-isopleths between neighbouring groups. Since badgers predominantly utilise latrines in close proximity to their active feeding areas, latrine usage patterns provide a reliable indicator of local foraging pressure, which in turn conveys information about local food resource depletion (Stewart *et al.*, 1997a).

While the evidence is not conclusive, so far it seems to accord more closely with the proposal that latrine sites form a line of individual-specific communication (as proposed in the passive range exclusion hypothesis by Stewart *et al.*, 1997a) rather than an aggressive territorial defence signal.

Scent-Marking at the Sett

All badgers from one social group at least sometimes share a communal denning site, the sett. Badger setts can be very large and complex with more than 60 active holes, persisting for several decades and being inhabited by generations of badgers (Neal and Cheeseman, 1996). Digging these extensive tunnel systems is energy-expensive and requires specific soil conditions (Roper *et al.*, 1991). Thus, setts are considered to be an important and often limiting resource for badgers (Doncaster and Woodroffe, 1993; but see Blackwell and Macdonald, 2000). Video observations have revealed that badgers mark their sett extensively with urine and subcaudal gland secretions, and also maintain latrines on top of the sett, which are especially active during the breeding season (Buesching, 2000, Buesching and Macdonald, submitted).

The subcaudal marking of objects has often been explained in the context of active territorial defence (e.g. Gorman *et al.*, 1984; Kruuk *et al.*, 1984). The scent-matching hypothesis (Gosling, 1982) proposes that extensive scent-marking of the territory with a scent-profile characteristic to the resident individual(s) will enable intruders to recognise resident animals immediately by matching the scent of the territory to the scent of the residents. The resulting asymmetry of contest thus reduces the risk of escalating aggressive encounters. However, badger setts are usually situated in the core area of the territory (Doncaster and Woodroffe, 1993) and are hence only rarely visited by intruders (Buesching, unpubl. data).

Consequently, the high scent-marking activity observed at badger setts (Kruuk *et al.*, 1984; Buesching and Macdonald, submitted) is likely to play an important role in intra-group communication rather than in territorial defence. Suitable setts are especially important resources for breeding females (Stewart *et al.*, 1999). Behavioural observations using remote video cameras revealed that throughout the year females scent-marked significantly more frequently in the vicinity of the sett than did males, but subcaudal marking activity of both sexes reached a maximum during the mating season in winter (Buesching and Macdonald, submitted). Since the chemical composition of subcaudal gland secretions is highly individual-specific (Gorman *et al.*, 1984; Buesching *et al.*, submitted) and scent-marking activity of sett entrances appears to be related to where individuals sleep (Buesching and Macdonald, submitted), subcaudal scent-marks are likely to be a reliable indicator of sett use (i.e. specifically which parts/ holes are used) by individual badgers. This is especially relevant in the context of reproduction. The chemical composition of subcaudal gland secretions is known to vary according to individual specific parameters, such as sex, age, and reproductive status of the marking individual, thus allowing subcaudal scent-marks to be used as reproductive advertisement signals (Buesching *et al.*, submitted). Further, because olfactory signals are pervasive over time and subcaudal secretions mature in a manner indicative of the age of the scent-mark (Buesching *et al.*, submitted) receivers may be able to use this information to pinpoint the current location of the marking individual.

Scent-Marking of Conspecifics

Two types of allo-marking (i.e. subcaudal scent-marking of conspecifics) can be distinguished (Kruuk, 1989): sequential allo-marking (one badger scent-marks the body of another individual), and mutual allo-marking (i.e. two badgers press the openings of their subcaudal pouches against each other simultaneously). Analyses of video observations revealed that both mutual and sequential allo-marking are performed most frequently during the mating season in winter and the cub rearing season in :ing, and significantly less often during summer and autumn. During all seasons, ma ⹁ mark significantly more with both types than do females. However, whereas mutual allo-marking proves to be influenced only by sex and season, sequential allo-marking is also dependent on individual-specific parameters (e.g. age, reproductive status), indicating different functions of the two types of allo-marking behaviour (Buesching *et al.*, submitted). The characteristic smell of subcaudal gland secretions is largely generated by pouch bacteria, which metabolise the primary gland products (Albone, 1984). Mutual allo-marking is performed exclusively between group-members and presumably facilitates the exchange of pouch bacteria, resulting in the generation of a common group-odour (Gorman *et al.*, 1984; Kruuk *et al.*, 1984). The ability to recognise individuals as (non-) group members is especially important in the context of active territorial defence (scent-matching hypothesis; Gosling, 1982) and in the context of reproduction. Badgers in the United Kingdom have a promiscuous mating system (Macdonald *et al.*, submitted), and whereas females should seek extra-group matings to avoid inbreeding, males should seek to expel competitors from other social groups to maximise their own reproductive success. Consequently, the existence of a common group-odour will benefit both sexes. However, levels of mutual allo-marking are very low, possibly indicating the minimal level required to maintain a common group-smell without obscuring individual-specific characteristics in the odour profile.

By contrast, sequential allo-marking is far more frequent and is strongly influenced by individual-specific parameters, indicating at least a partly different, and presumably two-fold function (Buesching *et al.*, submitted) to distribute the common group-odour generated by mutual allo-marking, and b) to serve as an individual advertisement signal by encoding reliable individual-specific and fitness-related information in the scent-marks.

CONCLUSION

The social organisation of the European badger is still poorly understood. Recent studies suggest that the social system has more structural layers than previously assumed and indicate the existence of sub-groups within badger clans (Woodroffe and Macdonald, 1992; Stewart, 1997; Buesching *et al.*, submitted; Macdonald and Newman, unpubl. data.). However, as badgers are solitary foragers and evidence of truly co-operative behaviour is scarce, the evolution of territorial defence at the group-level remains intriguing.

To consider badger scent-marking behaviour at the level of self-interested individuals, which advertise their presence in the area and simultaneously include information about individuality, group-membership, and individual-specific parameters in their scent-marks, offers an alternative mechanism to convey honest information on resource utilisation (Stewart *et al.*, submitted). Simultaneously, this behaviour can serve as an individual advertisement signal allowing badgers to communicate detailed information to other individuals in a social context. Specifically the evolution of the subcaudal gland and its secretion is of major interest as this structure fulfils a function in intra-specific communication beyond what is possible with faeces and urine alone. Unlike urine and faeces, subcaudal gland secretion involves the investment of energy, which is especially important during times of metabolic stress (e.g. mating, cub rearing), and conveys a subtle repertoire of individual- and context-specific information.

REFERENCES

Albone, E., 1984, *Mammalian Semiochemistry: The Investigation of Chemical Signals between Mammals*, Wiley and Sons, Chichester.
Blackwell, P., and Macdonald, D. W., 2000, Shapes and sizes of badger territories, *Oikos* **89**: 292-298.

Brown, R. E., and Macdonald, D. W. (eds.), 1985, *Social Odours in Mammals*, Volume 2, Clarendon Press, Oxford.

Buesching, C. D., 2000, The subcaudal gland of the European badger (*Meles meles*), chemistry and scent-marking behaviour, PhD-thesis, Dept. of Zoology, University of Oxford, England.

Cheeseman, C. L., Jones, G. W., Gallagher, J., and Mallinson, P. J., 1981, The population structure, density and prevalence of tuberculosis (*Mycobacterium bovis*) in badgers (*Meles meles*) from four areas in south-west England, *J. App. Ecol.* **18**: 795-804.

Cheeseman, C. L., Mallinson, P. J., Ryan, J., and Wilesmith, J. W., 1993, Recolonisation by badgers in Gloucestershire, in: *The Badger* (T. J. Hayden, ed.), Irish Royal Academy, Dublin, pp. 78-93.

Christian, S. F., 1993, Behavioural Ecology of the Eurasian badger (*Meles meles*): space use, territoriality, and social behaviour, PhD-thesis, School of Biological Sciences, University of Sussex, England.

Davies, J. M., Lachno, D. R., and Roper, T. J., 1988, The anal gland secretion of the European badger (*Meles meles*) and its role in social communication, *J. Zool. Lond.* **216**: 455-463.

Delahay, R. J., Brown, J., Mallinson, P. J., Spyvee, P. D., Handoll, D., Rogers, L. M., and Cheeseman, C. L., 2000, The use of marked bait in studies of the terrestrial organisation of the European badger (*Meles meles*), *Mamm. Rev.* **30**: 73-87.

Doncaster, C. P., and Woodroffe, R., 1993, Den site can determine shape and size of badger territories: implications for group living, *Oikos* **66**: 88-93.

Gorman, M. L., Kruuk, H., and Leitch, A., 1984, Social functions of the sub-caudal scent gland secretion of the European badger (*Meles meles*) (*Carnivora: Mustelidae*), *J. Zool. Lond.* **203**: 549-559.

Gosling, L. M., 1982, A reassessment of the function of scent marking in territories, *Z. Tierpsychol.* **60**: 89-118.

Heisterman, M., Möstl, M., and Hodges, J. K., 1994, Non-invasive endocrine monitoring of females reproductive status: methods and applications to captive breeding and conservation of exotic species, in: *Research and Captive Propagation* (U. Gansloßer, and W. Kaufmann, eds.), Filander Verlag, Erlangen, Germany.

Johnson, D. P., Macdonald, D. W., and Dickman, A. J., 2000, An analysis and review of the sociobiology of the Mustelidae, *Mamm. Rev.* **30**: 171-196.

Kruuk, H., 1978, Spatial organisation and territorial behaviour of the European badger (*Meles meles*), *J. Zool. Lond.* **184**: 1-19.

Kruuk, H., 1989, *The Social Badger: ecology and behaviour of a group living carnivore (Meles meles)*, Oxford University Press, Oxford.

Kruuk, H., Gorman, M. L., and Leitch, A., 1984, Scent-marking with the subcaudal gland by the European badger, *Meles meles* L., *Anim. Behav.* **32**: 899-907.

Macdonald, D. W., 1983, The ecology of carnivore social behaviour, *Nature* **301**: 379-384.

Macdonald, D. W., Stewart, P. D., Stopka, P., and Yamaguchi, N., 2000, Measuring the dynamics of mammalian societies: an ecologist's guide to ethological methods, in: *Research techniques in animal ecology: controversies and consequences.* (L. Boitani and T. K. Fuller, eds.), Columbia University Press, New York, pp. 332-388.

Murphy, O., Stopka, P., Stewart, P. D., and Macdonald, D. W., European badger (*Meles meles*) society: sociality without status?, *J. Zool. Lond.*, (in press)

Neal, E. G., 1977, *Badgers*, Blandford, Poole, Dorset.

Neal, E. G., and Cheeseman, C., 1996, *Badgers*, T. and A.D. Poyser, London.

Östborn, H., 1976, Doftmarkering hos graveling, *Zool. Revy.* **38**: 103-112.

Powell, R. A., 1979, Mustelid spacing patterns: variations on a theme by Mustela, *Z. Tierpsychol.* **50**: 153-165.

Rogers, L. M., Cheeseman, C. L., and Mallinson, P. J., 1997, The demography of a high-density badger (*Meles meles*) population in the west of England. *J. Zool. Lond.* **242**: 705-728.

Roper, T. J., Shepherdson, D. J., and Davies, J. M., 1986, Scent-marking with faeces and anal gland secretion in the European badger (*Meles meles*): seasonal and spatial characteristics of latrine use in relation to territoriality, *Behaviour* **97**: 94-117.

Roper, T. J., Tait, A. I., Christian, S. F., and Fee, D., 1991, Excavation of three badger (*Meles meles* L.) setts, *Z. Säugetierk.* **56**: 129-134.

Stewart, P. D., 1997, The social behaviour of the European badger *Meles meles*, PhD-thesis, Dept. of Zoology, University of Oxford, England.

Stewart, P. D., Anderson, C., and Macdonald, D. W., 1997a, A mechanism for passive range exclusion: Evidence from the European badger (*Meles meles*), *J. Theoret. Biol.* **184**: 279-289.

Stewart, P. D., Ellwood, S. A., and Macdonald, D. W., 1997b, Remote video-surveillance of wildlife - an introduction from experience with the European badger *Meles meles*, *Mamm. Rev.* **27**: 185-204.

Stewart, P. D., Bonesi, L., and Macdonald, D. W., 1999, Individual differences in den maintenance effort in a communally dwelling mammal: the Eurasian badger, *Anim. Behav.* **57**: 153-161.

Stewart, P. D., Macdonald, D. W., Newman, C., and Cheeseman, C., Boundary faeces matching in the European badger (*Meles meles*): a potential role in range exclusion, *J. Zool. Lond.* (in press)

Stubbe, M., 1971, Die analen Markierungsorgane des Dachses (*Meles meles*), *Zool. Garten N.F.* **40:** 125-135.

Tuyttens, F. A. M., and Macdonald, D. W., 2000, Social Perturbation: consequences for wildlife management and Conservation, in: *Behaviour and Conservation*, (L. M. Gosling and W. J. Sutherland, eds.), Cambridge University Press, Cambridge, pp. 315-329.

Woodroffe, R., 1993, Alloparental behaviour in the European badger, *Anim. Behav.* **46:** 413-415.

Woodroffe, R., and Macdonald, D. W., 1992, Badger clans: demographic groups in an antisocial species, *J. Zool. Lond.* **227:** 696-698.

Woodroffe, R., and Macdonald, D. W., 1993, Badger sociality - models of spatial grouping, *Symp. Zool. Soc. Lond.* **65:** 145-169.

Woodroffe, R., and Macdonald, D. W., 2000, Helpers provide no detectable benefits in the European badger (*Meles meles*), *J. Zool. Lond.* **250:** 113-119.

SCENT MARKING BEHAVIORS OF THE STRIPED MONGOOSE, *MUNGOS MUNGO*

Igor Ianovschi

Department of Vertebrate Zoology
St.-Petersburg State University
Universitetskaya nab., 7/9
St.-Petersburg, 199034, Russia

INTRODUCTION

In mammalian species that live in complex social organizations, such as striped mongooses, *Mungos mungo*, chemical signaling is often the result of visually prominent marking behaviors that might enhance the effectiveness of communication (Sokolov and Zinkevich, 1974). Three important sources of chemical signals include skin glands (Clark, 1982a,b), urine, and anal glands (Roeder, 1983). I describe here the visual components of marking behaviors that are associated with the deposition of such chemical signals by striped mongooses.

METHODS

I studied 17 male and 17 female striped mongooses born in captivity in four different family groups. The animals were housed as families in two-compartment cages of overall size 4x1.5x2 m, provisioned with logs, stones and shelter boxes. Meat, fish, curds, vegetables, fruits, nuts, sprouts and live mice, rats and chicken were included in the diet. Morning and evening observations took place after the animals had four or five days to familiarize themselves with the cages.

There were three observation series of marking behaviors, and all observation sessions were of 10 min duration. The first series was based on behavioral responses to the presentation of an unfamiliar object, e.g., plastic bottles, stones, rubber balls (d=20 cm), or a small board fastened to lattice at a height 1.5 m. All objects were used only once and they were placed in the center of the cage. In the second series, I observed the mongooses as they were allowed to enter and explore the adjacent half of the cage for the first time; in some cases, the cage was empty, in others it contained a wooden box, two stones and a wooden log. Observations began as soon as I opened the partition separating the two compartments. In the third series, I placed a portable, wire-mesh, cage (40x40x60 cm) containing other animals within 5 cm of the subjects' cages and observed responses; the caged „intruders" were: unfamiliar male and female *Mungos mungo* from another group, or *Nasua nasua*, or *Oryctolagus cuniculus*, or *Felis catus*, or *Helogale pervula*.

I observed four types of scent deposition: anal-gland marking, chin marking, ventral marking, and urine marking. To accomplish anal-gland marking, the animal moves apart its posterior paws, raises its tail, and then presses down its anus onto the substrate, dragging it forward using mainly the fore-legs. To leave the secretion of anal glands on a vertical surface, the animals clutched at the side of the cage using the fore-claws, arched their backs and rubbed the anogenital region on the surface (Figure 1). The ventral marking behaviors occur as the rear paws of the animal are moved apart and are not in use, at the same time the groin area and the back part of the ventrum are dragged along the substrate, powered by movement of the fore-legs (Figure 2). In chin marking, animals rub with their cheek or lower jaw, and sometimes their shoulders and chest against an object to be marked (Figure 3). Urine marking rarely occurred and only while the animals were moving along horizontal or inclined planes. The animal moves apart and bends its back paws, stretches out the tail parallel with the floor. Firstly it looks like the first steps of anogenital marking. However, instead of rubbing the animal stamps the object with its back paws rubbing the feet and simultaneously excretes a little of urine while the tail is raised up 45 degrees (Figure 4).

Figure 1. Marking by anal glands.

Figure 2. Marking with the belly.

Figure 3. Marking by head.

Figure 4. Marking by the use of urine.

RESULTS

Object Presentation

Upon presentation of an unknown object, the animals approached it, all adults simultaneously. They sniffed the object and then marked it. The marking was performed by turns: a male was the first, followed by a female, and then the male might take control of the object another time. This alternation of investigation and marking was occasionally supported by displays of threatening and aggressive behaviors. Young animals were allowed to investigate the object only after the adults were no longer interested in it.

The analysis of marking behaviors revealed that marking an object with the anal area was observed in 87% of the males, but only 29% of the female (Table 1). More than half of the females chin-marked the object. In total, males left more marks than females did in all experimental settings excerpt for those where a ball was presented. The latter kind of object stimulated females to actively display marking behaviour predominantly targeted not to the object presented but rather to fixed details of the cage environment. It was also noticed that the intensity of marking behaviour was strongly dependent on object size and increased when large objects were placed into cages (data not shown).

Occupation of New Territories

The animals entered the territory immediately after the door between the two cages was opened. During the first minute, their behaviors included pressing themselves to the floor, stretching their necks forward, and vocalizing. Adults usually entered unknown areas collectively, with the cubs close behind the parents. Males and females started to mark a new cage simultaneously irrespective of whether it was empty or contained items (log, stones, and box). The first target of odour gland secretions was always the border between two cages. In an experimental environment enriched with different object, the mongooses started to mark them and the floor around them from the first minute of an experiment.

The proportions of different marks in this series of experiments were basically the same as those observed in responses to objects, i.e., males made predominantly anal marks, whereas females made more chin marks (Table 1).

Intruder Arrival

On encountering an unknown animal belonging to their own or alien species, the mongooses displayed abundant scent marking. Responses to mongoose intruders were much stronger than those to animals of other species (Table 1).

Table 1. The proportion of different marks by striped mongoose, *Mungos mungo* in presence of unknown object, in new territory, or after introduction of an unknown animal.

	Anal marking (% of total)	Chin marking (% of total)	Ventral marking (% of total)	Urine marking (% of total
Object presentation				
Males	87	9	2	2
Females	29	55	16	0
Investigation of new territories				
Males	53	40	7	0
Females	39	45	16	0
Arrival of intruder mongooses				
Males	74	13	13	0
Females	43	51	6	0
Arrival of non-mongoose intruders				
Males	61	28	8	3
Females	61	50	4	0

In all experiments of this series mongoose males were noticed to use their anal area for odour marking of their kin, not only inanimate objects. Females marked their kin the ventral region. Reciprocal marking was observed only when mongoose females were presented. When a mongoose male was presented, females marked only their territory. It is noteworthy is that when a dwarf mongoose was presented to striped mongooses, alarm vocalising of the former immediately evoked watch-out responses of the latter, probably because natural habitats of *Mungos mungo* and *Helogale parvula* significantly overlap, although all animals tested were born in captivity.

DISCUSSION

This study of striped mongooses expands the diversity of marking behaviors known from studies on other mongoose species (Gorman, 1980; Gorman *et al.*, 1974; Roeder, 1983). For example, meerkats were noted to use only two modes of applying scent, e.g., with their paws or with their anal areas (Moran and Sorensen, 1986; Rasa, 1973). Anal marking was also described by Ewer (1963). The diversity of marking behaviors of species in the *Herpestidae* family might be limited by available ways of leaving their smell. This issue might be elucidated by histological studies of the skin and skin glands. Such studies are currently under way but, by now, they involve only few mongoose species and related *Viverridae* species, which have been shown to possess anal sacks and hypertrophied oil-glands and sweat-glands in lip corners, sex lips, and paw pads (Fiedler, 1975; Sokolov and Nekliudova, 1988).

The present work utilized three experimental settings that caused a high level of marking responses by striped mongooses. The four marking patterns we described were manifested by both males and females. However, the proportions of two behavioural patterns were different in males and females. Males always displayed anal marking at a rate above 50%, whereas females did it at a rate less than 50%. By contrast, the proportion of chin marking amounted to 9%-40% in males and 45%-55% in females.

Odour marking of the kin is a characteristic feature of the striped mongoose. It is more typical of males and is displayed only when intruders belonging to the same species are presented, which suggests that this behavioral pattern performs the function of group consolidation or its olfactory distinguishing from strangers. In his studies of *Viverra civetta*, Randall (1979) reported an increase in marking with anal-gland secretions when strangers were placed into the cage.

Another peculiarity of the striped mongooses is taking turns in odour marking an object. This is most evident with inanimate objects presented to the animals, and less evident with intruders. Taking turns in odour marking was also evident when a new territory was explored. In this case, the animals fairly rapidly focussed on objects used to enrich the environment and initially marked the larger objects first. Interestingly, despite the multitude of objects enriching the new environment, a group of animals always concentrated together around one of them and ignored the rest for a time. As their interest to an object weakened, the group turned to a smaller one in a rather consolidated way. The group was lead by a male.

When a new territory is entered, the mongooses place their marks not randomly but rather at definite zones around cage corners, which the animals visited most often. Because a male and a female move separately, they perform marking within these zones by turns. Thus, the reasons for taking turns in marking objects or an empty area (a barren landscape) are different.

REFERENCES

Clark, A. B., 1982a, Scent marks as social signals in *Galago crassicaudatus*. 1. Sex and reproductive status as factor in signal and responses, *J. Chem. Ecol.* **8**:1133.

Clark, A. B., 1982b, Scent marks as social signals in *Galago crassicaudatus*. 2. Discrimination between individual as scent, *J. Chem. Ecol.* **8**:1153.

Ewer, R. F., 1963, A note on the suckling behaviour of the viverrid, *Suricata suricatta*, *Anim.Behav.* **11**:599-601.

Fiedler, W., 1975, Beobachtungen zum Markierungsverhalten einiger Saugetier, *Ztschr. Saugetier.* **22**:57-76.

Gorman, M. L., 1980, Sweaty mongoose and other smelly carnivores, *Symp. Zool. Soc. Lond.* **45**:87-105.

Gorman, M. L., Nedwell, D.B., and Smith, R.M., 1974, An analysis of the contens of the anal scent pockets of *Herpestes auropunctatus*, *J. Zool.* **172**:389-399.

Moran, G., and Sorensen, L., 1986, Scent marking behaviour in a captiva group of meerkats (*Suricata suricatta*), *J. Mammal.* **55**:120-130.

Randall, R. M., 1979, Perineal gland marking free-ranging African civets (*Civettictis civetta*), *J. Mammal.* **60**:622-627.

Rasa, O. A. E., 1973, Marking behaviour and its social significance in the African dwarf mongoose, *Helogale undulata rufula*, *Ztschr. Tierpsychol.* **32**:293-318.

Roeder, J. J., 1983, Memorisation des margues olfactives cher la Genette (*Genetta genetta* L.): duree de reconnaissance par les femeles des margues olfacives des males, *Ztschz. Tierpsychol.* **61**:311-314.

Sokolov, V. E., and Zinkevich, A. P., 1974, Osnovnie principi kommunikacii nasemnih pozvonochnih, *II Megidunarodnii teriologhicheskii kongress*, T.2. Moskva. pp. 209-210.

Sokolov, V. E., and Nekliudova, T. I., 1988, Morfologia kogi i specificheskih koginih geles nekotorih predstavitelei semeistva viverrovih (Viverridae) Vietnama, *Ekologo-funkcionalnaja morfologia koginogo pokrova mlekopitajuschih*, Moskva, Nauka, pp. 4-62 (in Russian).

INCREASED SOCIAL DOMINANCE IN MALE RABBITS, *ORYCTOLAGUS CUNICULUS*, IS ASSOCIATED WITH INCREASED SECRETION OF 2–PHENOXY ETHANOL FROM THE CHIN GLAND

R. Andrew Hayes, Barry J. Richardson and S. Grant Wyllie

Centre for Biostructural and Biomolecular Research
University of Western Sydney, Hawkesbury
Richmond, NSW 2753, Australia

INTRODUCTION

The European rabbit, *Oryctolagus cuniculus*, is a gregarious and sedentary organism that commonly lives in defined social groups of between two and about twenty individuals (Cowan and Bell, 1986). This social group lives together in a series of interconnecting burrows known as a warren (Lockley, 1961; Mykytowycz, 1958; Southern, 1940). There are distinct social hierarchies within each social group (Cowan and Bell, 1986; Lockley, 1961; Mykytowycz, 1958). These hierarchies are strictly maintained during the breeding season, but are not as strong at other times of the year (Mykytowycz, 1964).

Rabbits are known to scent-mark their environment with urine (Bell, 1981), faeces (Myers *et al.*, 1994) and the secretions from several skin glands (Mykytowycz, 1968). The secretion of the submandibular cutaneous (chin) gland is strongly correlated with social status. Dominant, male rabbits have a higher secretory activity of the gland (Mykytowycz and Dudzinski, 1966), and show much more scent-marking behaviour than do other individuals within the social group (González-Mariscal *et al.*, 1993; Mykytowycz, 1968; Verberne and Blom, 1981).

The composition of the chin gland secretion has been analysed chemically and was found to be mainly protein, with only trace amounts of free lipids (Goodrich and Mykytowycz, 1972). There is a predominance of high molecular mass, non-volatile constituents, however rabbits can detect chin secretion olfactorially (Mykytowycz, 1975). Goodrich (1983) examined the headspace volatiles of the chin secretion, and found them to be primarily of an aromatic structure, including aldehydes, ketones and phenyl compounds, with only a few aliphatic compounds, aldehydes, ketones and *n*-butanoic acid. It could be that the volatile components are bound to higher molecular mass constituents that slowly release the message. The slow release may also allow time for the action of microorganisms to decompose these compounds and release the odorants (Goodrich, 1983).

COMPONENTS OF CHIN GLAND SECRETION

The present study was principally conducted at Hope Farm, Cattai National Park (150°52' E, 33°33' S) on the east bank of the Hawkesbury River, in the north-west of Sydney, Australia. Wild rabbits were trapped in treadle-operated wire box traps. A total of 3459 trapping events occurred between 21/8/95 and 24/12/97, with a total number of 24 450 trap-nights. Chin gland secretion was obtained from trapped rabbits by collection of extruded secretion onto a cotton bud, or into a glass, micro-capillary tube following external palpation of the gland. The swab was then placed into a headspace vial (10 mL), and the vial closed with an airtight teflon-faced rubber seal for analysis by dynamic headspace gas chromatography. All samples were stored in a freezer at $-20°C$, until analysed.

The vial was heated to 30°C in a heating block. Laboratory air (3.2 L) was pumped through a charcoal trap, then over the sample at a flow rate of 13 mL min^{-1}. After passing over the sample the air then flowed through gas chromatography injection port liners packed with the adsorbent agent Tenax TA (35/60 mesh) (90 ± 2 μg). At the end of the collection time (246 min), the liner was placed in the injector port of a gas chromatograph (Hewlett Packard 5890 Series II) equipped with a flame ionisation detector and fitted with a silica capillary column (SGE, model BP-1, 25 m length, 0.22 mm ID and 0.1 μm film thickness). The Tenax was thermally desorbed using a Programmable Temperature Vapouriser (PTV).

Data were acquired under the following GC conditions – inlet temperature: 200°C, carrier gas: nitrogen at 23 cm s^{-1}, split ratio: 1:20, detector temperature: 260°C, temperature program: initial temperature 45°C, initial time 5 min, rate 8°C min^{-1}, final temperature 250°C, final time 5 min. The temperature program for the PTV was - initial temperature: 40°C, ramp rate: 16°C s^{-1}, final temperature: 200°C.

The identities of peaks in the chromatogram were determined by use of another HP5890 GC coupled to a mass spectrometer (HP 5971A, Hewlett Packard) (GC-MS) with an attached PTV. Data was acquired under the following GC conditions – inlet temperature: 260°C, carrier gas: helium at 1.1 mL min^{-1}, split ratio 1:2, detector temperature 260°C, temperature program: initial temperature 40°C, initial time 5 min, rate 8°C min^{-1}, final temperature 250°C. The MS was held at 190°C in the ion source using an ionisation energy of 70 eV and a scan rate of 0.9 scans s^{-1}. The PTV program was – initial temperature: 35°C, ramp rate: 12°C s^{-1}, final temperature 220°C. Peak identities were assigned by comparison with the Wiley mass spectral library.

An external standard hydrocarbon mixture ($C_8 - C_{18}$) was injected into the GC and analysed under the same conditions to enable calculation of Kovats indices.

Sample blanks were run at regular intervals (approximately twice every week) to ensure that volatiles recorded were from the secretion and not due to the sampling procedure. The blanks consisted of an unused cotton bud analysed by the same procedure as for the samples.

A range of different volatile compounds was identified from the 120 samples of chin gland secretion examined, and these consist mainly of aromatic and aliphatic hydrocarbons. Especially common are a range of alkyl-substituted benzene rings, that provide most of the diversity in the compounds in the secretion.

Statistical analysis of the chromatograms obtained shows that, while there is variation within individuals, there are also significant differences among social classes of rabbits. These differences are in the relative areas of the peaks in the chromatogram, and are not primarily due to the presence or absence of components. Discriminant function analysis, using the relative areas under the peaks of the chromatogram as variables, allowed us to predict the status of an unknown individual with a reasonable degree of confidence ($r^2=0.77$).

Because of this finding, it was decided that a manipulative experiment might further elucidate the effect of a change in status on the chemical characteristics of the chin-gland secretion of the male, wild rabbit. The experiment was based on that of Reece (1985).

CHANGES IN DOMINANCE STATUS AND CHIN GLAND ACTIVITY

Nine wild, adult animals were captured in box-traps and returned to the campus of the University, where they were held in groups of three in floor-pen enclosures. The pens were constructed to the published specifications for group-housing of laboratory rabbits (BVAAWF *et al.*, 1993; Harris, 1994; Love, 1994). The side walls of the pens were solid aluminium, while the front wall was wire mesh. The pens were lined with pine-shavings over a concrete floor. Commercial rabbit pellets were provided in auto-feed hoppers that were re-filled daily and green lucerne hay was also provided daily. Water was available *ad libitum*.

The groups of three (triads) consisted of two males and one female, with a difference in mass of at least 200g between the two males. Chin swabs were collected from all three animals, as described above. The dominant male was determined in each triad from behavioural observations of agonistic interactions and scent-marking behaviour (after Mykytowycz and Dudzinski, 1966; Reece, 1985; Southern, 1940, 1948).

After animals had been held in triads for seven days, the dominant male was removed from the enclosure (Day 0). The remaining male and female were kept together for another fourteen days (Days 1 to 14). Swabs were collected from each animal throughout this period. Any change which was to occur was expected to take place rapidly. Reece (1985) found that behavioural modification to the dominant pattern in the originally subordinate animal occurred within one day, while circulating plasma testosterone increased to dominant levels within a week.

Data collected before manipulation began indicated that all male rabbits showed a signal with a similar complexity regarding the number of peaks detected. However, after removal of the dominant, the chromatograms of the previously subordinate males underwent significant change (ANOVA, F=23.32, $p < 0.0001$).

On Day 1, the number of peaks in the chromatogram dropped from approximately fifty to, at most, nineteen. A compound that was not previously found in the subordinate traces, peak 45, was now the largest peak in the trace (Figure 1). These changes occurred within twenty-four hours, the same time period noted in an earlier study for behavioural modification to occur (Reece, 1985).

After a further three days (*i.e.* on Day 4), the chromatograms looked the same, still with a reduced complement of compounds, and an extremely large peak 45 (Figure 1). By Day 7, the chromatogram had reverted to an appearance that is not significantly different to that seen before Day 0, but with the addition of peak 45. This compound was now found in the chromatogram of all the animals that had undergone a shift in social status, and persisted in the trace for at least another week.

Upon further examination of the chromatograms of the originally dominant animals it is seen that compound 45 was present in the secretions of these rabbits at the beginning of the experiment.

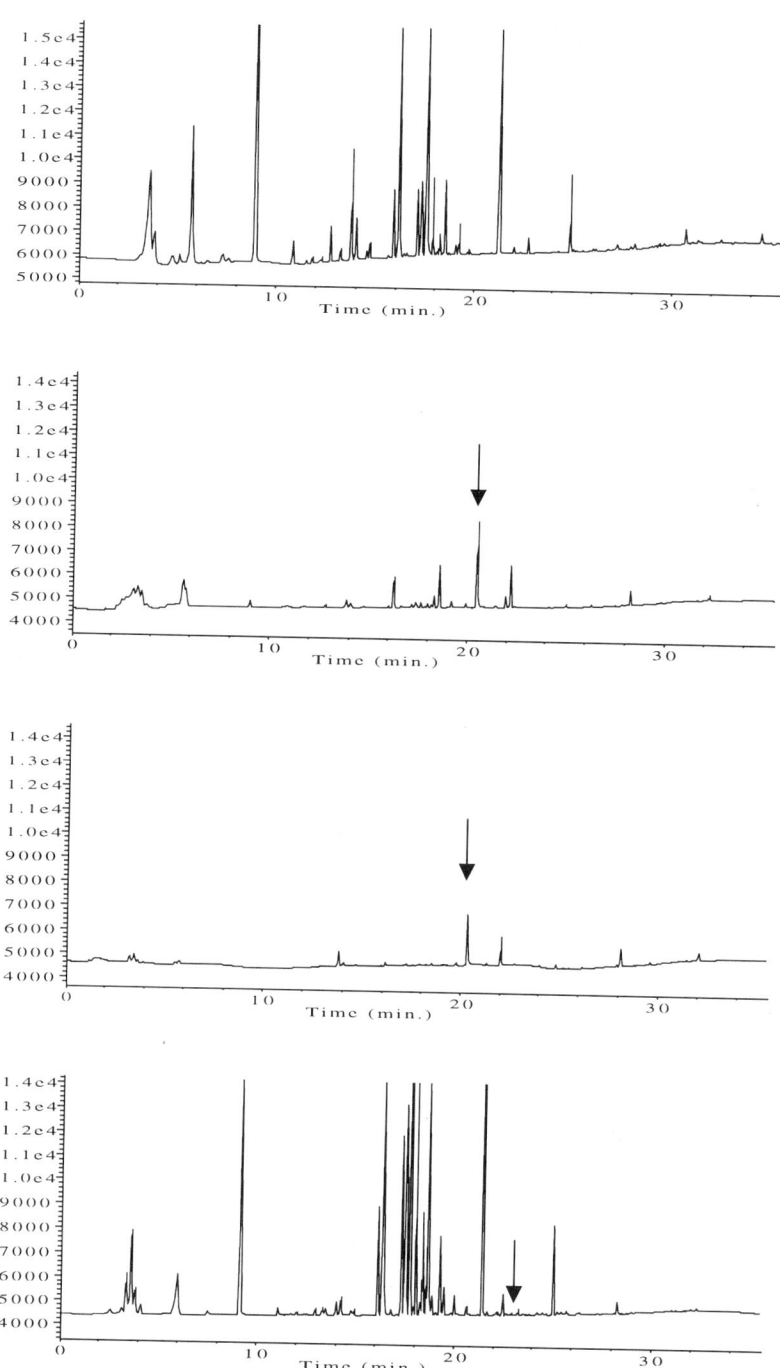

Figure 1. Chromatograms of swabs from an initially subordinate rabbit (Yellow tag). The chromatograms are (from top to bottom) Day –7, Day 1, Day 3 and Day 7 respectively. There is no change after Day 7. Peak 45 is marked with an arrow.

FIELD TESTS SUPPORT DOMINACE-RELATED SIGNAL USE

Having observed that there was a compound found in the secretion of dominant rabbits, and not found in that of subordinates, data from the field was examined to see if this held true in a more natural setting. Chromatograms from chin swabs of wild male rabbits trapped at Cattai National Park during the breeding season were compared with the social status of these individuals. The animals were all rabbits where the social status could be ascertained with confidence. The presence or absence of compound 45 in the secretion was then ascertained for each rabbit (Table 1). This substance was again never detected in the secretion of a subordinate animal. It was, however, identified in the secretion of all of the presumed dominant animals examined.

Mass spectral analysis of this compound shows that it is 2-phenoxy ethanol. It would seem to be a compound that the rabbits are using to signal dominance.

Table 1 Correlation of compound 45 in chromatograms of the chin swab with social status in wild rabbits. A ✓ indicates that the peak occurs in the chromatogram, an × indicates that it does not.

Tag combination	Social status	Presence of #45
Bk+N/W 3/7/96	Dominant	✓
Bk+R/W 17/7/96	Dominant	✓
Bl+R/W 20/8/96	Dominant	✓
N+N/W 12/6/96	Dominant	✓
Pi+R/W 21/8/96	Dominant	✓
R+Bk*Y 7/8/96	Dominant	✓
R+Bk/G 3/7/96	Dominant	✓
R+Y/G 22/8/96	Dominant	✓
Y+Bk/O 22/8/96	Dominant	✓
Bk+Bk/O 29/11/95	Subordinate	×
Bk+Y/Bk 10/12/96	Subordinate	×
G+W/G 5/10/96	Subordinate	×
N+Bk/G 19/11/96	Subordinate	×
O+Bk/O 15/11/95	Subordinate	×
P5160 27/11/95	Subordinate	×
Pu+Bk/O 1/11/95	Subordinate	×
R+Bk/O 14/11/95	Subordinate	×
W+Bk/O 13/12/95	Subordinate	×
W+R/W 5/9/95	Subordinate	×

GENERAL DISCUSSION

Stoddart (1973) noted that there were sometimes differences in the composition of the glandular secretions between the sexes and between the different social categories of animals in a natural population. He proposed that changes that occur in the secretion as an animal moves between these categories may involve the production of new chemical compounds as well as changes in the relative proportions of the other components.

2-phenoxy ethanol appears in the secretion of the rabbit soon after such a shift in social status. This compound is used as a fixative in the perfume industry (Winter, 1984). The addition of 2-phenoxy ethanol may not, then, merely be the addition of a new component that signals dominance directly, but the system may work in another way. While the change in the signal from a non-signaller (subordinate) to a signaller (dominant) is seen by the addition of this one compound, it may have an additional function – to make the signal last longer.

Acting as a fixative, the presence of this compound in the chin gland secretion ensures that the signal produced by the dominant male rabbit remains in the environment and does not rapidly dissipate. He, therefore, comes to „dominate" the semiochemical space in a similar fashion to the way he dominates the physical space. Is this, then, a general principle? Does the change in signal from a subordinate to a dominant status occur in many mammal species by the addition of a fixative that allows the signal to persist in the environment?

The work in the present study provides new insights into the way that rabbit semiochemical messages work. The difference between the secretion of a subordinate and a dominant rabbit appears to be due to the presence or absence of one compound, 2-phenoxy ethanol. This compound, with known fixative properties, supports the idea that the only difference between dominant and subordinate secretions is whether or not they persist in the environment after marking.

The addition of a fixative would seem to be a parsimonious way of signalling this change in status. It is far easier to understand the evolution of this than to explain the development of an entirely new communication system. The fixative hypothesis requires only a change in one category within the population, the signaller. The rest of the population is already pre-adapted to register this suite of compounds; they are attuned to this signal that will now persist in the environment instead of being transitory. An entirely new system would require the whole population to evolve, in order to recognise and respond to a new signal. We would predict that the approach of adding a fixative may occur much more commonly than would appear from its complete lack of reporting, due to its relative simplicity.

REFERENCES

Bell, D. J., 1981, Chemical communication in the European rabbit: urine and social status, in: *Proceedings of the World Lagomorph Conference, Guelph 1979*, (K. Myers, and C. D. MacInnes, eds.), Guelph, Ontario, pp. 271-279.

BVAAWF/FRAME/RSPCA/UFAW Joint Working Party on Refinement, 1993, Refinements in rabbit husbandry, *Lab. Anim.* **27**: 301-329.

Cowan, D. P., and Bell, D. J., 1986, Leporid social behaviour and social organisation, *Mammal Rev.* **16**: 169-179.

González-Mariscal, G., Melo, A. I., Zavala, A., Chirino, R., and Beyer, C., 1993, Sex steroid regulation of chin-marking behavior in New Zealand rabbits, *Physiol. Behav.* **54**: 1035-1040.

Goodrich, B. S., 1983, Studies of the chemical composition of secretions from the skin glands of the rabbit *Oryctolagus cuniculus*, in: *Chemical Signals in Vertebrates 3*, (D. Müller-Schwarze, and R. M. Silverstein, eds.), Plenum Press, New York, pp. 275-290.

Goodrich, B. S., and Mykytowycz, R., 1972, Individual and sex differences in the chemical composition of pheromone-like substances from the skin glands of the rabbit, *Oryctolagus cuniculus, J. Mammal.* **53**: 540-548.

Harris, I., 1994, The laboratory rabbit, *ANZCCART News* **7**(4): 1-8.

Lockley, R. M., 1961, Social structure and stress in the rabbit warren, *J. Anim. Ecol.* **30**: 385-423.

Love, J. A., 1994, Group-housing: meeting the physical and social needs of the laboratory rabbit, *Lab. Anim. Sci.* **44**: 5-11.

Myers, K., Parer, I., Wood, D., and Cooke, B. D., 1994, The rabbit in Australia, in: *The European Rabbit. The History and Biology of a Successful Colonizer*, (H. V. Thompson, and C. M. King, eds), Oxford University Press, Oxford, pp. 108-157.

Mykytowycz, R., 1958, Social behaviour of an experimental colony of wild rabbits, *Oryctolagus cuniculus* (L.). 1. Establishment of the colony, *CSIRO Wildl. Res.* **3:** 7-25.

Mykytowycz, R., 1964, Territoriality in rabbit populations, *Aust. Nat. Hist.* **14:** 326-329.

Mykytowycz, R., 1968, Territorial marking by rabbits, *Sci. Amer.* **218:** 116-126.

Mykytowycz, R., 1975, Activation of territorial behaviour in the rabbit, *Oryctolagus cuniculus*, by stimulation with its own chin gland secretion, in: *Olfaction and Taste V*, (D. A. Denton, and J. P. Coghlan, eds.), Academic Press, New York, pp. 425-432.

Mykytowycz, R., and Dudzinski, M. L., 1966, A study of the weight of odoriferous and other glands in relation to social status and degree of sexual activity in the wild rabbit, *Oryctolagus cuniculus* (L.), *CSIRO Wildl. Res.* **11:** 31-47.

Reece, C., 1985, *Aspects of reproduction in the European rabbit*, Ph.D. Thesis, University of East Anglia, UK.

Southern, H. N., 1940, The ecology and population dynamics of the wild rabbit (*Oryctolagus cuniculus*), *Ann. Appl. Biol.* **27:** 509-526.

Southern, H. N., 1948, Sexual and aggressive behaviour in the wild rabbit, *Behaviour* **1:** 173-194.

Stoddart, D. M., 1973, Preliminary characterisation of the caudal organ secretion of *Apodemus flavicollis*, *Nature* **246:** 501-503.

Verberne, G., and Blom, F., 1981, Scentmarking, dominance and territorial behaviour in male domestic rabbits, in: *Proceedings of the World Lagomorph Conference, Guelph 1979*, (K. Myers, and C. D. MacInnes, eds.), Guelph, Ontario, pp. 280-290.

Winter, R., 1984, *A Consumer's Dictionary of Cosmetic Ingredients*, Crown Publishers, Inc., New York, pp. 201.

THE RESPONSE OF INDIVIDUALS TO OVER-MARKS OF CONSPECIFICS DIFFERS BETWEEN TWO SPECIES OF MICROTINE RODENTS

Michael H. Ferkin

Department of Biology
Ellington Hall
University of Memphis
Memphis, TN 38152

INTRODUCTION

Animals communicate to conspecifics their identity, sexual condition, and features of their quality through displays and signals. In general, these signals will be used to attract opposite-sex conspecifics and to compete with same-sex conspecifics. For mammals, these signals are deposited as scent marks. In that these scent marks are usually deposited along well-traversed paths and on prominent objects in an area, conspecifics may deposit their scent marks on top of those deposited previously by conspecifics. The literature is replete with accounts of over-marking and its potential functions for a wide array of mammals (Lazaro-Perea *et al.*, 1999; Hurst and Rich, 1999; Johnston, 1999; Roberts and Dunbar, 2000), but the literature does not provide many details of how conspecifics respond to over-marks.

The handful of detailed studies of over-marking, conducted principally on golden hamsters and voles, led to the hypothesis that over-marking may be a competitive form of olfactory communication (Johnston *et al.*, 1994). This competition hypothesis was based on the finding that golden hamsters (*Mesocricetus auratus*), meadow voles (*Microtus pennsylvanicus*), and prairie voles (*M. ochrogaster*) exposed to an over-mark of two same-sex scent donors later display a better selective memory and show a preference for the top-scent mark than the bottom-scent mark (Johnston *et al.*, 1994, 1995, 1997a,b; Woodward *et al.*, 1999). Johnston and colleagues concluded that individuals exposed to an over-mark behave as if the mark of the top-scent donor was more important than that of the bottom-scent donor and that the top-scent donor may be more likely than the bottom-scent donor to communicate its chemical message to conspecifics (Johnston *et al.*, 1994, 1997a,b; Ferkin *et al.*, 1999). These conclusions have been supported and extended by recent studies on golden hamsters, meadow voles, and prairie voles, and has led to the hypothesis that over-marking provides advantages in information transfer to conspecifics by the top-scent donor but not by the bottom-scent donor (Wilcox and Johnston, 1995; Ferkin, 1999a,b; Johnston, 1999).

In this paper, I extend the competition hypothesis by suggesting that 1) individuals investigating an over-mark may be provided with a record of an interaction between two conspecifics in a given area and 2) that the response of an individual to an over-mark is associated with its social biology. I do so by comparing the responses to conspecific over-marks by meadow voles, a promiscuous rodent that has infrequent interactions with the same opposite-sex conspecific, and prairie voles, a monogamous rodent that has frequent interactions with the same-opposite-sex conspecific, presumably its mate (Boonstra et al., 1993; Carter and Getz, 1993).

HOW DO PRAIRIE AND MEADOW VOLES RESPOND TO OVER-MARKS OF TWO SAME-SEX CONSPECIFICS?

Recent experiments on prairie voles and meadow voles suggest that responding to an over-mark may be more complex than originally thought (Johnston et al., 1995). For example, meadow voles and prairie voles differ in how they respond to the bottom-scent of an over-mark. Meadow voles exposed to an over-mark of two same-sex scent donors spent similar amounts of time investigating the mark of the bottom-scent donor as compared to that of a novel-scent donor, a donor whose scent was not part of the over-mark (Ferkin et al., 1999). In contrast, prairie voles spent more time investigating the novel-scent donor's mark than the bottom-scent donor's mark of this type of over-mark (Woodward et al., 1999). These results suggested the following. For meadow voles, the bottom-scent mark and the novel-scent mark had roughly a similar value attached to them. For prairie voles, the bottom-scent mark appeared to have a lower value attached to it relative to that of the novel-scent mark. Taken together, it appears that it may be more costly for prairie voles than it is for meadow voles to be the bottom-scent donor of an over-mark (Woodward et al., 1999).

HOW LONG DO PRAIRIE AND MEADOW VOLES PREFER THE TOP-SCENT OF AN OVER-MARK?

Species differences existed in the length of time that meadow voles and prairie voles maintained a preference for the mark of the top-scent donor relative to that of the bottom-scent donor. After a single exposure to an over-mark, male and female meadow voles showed a preference for the top-scent mark over the bottom-scent mark up to 48 hours later. The duration of preference for the top-scent donor's mark over the bottom-scent donor's mark was considerably shorter for prairie voles than it was for meadow voles. Male prairie voles showed this type of preference for 24 hours, whereas female prairie voles showed this preference for only 12 hours (Ferkin, Leonard, Bartos, and Schmick, unpublished data). The difference between meadow voles and prairie voles in how long they maintained a preference for the mark of the top-scent donor over that of the bottom-scent donor may be associated with its selective memory for the top-scent donor (Johnston and Bhorade, 1998; Ferkin et al., 1999), and the likelihood that an individual has of having repeated encounters with that donor's scent marks (Ferkin, 1999a,b). In that male and female prairie vole have regular and frequent encounters with the scent marks of their mate (Carter and Getz, 1993), there may be no need for them to maintain a relatively long selective memory for the scent marks of that scent donor. A short selective memory by prairie voles for the top-scent donor's mark may allow them to pay particular attention to the scent marks of their mates and other opposite-sex conspecifics that they may encounter in their territories. If its mate's marks are over-marked by an opposite-sex conspecific, it may indicate to a pairbonded prairie vole that its mate is absent. Thus, the relatively short selective memory by female prairie voles for the top-scent male's mark may discourage that male from leaving its territory for long periods of time (Ferkin *et al.*, unpublished data). In contrast to prairie voles, meadow voles maintained a relatively long selective memory for the top-scent donor of an opposite-sex over-mark. A long selective memory for the top-scent donor's marks may allow meadow voles to distinguish between that individual's marks and those of another opposite-sex conspecific, thereby increasing the likelihood that they can mate with multiple partners.

HOW DO PRAIRIE AND MEADOW VOLES RESPOND TO MIXED SEX OVER-MARKS?

Species differences existed in the manner in which meadow voles and prairie voles responded to an over-mark containing the marks from a male donor and a female donor (mixed-sex over-mark). Experiments by Woodward *et al.* (1999, 2000) showed that after exploring an over-mark in which the top-scent donor was the same-sex as the investigating animal and the bottom-scent donor was the opposite-sex of the investigating animal, male and female meadow voles spent significantly more time with the odor of the bottom-scent donor than that of the top-scent donor. Prairie voles showed no preference for either the top scent or the bottom scent. Thus, meadow voles and prairie voles differ in how they respond to the top and bottom-scent donors of a mixed-sex over-mark. Apparently, meadow voles use the sex of the donor rather than the position of its mark in the over-mark to respond preferentially to it. Meadow voles prefer the scent mark of the opposite-sex conspecific, independent of whether it is the top or bottom-scent mark in an over-mark. In that meadow voles also prefer the mark of an opposite-sex bottom-scent donor suggests that a mixed-sex over-mark may be used to attract multiple mates (Woodward *et al.*, 2000). The opposite inference may be drawn after we examine the responses of prairie voles to such an over-mark. Prairie voles seem to rely on both the position and the sex of the donor's mark in an over-mark. That is, prairie voles respond preferentially to a particular scent donor if it is both the opposite-sex and if its mark is on top. In addition, male prairie voles prefer the mark of an opposite-sex novel-scent donor as compared to that of an opposite-sex bottom-scent donor. This response suggests male prairie voles may use this type of over-marking as a form of mate guarding (Woodward *et al.*, 2000). Apparently, the manner in which conspecifics respond to the marks of the top and bottom-scent donors may be affected by both the sex of the investigating animal and the sex of the top and bottom-scent donors.

SPECIES DIFFERENCES IN OVER-MARKING - A VOLE'S TALE

The species differences that exist in the manner in which meadow voles and prairie voles differ respond to over-marks may reflect the unique social biology of each species. Prairie voles form a pairbond and share in nest defense with their mates (Carter and Getz 1993). Within a given area, their own scent marks or those of their mate's should be the top-scent mark of a same sex or a mixed-sex over-mark. If so, the top-scent donor may be able to signal its residency in an area to intruders (Woodward *et al.*, 1999). This hypothesis is supported by the fact that prairie voles display relatively short-term memory for the top-scent donor and that they treat the bottom-scent donor as if its mark has been devalued relative to that of a novel donor. If one of the residents' scent marks are over-marked and not attended to, it may signal to the other resident that its mate may no longer be present (Woodward *et al.*, 2000). For prairie voles, mixed-sex and same-sex over-marking may be associated with the formation and maintenance of the male-female pair bond and in territory defense. In contrast to prairie voles, for meadow voles same-sex and mixed-sex over-marking may be closely aligned with indicating recency in an area and attracting multiple partners (Woodward et al., 2000). Meadow voles do not form a pairbond and opposite-sex conspecifics do not share in nest defense (Boonstra et al., 1993). Within a given area, meadow voles may likely encounter over-marks containing the scent marks of past and present visitors. Previous mates or competitors may have had their scent marks over-marked by more recent visitors to that area. A relatively long selective memory for the top-scent donor suggests that meadow voles may selectively attend to the scent marks of potential mates and competitors over those of previous mates and competitors.

ACKNOWLEDGMENTS

I thank R. Johnston, S. Leonard, R. Woodward, and J. Wolff for commenting on earlier drafts of this manuscript. This research was supported by National Science Foundation grant IBN 9421529 and National Institutes of Aging grant AG16594-01.

REFERENCES

Boonstra, R., Xia, X., and Pavone, L., 1993, Mating system of the meadow vole, *Microtus pennsylvanicus*, *Behav. Ecol.* **4**:83-89.
Carter, C. S., and Getz, L. L., 1993, Monogamy and the prairie vole, *Sci. Amer.* **268**:100-106.
Ferkin, M. H., 1999a, Scent over-marking as a competitive form of chemical communication in voles, in: *Advances in Chemical Signals in Vertebrates, Number 8*. (R. E. Johnston, D. Müller-Schwarze, and P. Sorenson, eds.), Plenum Press, New York, pp. 239-246.
Ferkin, M. H., 1999b, Meadow voles, *Microtus pennsylvanicus* (Rodentia), over-mark and adjacent mark the scent marks of same-sex conspecifics, *Ethology* **105**:825-837.
Ferkin, M. H., Dunsavage, J., and Johnston, R. E., 1999, What kind of information do meadow voles, *Microtus pennsylvanicus*, use to distinguish between the top and bottom scent of an over-mark? *J. Comp. Psych.* **113**:43-51.
Hurst, J. L., and Rich, T. J., 1999, Scent marks as competitive signals of mate quality, in: *Advances in Chemical Signals in Vertebrates, Number 8*. (R. E. Johnston, D. Müller-Schwarze, and P. Sorenson, eds.), Plenum Press, New York, pp. 209-225.
Johnston, R. E., 1999, Scent over-marking: how do hamsters know whose scent is on top and why should it matter? in: *Advances in Chemical Signals in Vertebrates, Number 8*. (R. E. Johnston, D. Müller-Schwarze, and P. Sorenson, eds.), Plenum Press, New York, pp. 227-238.
Johnston, R. E., and Bhorade, A., 1998, Perception of scent over-marks: novel mechanisms for determining which individual's mark is on top, *J. Comp. Psych.* **112**:1-14.
Johnston, R. E., Chiang, G., and Tung, C., 1994, The information in scent over-marks of golden hamsters, *Anim. Behav.* **48**:323-330.
Johnston, R. E., Munver, R., and Tung, C., 1995, Scent counter marks: selective memory for the top scent by golden hamsters, *Anim. Behav.* **49**:1435-1442.
Johnston, R. E., Sorokin, E. S., and Ferkin, M. H., 1997a, Scent counter-marking by male meadow voles: Females prefer the top-scent male, *Ethology* **103**:443-453.
Johnston, R. E., Sorokin, E. S., and Ferkin, M. H., 1997b, Female voles discriminate males' over-marks and prefer top-scent males, *Anim. Behav.* **54**:679-690.
Lazaro-Perea, C., Snowdon, C. T., and de Fatima Arruda, M., 1999, Scent marking behavior in wild groups of common marmosets (*Callithrix jacchus*), *Behav. Ecol. Sociobiol.* **46**:313-324.
Roberts, S. C., and Dunbar, R. I. M., 2000, Female territoriality and the function of scent marking in a monogamous antelope (*Oreotagus oreotagus*), *Behav. Ecol. Sociobiol.* **47**:417-423.
Wilcox, R. M., and Johnston, R. E., 1995, Scent counter-marks: specialized mechanisms of perception and response to individual odors in golden hamsters (*Mesocricetus auratus*), *J. Comp. Psych.* **109**:349-356.
Woodward, R. L., Jr., Schmick, M. K., and Ferkin, M. H., 1999, Response of prairie voles, *Microtus ochrogaster*, to scent over-marks, *Ethology* **105**:1009-1017.
Woodward, R. L., Jr., Bartos, K., and Ferkin, M. H., 2000, Meadow voles and prairie voles differ in their responses to over-marks from opposite- and same-sex conspecifics, *Ethology* **106**:979-992.

THE SECRETION OF THE SUPPLEMENTARY SACCULI OF THE DWARF HAMSTER *PHODOPUS CAMPBELLI*

Raimund Apfelbach,[1] Urte Schmidt,[1] and Nina Y. Vasilieva[2]

[1]Institute of Zoology, University of Tübingen
Auf der Morgenstelle 28, 72076 Tübingen, Germany
[2]Institute for Ecology and Evolution, Russian Academy of Sciences
Leninsky pr. 33, Moscow 117071, Russia

INTRODUCTION

Supplementary sacculi (sacs) at the openings of the cheek pouches are a unique anatomical structure found only in two species of *Phodopus* hamsters: *Phodopus sungorus* and *P. campbelli*, the subject of this report. Vasilieva and Feoktistova (1993) demonstrated the importance of the creamy secretion from these sacs in *P. campbelli* for normal growth, development, and survival during the first month of life. Aside from its possible role in nutrition, the strong smelling secretion is suspected to play roles in intraspecific communication (Feoktistova and Vasilieva, 1995) and in defensive interactions (Krischke, 1986). We propose here that this odorous secretion is also used to scent mark cached foods with odors that function to facilitate the food's retrieval.

Our hypothesis is based on the observation that as foraging hamsters collect food items in their cheek pouches and transport them to food caches, the transported foods come in contact with the secretion of the supplementary sacs and thus become odor-marked. When searching to retrieve the cached foods, the deposited scent might help guide the searching animals in finding the foods. To address this hypothesis, we conducted behavioral tests using foods „marked" with the secretion of the sacs. In addition, we conducted a chemical analysis of the secretion in order to identify its volatile components, which we then used in behavioral tests.

MATERIAL AND METHODS

The subjects in these experiments were 6 adult males and 7 adult females. Animals were from laboratory stock. Animal were housed in groups of 2 to 5 individuals in cages of size 35 x 55 x 20 cm. Cages were equipped with small tubes and other shapes for refugia, and we used wood chips as the cage substrate. To exclude dietary effects on the composition of the secretion of the supplementary sacs (Salamon, 1995), all animals were fed the same diet, consisting of special hamster chow, fresh fruits and vegetables, and ad libitum access to water. The colony was maintained on a 14L:10D light:dark cycle with temperature maintenance at 20±2°C.

The supplementary sacs are easily accessible. Samples of the secretion were obtained by carefully inserting a cotton-tipped applicator stick into the opening of the sac and slowly rotating it inside the sac as the secretion was absorbed into the cotton.

For the behavioral experiments, subjects were placed into the middle of a 6-arm-maze (Figure 1) containing a substrate of wood chips. Into four of the six arms we placed sunflower seeds which had been „scented" by mixing them with the secretions of: 1) the subject being tested, 2) a cagemate of the subject, 3) a conspecific that was not familiar to the subject, and, 4) an unfamiliar heterospecific (*P. sungorus*). Into the fifth arm we placed unscented sunflower seeds, and the sixth arm was left without seeds. For each trial the scented seeds were randomly distributed between the arms. In the behavioral tests, we noted the percentage of each trial spent in each arm as an indicator of odor discriminations by the subjects.

For the chemical analyses of the secretions, we used a GC/MS-System (Headspace-GC/MS, Headspace autosampler HP 7694). The system consisted of the gas chromatograph HP 6890 with the mass-selective detector HP 5973, fitted with an HP-VOC column. The chemical constituents identified in this way were later used as odorants in the 6-arm-maze for behavioral testing.

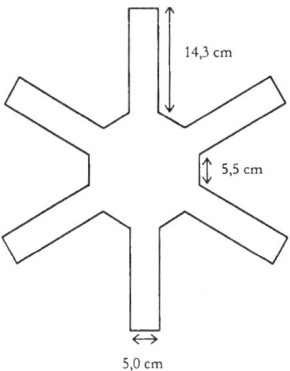

Figure 1. 6-arm-maze used for behavioral tests.

RESULTS

The results of the first set of six-arm maze tests are shown in Figure 2. When exposed simultaneously to maze arms containing sunflower seeds scented with different secretions versus an empty arm or one with unscented sunflower seeds, the subjects spent most of the testing time in the arms containing the scented food. The subjects appeared to discriminate between food scented with the secretion of their own species versus that scented by secretion from a heterospecific, i.e., *P. sungorus* ($p<0.001$; Man-Whitney U-test); in fact the latter was visited for approximately the same percentage of the test as unscented seeds. In spite of the trend that seeds marked with the secretion of a known individual tended to be most attractive, the statistical analysis forces us to conclude that the subjects did not discriminate between arms containing food marked with their own secretion from arms with food marked with the secretions of family members or unfamiliar conspecifics.

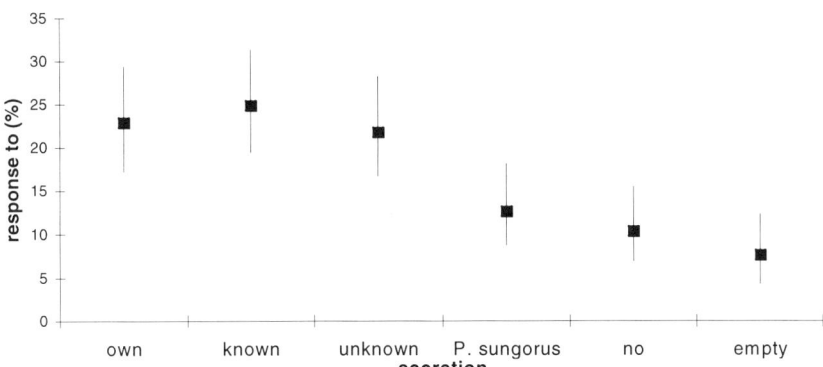

Figure 2. Subjects spent a greater percentage of the testing time in the three arms of the maze which contained seeds scented by conspecific sac secretions, compared to time spent in arms containing food.

Chemical analyses of the odorous secretions from several individuals showed they had a similar qualitative composition. There were six compounds, listed in Table 1 and Figure 3, which were found reliably between sessions of chromatographic analysis. Table 1 reveals some of the interindividual variation in the relative proportions of these six odorous substances, and it is possible that such interindividual differences might be the basis for individual odor discrimination.

Subsequent behavioral tests revealed that none of the pure odorants elicited the same evidence of discriminations that had been evidenced in the subjects' responses to native secretions (Figure 4). We, therefore, suspect that these compounds only elicit such discriminations when offered in the appropriate mixture.

Table 1. Retention times for the different odorants found in the secretion of the supplementary sacculi of *P. campbelli*. Data were obtained from three different adult males (No 1 - o 3).

Peak	Retention time (min)	Odorant	Intensity (%) of the total area		
			Male 1	Male 2	Male 3
1	13.13	Acetic acid	7.87	5.96	5.21
2	16.27	Isobutyric acid	10.57	9.90	10.46
3	17.20	Butyric acid	45.21	45.58	48.02
4	18.89	Isovaleric acid	11.62	13.39	13.81
5	19.19	2-Methylbutyric acid	12.57	15.41	12.24
6	23.46	Phenol	5.63	9.44	9.49

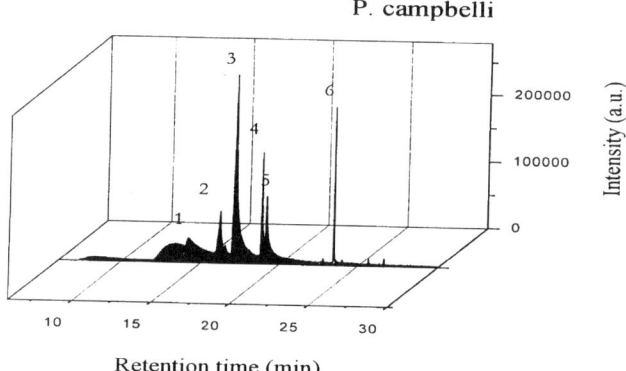

Peak 1: Acetic acid
Peak 2: Isobutyric acid
Peak 3: Butyric acid
Peak 4: Isovaleric acid
Peak 5: 2-Methylbutyric acid
Peak 6: Phenol

Figure 3. Chromatogram of the secretion from the supplementary sacs (a.u. = arbitrary units).

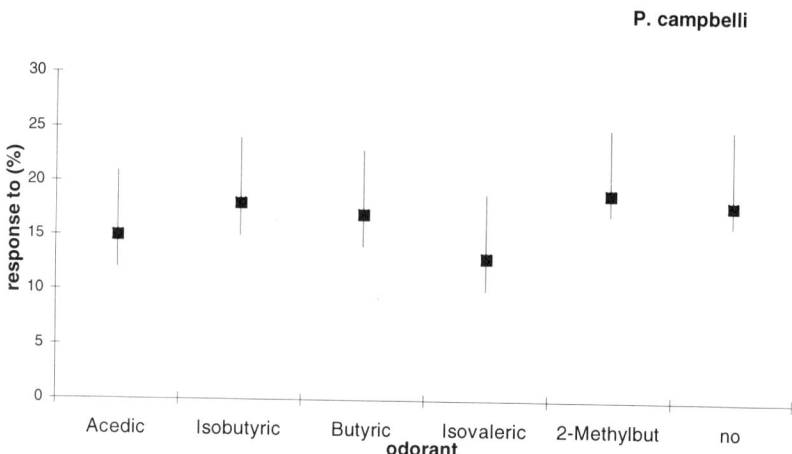

Figure 4. When simultaneously presented in the 6-arm-maze, none of the pure odorants elicited evidence of discrimination, ie., statistically, responses to all substances were equivalent. Mean values and confidence intervals are given.

DISCUSSION

Our data bring new understanding to the role of the supplementary sacs at the opening of their cheek pouches in male and female *Phodopus campbelli* (Sokolov et al.,1991). Both sexes feed their young with the secretions from these unique structures. Since hamsters collect food items in their cheek pouches and transport them to food caches, we suspected an additional function of the strong smelling secretion. While in the cheek pouch food items might become odorized in ways that function to guide the animals back to the cache of scented food. Indeed, we demonstrated that *P. campbelli* subjects spent proportionately more time in proximity to sunflower seeds odorized with the secretion of their own species, compared to unscented seeds or those scented with heterospecific odors. This conspecific discrimination occurs in spite of the fact that the closely related species *P. sungorus* (Wynn-Edwards and Liske, 1986; Sokolov and Vasilieva, 1993) also produces a similar secretion in its supplementary sacs.

Our chemical analyses (headspace chromatography) revealed qualitative similarities in the composition of the sac secretions from different animals. We suggest that the small, quantitative, inter-individual, differences in proportions of constituent volatiles might provide the basis for individual odor recognition. However, the chemical analyses failed to provide reliable sex or individual identity differences in composition that would allow us to identify the sex or individual identity based on GC/MS data. Individual odors, especially warning odors, often consist of many components (Li *et al.*, 1997). To our surprise, the chromatograms revealed only six peaks, attributable to six different odorants. Five of these six odorants are known to be involved in the chemical communication of vertebrates (Stoddart, 1980) while the sixth compound, phenol, is regarded as a defensive odorant.

Our work on the chemical composition of secretions is as yet preliminary. We do not know what role is played by the microflora (Sokolov *et al.*, 1995; Stoddart, 1995) associated with the glandular region in producing odorous compounds.

ACKNOWLEDGMENTS

We want to acknowledge the valuable help given to us by our colleagues Dr. W. Göpel and Dr. U. Weimar of the Institute of Physical and Theoretical Chemistry, Tübingen University.

REFERENCES

Feoktistova, N. Yu., and Vasilieva, N. Yu., 1995, The communicative function of supplementary sacculi at the openings of cheek pouches of Campbell's hamster (*Phodopus campbelli*) and dwarf hamster (*Phodopus sungorus*), *Zoologicheskii Zhurnal* **74**:136-143 (in Russian).

Krischke, N., 1986, Die Mundwinkeldrüsen des Dschungarischen Zwerghamster *Phodopus sungorus*: Struktur und Funktion, *Säugetierkundliche Mitteilungen* **33**:195-204.

Li, G., Roze, U., and Locke, D. C., 1997, Warning odor of the North American porcupine (*Erethizon dorsatum*), *J. Chem. Ecol.* **23**:2737-2754.

Salamon, M., 1995, Seasonal, sexual and dietary induced variations in the sternal scent secretion in the brushtail possum (*Trichosurus vulpecula*), in: *Chemical Signals in Vertebrates VII*, (R. Apfelbach, D. Müller-Schwarze, K. Reutter, and E. Weiler, eds.), Elsevier Science Ltd., Oxford, pp. 211-221.

Sokolov, W. E., Vasilieva, N. Yu., Demina, N. J., and Feoktistova, N. Yu., 1991, Supplement saccules at the entrance of the cheek pouches in the Campbell hamster (*Phodopus campbelli*), Thomas, 1905; Cricetidae, Rodentia, *Dokl. Akad. Nauk SSSR* **316**:479-483 (in Russian).

Sokolov, W. E., and Vasilieva, N. A., 1993, Hybridological analysis confirms species independence of Phodopus sungorus (Pallas, 1773) and Phodopus campbelli (Thomas, 1905), *Dokl. Akad. Nauk* **332**:120-123 (in Russian).

Sokolov, W. E., Ushakova, N. A., Feoktistova, N. Yu., and Vasilieva, N. Yu., 1995, Bacteria in the secrete of auxiliary sacs in the opening of buccal pockets of Campbell hamsters, *Microbiologia* **64**:216-221 (in Russian).

Stoddart, D. M., 1980, The Ecology of Vertebrate Olfaction, Chapman and Hall.

Stoddart, D. M., 1995, Olfactory biology, social behaviour and ecophysiology: an integrated marsupial study, in: *Chemical Signals in Vertebrates VII*, (R. Apfelbach, D. Müeller-Schwarze, K. Reutter, and E. Weiler, eds., Elsevier), Science Ltd., Oxford, pp. 201-210.

Vasilieva, N. Yu., and Feoktistova, N. Yu., 1993, The function of supplementary sacculi at the openings of cheek pouches of Campbell's hamster, *Phodopus campbelli* (Cricetidae, Rodentia), *Zoologicheskii Zhurnal* **72**:103-113 (in Russian).

Wynne-Edwards, K. E., and Liske, R. D., 1986, Behavioral interactions differentiate Djungarian (*Phodopus campbelli*) and Siberian (*Phodopus sungorus*) hamsters. *Can. J. Zool.* **65**:2229-2235.

THE ROLE OF URINARY PROTEINS AND VOLATILES IN COMPETITIVE SCENT MARKING AMONG MALE HOUSE MICE

Rick E. Humphries,[1] Duncan H.L. Robertson,[2] Charlotte M. Nevison,[1] Robert J. Beynon,[2] and Jane L. Hurst[1]

[1] Animal Behaviour Group and [2] Protein Function Group
Faculty of Veterinary Science
University of Liverpool
Leahurst, Neston, CH64 7TE, UK

1. INTRODUCTION

Male house mice (*Mus domesticus*), like many other male mammals, advertise their competitive dominance and ability to defend territories by depositing numerous urinary scent marks throughout their territory (reviewed by Ralls, 1971; Johnson, 1973; Gosling, 1982, 1990; Hurst, 1987). Male mice also increase their rate of marking near any competing scent marks from other males, a behaviour termed counter-marking (Hurst, 1990, 1993; Hurst and Rich, 1999). Because only those males successfully dominating their territory can ensure that their own marks are always the freshest and predominant in the area, other males can use the temporal and spatial deposition dynamics of male scent marks to assess territory ownership and competitive ability (see Hurst *et al.*, this volume). Perhaps more importantly, female mice can also use these scent marks to assess the quality of potential mates, preferring dominant male territory owners that counter-mark scent mark challenges from competitors and which ensure that their own marks are always the freshest (Rich and Hurst, 1999).

Both inbred (Finlayson *et al.*, 1963) and wild (Robertson *et al.*, 1997; Pes *et al.*, 1999) male mice produce a high concentration of proteins termed major urinary proteins (MUPs) in their urine. MUPs bind male signalling volatiles, chiefly 2-sec-butyl-4,5-dihydrothiazole (thiazole) and 3,4-dehydro-exo-brevicomin (brevicomin; Bacchini *et al.*, 1992; Robertson *et al.*,1993), and slowly release these volatiles from urinary scent marks (Hurst *et al.*, 1998). These protein-ligand complexes have a number of reproductive priming effects (Novotny *et al.*, 1999; Brennan *et al.*, 1999).

Humphries *et al.*, (1999) examined the roles of urinary proteins, male signalling volatiles bound to proteins, and unbound urinary volatiles in stimulating competitive scent counter-marking among territorial male mice. This study suggested that the urinary proteins in an intruder's scent mark, largely MUPs, stimulate competitive scent marking between male mice. Surprisingly, though, this study also suggested that there was an increase rather than a decrease in competitive counter-marking towards aged protein scent marks that had lost their volatile ligands. Here we review this study and report a further study that investigates the effect of scent mark age on competitive scent counter-marking.

2. METHODS

2.1. Urine Fractionation and Analysis

We tested wild-caught males using pooled urine collected from 10 singly housed male inbred BALB/c mice (Humphries et al., 1999). Test urine was split into three fractions according to molecular weight by size-exclusion chromatography. These three fractions were then analysed for protein, creatinine, thiazole and brevicomin. The protein concentration in the three pools was measured by using the Coomassie Plus assay (Pierce Chemicals, Chester, U.K.). Creatinine concentration, used to determine the elution position of low molecular weight urinary components, was measured by the alkaline picrate assay (Sigma Chemicals, Poole, Dorset, U.K.). To determine the concentration of thiazole and brevicomin in the fractionated urine, and also to measure the loss of these MUP ligands from deposited samples of urine, we used static headspace sampling and GC/MS analysis of volatiles.

Pool 1 was found to contain all the urinary proteins (over 99% of which are MUPs) and most of the thiazole and brevicomin bound to the proteins as ligands. It had a relatively weak 'mousey' odour to the human nose. Pool 2 contained the urinary creatinine and the unbound portion of the thiazole and brevicomin and had an acrid 'mousey' smell. Pool 3 had no detectable proteins or ligands, but contained the lowest molecular weight components including unbound signalling volatiles. It possessed a weaker acrid 'mousey' smell. The amount of free thiazole and brevicomin was much lower than in fresh wet urine (Robertson et al., this volume), probably because these molecules bind to the gel filtration matrix.

2.2. Counter-marking Trials

In each experiment, 11 or 12 wild-caught male house mice were tested. These were housed singly in enclosures (1.2 x 1.1 x 0.8 m) containing a nestbox and a food and water station. We gave the males a choice of four stimuli introduced simultaneously into their home enclosures at the beginning of the dark period when they were most active. These four stimuli consisted of three different urine stimuli and a column buffer control. For each stimulus, two 50µl streaks were deposited in the centre of a square of Benchkote (three 50µl streaks were used in the age of marks experiment). The four stimuli were each placed against a different side wall of a male's enclosure, allocated randomly. Investigatory behaviour was video recorded under infra-red lighting over the first 30 minutes, and the number of scent marks deposited on each of the stimulus squares was counted after 10 h. Counter-marking was measured as the number of marks deposited in response to a urine stimulus minus response to the column buffer control. This took into account scent marking in response to an introduced surface. Thus, counter-marking was assessed as a significant increase in the number of marks deposited in response to a urine stimulus.

3. RESULTS AND DISCUSSION

3.1. Response to Different Urine Pools

To establish whether protein-ligand complexes or free volatiles stimulate competitive counter-marking among male mice, we presented males with a choice between the three separate urine pools and a column buffer control (Humphries et al., 1999). Each male was tested separately with fresh stimuli, and with stimuli deposited 24 h previously.

Only pool 1 (protein) stimulated significantly more counter-marking than the column buffer control, when tested with either fresh or aged stimuli (Figure 1). Neither of the two free-volatile fractions (low molecular weight components) stimulated marking that was significantly different from the control. In total, mice deposited fewer marks when tested with urine fractions that had been aged by 24 h compared to fresh test stimuli (306 ± 75 marks in fresh trials, 240 ± 70 marks in aged trials, Wilcoxon matched sets test, $Z = -2.0$, $P < 0.025$). Corresponding to this bias in urine-marking behaviour, males showed a strong initial attraction to investigate pool 1, visiting significantly more quickly than control in the first 30 min of exposure, even when the stimulus had been deposited 24 h previously (see Humphries et al., 1999). All males consistently visited pool 1 within a few seconds of the start of the test in contrast to the very high variance in latency to visit the control stimulus. This suggests that the mice were attracted from a distance by volatiles emanating from the proteins, and that volatiles continued to be released and were clearly detected 24 h after the urinary protein fraction had been deposited.

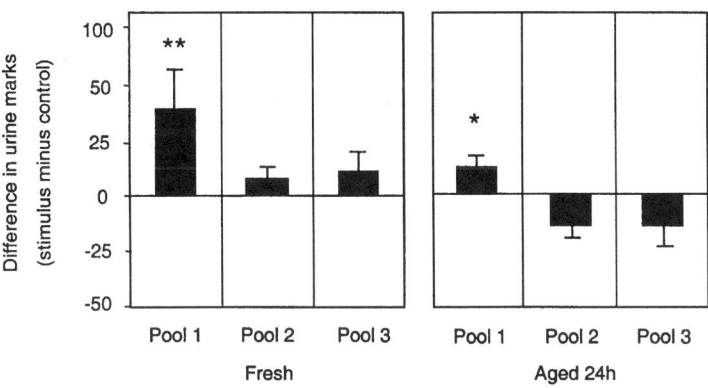

Figure 1. The number of counter-marks in response to urine pools 1, 2 and 3 minus control, when fresh or aged 24 h. Wilcoxon signed-ranks exact test p-values: * $p < 0.05$, ** $p < 0.025$. Data are means $\pm$ 1 SE ($N=11$). From data Humphries et al. (1999).

3.2. Effect of Protein Ligands

To test whether competitive counter-marking is stimulated by male signalling volatiles bound to the MUPs or by the proteins themselves, we manipulated the concentration of ligands in the protein fraction either by natural evaporation over 7 days, or by displacement with menadione (Hurst et al., 1998; Robertson et al., 1998). Mice were then presented with three different protein stimuli (fresh pool 1, pool 1 aged by 7 days, or pool 1 mixed with menadione) along with a column buffer control (Humphries et al., 1999).

The number of urine marks deposited over a 10 h period confirmed that the three urinary protein stimuli were counter-marked, stimulating significantly more scent marking than the control (Figure 2; Wilcoxon matched-sets test, $Z = -2.32$, $P < 0.025$). There was also a difference in the level of counter-marking on the three protein stimuli, suggesting that the presence of protein ligands in the marks play a role in stimulating counter-marking. However, this was not in the direction expected (Figure 2). Males deposited more marks when the protein had been air-dried and aged for 7 days than when it was fresh. Males also tended to deposit more marks near the protein treated with menadione to displace the ligands. These results suggest that males were able to detect and recognize the protein fraction of an intruder's scent mark even when few, if any, volatile ligands continued to be released from the proteins, and that the males preferred to deposit more counter-marks near protein stimuli that contained very few volatile ligands.

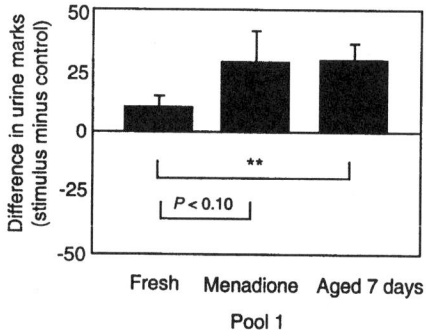

Figure 2. The number of counter-marks in response to urine pool 1, when presented fresh, mixed with menadione, or aged 7 days (stimulus minus control). Wilcoxon signed-ranks exact test p-values: * $p < 0.05$, ** $p < 0.025$. Data are means ± 1 SE ($N=12$). From data Humphries et al. (1999).

3.3. Age of Marks

We were surprised that the aged intruder scents stimulated a stronger counter-marking response than the fresh scents. This was not because males were too intimidated to counter-mark a fresh scent mark challenge from an intruder since fresh urinary protein clearly stimulated a significant increase in scent marking, especially when presented alone (see Figure 1). It seems very unlikely that the aged signal presented a greater competitive challenge to the resident's own scent marks than a fresh signal and thus stimulated a stronger counter-marking response. However, Humphries et al., (1999) presented both the aged protein fraction and protein from which ligands had been displaced simultaneously with a fresh protein stimulus from an intruding male. The presence of the fresh stimulus, replete with ligands (see Robertson et al., this volume), might have influenced their response. Further, Humphries et al. (1999) only tested the effect of ageing on the response to the protein fraction of an intruder's urine. It is possible that other labile components in whole urine provide information on the age of a scent mark that modulates the response to the protein fraction. We thus conducted another experiment to examine whether aged urine stimulates counter-marking in the absence of fresh scent, and whether the response to the protein fraction is modulated by other urinary components.

To establish whether mice respond strongly to aged scents in the absence of fresh stimuli, we presented males with urine stimuli that were all aged by 24 h, or all aged by 7 days. In case low molecular weight components in whole urine provide information on the age of the mark and modulate response to the protein fraction, mice were presented with a mixture of the high and lower molecular weight urine pools (pools 1+2+3) as well as the protein fraction alone (pool 1). Hence four stimuli were presented simultaneously; a mixture of all pools (1+2+3), proteins and their bound ligands (pool 1), lower molecular weight components alone (pools 2+3), and a column buffer control.

When presented with urine stimuli aged by 24 h, males counter-marked those stimuli that contained protein (i.e. pool 1 and pools 1+2+3) significantly more than the column buffer control (Figure 3). However, when the urine stimuli were aged by 7 days, none stimulated significantly more counter-marking than the control (Figure 3) and relatively few scent marks were deposited in this trial. Mice deposited significantly fewer marks in total when tested with urine stimuli aged by 7 days, compared with urine stimuli aged by 24 h (292 ± 55 marks on all four stimuli combined in aged 24 h trials, 242 ± 46 marks in aged 7 days trials; Wilcoxon signed-ranks test: $Z = -2.12$, $P < 0.05$). It was clear that marking on each of the urine stimuli aged by 7 days was low, with means of 50-69 marks deposited on each stimulus. This was very similar to the number of marks deposited on control stimuli in other tests (means of 59-60 marks per control stimulus). Where mice deposited significantly more marks on a urine stimulus than on a column buffer control, the level of marking was in the region of 80-100 marks per urine stimulus. The relatively low marking response shown to urine stimuli aged by 7 days in this experiment (Figure 3) thus contrasts with the very strong response to pool 1 aged by 7 days when this was presented alongside a fresh protein stimulus from an intruder (Figure 2).

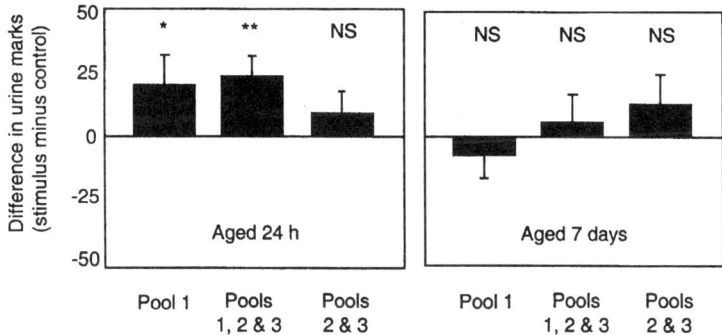

Figure 3. The number of counter-marks in response to urine pools 1 (protein), 1+2+3 (protein and lower molecular weight fractions), and 2+3 (lower molecular weight fractions only) minus response to control column buffer, when urine deposits were aged by 24 h or 7 days. Wilcoxon signed-ranks exact test p-values: * $p < 0.05$, ** $p < 0.025$. Data are means $\pm$ 1 SE ($N=12$).

Although the mice failed to counter-mark urine stimuli aged by 7 days, males visited pools 1+2+3 (mixture of proteins and lower molecular weight components, including volatiles) significantly more times than the column buffer control ($Z = -1.87$, $P < 0.05$). They also visited this stimulus significantly quicker than the control (latency to first visit pools 1+2+3: 24 ± 9 s; control: 71 ± 27 s; $Z = -2.12$, $P < 0.05$), suggesting that they could detect volatiles emanating from this stimulus at a distance even after 7 days. However, the protein alone (pool 1) no longer stimulated such attraction.

4. CONCLUSIONS

The protein fraction of male mouse urine contains the chemical signals that stimulate competitive counter-marking. This fraction consists predominantly of the major urinary proteins (MUPs) and a number of volatile male signalling molecules that are bound in the central cavity of the proteins (Bacchini et al., 1992; Robertson et al., 1993). MUPs provide a slow release of these male signalling volatiles (mostly a thiazole and a brevicomin) from male scent marks (Hurst et al., 1998). These volatiles attract attention to the scent marks, even 24 h after deposition, and provide information that the scent mark is from another male mouse. The protein fraction also binds and releases other volatiles that provide information such as the MHC type of the scent owner (Singer et al., 1993, 1997). Once approached, close investigation of the non-volatile proteins or protein-ligand complexes in the intruder's scent mark stimulates a local increase in the number of scent marks deposited by the resident male to counter-mark the stimulus. This is presumably because the male has detected that the scent mark is not one of his own but has been deposited within his territory and thus represents a challenge to his own scent mark advertisement of competitive ability (see Hurst et al., this volume).

In our earlier experiments (Humphries et al., 1999), we found that males counter-marked the fresh urinary protein scents of intruders but preferred to deposit more of their counter-marks near proteins that had lost most of their ligands through prolonged ageing or through displacement with menadione. However, the additional experiment reported here shows that males were only stimulated to counter-mark (i.e. increase their rate of scent marking relative to a clean control stimulus) in the presence of a fresh or 24 h aged protein stimulus. Proteins aged by 7 days did not stimulate any increase in marking, although these were detected and investigated, especially in the presence of the lower molecular weight fractions of urine. To check that the males used had not habituated to the stimulus, we subsequently tested their response to a fresh stimulus and found normal counter-marking behaviour (unpublished data).

Thus males only respond competitively in the presence of a relatively fresh scent mark challenge from an intruder. It is likely that such aged scents of intruders present little competitive challenge within a male's scent-marked home territory since the territory will already contain much resident's scent that is fresher than the intruder's scent through normal scent mark replenishment (Hurst, 1989). The intruder's scent is thus already effectively 'counter-marked' by the resident's fresher scent. Perhaps more importantly, females can easily discriminate between scent marks from two competing males that differ in age by 24 h, preferring males whose counter-mark scents are fresher than those of the intruder (Rich and Hurst, 1999). Hence females should discriminate in favour of the fresher resident's scent marks without any need for the resident to increase his own scent marking rate and counter-mark such aged intruder scent marks. Since the intruder's scent will have lost its volatile male signalling ligands after 7 days (Humphries et al., 1999), these scent marks are unlikely to attract much attention from females.

In contrast, the strong counter-marking of urinary proteins aged by 7 days in the presence of fresh urinary proteins from an intruder (compared to their response to an unmarked substrate) indicates that they clearly detected the presence of urinary protein even though this was no longer releasing volatile ligands. This suggests that the proteins themselves stimulated deposition of counter-marks in that location. Mouse major urinary proteins are highly polymorphic (see Payne et al., and Veggerby et al., this volume) and have the capacity to signal the individual identity of the scent mark owner (see Beynon et al., and Hurst et al., this volume). Moreover, resident males counter-marked an aged intruder scent in preference to a fresh scent mark challenge. This preference for depositing more marks near the aged scent may be because the aged intruder scent provides a better contrast to their own fresh scent. Female mice failed to discriminate between the scent marks of males when one scent was deposited on top of the other but there was no age difference between the bottom scent mark and the over-mark (Rich and Hurst, 1999). This might be because females are incapable of discriminating between scent marks and countermarks when there is no age difference between them, or because the competitive challenge has not been clearly resolved until a defeated challenger is no longer able to deposit fresh competing scent marks in the territory. In either case, although it is important to countermark fresh scent mark challenges to ensure that an intruder's scent is no fresher than their own, males will gain a greater advantage from depositing their own fresh scent marks near to older scents from intruders. The age difference between the scents will also maximise the difference in concentration of male signalling volatiles, that are positively associated with the male's aggressive status (Apps *et al.*, 1988; Harvey *et al.*, 1989; Novotny *et al.*, 1990; Jemiolo *et al.*, 1992) but which will be lost as the scent mark ages. Thus the age difference will allow their scents to be clearly discriminated as fresher or stronger than the intruder's, and allow them to be identified as the successful competitor by females. It thus appears that males deposit scent marks strategically to provide the most advantageous competitive signal in the face of scent mark challenges from other males, and thus communicate their superior competitive ability to females and to other male competitors.

ACKNOWLEDGMENTS

We thank all the members of the Animal Behaviour Group, Amr Marie for collecting stimulus urine, Edward Smith for analysing the videos, and the farmers for allowing us to catch their mice. This work was supported by research grants from BBSRC, and a travel award from the Skeath-Hughes Travel Fund.

REFERENCES

Apps, P. J., Rasa, A., and Viljoen, H. W., 1988, Quantitative chromatographic profiling of odours associated with dominance in male laboratory mice, *Aggr. Behav.* **14**:451-461.

Bacchini, A., Gaetani, E., and Cavaggioni, A., 1992, Pheromone binding proteins of the mouse, Mus musculus, *Experientia* **48**:419-421.

Brennan, P. A., Schellinck, H. M., and Keverne, E. B., 1999, Patterns of expression of the immediate-early gene egr-1 in the accessory olfactory bulb of female mice exposed to pheromonal constituents of male urine, *Neuroscience* 90 **4**:1463-1470.

Finlayson, J. S., Potter, M., and Runner, R. C., 1963, Electrophoretic variation and sex dimorphism of the major urinary protein complex in inbred mice: a new genetic marker, *J. Nat. Cancer Inst.* **31**:91-107.

Gosling, L. M., 1982, A reassessment of the function of scent marking in territories, *Z. Tierpsychol.* **60**:89-118.

Gosling, L. M., 1990, Scent marking by resource holders: alternative mechanisms for advertising the costs of competition, in: *Chemical Signals in Vertebrates V* (D. W. Macdonald, D. Müller-Schwarze, and S. E. Natynczuk, eds.), Oxford University Press, Oxford, pp. 315-328.

Harvey, S., Jemiolo, B., and Novotny, M., 1989, Pattern of volatile compounds in dominant and subordinate male-mouse urine, *J. Chem. Ecol.* **15**:2061-2072.

Humphries, R. E., Robertson, D. H. L., Beynon, R. J., and Hurst, J. L., 1999, Unravelling the chemical basis of competitive scent marking in house mice, *Anim. Behav.* **58**:1177-1190.

Hurst, J. L., 1987, The functions of urine marking in a free-living population of house mice, Mus domesticus Rutty, *Anim. Behav.* **35**:1433-1442.

Hurst, J. L., 1989, The complex network of olfactory communication in populations of wild house mice *Mus domesticus* Rutty: urine marking and investigation within family groups, *Anim. Behav.* **37**:705-725.

Hurst, J. L., 1990, Urine marking in populations of wild house mice *Mus domesticus* Rutty. I. Communication between males, *Anim. Behav.* **40**:209-222.

Hurst, J. L., 1993, The priming effects of urine substrate marks on interactions between male house mice, *Mus musculus domesticus* Schwarz and Schwarz, *Anim. Behav.* **45**:55-81.

Hurst, J. L., and Rich, T. J., 1999, Scent marks as competitive signals of mate quality, in: *Advances in Chemical Communication in Vertebrates* (R. E. Johnson, D. Müller-Schwarze, and P. Sorensen, eds.), Plenum Press, New York, pp. 209-226.

Hurst, J. L., Robertson, D. H. L., Tolladay, U., and Beynon, R. J., 1998, Proteins in urine scent marks of male house mice extend the longevity of olfactory signals, *Anim. Behav.* **55**:1289-1297.

Jemiolo, B., Xie, T. M., and Novotny, M., 1992, Urine marking in male mice: Responses to natural and synthetic chemosignals, *Physiol. Behav.* **52**:521-526.

Johnson, R. P., 1973, Scent marking in mammals, *Anim. Behav.* **21**:521-535.

Novotny, M., Harvey, S., and Jemiolo, B., 1990, Chemistry of male dominance in the house mouse, *Mus domesticus, Experientia* **46**:109-113.

Novotny, M., Ma, W., Wiesler, D., and Zidek, L., 1999, Positive identification of the puberty-accelerating pheromone of the house mouse: the volatile ligands associating with the major urinary protein, *Proc. R. Soc. Lond. B.* **266**:2017-2022.

Pes, D., Robertson, D. H. L., Hurst, J. L., Gaskell, S. J., and Beynon, R. J., 1999, How many major urinary proteins are produced by the house mouse Mus domesticus? in: *Advances in Chemical Communication in Vertebrates,* (R. E. Johnston, D. Müller-Schwarze, and P. Sorensen, eds.), Plenum Press, New York, pp. 149-161.

Ralls, K., 1971, Mammalian scent marking, *Science* **171**:443-449.

Rich, T. J., and Hurst, J. L., 1999, The competing countermarks hypothesis: reliable assessment of competitive ability by potential mates, *Anim. Behav.* **58**:1027-1037.

Robertson, D. H. L., Beynon, R. J., and Evershed, R. P., 1993, Extraction, characterization, and binding analysis of two pheromonally active ligands associated with major urinary protein of House Mouse (Mus-Musculus), *J. Chem. Ecol.* **19**:1405-1416.

Robertson, D. H. L., Hurst, J. L., Bolgar, M. S., Gaskell, S. J., and Beynon, R. J., 1997, Molecular heterogeneity of urinary proteins in wild house mouse populations, *Rapid Commun. Mass Spectrom.* **11**:786-790.

Robertson, D. H. L., Hurst, J. L., Hubbard, S. J., Gaskell, S. J., and Beynon, R. J., 1998, Ligands of urinary lipocalins from the mouse: uptake of environmentally derived chemicals, *J. Chem. Ecol.* **24**:1127-1140.

Singer, A. G., Tsuchiya, H., Wellington, J. L., Beauchamp, G. K., and Yamazaki, K., 1993, Chemistry of Odortypes in Mice - Fractionation and Bioassay, *J. Chem. Ecol.* **19**:569-579.

Singer, A. G., Beauchamp, G. K., and Yamazaki, K., 1997, Volatile signals of the major histocompatibility complex in male mouse urine, *Proc. Nat. Acad. Sci. USA* **94**:2210-2214.

INVADING PEST SPECIES AND THE THREAT TO BIODIVERSITY: PHEROMONAL CONTROL OF GUAM BROWN TREE SNAKES, *BOIGA IRREGULARIS*

Robert T. Mason[1] and Michael J. Greene[1,2]

[1]Department of Zoology
Oregon State University
Corvallis, OR 97331-2914

[2]Department of Biological Sciences
Stanford University
Stanford, CA 94305-5020

1. INTRODUCTION

1.1 Natural History of the Brown Tree Snake

The tropical colubrid brown tree snake, *Boiga irregularis*, is native to Australia, Papua-New Guinea, and the Solomon Islands (Cogger, 1992). The snake is an arboreal, rear-fanged colubrid that specializes in bird predation. The primary habitat of this species is in the tropical rain forests where they are nocturnal and actively forage for a generalized diet of lizards, small mammals, birds and bird eggs (Greene, 1989; Savidge, 1988). Their prey is killed by a combination of constriction and envenomation with the venom injected into the prey through grooves in their enlarged rear fangs (Weinstein *et al.*, 1991). Brown tree snakes reach snout-vent lengths of up to approximately three meters, masses of up to approximately two kilograms and have a relatively slender body plan characteristic of arboreal snakes (Lillywhite and Henderson, 1993; Rodda *et al.*, 1997). The species is sexually dimorphic with males being longer and heavier than females as mature adults (Shine, 1996).

During, or shortly after World War II, this snake was accidentally introduced onto the island of Guam (Rodda *et al.*, 1992). Like many other Pacific islands, no native snakes have ever existed on Guam. Currently, the flowerpot snake, *Rhamphotyphlops braminus*, is native to the island and is thought to have been introduced by the early Polynesian ancestors who first colonized these islands. These small subterranean snakes live almost exclusively underground feeding on ant and termite larvae. Therefore, the brown tree snake was introduced into an environment that had no snake predators, little competition for their prey base and a number of potential prey species that were naive to snake predation. As a result, this species flourished on Guam over the next 50 years until a recent study found estimated densities of brown tree snakes to be nearly 13,000 snakes per square mile, one of the highest densities of snakes anywhere in the world (Rodda *et al.*, 1992).

As Guam's brown tree snake population exploded, it became a serious menace to humans on the island. The snakes climb power lines in search of prey and in many instances, will cause power outages when the snake crosses a wire and grounds out. This has been a major problem for the island (there were over 1200 snake caused power outages between 1978 and 1994) resulting in losses of millions of dollars in lost revenues and destroyed equipment (Rodda *et al.*, 1997). The snakes also enter houses where they can be very aggressive towards people if forced into a defensive position and will repeatedly strike at the threat (Johnson, 1975). The snakes have also attacked the hands and feet of numerous sleeping humans. Most tragic of all are the attacks on small infants in a manner that suggests that the snakes were attempting to eat the baby (Fritts *et al.*, 1990). The brown tree snake also preys on domestic animals including puppies, kittens, rabbits and poultry such as chickens and Japanese quail (Fritts and McCoid, 1991).

The most dramatic result of the colonization of brown tree snakes on Guam has been the decimation of the island's avifauna (Savidge, 1987; Rodda *et al.*, 1997). As a result of snake predation, nine out of twelve species of native forest birds have been completely extirpated from the wild on Guam. Three of these species were endemic to Guam and are now extinct (Savidge, 1987). With such a major loss in prey, the snakes now survive on rodents, lizards, small domesticated animals and the endangered Mariana fruit bat (Wiles, 1987; Rodda and Fritts, 1992; Savidge, 1987).

Unfortunately, the brown tree snake problem is not limited to Guam. Individual brown tree snakes have been discovered on many other islands including Wake Island, Diego Garcia, Okinawa, Saipan and Oahu, Hawaii. In addition, there has been at least one confirmed introduction into the mainland United States in the state of Texas. However, of particular concern is the island state of Hawaii, where there have been at least nine brown tree snake sightings since the mid-1980's (Rodda *et al.*, 1997). Current management and research efforts have focused on controlling the spread of the brown tree snake while also controlling the population levels on Guam itself to avoid another invasion and potential colonization.

1.2 Pheromones as Biological Control Agents

Current management tools are limited to trapping brown tree snakes using live prey and searching cargo leaving Guam to other areas in the Pacific using trained dogs (Rodda *et al.*, 1997). However, this requires an expensive breeding and maintenance program. In addition, trapping snakes is only effective in the near vicinity of the trap where a snake could either see the prey item in the trap or detect volatile prey odors emanating from the trap. The brown tree snake control plan (Brown Tree Snake Control Committee, 1996) calls for the development of new technologies that can act as chemical attractants, repellents and mate disruption agents. Particular attention has been paid to the potential use of synthetic brown tree snake pheromones for this purpose (Mason, 1998).

The main pest management uses of pheromones are: mass trapping, disruption of communication, and repellency. The use of pheromones for attraction (mass trapping) and mating disruption has proven to be quite effective in selected examples of insect pest management (Ridgway *et al.*, 1990). Attraction and mating disruption appear at this time to hold the most promise as biological control agents because: behavioral responses are elicited at very low pheromone concentrations, the pheromones produce highly specific behaviors, and they have the potential for immediate use in pest management. Pheromones may indirectly control snakes by luring them into traps, to toxicants or to pathogens, or they may alter normal reproductive or aggregative behavior. In some cases, such as mating behavior, pheromones may affect only one sex. However, aggregation and trailing pheromones will affect both sexes. Perhaps most importantly in this more conservation-minded era, no cases are known in which pheromones leave toxic residues or directly affect species other than the one being controlled.

One novel strategy that we are pursuing is the use of pheromones with naturally occurring pathogens. In this case, pheromones are used to attract snakes to pathogen-contaminated „traps" (Mason, 1998). The inoculated snakes are then allowed to escape. During aggregations or mating, the pathogen is passed on to additional individuals that in turn pass it on yet again. Guam presents an excellent opportunity to test these ideas as it is a small, isolated island with little chance of spreading the vector to mainland populations. In addition, there are no other closely related snakes on the island that would be affected. The Typhlopid flowerpot snake is completely subterranean living in ant mounds and termite nests. The probability of any significant contact between the arboreal brown tree snake and the subterranean flowerpot snake is exceedingly small.

The role of pheromonal agents in integrated pest management strategies as applied to reptilian pests is still in its infancy. Indeed, the same case could be made for pheromonal control of any vertebrate species. Until recently, attempts to isolate, characterize and chemically identify vertebrate pheromones have met with limited success in comparison to the enormous body of research conducted on insects (Albone, 1984; Macdonald et al., 1990). This is due in part to the many advantages that insect systems offer, but also to several constraints associated with vertebrate species. It is estimated that less than 20 vertebrate pheromones have been positively identified to date as compared to over a thousand pheromone identifications in insects alone (Abelson, 1985).

Previous research from our laboratory has shown that the brown tree snake displays a diverse set of pheromonally mediated behaviors including courtship in which females produce a sex attractiveness pheromone in their skin lipids. This pheromone serves to attract males and initiate courtship and mating from them. In addition, both male and female brown tree snakes will follow pheromone trails on the substrate. Males follow trails of females to gain access to potential mates. Males follow males in order to engage in combat behavior with other males in order to establish temporary hierarchies and thus gain access to females during the breeding season. Females may follow females in order to find suitable aggregation sites during the daylight hours or to find suitable egg-laying sites.

The brown tree snake provides an excellent system for adding to the current body of knowledge concerning the mediation of vertebrate behavior by pheromones. In addition, the compelling applied aspect of this research will be critical to help stem the increasing frequency of accidental introductions onto other islands.

2. PHEROMONE SYSTEMS OF THE BROWN TREE SNAKE

2.1 Courtship Behavior - Male

Courtship in this species begins when a male tongue-flicks the dorsal integument of the female. In snakes, the detection of sex pheromones occurs in the vomeronasal organ via delivery by tongue-flicking (Halpern, 1992). After tongue-flicking the female's dorsum, the male displays head-jerking whereby the male rapidly tongue-flicks the female while rhythmically jerking his head in a lateral direction (Greene and Mason, 2000). The male will soon mount the female by placing his chin on the female's dorsum and advancing along the female's body while pressing his chin to the female's dorsum (Figure 1).

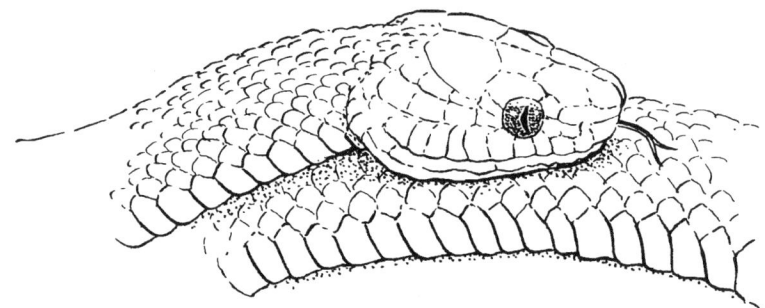

Figure 1. A male chin-rubbing the dorsum of a sexually attractive female during courtship. Chin-rubbing is accompanied by forward body jerks.

After several minutes of chin-rubbing, the male will periodically lift his head, coil his head and neck region in an „S" configuration and bob it up and down perpendicularly to the ground (Greene and Mason, 2000). Subsequent to these behaviors, males will chase a retreating female or chase-mount a female by pursuing her while remaining mounted on the female's body and displaying chin-rubbing. A period of chase-mount behavior eventually leads to body alignment where the bodies of the courting pair are aligned side by side or the male's body is mounted on the female's body (Figure 2). Tail-search copulatory attempts follow in which the male repeatedly attempts to align his cloaca with that of the female by wrapping his tail around hers. If the female is receptive to the male's courtship, the male will intromit one of his hemipenes into the female's gaped cloaca. During copulation, the male remains motionless except for periodic tongue-flicking. No copulatory plugs are deposited into the female's cloaca at the termination of copulation (Greene and Mason, 2000).

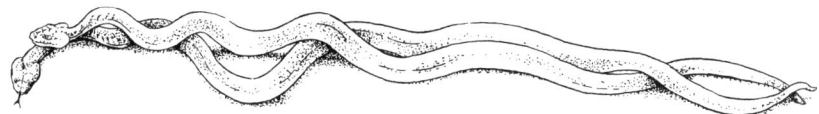

Figure 2. A male and female during body alignment. The male is mounted on the dorsum of the female and is displaying chin-rubbing behavior.

2.2 Courtship Behavior – Female

Reproductively active and receptive females appear to solicit mating behavior from males. After tongue-flicking a male, females display a brief bout of head-jerking behavior (Greene and Mason, 2000). Females displaying head-jerking will mount the male by moving her head along the male's back or slide along side the male's body towards his head. When the female reaches the male's head, she maneuvers her body under or directly in front of the male's snout in an apparent attempt to elicit tongue-flicking from the male. Males then will usually begin to court the female. The female then retreats from the male while exhibiting head-lifting behavior remarkably similar to that shown by males. Eventually the male will align with the female and mating will ensue. Male snakes cannot force copulation, thus mating only proceeds if the female is receptive and gapes her cloaca allowing the male to intromit one of his hemipenes. During copulation, the female remains motionless in the same fashion as the male.

2.3 Female Sex Attractiveness Pheromones

Bioassays utilizing the courtship and mating behaviors of male brown tree snakes were developed in order to elucidate the female sex attractiveness pheromone source. Filter papers treated with hexane extracts of sexually attractive female skin elicited significantly more courtship behaviors than extract and solvent controls (Greene and Mason, 1998). Subsequent fractionation of these hexane extracts on alumina activity III columns with hexane and diethyl ether as the mobile phase yielded 20 fractions. Significant differences in male sex behaviors exhibited in response to these fractions indicated that the sex attractiveness pheromone was present only in the first four fractions, with fraction four being the most active. The other 16 fractions elicited no sex behavior from males. The active fractions contained only hydrocarbons and not methyl ketones, which have previously been isolated as the sex attractiveness pheromone of another colubrid, the red-sided garter snake, *Thamnophis sirtalis parietalis* (Mason *et al.*, 1989). Future studies will concentrate on further fractionation and elucidation of the chemical constituents of the first four fractions of female brown tree snake skin lipids.

2.4 Courtship Inhibition

When the female is unreceptive to male courtship behavior, she will flee from the male and become agitated with body jerking and assume a defensive posture. If the male continues his courtship advances, the female will lift her tail and release cloacal secretions (Greene and Mason, 2000). Prior to this event, all the males had displayed vigorous courtship behavior. After the female releases these cloacal secretions, males immediately cease all courtship behavior but continue to stay near the female, following her around the enclosure. In only one instance did we observe the males actually tongue-flicking the cloacal secretions. This suggests that the behavior may be being induced by a volatile pheromone. In experiments with breeding males and females, female cloacal secretions alone cause a significant decrease in the amount of time spent courting females and a decrease in the intensity of male courtship as compared to controls.

2.5 Combat Behavior

Male brown tree snakes will exhibit ritualized combat behavior after tongue-flicking the dorsal integument of his male opponent. After tongue-flicking each other, both males display head-jerking behavior. Both males will try and mount the other while chin-rubbing and beginning forward-body jerking behavior. During this phase of combat, males will often display body-bridging in which a male will raise a loop of his body at that point where his opponent is attempting to chin-rub. At this point males will display vigorous bouts of head-lifting. Head-lifting by one male usually elicits a head-lifting response by his opponent. Eventually, both males will intertwine around each other with their heads lifted off the substrate (Figure 3). At this point males begin to display head-pinning behavior where one male pins his opponent's head to the ground using his head or a U-shaped coil of the neck. This behavior can continue for several hours until finally one of the males is victorious. The losing male will attempt to flee from the winner frequently lashing the body and tail. Winning males will subsequently exhibit vigorous courtship behavior to a female if present.

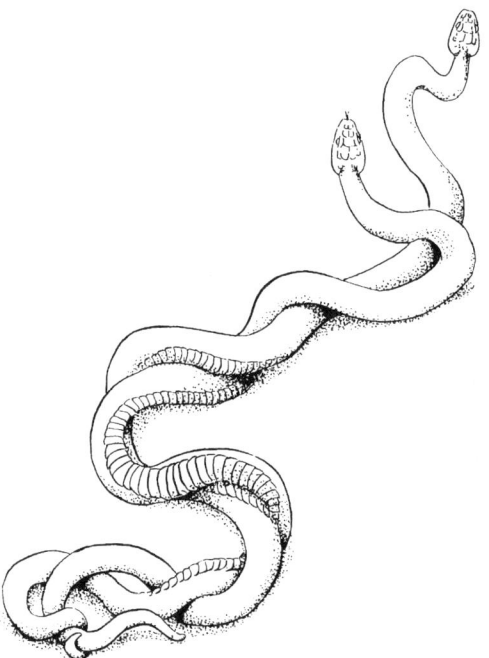

Figure 3. Two males in a typical combat posture with head-lifting and body-coiling displayed.

2.6 Trailing Behavior

Trailing behavior is common to many species of snakes and has significant consequences to their survival and reproduction. Most studies on other species of snake have concentrated on the male's ability to trail conspecifics females during the breeding season. Reproductively active brown tree snakes were tested in a Y-maze for their ability to trail conspecifics. Briefly, the results indicated that males were very adept at following trails of both females and males. It is not hard to imagine that it would be selectively advantageous for males to follow females in order to gain access to mates. However, males also trail rival males. This may also be advantageous such that when two males encounter each other, combat behavior will ensue and the winning male will then have a temporary dominance allowing him exclusive access to any females in that area (Shine et al., 1981). Unlike males, most of the females would not trail at all when they were reproductively active. Instead, they remained fairly sedentary and most chose not to trail. However, of those females that did trail, all chose to trail conspecifics.

Field data from studies in the brown tree snake's native range confirm these findings. Males were more active than females and were encountered far more frequently during months where mating behavior had been observed and in months where females were in the late stages of ovarian development or gravid (Bull and Whittier, 1996; Whittier and Limpus, 1996). In other months, the encounter rate of both males and females was equal. These results suggest that females remain fairly sedentary during the breeding season and thus males must be much more active in searching out the females. Thus, trailing behavior would be expected to be a sexually dimorphic character with predominately males expressing the trait.

CONCLUSION

The use of pheromones as control agents for vertebrates is only just beginning to receive attention. The introduced population of the brown tree snake on Guam presents an ideal opportunity to use a vertebrate pheromone to manage an introduced pest species. If successful, this would be the first case in which vertebrate pheromones were used to stem the spread of an injurious pest species. As humankind becomes more and more environmentally conscious, we are reluctant to use pesticides and other toxicants in the environment. Thus, it seems inevitable that pheromone technology will continue to provide safe alternatives to the current use of deleterious synthetic compounds. Only by integrating several control strategies will there be any hope of a feasible control of problem snake populations.

ACKNOWLEDGMENTS

The authors thank S. Stark for her help in maintaining the captive colony of snakes and for her help conducting courtship trials. R. Jones drew the figures included in the text. This work was supported by grants from the National Science Foundation (INT-9114567) as well as an NSF National Young Investigator Award (IBN-9357245) to RTM. A grant-in-aid was awarded to MJG by the Sigma-Xi Scientific Research Society. This research was conducted under the authority of U.S. Fish and Wildlife Permit PRT-769753 and Oregon State University Institutional Animal Care and Use Committee Protocols LAR 932 and 2181-B.

REFERENCES

Albone, E. S., 1984, *Mammalian Semiochemicals*, John Wiley and Sons, New York.
Abelson, P. H., 1985, Use of and research on pheromones, *Science* **229**:1342.
Brown Tree Snake Control Committee, 1996, *Brown Tree Snake Control Plan*, Aquatic Nuisance Species Task Force, USA.
Bull, K. H., and Whittier, J. M., 1996, Annual patterns of activity of the brown tree snake (*Boiga irregularis*) in southeastern Queensland, *Mem. Queensland Mus.* **39**:483-486.
Cogger, H. G., 1992, *Reptiles and Amphibians of Australia*, Reed Books, Chatswood, New South Wales, Australia.
Fritts, T. H., and McCoid, M. J., 1991, Predation by the brown tree snake, *Boiga irregularis,* on poultry and other domesticated animals on Guam, *Snake* **23**:75-80.
Fritts, T. H., McCoid, M. J., and Haddock, R. L., 1990, Risks to infants on Guam from bites of the brown tree snake (*Boiga irregularis*), *Am. J. Trop. Med. Hyg.* **42**:607-611.
Greene, H. W., 1989, Ecological, evolutionary and conservation implications of feeding biology in old world cat snakes, genus *Boiga* (Colubridae), *Proc. Calif. Acad. Sci.* **46**:193-207.
Greene, M. J., and Mason, R. T., 1998, Chemically mediated sexual behavior of the brown tree snake, *Boiga irregularis*, *Ecoscience* **5**:405-409.
Greene, M. J., and Mason, R. T., 2000, Courtship, mating, and male combat of the brown tree snake, *Boiga irregularis*, *Herpetologica* **56**:166-175.
Halpern, M., 1992, Nasal chemical senses in reptiles: structure and function, in: *Biology of the Reptilia, Vol. 18*, (C. Gans, and D. Crews, eds.), University of Chicago Press, Chicago, pp. 424-498.

Johnson, C. R., 1975, Defensive display behavior in some Australian and Papuan-New Guinean pygopodid lizards, boid, colubrid, and elapid snakes. *J. Linn. Soc.* **56**:265-282.

Lillywhite, H. B., and Hendersen, R. W., 1993, Behavioral and functional ecology of arboreal snakes, in: *Snakes: Ecology and Behavior*, (R. A. Seigel and J. T. Collins, eds.), McGraw-Hill, New York, pp. 12-15.

Macdonald, D. W., Müller-Schwarze, D., and Natynczuk, S. E., 1990, *Chemical Signals in Vertebrates 5*, Oxford University Press, Oxford.

Mason, R. T., 1998, Integrated pest management: the case for pheromonal control of habu (*Trimeresurus flavoviridis*) and brown tree snakes (*Boiga irregularis*), in: *Problem Snake Management, Habu and Brown Tree Snake Examples*, (G. H. Rodda, Y. Sawai, D. Chiszar, and H. Tanaka, eds.), Cornell University Press, Ithaca, New York, pp. 196-206.

Mason, R. T., Fales, H. M., Jones, T. H., Pannell, L. K., and Crews, D., 1989, Sex pheromones in snakes, *Science* **245**:290-293.

Ridgway, R. L., Silverstein, R. M., and Iscoe, M. N., (eds.), 1990, *Behavior Modifying Chemicals for Insect Management*. Marcel Dekker, Inc., New York.

Rodda, G. H., and Fritts, T. H., 1992, The impact of the introduction of the colubrid snake *Boiga irregularis* on Guam's lizards, *J. Herpetol.* **26**:166-174.

Rodda, G. H., Fritts, T. H., and Vonty, P. J., 1992, Origin and population growth of the brown tree snake, *Boiga irregularis*, on Guam, *Pacific Sci.* **46**:46-57.

Rodda, G. H., Fritts, T. H., and Chiszar, D., 1997, Disappearance of Guam's wildlife, *Bioscience* **47**:565-574.

Savidge, J. A., 1988, Food habits of *Boiga irregularis*, an introduced predator on Guam, *J. Herpetol.* **22**:275-282.

Savidge, J. A., 1987, Extinction of an island avifauna by an introduced snake, *Ecol.* **68**:660-668.

Shine, R., 1996, Sexual dimorphism in snakes revisited, *Copeia* **1996**:326-346.

Shine, R., Grigg, G. C., Shine, T. G., and Harlow, P., 1981, Mating and male combat in Australian blacksnakes, *Pseudachis porphyriacus*, *J. Herp.* **15**:101-107.

Weinstein, S. A., Chiszar, D., Bell, R. C., and Smith, L. A., 1991, Lethal potency and fractionation of Duvernoy's secretion from the brown tree snake, *Boiga irregularis*, *Toxicon* **29**:401-407.

Whittier, J. M., and Limpus, D., 1996, Reproductive patterns of a biologically invasive species: the brown tree snake (*Boiga irregularis*) in eastern Australia, *J. Zool. Soc. Lond.* **238**:591-597.

Wiles, G. J., 1987, Current research and future management of Marianas fruit bats (Chiroptera: Pteropodidae) on Guam, *Austr. Mammal.* **10**:93-95.

ANNUAL AND SEASONAL VARIATION IN THE FEMALE SEXUAL ATTRACTIVENESS PHEROMONE OF THE RED-SIDED GARTER SNAKE, *THAMNOPHIS SIRTALIS PARIETALIS*

Michael P. LeMaster and Robert T. Mason

Department of Zoology
Oregon State University
Corvallis, Oregon 97331-2914

1. INTRODUCTION

Reptiles are excellent models for vertebrate pheromone research because the initiation of many behaviors is dependent on pheromone production and expression (reviewed in Madison, 1977; Mason, 1992; Mason *et al.*, 1998). For example, the reproductive success of snakes depends on the production and perception of specific sex pheromones (Noble, 1937; Gillingham and Dickinson, 1980; Andrén, 1986; Mason *et al.*, 1989). In the absence of these pheromones, sexual behavior will not be initiated and mating will not occur.

Garter snakes (genus *Thamnophis*) are one of the most commonly encountered snakes in North America with a continuous range stretching from central Canada through Mexico (Rossman *et al.*, 1996). Since the 1930's, investigators have demonstrated that within this genus initiation of male courtship behavior depends on the production and expression of a female sexual attractiveness pheromone (e.g. Noble, 1937; Garstka and Crews, 1981; Mason *et al.*, 1989). When a male comes in contact with a female expressing such a pheromone, the male displays courtship behavior characterized by increased tongue-flick rate, chin-rubbing along the dorsum of the female and alignment of cloacal openings (Noble, 1937). These behaviors are simultaneously expressed only in a reproductive context in response to the sexual attractiveness pheromone (Camazine *et al.*, 1980).

The sexual attractiveness pheromone of the red-sided garter snake *(Thamnophis sirtalis parietalis)* is the first, and to date only, reptilian sex pheromone isolated and identified. The female sexual attractiveness pheromone of this species consists of a homologous series of long-chain saturated and ω–9 cis-unsaturated methyl ketones contained within the skin lipids of the female (Mason *et al.*, 1989; Mason *et al.*, 1990). Although both the saturated and unsaturated methyl ketones are required to elicit full male courtship behavior, unsaturated methyl ketones appear to be the more biologically active of the two groups. When presented in isolation, unsaturated methyl ketones elicit a five fold increase in male response over saturated methyl ketones (Mason *et al.*, 1989).

As part of our laboratory's ongoing work towards deciphering the pheromone system of the red-sided garter snake, we present here a study examining annual and seasonal variation in the composition of the female sexual attractiveness pheromone for this species. Because garter snake reproductive success depends on the production and expression of the female sexual attractiveness pheromone, we predict that the composition of this pheromone will remain relatively stable across breeding seasons for the red-sided garter snake. Additionally, we predict that there will exist observable differences in the composition of this pheromone between the breeding season and the non-breeding season, representing a mechanism for regulating male courtship behaviour out of the breeding season. Currently, there is limited information available concerning temporal variation in the production and expression of vertebrate reproductive pheromones.

2. MATERIALS AND METHODS

2.1 Study Population

The red-sided garter snake (*Thamnophis sirtalis parietalis*) is the most northerly living reptile in the Western Hemisphere (Logier and Toner, 1961). The annual aggregation of red-sided garter snakes at underground hibernacula in Manitoba, Canada is a unique natural phenomena representing the highest concentration of snakes in the world (Gregory, 1984). Marshes, shallow lakes and poor drainage offer good summer feeding grounds for the snakes while limestone bedrock provides hibernaculum sites where snakes are constrained to spend up to eight months of the year to avoid the harsh winters (Aleksiuk and Stewart, 1971).

Red-sided garter snakes utilized for this study were obtained from a field site at the Narcisse Wildlife Management Area in the Interlake region of Manitoba, Canada (50°44'N, 97°34'W). The Narcisse Wildlife Area contains three over-wintering hibernacula in close proximity to each other, with each hibernaculum possessing in excess of twenty thousand red-sided garter snakes during the winter months (Mason, unpublished data).

2.2 Skin Lipid Collection and Fractionation

For the annual variation samples, adult, sexually attractive female red-sided garter snakes were randomly collected immediately following spring emergence from the hibernacula, mean (±SEM) anatomical parameters were recorded, in May of 1997 (n = 19; snout-vent length (SVL) = 62.2 (± 8.1) cm; mass = 93.6 (± 41.8) g), 1998 (n = 20; SVL = 62.7 (± 6.7) cm; mass = 93.0 (± 31.1) g), and 1999 (n = 12; SVL = 65.7 (± 4.5) cm; mass = 103.4 (± 19.6) g). For the seasonal variation samples, adult, attractive female red-sided garter snakes were randomly collected from the hibernacula in May 1997 (n =5; SVL =58.3 (± 4.1) cm; mass = 72.1 (± 13.6) g) and adult, unattractive females were randomly collected from the hibernacula in September 1997 (n = 5; SVL = 57.8 (± 4.2) cm; mass = 74.2 (± 16.0) g).

Following each sampling period, the animals were briefly washed (1 minute per snake) in 300 ml hexane (C_6H_{14}) to remove a pooled sample of their skin lipids. The excess solvent was then evaporated off the skin lipid extract and the resulting residue was resuspended in fresh hexane and sealed in a 9-ml glass vial with a polyethylene-lined cap for storage at –20°C. The animals, following a brief recovery at the Chatfield research station, Chatfield, Manitoba, were returned to the point of capture.

Fractionation of the extracted skin lipids was accomplished in the laboratory using alumina activity III columns with hexane and ether as the mobile phases (Mason *et al.*,

1989). For each sample, the fraction containing the methyl ketones composing the female sexual attractiveness pheromone was removed (fraction 5 - Mason *et al.*, 1989) and the excess solvent evaporated off. The remaining residue was then resuspended in fresh hexane and sealed in a 9-ml glass vial with a polyethylene-lined cap for storage at –20°C.

2.3 Chemical Analysis

Analysis of the methyl ketones composing the sexual attractiveness pheromone was conducted on a Hewlett Packard 5890 Series II gas chromatograph fitted with a split injector (280°C) and a Hewlett Packard 5971 Series mass selective detector. A fused-silica capillary column (HP-1; 12 m x 0.22 mm ID; Hewlett Packard, California, USA) was used with helium as the carrier gas (5 cm/sec). All injections were made in the splitless mode with the split valve closed for 60 sec. Oven temperature was initially held at 70°C for 1 min, increased to 210°C at 30°C/min, held at 210°C for 1 min, increased to 310°C at 5 C/min, and held at 310°C for 5 min.

Quantification of the saturated and unsaturated methyl ketones was accomplished using peak integration to calculate the percent relative concentrations of each to the overall pheromone profile for a sampling period (e.g. percent relative concentration of saturated methyl ketones = area under saturated methyl ketone peaks / total area of all methyl ketone peaks x 100). Retention times and peak areas were determined using ChemStation software (Version B.02.05; Hewlett Packard, California, USA) interfaced with the gas chromatograph/mass spectrometer (GC/MS).

3. RESULTS

3.1 Annual Variation

When compared across years, there was no difference in the average snout-vent length (ANOVA, $F = 1.030$, $P > 0.25$) or mass (ANOVA, $F = 0.416$, $P > 0.50$) of females sampled.

Complete GC/MS analysis of the methyl ketone fractions from the three years revealed the presence of eighteen unique long-chain methyl ketones. Nine of these were identified as saturated methyl ketones while the other nine had mass spectra in accord with unsaturated methyl ketones (Mason *et al.*, 1990). Overall, there were only minor differences observed in the percent relative contribution of saturated and unsaturated methyl ketones to the overall pheromone profiles across the years (Figure 1). In all three years, unsaturated methyl ketones dominated the pheromone profiles, accounting for 81.3% of the pheromone profile in 1997, 82.4 % in 1998, and 86.1 % in 1999.

3.2 Seasonal Variation

When compared across seasons, there was no difference in the average snout-vent length (ANOVA, $F = 0.042$, $P > 0.50$) or mass (ANOVA, $F = 0.052$, $P > 0.50$) of females sampled.

Complete GC/MS analysis of the methyl ketone fractions from the two seasons revealed the presence of the eighteen unique long-chain methyl ketones previously described. Overall, there was a distinct difference in the percent relative concentrations of saturated and unsaturated methyl ketones composing the pheromone between seasons (Figure 2). Unlike the spring sample, which displayed a profile rich in unsaturated methyl ketones (72.0%), the fall sample shows a marked decrease in the percent relative concentration of unsaturated methyl ketones (63.0%).

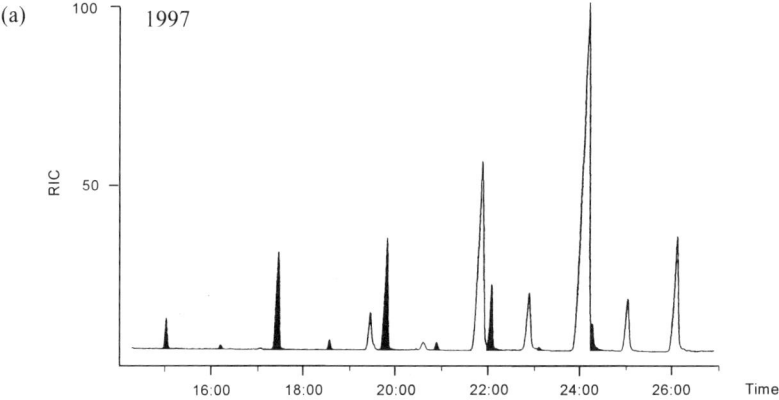

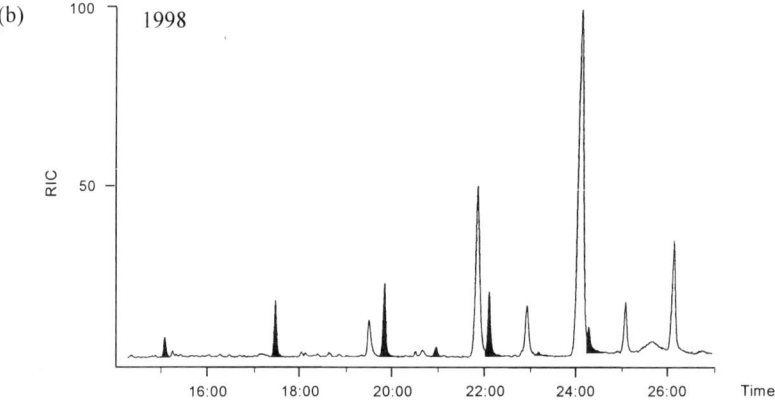

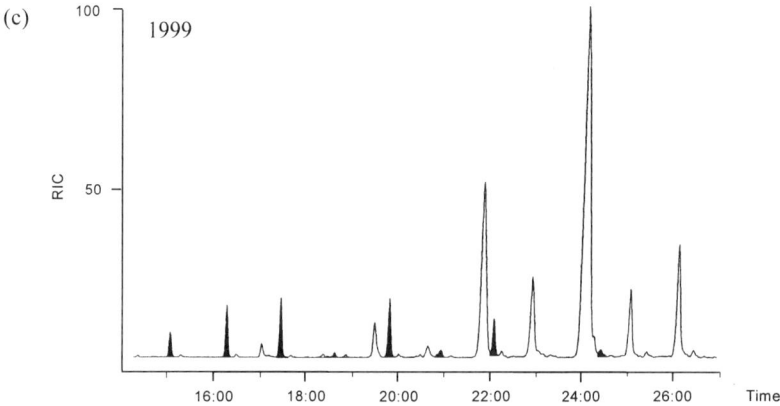

Figure 1. Gas chromatograms of the female sexual attractiveness pheromone profile for red-sided garter snakes sampled in (a) 1997, (b) 1998 and (c) 1999. Pheromone profiles are composed of saturated (shaded peaks) and unsaturated (open peaks) methyl ketones.

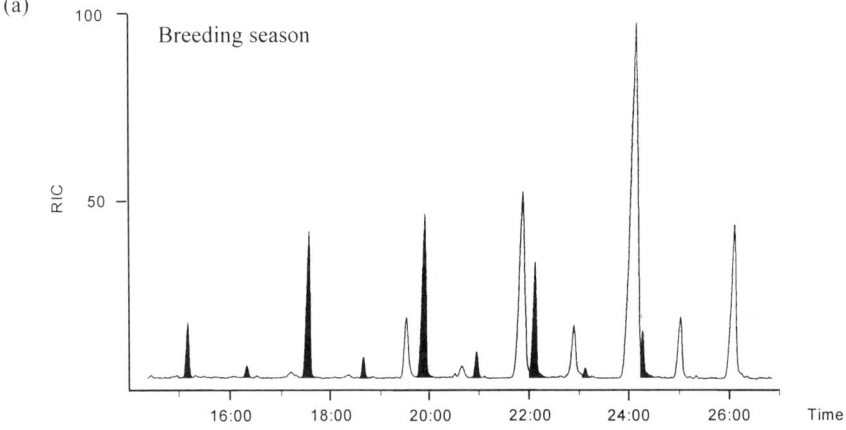
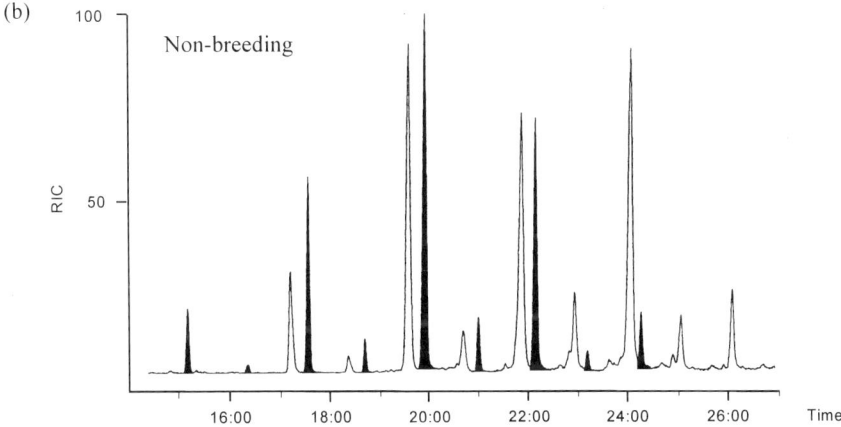

Figure 2. Gas chromatograms of the female sexual attractiveness pheromone profile for red-sided garter snakes sampled during the (a) breeding season and (b) non-breeding season. Pheromone profiles are composed of saturated (shaded peaks) and unsaturated (open peaks) methyl ketones.

4. DISCUSSION

The results of this study demonstrate that the composition of the female sexual attractiveness pheromone does exhibit temporal variation. As predicted, complete GC/MS analysis revealed only minor fluctuations in the relative concentrations of saturated and monounsaturated methyl ketones to the overall pheromone profiles across years, but revealed distinct differences in the pheromone profiles between the seasons. To the best of our knowledge, this is the first study to examine annual and seasonal variation in the expression of a characterized vertebrate pheromone.

The lack of variation across the three years of this study suggests that the composition of the female sexual attractiveness pheromone is tightly regulated during the breeding season. It is not surprising that some degree of regulation was observed. Communication systems depend on the production of a signal by the signaler and the perception of the signal by a receiver. If the variation of a signal extends beyond the range of what the receiver is tuned for, then the signal will no longer function as intended (Bradbury and Vehrencamp, 1998). Thus, female garter snakes expressing methyl ketone profiles that fall out of the tuning range of males would not elicit courtship behavior nor successfully mate.

Insects have been noted to display precise control over sex pheromone expression (e.g. Miller and Roelofs, 1980), similar to our observations. This precise regulation is often regarded as a mechanism by which closely related species, living in sympatry and utilizing similar pheromone components, can remain reproductively isolated (Cardé and Baker, 1984; Attygalle et al., 1986). The red-sided garter snake lives in sympatry with a closely related species, the plains garter snake (*Thamnophis radix*), in Manitoba, Canada. Initial examinations of the methyl ketone profile for the plains garter snake reveals a profile similar in composition, but distinct in the relative concentrations of individual components, to that for the red-sided garter snake (Mason, unpublished data). Future studies will examine the role of the female sexual attractiveness pheromone in the reproductive isolation of these two species.

The observed variation in the pheromone profiles between the breeding season and the non-breeding season suggests that the integrity of the female sexual attractiveness pheromone of the red-sided garter snake is not maintained throughout the year. Most likely, the females are halting production of the pheromone following the breeding season, similar to what occurs in most insects expressing sex pheromones contained within the cuticular skin lipids during the breeding season (reviewed in Tillman et al., 1999). The change in composition observed in the garter snake may result from differences in the production rates of the saturated and unsaturated methyl ketones. Unlike unsaturated methyl ketones, saturated methyl ketones are present in the skin lipids of males as well as females (Mason, 1993) suggesting a possible biological function other than sex recognition. Thus, females most likely are shifting the levels of unsaturated methyl ketones, the more active of the two pheromone components (Mason et al., 1989), between the breeding and non-breeding season while maintaining a similar saturated methyl ketone level. Although the exact physiological mechanisms involved in regulating pheromone production in garter snakes are not known, the expression of the sexual attractiveness pheromone appears to be under endocrine regulation. Estrogen treatment has been shown to induce pheromone production in female garter snakes (Crews, 1985; Kubie et al., 1978; Garstka and Crews, 1981).

Because male red-sided garter snakes have an obligate reliance on the pheromone to initiate courtship behavior, this seasonal shift in pheromone composition may explain the lack of male courtship interest displayed to females during the non-breeding season. Insects often rely on shifts in pheromone composition to regulate courtship behavior (e.g. Schal et al., 1994; Hurd and Perry, 1991). In support of this hypothesis is the fact that a small subset of female garter snakes do elicit male courtship behavior out of the breeding season (LeMaster, personal observation). This indicates that males are still receptive to the pheromone outside the breeding season, unlike males of other vertebrate species which may be physiologically unreceptive to pheromone cues in the non-breeding season (e.g. newts, Moore et al., 2000). Additional studies are necessary to demonstrate whether the small subset of females attracting courtship attention outside the breeding season are displaying a pheromone profile similar to that observed in the breeding season, or whether there are some other mechanisms at work which still remain to be elucidated.

ACKNOWLEDGMENTS

We thank the Manitoba Department of Natural Resources, Dave Roberts and Al and Gerry Johnson for assistance in the field. We also thank Tom Roberts and Julie LeMaster for useful comments on the manuscript, and William Gerwick for the use of his GC/MS system. This research was supported by a Sigma Xi Grants-in-Aid of Research grant and Oregon State University Zoology Research Funds to M.P.L., and the National Science Foundation (INT-9114567), NSF National Young Investigator Award (IBN-9357245), and the Whitehall Foundation (W95004) to R.T.M. The research presented here was conducted under the authority of Manitoba Wildlife Scientific Permits No. WSP 02-97 and in accord with the Oregon State University Institutional Animal Care and Use Committee Protocol No. LAR-1848B.

REFERENCES

Aleksiuk, M., and Stewart, K. W., 1971, Seasonal changes in the body composition of the garter snake (*Thamnophis sirtalis parietalis*) at northern latitudes, *Ecology* **52**:485-490.

Andrén, C., 1986, Courtship, mating and agonistic behavior in a free-living population of adders, *Vipera berus*, *Amphibia-Reptilia* **7**:353-383.

Attygalle, A. B., Schwarz, J., Vostrowsky, O., and Bestmann, H. J., 1986, Individual variation in the sex pheromone components of the false codling moth, *Cryptophlebia leucotreta* (Lepidoptera: Tortricidae), *J. Naturforsch.* **41**:1077-1081.

Bradbury, J. W., and Vehrencamp, S. L., 1998, *Principles of Animal Communication*, Sinauer Associates, Inc., Sunderland, Massachusetts.

Camazine, B., Garstka, W., Tokarz, R., and Crews, D., 1980, Effects of castration and androgen replacement on male courtship behavior in the red-sided garter snake (*Thamnophis sirtalis parietalis*), *Horm. Behav.* **7**:451-460.

Cardé, R. T., and Baker, T. C., 1984, Sexual communication with pheromones, in: *Chemical Ecology of Insects* (W. J. Bell, and R. T. Cardé, eds.), Chapman and Hall, London, pp. 355-383.

Crews, D., 1976, Hormonal control of male courtship behavior and female attractivity in the garter snake (*Thamnophis sirtalis parietalis*), *Horm. Behav.* **7**:451-460.

Crews, D., 1985, Effects of early sex steroid hormone treatment on courtship behavior and sexual attractivity in the red-sided garter snake, *Thamnophis sirtalis parietalis*, *Physiol. Behav.* **35**:569-575.

Garstka, W., and Crews, D., 1981, Female sex pheromone in the skin and circulation of a garter snake, *Science* **214**:681-683.

Gillingham, J. C., and Dickinson, J. A., 1980, Postural orientation during courtship in the eastern garter snake, *Thamnophis sirtalis sirtalis*, *Behav. Neural Biol.* **28**:211-217.

Gregory, P. T., 1984, Communal denning in snakes, in: *Vertebrate Ecology and Systematics – A Tribute to Henry S. Fitch* (R. Seigel, L. Hunt, J. Knight, L. Malaret, and N. Zuschlag, eds.), University of Kansas Press, Lawrence, Kansas, pp. 57-75.

Hurd, H., and Parry, G., 1991, Metacestode-induced depression of the production of, and response to, sex pheromone in the intermediate host *Tenebrio molitor*, *J. Invert. Path.* **58**:82-87.

Kubie, J. L., Vagvolgyi, A., and Halpern, M., 1978, The roles of the vomeronasal and olfactory systems in the courtship behavior of male snakes, *J. Comp. Physiol. Psychol.* **92**:627-641.

Logier, E. B. S., and Toner, G. C., 1961, Checklist of amphibians and reptiles in Canada and Alaska, Royal Ontario Museum Contribution No. 53, University of Toronto Press, Toronto.

Madison, D. M., 1977, Chemical communication in amphibians and reptiles, in: *Chemical Signals in Vertebrates* (D. Müler-Schwarze, and M. M. Mozell, eds.), Plenum Press, New York, pp. 135-168.

Mason, R. T., 1992, Reptilian pheromones, in: *Biology of the Reptilia*, Volume 18 (C. Gans, and D. Crews, eds.), The University of Chicago Press, Chicago, pp. 115-216.

Mason, R. T., 1993, Chemical ecology of the red-sided garter snake, *Thamnophis sirtalis parietalis*, *Brain Behav. Evol.* **41**:261-268.

Mason, R. T., Chivers, D. P., Mathis, A., and Blaustein, A. R., 1998, Bioassay methods for amphibians and reptiles, in: *Methods in Chemical Ecology*, Volume 2 (K. F. Haynes, and J. G. Millar, eds.), Kluwer Academic Publishers, Norvell, Massachusetts, pp. 271-325.

Mason, R. T., Fales, H. M., Jones, T. H., Pannel, L. K., Chinn, J. W., and Crews, D., 1989, Sex pheromones in snakes, *Science* **245**:290-293.

Mason, R. T., Jones, T. H., Fales, H. M., Pannell, L. K., and Crews, D., 1990, Characterization, synthesis, and behavioral responses to sex attractiveness pheromones of red-sided garter snakes (*Thamnophis sirtalis parietalis*), *J. Chem. Ecol.* **16**:2353-2369.

Miller, J. R., and Roelofs, W. L., 1980, Individual variation in sex pheromone component ratios in two populations of the redbanded leafroller moth, *Argyrotaenia velutinana*, *J. Chem. Ecol.* **9**:359-363.

Moore, F. L., Richardson, C., and Lowry, C. A., 2000, Sexual dimorphism in numbers of vasotocin-immunoreactive neurons in brain areas associated with reproductive behaviors in the roughskin newt, *Gen. Comp. Endocr.* **117**:281-298.

Noble, G. K., 1937, The sense organs involved in the courtship of *Storeria*, *Thamnophis*, and other snakes, *Bull. Am. Mus. Nat. Hist.* **73**:673-725.

Rossman, D. A., Ford, N. B., and Seigel, R. A., 1996, *The Garter Snake: Evolution and Ecology*, University of Oklahoma Press, Norman, Oklahoma.

Schal, C., Gu, X., Burns, E. L., and Blomquist, G. J., 1994, Patterns of biosynthesis and accumulation of hydrocarbons and contact sex pheromone in the female German cockroach *Blattella germanica*, *Arch., Insect Biochem. Physiol.* **25**:375-391.

Tillman, J. A., Seybold, S. J., Jurenda, R. A., and Blomquist, G. J., 1999, Insect pheromones – an overview of biosynthesis and endocrine regulation, *Insect Biochem. Molec. Biol.* **29**:481-514.

HYPOTHALAMIC AND OVARIAN RESPONSE TO PHEROMONE APPLICATION IN SEASONAL ANOESTROUS GERMAN MUTTON MERINO SHEEP

Karl-Heinz Kaulfuß, Reinhard Süß, Paul Schenk, Elke Berger, and Eberhard von Borell

Institute of Animal Breeding and Husbandry with Veterinary Clinic
Martin-Luther-University Halle-Wittenberg
Emil-Abderhalden-Str. 27
D-06108 Halle/Saale, Germany

INTRODUCTION

Merino sheep normally undergo a prolonged anestrous period during spring and summer, thus preventing the continuous production of lambs for the marketplace. To meet this challenge, it has been demonstrated that off-season estrus can be artificially induced by the delivery of exogenous hormones, including the intravaginal application of progestagen sponges in combination with pregnant mare serum gonadotropin (PMSG), and melatonin. However, questions about possible hormone persistence in lamb meat have resulted in the prohibition of such hormone treatments in the agricultural practices of many countries, including Germany. The development of a natural and safe method of estrus induction in sheep has been advanced by the discovery of the „ram effect," in which the introduction of a ram to a group of anoestrous ewes induces estrus (Martin et al., 1986).

The ram effect is largely due to pheromones produced by the ram's sebaceous glands, distributed over much of the body surface (Knight and Lynch, 1980a, b). Pheromone synthesis is believed to be androgen dependent, since the ram's potency in activating estrus is lost following castration but can be restored by testosterone treatment (Claus, 1979; Croker et al., 1982; Knight, 1983; Fulkerson et al., 1981). Because of seasonal variation in testosterone concentrations in the blood, it is likely that there is also seasonal variation in the potency of pheromones inducing the ram effect, and these changes likely correspond with the seasonal breeding cycle of the ewes (Al-Merestani, 1989; Boland et al., 1985; D`Occhio and Brooks, 1983; Haynes and Haresign, 1987; Knight, 1983).

We propose that off-season estrus induction can be safely and reliably promoted by the use of pheromone-rich wool fat, extracted from the wool of rams during the breeding season. Existing methods of pheromone delivery to ewes, including the evaporation of wool fat extracts, the dispersion of aerosol, and the use of wool-containing masks, are labor-intensive and provide variable doses of pheromones (Knight and Lynch, 1980b; Al-Merestani, 1989; Opitz, 1991; Cohen-Tannoudji et al., 1994; Ichimaru et al., 1999). The aim of the present study was to assess the estrus-induction efficacy of a single nasal application of ram-pheromone to non-breeding ewes. Correspondingly, we measured ovarian and hormonal responses of anoestrous German Mutton Merinos presented with wool fat.

MATERIAL AND METHODS

Production and Processing of the Wool

To produce the pheromone containing wool fat, the wool of one sexually vigorous ram was collected. The wool had grown during one breeding season and the ram was sheared in September and February. The wool fat was cold extracted with methylenchloride, as described by Over (1992). After filtration and multiple washing with the same solution, the extract was condensed in a vacuum rotation vaporizer until a paste was obtained.

Pre-selection of the Subject Ewes

We used German Mutton Merinos ewes that had lambed during a 3-week period in February, with weaning of the lambs after 60 days. Ovarian activity was monitored by transrectal ultrasonography three times: 27 days, 7 days, and roughly 4 hours before the beginning of the main experiment on May 13th (day zero). Of the 80 ewes screened, ten were selected as subjects, based on the fact that they lacked corpora lutea (CL), and had no ovarian follicles greater than 4 mm in diameter, although we did allow a maximum of one follicle with diameter of 4 mm. The ten ewes were subdivided into two groups based on the number of tertiary follicles with a diameter of ≤ 4 mm. Group A was composed of five sheep with fewer than 5 diagnosed follicles, and Group B had five sheep with 5 or more diagnosed follicles. Progesterone concentrations in blood, sampled weekly prior to the experiment, were consistently lower than 0.1 ng/ml.

Experimental Design

Starting 2 hours prior to pheromone application, blood samples were collected every 20 minutes via the vena jugularis externa to characterize initial hormone levels. The pheromone treatment consisted of rubbing 2 ml of the wool fat onto the nose of every sheep (this occurred at 9.00 a.m.). Afterwards, blood samples were taken every 10 minutes during the first hour, and then every 20 minutes during hours 1 through 7. Then, on days 1 and 2 post application, blood samples were collected every two hours; from day 3 until day 25, blood samples were taken every two days. On day 1, 2, 6, 12, 17 and 25 an ovarian diagnosis was conducted by ultrasonography and the number and size of all visible tertiary follicles as well as CL were recorded. The animals were housed in a pen until day 25 after stimulation and had no contact with rams. The serum concentrations of gonadotropins were determined by an ELISA method, and progesterone was determined by an RIA method.

RESULTS

Serum Gonadotropins

Ten minutes after the pheromone application, all sheep exhibited a significant increase in luteinizing hormone (LH) concentration in the blood; and, after ten minutes more, the concentration of follicle-stimulating hormone (FSH) was significantly increased ($p < 0.001$; Figure 1 and 2). After 180 minutes, LH and FSH concentrations returned to basal levels. A second increase in gonadotropin concentration occurred between 9 and 26 hours post application ($p < 0.001$). However, this second increase did not reach the maximum levels of the first secretion. Further increases in serum LH and FSH were determined to be simultaneous in the sequence of increase, plateau and decrease, until 26 hours post application. From this time on, simultaneous secretion was no longer found, and the animals showed individual differences in gonadotropin levels.

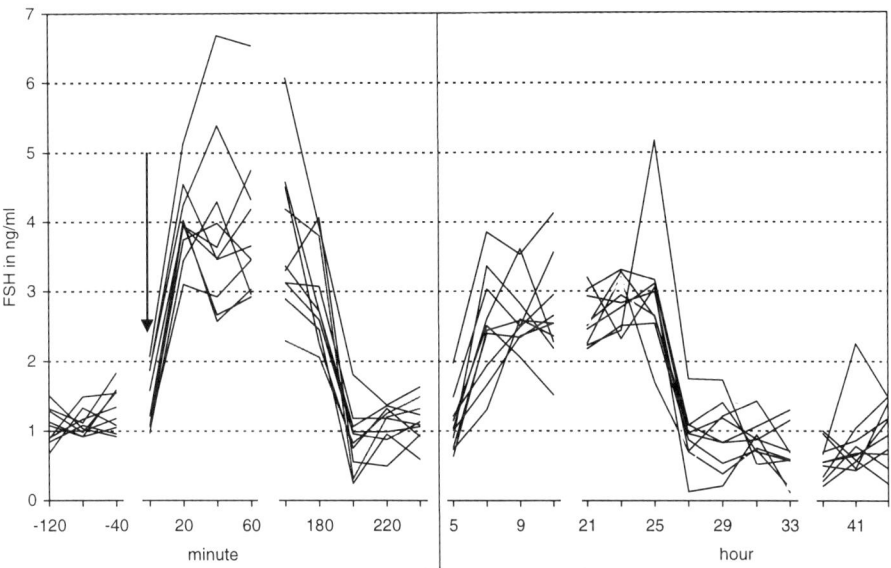

Figure 1. FSH concentrations of 10 German Mutton Merinos after pheromone application (arrow).

Ovarian Follicle Dynamics

There were detectable increases in follicle number and size in all sheep, and the specific pattern of follicle dynamics differed depending on the number of small follicles before the start of the experiment. Ewes in group A, which originally had less than 5 small follicles, had significant increases in total follicle number and in the follicle number of all three diameter-classes within 24 hours after pheromone application ($p < 0.05$, Figure 3). In comparison, among the ewes in group B, which originally had 5 or more small follicles, only the follicle diameter-class changed (Figure 3).

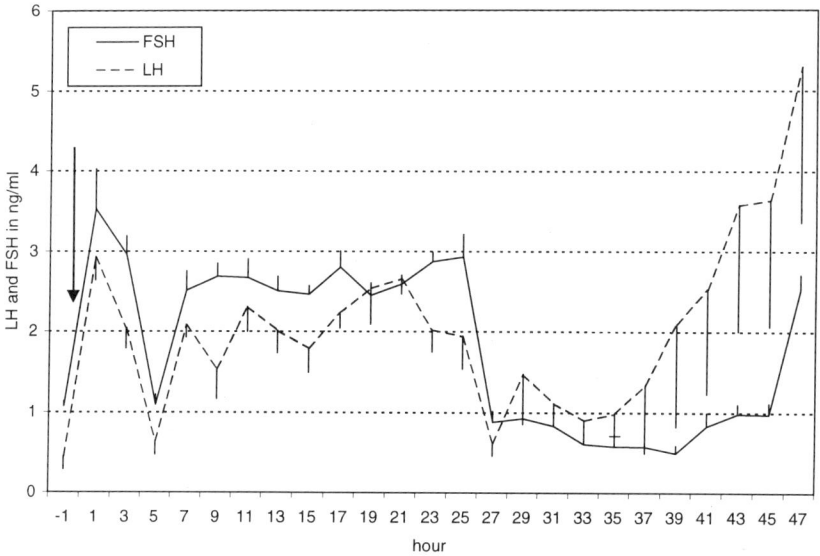

Figure 2. FSH and LH (mean ± SEM) during the first two days after pheromone application (arrow).

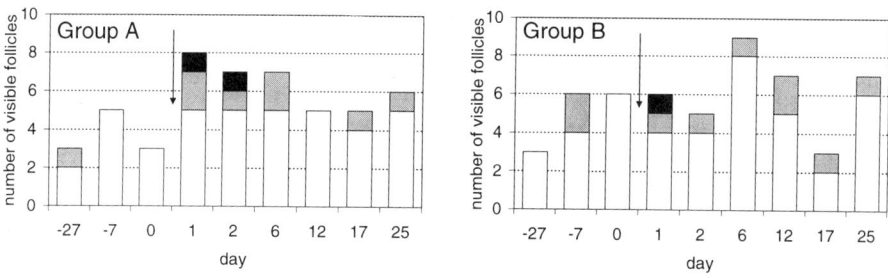

Figure 3. Development of the ultrasonically visible follicles (diameter:☐< 4 mm; ▨ 4 –5 mm ; ■ > 5 mm) in group A and B after pheromone application (arrow).

Ovulation, Corpora Lutea, and Progesterone

All 10 ewes of the main experiment ovulated, as evidenced by ultrasonography and by increased progesterone concentrations (Table 1, Figure 4). In seven sheep, an ovulation-type LH peak was detected during the first two days post application. The CL diameters that were measured post ovulation corresponded to the blood plasma progesterone concentrations. Four animals showed large, functional CL, with diameters of more than 10 mm, and had blood progesterone levels of more than 4 ng/ml, similar to our observations of cyclic ewes. The progesterone values of six ewes were below 4 ng/ml and in five cases accompanied by insufficient CL (diameter 4-6 mm). A second ovulation occurred as a result of the insufficiency of the first CL. This was ascertained by the ultrasonic examinations of the CL as well as by the progesterone concentrations. In this case, two peaks were apparent in the progesterone curve (Table 1, Figure 5). Within group B, progesterone concentrations generally remained at an increased level over 19 days and showed two peaks during this period, and a new CL developed in 3 ewes. In group A, however, a higher progesterone level was diagnosed for 11 days and only one ewe showed two peaks in the curve.

Table 1. Duration of progesterone secretion (mean ± SD) in group A and B.

	Group A	Group B
start of progesterone increase (day post application)	4.80 ± 1.79	4.4 ± 0.89
start of progesterone decrease (day post application)	16.40 ± 4.93	22.80 ± 1.79*
duration of the progesterone secretion (in days)	11.60 ± 4.51	18.40 ± 1.34*
number of progesterone peaks	1.20 ± 0.45	2.00 ± 0.00*

* Values differ $p < 0.05$

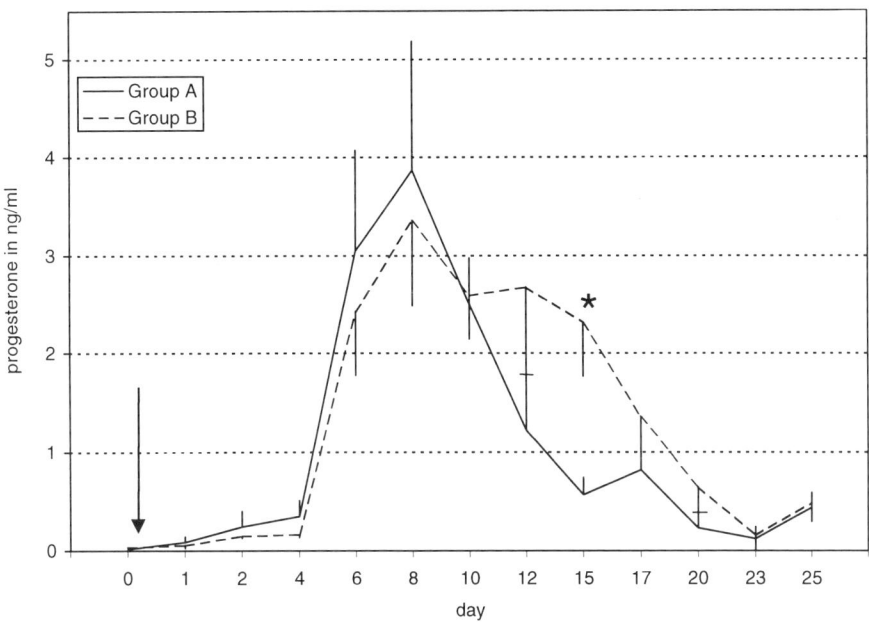

Figure 4. Progesterone concentrations (mean ± SEM) in group A and B after pheromone application (arrow, * Values differ $p < 0.05$).

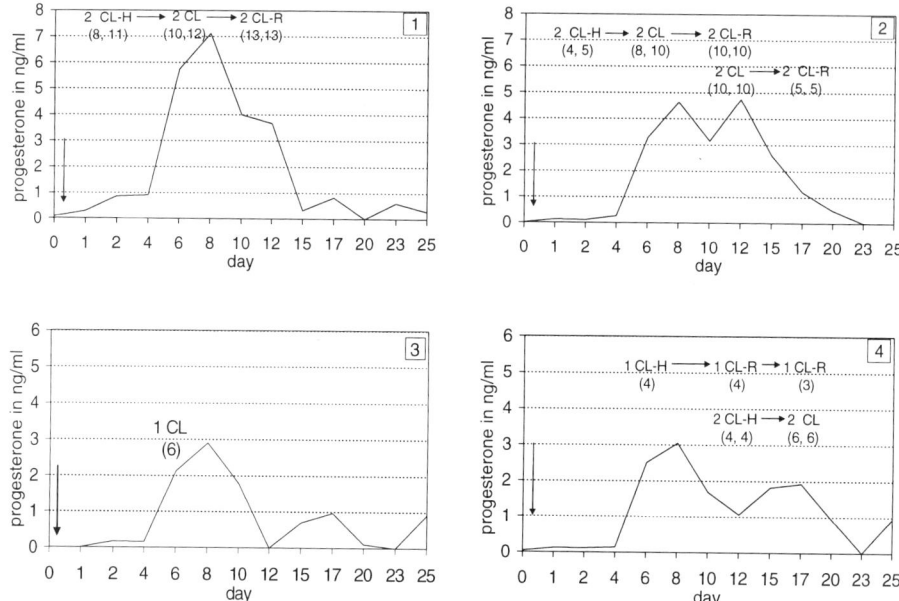

Figure 5. Progesterone concentrations and CL diagnosis (CL diameter in brackets) in sheep of group A (1, 3) and B (2, 4) with one (1, 3) or two (2, 4) progesterone peaks and physiological (1, 2) and insufficient (3, 4) CL (CL-H = corpus haemorrhagicum; CL-R = corpus luteum in regression)

DISCUSSION

A single pheromone application was sufficient to induce an ovulation in all ewes. Consequently, we suggest that visual and physical components of the ram-effect are not crucial factors for the stimulation of the gonadotropin release and the follicle development in seasonal anoestrous German Mutton Merinos. Similar to results with the presentation of whole rams, the LH concentrations in our wool fat-exposed subjects were elevated within 10 minutes of exposure, and during the next 48 hours, we recorded preovulatory LH peaks followed by ovulations (Martin et al., 1980a; Martin and Scaramuzzi 1983; Atkinson and Williamson, 1985; Al-Merestani, 1989; Scaramuzzi and Downing, 1999). We also observed a parallel increase of FSH, responsible for follicle development, which was not found by Atkinson and Williamson (1985) or Martin et al. (1980b).

The hypothalamic-pituitary endocrine reactions to pheromone exposure were not influenced by the two categories of ovary status on which we based our subject selection. However, ovary status influenced follicle dynamics and the development of the CL (size and progesterone secretion), although in both groups there were no follicles > 4 mm in diameter prior to pheromone exposure. This underlines the importance of knowing ovary status in subjects used in studies of the ram effect or pheromone effects. Therefore, in breeds with aseasonal fertility (like in German Mutton Merino: Kaulfuß et al., 1999; Kaulfuß and Berger, 2000) it is not possible to identify the reaction of the manipulated ewes as either real estrous induction of an evidently anoestrous female (no estrus behavior, no CL, no follicle > 4 mm), or stimulation of latent estrous animal (no estrus behavior but CL and/or follicles > 4 mm), or simple estrus synchronization of normal cyclic ewes.

In six of ten ewes, insufficient CL and low progesterone concentrations were diagnosed during the first 10 days post application. The reason for this might be inadequate numbers or activity of LH and FSH receptors in the follicles with < 4 mm in diameter (Braden et al., 1989). Even before the complete regression of the first CL, a second ovulation took place and the progesterone concentration failed to return to basal levels before increasing again (Knight et al., 1981). Prolonged periods of progesterone secretion were mainly observed in group B.

CONCLUSIONS

The nasal application of 2 ml pheromone containing wool fat causes a hypothalamic reaction, which is analogous to the natural ram effect. The advantage of this method to apply pheromones, which are contained in the wool extract, is the exact and easily handled dosage of the pheromone per animal. Consequently, the application of pheromones seems to be an appropriate method for estrus induction in ewes under field conditions.

REFERENCES

Al-Merestani, M. R., 1989, Examinations to influence the endocrine function of reproduction during and outside the breeding season by pheromones, PhD thesis, University of Leipzig.

Atkinson, S., and Williamson, P., 1985, Ram-induced growth of ovarian follicles and gonadotrophin inhibition in anoestrous ewes, *J. Reprod. Fert.* 73:185-189.

Boland, M. P., Al-Kamali, A. A., Crosby, T. F., Haynes, N. B., Howles, C. M., Kelleher, D. L., and Gordon, I., 1985, The influence of breed, season and photoperiod on semen characteristics, testicular size, libido and plasma hormone concentrations in rams, *Anim. Reprod. Sci.* 9:241-252.

Braden, T. D., King, M. E., Odde, K. G., and Niswender, G. D., 1989, Development of preovulatory follicles expected to form short-lived corpora lutea in beef cows, *J. Reprod. Fert.* 85:97-104.

Claus, R., 1979, *Pheromone bei Säugetieren unter besonderer Berücksichtigung des Ebergeruchsstoffes und seiner Beziehung zu anderen Hodensteroiden*, Paul Paray, Verlag, München.

Cohen-Tannoudji, J., Einhorn, J. and Signoret, J. P., 1994, Ram sexual pheromone: First approach of chemical identification, *Physiol. Behav.* 56:955-961.

Croker, K. P., Butler, L. G., Johns, M. A., and McColm, S. C., 1982, Induction of ovulation and cyclic activity in anestrous ewes with testosterone treated wethers and ewes, *Theriogenology* 17:349-354.

D'Occhio, M. J., and Brooks, D. E., 1983, Seasonal changes in plasma testosterone concentration and mating activity in Border leicester, Poll Dorst, Romney and Suffolk rams, *Austr. J. Exper. Agric. Anim. Husb.* 13:248-253.

Fulkerson, W. J., Adams, N. R., and Gherardi, P. B., 1981, Ability of castrate male sheep treated with oestrogen or testosterone to induce and detected oestrus in ewes, *Appl. Anim. Ethology* 7:57-66.

Haynes, N. B., and Haresign, W., 1987, Endocrine aspects of reproduction in the ram important to the male effect, *World Rev. Anim. Prod.* 23:21-28.

Ichimaru, T., Takeuchi, Y., and Mori, Y., 1999, Stimulation of the GnRH pulse generator activity by continuous exposure to the male pheromones in the female goat, *J. Reprod. Develop.* 45:243-248.

Kaulfuß, K.-H., and Berger, E., 2000, Ultrasonic examination of the ovary activity in sheep before and after weaning, *Reprod. Dom. Anim.* 35:29.

Kaulfuß, K.-H., Süß, R. and Moritz., 1999, Examinations about the seasonality of reproduction in German Mutton Merino and German Blackheaded sheep, in: *II. VDL-Fachtagung Forschung im Schafsektor*, pp. 115-122.

Knight, T. W., 1983, Ram induced stimulation of ovarian and oestrous activity in anoestrous ewes - a review, *Proc. New Zealand Soc. Anim. Prod.* 43:7-11.

Knight, T. W., and Lynch, P. R., 1980a, Source of ram pheromones that stimulates ovulation in the ewe, *Anim. Reprod. Sci.* 3:133-136.

Knight, T. W., and Lynch, P. R., 1980b, The pheromone from rams that stimulates ovulation in the ewe, *Proc. Aust. Soc. Anim. Prod.* 13:74-76.

Knight, T. W., Tervit, H. R., and Fairclough, R. J., 1981, Corpus luteum function in ewes stimulated by rams, *Theriogenology* 15:183-190.

Martin, G. B., and Scaramuzzi, R. J., 1983, The induction of oestrus and ovulation in seasonally anovular ewes by exposure to rams, *J. Steroid Biochem.* **19**:869 - 875.

Martin, G. B., Oldham, C. M., and Lindsay, D. R., 1980a, Increased plasma LH levels in seasonally anovular merini ewes following the introduction of rams, *Anim. Reprod.* Sci. **3**:125 - 132.

Martin, G. B., Cognie, Y., Gayerie, F., Oldham, C. M., Poindron, P., Scaramuzzi, R. J., and Thierry, J. C., 1980b, The hormonal response of teasing, *Proc. Aust. Soc. Anim. Prod.* **13**:77-79.

Martin, G. B., Oldham, C. M., Cognie, Y., and Pearce, D. T., 1986, The physiological responses of anovulatory ewes to the introduction of rams: a review, *Livest. Prod. Sci.* **15**:219-247.

Opitz, H., 1991, Pheromones in mammals and the possibility of the application of ram pheromones to induce and stimulate oestrous, PhD thesis, University of Leipzig.

Over, R., 1992, Physiological reactions and operations to isolate pheromones of rams and he-goats, PhD thesis, University of Hohenheim.

Scaramuzzi, R. J., and Downing, J. A., 1999, Effect of progesterone on the GnRH-induced secretion of oestradiol and androstenedione from the autotransplanted ovary of the anoestrous ewe, *J. Reprod. Fert.* **116**:127-132.

MALES' OLFACTORY DISCRIMINATION OF RECEPTIVE STATE OF FEMALES IN RAT-LIKE HAMSTERS (*CRICETULUS TRITON*)

Jian-Xu Zhang, Zhi-Bin Zhang, and Zu-Wang Wang

State Key Laboratory of Integrated Management of Pest Insects
and Rodents in Agriculture
Institute of Zoology
Chinese Academy of Sciences
Beijing, 100080, P.R. China

INTRODUCTION

The odors of female mammals varies with reproductive states. Males can thus detect a female's receptive status by the odors from the whole body odor, urine, specialized skin glands, or vaginal discharge. Generally, the odors of estrous females are more attractive to males than are the odors of diestrous females (Johnston, 1983; 1985; Lai *et al.*, 1996; Ordinola *et al.*, 1997). In many species, estradiol plays a role in regulating the attractiveness to males of some odor sources (Ferkin and Johnston, 1993; Ferkin *et al.*, 1991).

Estrogen-dependent chemical signals seem to be especially important cues for coordinating reproduction in rodents (Müller-Schwarze, 1983; Ebling, 1977; Ferkin *et al.*, 1993). As a result, changes in estrogen levels could be important for shifts in odors signaling sexual receptivity (Wynne-Edwards *et al.*, 1987). Several studies of rodents have shown that males prefer odors of the whole body, urine, saliva, skin, and scent glands of estrous females over those of diestrous females, and that exposure to such odors can result in increased sex-related scent marking and reduced aggression-related scent marking (Ferkin *et al.*, 1993; Lai *et al.*, 1996; Johnston, 1983; 1985).

Rat-like hamsters lead mostly solitary lives, except when mating, which appears to follow a promiscuous pattern (Wang *et al.*, 1996; Zhang *et al.*, 1999a). During the breeding season, males regularly visit several neighboring females' burrow entrances (Wang *et al.*, 1996). Except on the day of sexual receptivity, which is marked by copulation, females aggressively attack approaching males and chase them away (Zhang *et al.*, 1999a). In addition to urine, vaginal secretions, saliva, and feces, female rat-like hamsters appear to use a pair of specialized flank glands, slightly smaller than those of males, in chemosensory communication (Zhang, 1997; Zhang *et al.*, 1999b). We report here that males closely investigate odors of females, and that estradiol treatments to females increases the males' preference for females' odors.

METHODS

Subjects

About 30 male and 30 female adults of rat-like hamsters were captured by live-traps made of wire meshes in the farmland of the center of Hebei Province, North China, in April (breeding season) of 1999. The males and females were trapped in different sites, among which the distances were far enough to ensure them to be socially inexperienced. Those with body mass greater than 100g were assumed to have had sexual experience, an assumption supported by observations of Yang et al. (1996). All of our subjects in this study were sexually mature animals, ie., scrotal males and females with perforate vagina. Estrous cycles of females were determined by extra-vaginal examination and behavioral test of sexual receptivity with males. Diestrus occurred 1 to 2 days after the day of sexual receptivity.

The hamsters were housed individually in wire-mesh cages (40×25×15 cm) containing wood shavings and cotton nesting materials with rat chow and water provided continuously in every dark phase. The colony was maintained on a 14L : 10D light cycle (lights on at 1700 h) at approximately 20^0C for at least two weeks prior to surgery and behavioral testing. Ten females were ovariectomized and treated as described below; these females served as odor donors 5 weeks after surgery.

Surgical Procedure

The overiectomies were performed through bilateral incisions in the abdominal region above the ovaries. Five females received a subcutaneous implant containing estradiol. The females were anesthetized with sodium pentobarbital (60mg/kg). Every estradiol capsule was constructed from 5 mm of Silastic tubing (o.d.0.125 in, i.d. 0.062 in, No. 602-285, Dow Corning Corp., Midland, MI), which was packed with 5-mm lengths of crystalline 17β-estradiol-3-benzoate (Sigma, St. Louis) and sealed with 2.5- mm lengths of Medical Adhesive Silicone Type A (Dow Corning Corp.) in two ends. The wound was closed with sterile sutures and treated with 5% tincture of iodine and crystalline sulphanilamide.

Measurement of Testosterone Level

On the day after the completion of odor collection, scent donor females were anesthetized and trunk blood was collected. Blood was centrifuged at 4000 r/min and serum was collected and stored at -20^0C until radioimmunoassay. Radioimmunoassays were run using the reported methods (Zeng, et al., 1980) in the Laboratory of Reproduction Biology, Institute of Zoology, Chinese Academy of Sciences. ^{3}H-estradiol was from Amersham Co., UK; antibody was produced by Institute of Zoology, Chinese Academy of Sciences; Standard estradiol was from Organ Co., Holland. The detectable range of the assay was 62.5 to 2000 pg/ml. The intra-assay coefficient of variation was 8.64%.

Odor Collection

Unless otherwise noted, all scents were collected fresh for each trial from scent donor females by rubbing a piece of filter paper (6.0 mm diameter) against a flank gland surface or anogenital area. We placed a piece of filter paper into the front of the mouth and moved it around on the sides and then over the tongue to obtain saliva. We put a piece of filter paper into vagina to obtain vaginal secretion. We shaved the fur around the donor hamsters' flank glands, so scent could easily be obtained direct from the gland surface with filter paper. We collected urine and feces samples by placing a hamster in a metabolic chamber for about 6 h during its dark phase, then the urine or feces samples was placed into a clean vial which was frozen at 20^0. On the day of testing, we thawed the urine or feces samples at room temperature.

Odors Testing Procedure

Fifty pieces of filter paper pooled from the same odor source on 5 individuals were soaked in a glass vial containing 2 ml water; 10 fecal pellets were placed in a glass vial with 2 ml water; and urine was used without dilution. Glass capillary tubes (1.8 mm diameter) were used to absorb the odorous solutions, and these were presented to male subjects in their home cages as pairs of stimuli, as described below.

50 pieces of the filter papers pooled from same odor sources of every group with 5 individuals were soaked into a glass vial with 2-ml water. 10 pieces of feces were put into a glass vial with 2 ml water. Urine was directly used. We used a glass capillary (1.8 mm diameter) to absorb every odor solution or urine, and then present to experimental males in their home cage

We recorded the time spent sniffing and licking the scented ends of two glass capillaries continuously during a 3-min two-choice test with two stopwatches, respectively. Testing was initiated when the experimental hamster did the first investigation. Wilcoxon matched-pairs tests were used in this study to test for differences between the total time hamsters spent investigating each stimulus pair. Significant differences were accepted at $P<0.05$.

Experimental Procedure

In every test, an experimental male was presented in its home cage with two odor stimuli in a glass capillary from each of following pairs: (1) estrous female (E) versus diestrous female (DE); (2) estradiol-treated ovariectomized female (EOC) versus ovariectomized female (OE); (3) estradiol-treated ovariectomized (EOC) versus estrous female (E); (4) estradiol-treated ovariectomized female (EOC) versus diestrous female (DE); (5) ovariectomized female (OC) versus diestrous female (DE); (6)ovariectomized female (OE) versus estrous females (E).

We used five different scents (female urine, vaginal secretion, flank gland, saliva, and feces) to test 10 male subjects, with at least 4 days between successive tests.

RESULTS AND DISCUSSION

Male rat-like hamsters spent more time in sniffing and licking the odors of estrous females' urine, vagina, flank gland, saliva and feces than those of diestrous female (Figure 1). These results suggest that the five scents from females in estrus were all attractive to males and can be used to detect female's estrus. The five scents should be correlated with estrogen.

Male rat-like hamsters displayed preferences for these odors of EOC females over of OC females (Table 1, Figure 2). These implied that estradiol stimulated these odors' attractiveness to males, and males can detect the difference in estradiol level by these odors. Chronic changes in reproductive status and estradiol levels affected the attractiveness of these five scents of females to males.

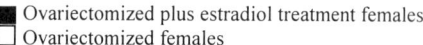
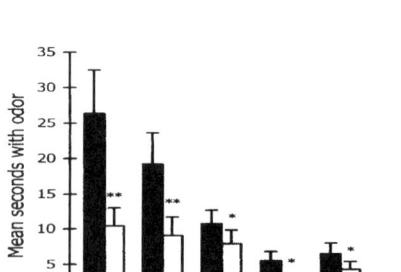 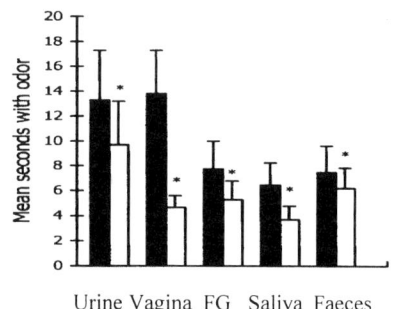

Figure 1. Mean time (sec) that males investigated the odors from estrous and diestrous females. *p<0,05; **p<0,01.

Figure 2. Mean time (sec) that males investigated the odors from ovariectomized plus estradiol replacement and ovariectomized females. *p<0,05.

Table 1. Circulating levels of estradiol in ovariectomized, intact and (ovariectomized plus estradiol treatment female rat-like hamsters ($X \pm SD$, Sample size=5).

	Ovariectomized females	Intact females	Ovariectomized plus estradiol treatment females
Estradiol level (pg/ml)	242.67±279.49	778.33±189.41	1603.80±978.20
Body weight (g)	166.3±15.0	175.3±22.4	180.3±18.4

As expected, males spent significantly more time in close proximity to the odors of estradiol-treated ovariectomized females than they did near the odors of ovariectomized females not treated with steroids (Figure 2). There were no differences in any of the stimulus pairs from estrous females and estradiol-treated ovariectomized females (Figure 3). Males spent more time in close proximity to urine, vaginal secretion, saliva and feces of diestrous females than near the same odor sources from ovariectomized females not treated with steroids, but there was no significant difference in proximity near flank gland odors (Figure 4). All odors from estrous females were preferred over those from ovariectomized females not treated with steroids (Figure 5). Likewise, all of the odors of estradiol-treated ovariectomized females were more attractive than those of diestrous females (Figure 6). All of these results are generally consistent with the conclusion that ovarian steroids in rat-like hamster females increase the attractiveness of their odors to males.

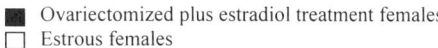
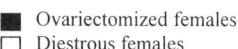

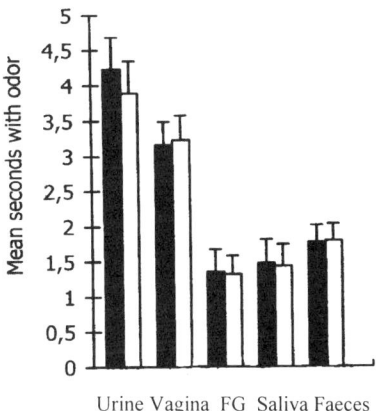
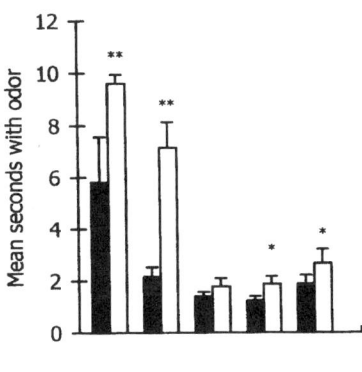

Figure 3. Mean time (sec) that males investigated the odors form ovariectomized plus estradiol treatment and estrous females.

Figure 4. Mean time (sec) that males investigated the odors from ovariectomized and diestrous females. *$p<0,05$; **$p<0,01$.

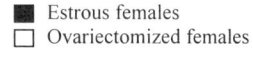
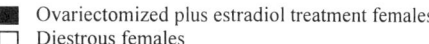

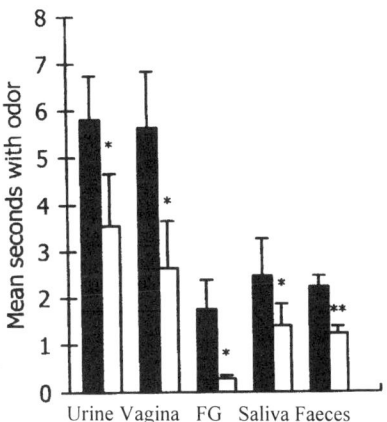
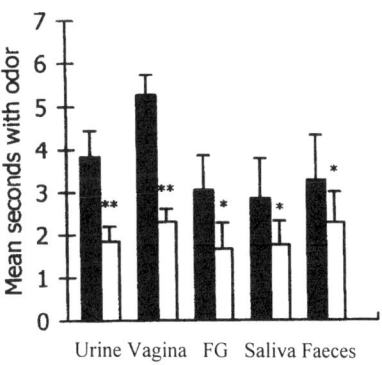

Figure 5. Mean time (sec) that males investigated the odors from estrous and ovariectomized females. *$p<0,05$; **$p<0,01$.

Figure 6. Mean time (sec) that males investigated the odors from ovariectomized plus estradiol treatment and diestrous females. *$p<0,05$; **$p<0,01$.

ACKNOWLEDGMENTS

We express our gratitude to Dr. D. Ananie for carefully commenting on this manuscript, S. Wang, S. Hao, F. Wang, X. Cao, T. Xu , G. Han and D. Guo for their many field and laboratory assistants. The work was supported by Chinese Academy of Sciences (KSCX2-1-03) and Chinese NSF (39770103, 39730090 and 39825105).

REFERENCES

Ebling, F. J., 1977, Hormonal control of mammalian skin glands, in: *Chemical Signals in Vertebrates* (D. Müller-Schwarze, and M. M. Mozell, eds.), Plenum Press, New York and London, pp. 17-34.

Ferkin, M. H., and Johnston., R. E., 1993, Roles of gonadal hormones in control of five sexually attractive odors of Meadow voles (*Microtus pennsylvanicus*), *Horm. Behav.* **27**:523-538.

Ferkin, M. H., Gorman, M. R., and Zucker, I., 1991, Ovarian hormones influence odor cues emitted by female meadow voles (*Microtus pennsylvanicus*), *Horm. Behav.* **25**:572-581.

Johnston, R. E., 1983, Chemical signals and reproductive behavior. in: *Pheromone and Reproduction in Mammals* (J. G. Vandenbergh, ed.), Academic Press Inc. Orlando, Florida, pp. 3-37.

Johnston, R. E., 1985, Communication. in: *The Hamster-Reproduction and Behavior.* (H. I. Siegel, ed.), Plenum Press, New York, pp. 121-154

Lai, S., Vasilieva, N. Y., and Johnston, R. E., 1996, Odors providing sexual information in Djungarian hamsters: evidence for an across-odor code, *Horm. Behav.* **30**:26-36.

Müller-Schwarze, D., 1983, Scent gland in mammals and their functions. in : *Advances in the Study of Mammalian Behavior* (J. F. Eisenberg, and D. G. Kleiman, eds), Special Publications No. 7, The American Society of Mammalogist, pp. 147-191.

Ordinola, P., Martinez-Gomez, M., Manzo, J., and Hudson, R., 1997, Response of male domestic rabbits (*Oryctolagus cuniculus*) to inguinal gland secretion from intact and ovariectomized females, *J. Chem. Ecol.*, **23**:2079-2091.

Wang, S., Yang, H., and Hao, S., 1996, Activity rang, activity rhythm and food preference in ratlike hamsters (*Cricetulus triton*), *Chinese J. Zool.* **31**:28-31.(in Chinese)

Wynne-Edwards, K. E., Terranova, P. F., and Lisk, R. D., 1987, Cyclic Djungarian hamsters (*Phodopus campbelli*) lack the progesterone surge normally associated with ovulation and behavioral receptivity, *Endocrinology* **120**:1308-1316.

Yang, H., Wang, S., and Hao, S., 1996, An investigation on populations of ratlike hamsters (*Criccetulus triton*), their predication and the integrated management in the non-irrigated area on Huabei Plain, China. in: *Theory and Practice of Rodent Pest Management* (Z. Wang and Z. Zhang eds.), Science Press, Beijing, China, pp.229-246 (in Chinese).

Zeng, G., Jiang, G., and Wang, H., 1980, Changes of serum LH, estradiol-17β and progesterone levels in Hu-Yang and Chinese Karakul ewes during the oestrous cycle, *Acta Veterinaria Et Zootechnica Sinica* **11**:147-154 (in Chinese).

Zhang, J., 1997, Chemical communication in ratlike hamsters (*Cricetulus triton*), Unpubl. Ph.D. dissert., Institute of Zoology, Chinese Academy of Sciences, Beijing, China (in Chinese).

Zhang, J., Zhang, Z., and Wang, Z., 1999a, Behavioral interactions and mating behavior of ratlike hamsters (*Cricetulus triton*) during breeding season, *Acta Theriologica Sinica* **19**(2): 132-142 (in Chinese).

Zhang, J., Zhang, Z., and Wang, Z., 1999b, Development and sexual differences in flank glands of rat-like hamsters (*Cricetulus triton*), *Acta Zool. Sinica* **45**:390-397.

PHEROMONAL REGULATION OF BANK VOLE (*CLETHRIONOMYS GLAREOLUS*) REPRODUCTION

Anna Marchlewska-Koj

Institute of Environmental Sciences
Jagiellonian University
Ingardena 6, 30-060 Kraków, Poland

INTRODUCTION

The bank vole *(Clethrionomys glareolus)* is a common European rodent, which is relatively easy to breed in laboratory conditions. In this species, similarly to other mammals, chemical compounds present in urine, faces or skin gland secretions can serve as olfactory signals for conspecifics (Marchlewska-Koj, 1984; Stoddard and Sales, 1985). In a natural population, adult females are territorial, and during the breeding season each female's home range overlaps with the home range of several males (Bujalska, 1973). The results of laboratory experiments indicate that marking behavior in bank vole males is correlated with the hierarchical status of the animals, and it has been suggested that the urine of adult males contains chemical signals involved in maintaining their social organization (Rozenfeld *et al.*, 1987). This is consistent with observations of bank vole females, which show an increase of behavioral activity in response to chemosignals from dominating adult males (Kruczek and Pochroń, 1997).

The bank voles reared in our laboratory came from Northeast Poland. They are maintained as an outbred stock (heterogenic but closed population), and we assume that the animals from such a colony are comparable to rodents living in a natural population.

This paper reviews evidence supporting the idea that the reproductive activity of bank voles is significantly influenced by olfactory cues. We began a series of experiments to take a systematic approach by examining a number of different endocrine and behavioral responses influenced by chemosignals.

INFLUENCE OF CHEMOSIGNALS ON HORMONAL ACTIVITY OF CONSPECIFICS

A number of investigators have reported that in natural populations of *Clethrionomys rufocanus* (Kalela, 1957; Saitoh, 1981) and *Clethrionomys glareolus* (Bujalska, 1970,

1973), females born into a high-density population mature later than those born into low-density populations. This was confirmed by experiments in seminatural conditions. The onset of puberty in bank vole females, measured by uterine weight, ovarian weight and the number of Graafian follicles in ovaries, supported the hypothesis that maturation of females born at the beginning of breeding season is suppressed by the presence of conspecific adult females (Kruczek and Marchlewska-Koj, 1986). A number of different stimuli – tactile, auditory, visual, and olfactory – could have operated in these experiments. Animals from the same colony were used for experiments to evaluate the influence of pheromones on maturation of bank vole females.

The animals were maintained in cages at $18\pm2°C$, under a 14 h light : 10 h dark regime (light on at 06:00 h). Standard pelleted chow for rabbits and water were available *ad libitum*. Wood shavings were provided as bedding material. Females after weaning at 21 days of age were reared in unisexual groups of 2–3 animals per cage and exposed to urine of an intact male, intact female or ovariectomized (OVX) female for 3 weeks (Kruczek *et al.*, 1989). The number of Graafian follicles in ovaries was used as an indicator of sexual maturation of females (Figure 1). The presence of intact male urine increases the number of mature ovarian follicles, whereas females exposed to urine of hormonally active or OVX females have fewer mature follicles than non–exposed animals. The results of the experiments clearly show that delayed puberty in females is evoked by the urine of adult females and that the presence of chemosignals is not controlled by ovarian hormones. In this respect, there is a strong similarity between *Mus musculus* and *Clethrionomy glareolus* females.

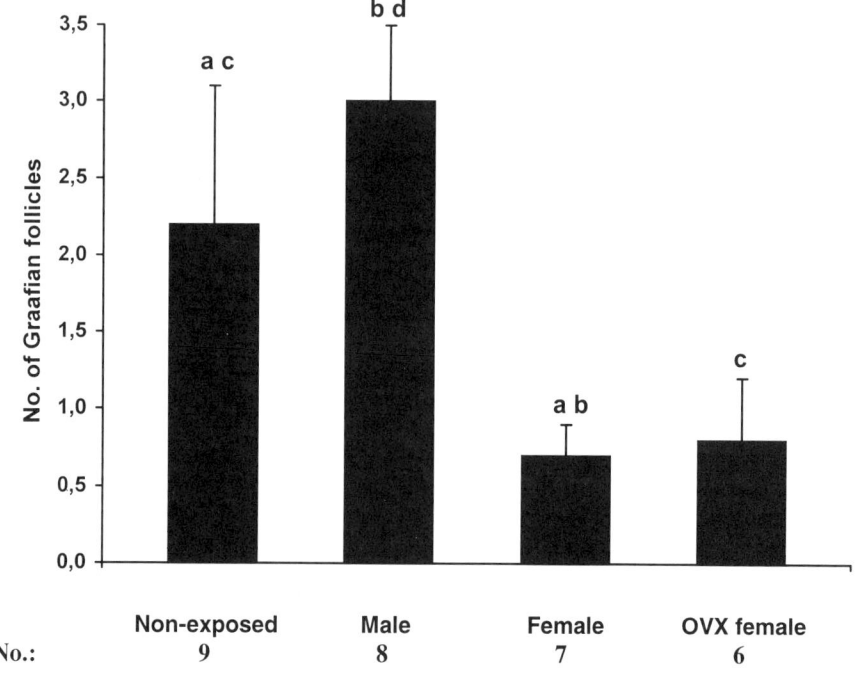

Figure 1. Number of Graafian follicles in ovaries of 6-week-old bank voles maturated in presence of adult conspecifics. Means (±SE) with the same symbols (letter) differ significantly; ANOVA Duncan's new multiple range test at p<0,05 (Data from Kruczek *et al.*, 1989).

Clethrionomys glareolus female, similarly to other Microtidae, is characterized by provoked ovulation (Clarke, 1985; Marchlewska-Koj, unpublished data), ovulation is induced by mating, and occurs 6 to 14 hours after coitus (Clarke *et al.*, 1970). However, in laboratory conditions a female kept in the presence of a male (located in a small wire-net cage) showed spontaneous ovulation during the next 4 to 6 days after introduction of the male. Copulation did not take place but tactile, olfactory, visual, and auditory stimuli passed between them (Clarke and Hellwing, 1977). A similar male effect was observed in female field voles (*Microtus agrestis*), another induced–ovulating species (Milligan, 1974). There is some evidence that the reproductive activity of adult bank vole females is stimulated by male pheromones (Jemioło *et al.*, 1980). This appears as an increased number of Graafian follicles in 4-month-old females reared in the presence of males in the same cage but separated by a net, or in females exposed to male urine (Figure 2).

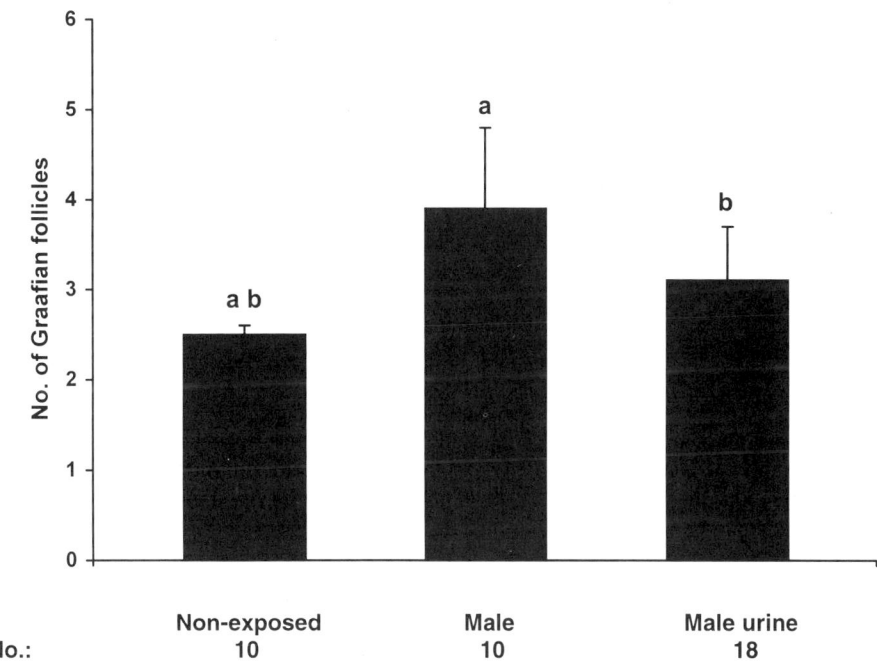

Figure 2. Number of Graafian follicles in ovaries of 4-month-old bank voles exposed to adult male or male's urine. Means (±SE) with the same symbols (letter) differ significantly; ANOVA Duncan's new multiple range test at p<0,05 (Data from Jemiolo *et al.*, 1980).

Housing a recently inseminated female with an alien male, or even the presence of its urine, terminates pregnancy, and the female returns to estrus during the next 3 to 4 days. This typical male pheromonal effect (Bruce effect) was first described in laboratory mice (Bruce, 1960; Marchlewska-Koj, 1983) and later reported in a few species of *Microtidae* including *Clethrionomys glareolus* (Clarke, 1985). Gray–tailed voles (*Microtus canicaudus*) recently were used as an experimental model to investigate the Bruce effect in *Microtus* living in natural conditions (de la Maza *et al.*, 1999). In controlled field studies, reproductive females exposed to unfamiliar males or females did not show significant differences in intervals between parturitions. In experiments in our laboratory, receptive female bank voles were mated with males, and the day after copulation they were transferred to an empty cage or a cage with a strange male. Two of 10 females reared with strange males copulated again on the third day after pairing. Moving a recently inseminated female to a new cage did not disturb pregnancy, and all 16 females transferred to empty cages delivered pups between the 17th and 19th days after copulation. The laboratory observations indicate that chemosignals released by males and females affect the hormonal activity of juvenile and adult female bank voles.

MODIFICATION OF BANK VOLE BEHAVIOR BY CHEMOSIGNALS

In *Clethrionomys glareolus*, as in other rodents, olfactory signals have been suggested as important modifiers of behavior. Female bank voles use the olfactory system to discriminate between sexual partners. During a two–choice preference test, adult females selected the sexually active male over a castrated one. Consequently, females are able to recognize the social status of males on the basis on their odor, and they strongly prefer dominant over subordinate male (Kruczek, 1997). However, olfaction is not essential for reproduction of bank vole females; removal of both olfactory systems in females did not affect their mating or maternal behaviors. Six of 9 OBX females paired with adult males copulated, gave birth, and nursed pups until weaning (Kruczek, 1998).

Sexually active bank vole males are much more dependent on functional olfactory systems. Chemosignals may affect ultrasonic vocalization, which can be a part of courtship in bank voles. Introduction of an anaesthetized female to a vivarium with a male elicited ultrasound vocalization in 6 tested animals. On the other hand, none of the 15 tested adult females exposed to anaesthetized males emitted ultrasounds. Bank vole males emitted ultrasounds at frequencies of 30 to 45 kHz, responding to the presence of an awake or anaesthetized female and even of the bedding of an adult female. The stimulating effect of freely moving females was noticed in non–operated males, but in bulbectomized animals ultrasounds were not recorded (Kapusta *et al.*, 1996). This indicates that female olfactory signals were the stimulus.

Bank vole males with functional olfactory systems exhibited aggression only toward other adult males (Marchlewska-Koj *et al.*, 1989). Adult males are non-aggressive during encounters with non-receptive females but the behavior changes after bulbectomy (Kapusta *et al.*, 1996). Total activity estimated by the number of non-aggressive and aggressive approaches was significantly lower in OBX males than in sham-operated animals. However, in the OBX males aggressive behavior toward tested females, significantly increased and they attacked female partners (Figure 3a, b).

Removal of both the main and accessory olfactory systems impaired the sexual behavior of males. OBX males paired with receptive females showed not only significantly fewer non–aggressive approaches to females but also impaired mounting behavior. Only 2 of 6 OBX males paired with receptive females copulated, whereas all 8 sham–operated males showed mounting behavior during 3–hour encounters. Copulation was confirmed by the presence of spermatozoa in the vagina. This indicates that olfaction plays a crucial role in identification of sex in the bank vole, and that this system directly controls the reproductive activity of males (Marchlewska-Koj, unpublished data).

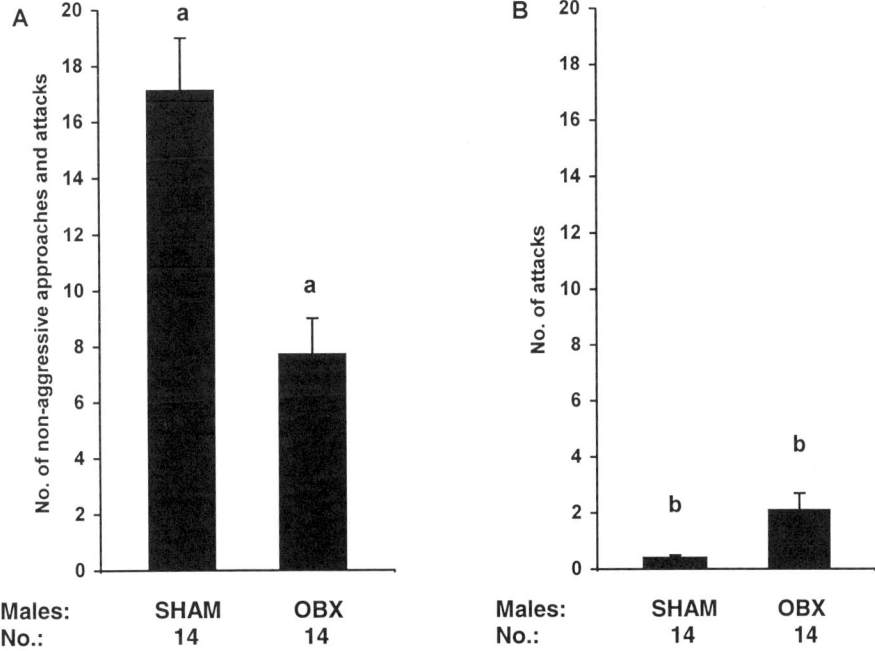

Figure 3. Behavior of sham-operated (SHAM) and bulbectomized (OBX) bank vole males in the presence of non-receptive females during 10 min encounters. **A**, Total activity: number of non-aggressive approaches and attacks. **B**, Aggressive behavior: number of attacks. Means (±SE) with the same symbols (letter) differ significantly; ANOVA at p<0,05 (Data from Kapusta et al., 1996).

CONCLUDING REMARKS

Bank voles occupy the litter zone of the forest floor or underground burrows, and live mostly non-visual lives. They have a very complicated and active social life, but because they are subject to predator attacks they have developed intraspecific communication understood only by members of this species. There is no doubt that the olfactory system is involved in many area of bank vole biology including reproductive activity, and that it serves same–sex or male–female communication. Females are able to differentiate dominant from subordinate males by odor. This ability may play an important role in sexual selection in this species. However, olfaction is not essential for reproduction and removal of olfactory systems does not affect mating or maternal behavior. On the other hand, chemical signals released by females affect male behavior. In males the olfactory system plays a crucial role in identification of sex and is directly involved in reproduction, including copulatory behavior. Moreover, bank vole males respond to female odor by production of ultrasounds during sexual encounters, but the biological function of this phenomenon is not clear.

Laboratory observations indicate that the presence of sexually active males affects hormonal activity of female bank voles. However, both male effects: – acceleration of puberty and pregnancy termination – were observed only in laboratory colony. Field verification of these results is required. A much more pronounced effect is observed during chemical interaction among bank vole females. Suppression of the maturation of juvenile females by chemosignals of adult females can affect the reproduction of animals living in natural conditions. Data obtained in seminatural and laboratory conditions can be referred to the density–dependent fluctuation of the reproductive activity of bank voles observed in natural populations.

ACKNOWLEDGMENTS

The work was supported by a grant from the State Committee for Scientific Research, Warsaw, Poland, KBN PB 0919/PO4/98/15.

REFERENCES

Bruce, H. M., 1960, A block to pregnancy in the mouse caused by proximity of strange males, *J. Reprod. Fert.* **1**:96-103.
Bujalska, G., 1970, Reproduction stabilising elements in an island population of *Clethrionomys glareolus* (Schreber 1980), *Acta theriol.* **15**:381-412.
Bujalska, G., 1973, The role of spacing behaviour among females in the regulation of reproduction in the bank vole, *J. Reprod. Fert.* (Suppl), **19**:465-474.
Clarke, J. R., 1985, The reproductive biology of the bank vole *(Clethrionomys glareolus)* and the wood mouse *(Apodemus sylvaticus), Symp. Zool. Soc. Lond.* **55**:33-59.
Clarke, J. R., and Hellwing, S., 1977, Remote control by males of ovulation in bank voles *(Clethrionomys glareolus), J. Reprod. Fert.* **50**:155-158.
Clarke J. R., Clulow, F. V., and Greig, F., 1970, Ovulation in the bank vole, *Clethrionomys glareolus, J. Reprod. Fert.* **23**:531.
de la Maza H., Wolff J. O., and Lindsey A., 1999, Exposure to strange adults does not cause pregnancy disruption or infanticide in the gray-tailed vole, *Behav. Ecol. Sociobiol.* **45**:107-113.
Jemioło B., Marchlewska-Koj, A., and Buchalczyk, A., 1980, Acceleration of ovarian follicle maturation of female caused by male in *Microtus agrestis* and *Clethrionomys glareolus, Folia Biol.* (Kraków) **28**:269-277.
Kalela, O., 1957, Regulation of reproductive rate in subarctic populations of the vole, *Clethrionomys rufocanus (Sund), Ann. Acad. Sci. Fenn. Ser. A IV Biol.* **34**:1-60.
Kapusta, J., Marchlewska-Koj, A., Olejniczak, P., and Kruczek M., 1996, Removal of the olfactory system modifies male bank vole behaviour in the presence of females. *Behav. Process.*, **37**:39-45.
Kruczek, M., 1997, Male rank and female choice in the bank vole, *Clethrionomys glareolus, Behav. Process.* **40**:171-176.
Kruczek, M., 1998, Male chemical signals and female choice in the bank vole *Clethrionomys glareolus* - Jagiellonian University - Kraków - Monography no **334**:1-52.
Kruczek M., and Marchlewska-Koj, A., 1986, Puberty delay of bank vole females in a high-density population, *Biol. Reprod.* **35**:537-541.
Kruczek, M., Marchlewska-Koj, A., and Drickamer, L. C., 1989, Social inhibition of sexual maturation in female and male bank voles *(Clethrionomys glareolus)* - *Acta theriol.* **34**:479-485.
Kruczek M., and Pochroń, E., 1997, Chemical signals from conspecifics modify the activity of female bank voles *(Clethrionomys glareolus), Acta theriol.* **42**:71-78.
Marchlewska-Koj, A., Kołodziej, B., and Filimowska, A., 1989, Aggressive behavior of adult bank voles *(Clethrionomys glareolus)* towards conspecifics, *Aggr. Behav.* **15**:381-387.
Marchlewska-Koj, A., 1983, Pregnancy blocking by pheromones, in: *Pheromones and Reproduction in Mammals*, (J. G. Vandenbergh ed.), Acad. Press Inc., New York, pp. 151-174.
Marchlewska-Koj, A., 1984, Pheromones and mammalian reproduction, in: *Oxford Reviews of Reproductive Biology*, Vol. 6, (J. R. Clarke ed.), Oxford Univ. Press, Oxford, pp. 266-302.
Milligan, S. R., 1974, Social environment and ovulation in the vole, *Microtus agrestis, J. Reprod. Fert.* **41**:35-47.
Rozenfeld, F. M., Boulange, E., and Rasmont, R.,1987, Urine marking by male bank voles (*Clethrionomys glareolus* Schreber, 1780; Microtidae, Rodentia) in relation to their social rank, *Can. J. Zool.* **65**:2594-2601.
Saitoh, T., 1981, Control of female maturation in high density populations of the red-banked vole, *Clethrionomys rufocanus bedfordial, J. Anim. Ecol.* **50**:79-87.
Stoddart, D. M., and Sales, G. D., 1985, The olfactory and acoustic biology of wood mice, yellow-necked mice and bank voles, *Symp. Zool. Soc. Lond.* **55**:117-139.

OLFACTORY COMMUNICATION IN BRANDT'S VOLE (*MICROTUS BRANDTI*)

Li Zhang, Jiming Fang, and Ruyong Sun

Institute of Ecology
College of Life Sciences
Beijing Normal University
Beijing, 100875, P. R. China

INTRODUCTION

Chemical signals play an important role in social behaviors of many rodent species, mediating sexual, aggressive, parental and spacing behaviors, as well as influencing an animal's internal hormonal milieu (Halpin, 1986; Johnston, 1990, 1993). Studies of conspecific odor preferences in voles have provided insight on chemosensory communication mechanisms in the diverse social systems among different *Microtus* species (Sawrey and Dewsbury, 1994).

The odor preferences of Brandt's vole (*Microtus brandti*) in behavioral development, mate choice, and in individual and group discrimination, were evaluated in a Y-shaped testing apparatus and in home cages (Zhang and Fang, 1996, 1999; Zhang *et al.*, 2000). We also compared the odor preferences of voles housed in long and short photoperiods, responses to odors from other voles housed in long and short photoperiods, and the effects of social odors on testosterone in males. These data attest to a prominent role for olfactory communication in the social life of Brandt's vole.

ODOUR DISCRIMINATION OF PUPS

Animals appear to base their discrimination of conspecific odors on their own social experiences during development and in adulthood (Brown, 1991). Studies of the development of odor preferences were conducted to help us understand the chemosensory mechanisms of social relationship parents and their pups, as well as documenting the activities of the developing pups in their colonies (Zhang *et al.*, 2000). During the weaning period (15 through 30 days of age; see Table 1), we examined the odor preference of pups by observing their investigation of four categories of soiled bedding, taken from the cages of their mothers, fathers, and unfamiliar males and females.

Pups investigated the odors of unfamiliar adults significantly more than those of their parents. Beginning at 24 days of age, pups spent more time close to the odors of their mothers than near that of their fathers (Table 1). These data show age-related changes in olfactory discriminations.

Table 1. Comparison between the duration (mean±SE) of sniffing experimental odors, n=8.

Odor	Duration of sniffing substrates (s)					
	15d	18d	21d	24d	27d	30d
Mother	6.9±2.7	8.0±0.7cd.	6.0±0.5cd.	6.2±0.4bcd	5.0±0.3bcd	7.0±1.1bcd
Father	3.3±0.3cd.	8.2±0.7cd.	5.4±0.6cd	4.9±0.4acd	4.4±0.6acd	4.7±0.3acd
Novel female	6.8±1.1b	14.7±1.9ab	12.6±0.9ab	16.1±2.0ab	11.7±1.1ab	11.4±0.5ab
Novel male	4.9±0.5b	11.1±1.2ab	13.6±1.4ab	12.9±0.9ab	9.6±0.7ab	9.2±0.8ab

Mann-Whitney test: a, b, c, d: $p<0.05$; a: comparing with mother's odor, b: comparing with father's odor, c: comparing with novel female odor, d: comparing with novel male odor

The ability of Brandt's vole pups to discriminate colony odors was also studied. Soiled bedding substrates from pups' own nests and from nests of unfamiliar voles were presented in a wooden Y-maze. Beginning at age 27 days, pups spent significantly more time near their own nest odors (Table 2), so it is suggested that they develop their odor discrimination at this age.

Table 2. Comparison on the duration of sniffing (mean±SE) of pups to different colony odors, n=8.

Colony odor	Duration of sniffing substrate (s)					
	15d	18d	21d	24d	27d	30d
Own colony	7.0±1.2	9.4±1.4	12.6±1.9	13.4±2.8	9.1±1.1	11.8±0.8
Novel colony	8.1±2.1	7.1±1.4	9.4±1.4	9.7±1.2	5.5±0.5	6.7±0.7
Wilcoxon test	ns	ns	ns	ns	**	***

** $p<0.01$, *** $p<0.001$, ns: no significance

ODOR PREFERENCES OF ADULT VOLES

Whole body odors, presumed to represent a combination of odors from different sources, may be individually distinctive (Brown, 1979; Halpin, 1986), and the discrimination of an individual's species and sex is the first step in a series of behavioral interactions (Ferkin and Johnston, 1995). We have studied voles during their breeding period, making observations of agonistic behavior, parental behavior, and other social behaviors, but the role of olfaction has not been determined. The ability of female voles to discriminate the odors of their mates from those of unfamiliar males was investigated by presenting female voles with a choice between bedding soiled by their mates or by unfamiliar males during a 30 min test. We observed that female voles spent more time near the odors of their mates than near those of unfamiliar males (Table 3). These results suggest that female Brandt's voles can discriminate males by their scents. It is suggested that females pair-bonded to their mates, which could enhance the survival of pups.

Table 3. Comparison of female behaviors (mean±SE) in different odor locations, n=15.

Odor	Entrance Frequency (number)	Staying Duration (s)	Self-grooming		Resting		Flank marking	
			Frequency (number)	Duration (s)	Frequency (number)	Duration (s)	Frequency (number)	Duration (s)
Spouse male	23.7±2.8	802.0±59.5	11.1±1.5	69.1±12.4	3.2±0.7	56.2±36.	56.2±36.	5.7±1.5
Novel male	5.5±4.2	360.0±31.8	2.2±0.4	6.0±1.9	0.1±0.1	0.2±0.2	2.9±0.6	3.4±1.2
Wilcoxon test	**	*	*	**	*	*	*	ns

* $p<0.05$; ** $p<0.01$; ns: no significance

The responses of adult male Brandt's voles to the odors of their mates and unfamiliar adult females were also studied. Four pairs of odor combinations (soiled bedding) were tested: estrous mates with estrous unfamiliar females, estrous mates with diestrous unfamiliar females; diestrous mates with estrous unfamiliar females; diestrous mates and diestrous unfamiliar females. In three of these four tests, males responded by spending more time near the odors of the unfamiliar females than near that of their mates (Table 4). Only when the mate was in estrous condition and the unfamiliar female was diestrous did the males show a preference for their mates. These results indicate that male voles prefer odors of estrous females over those of diestrous females, and that they prefer the odor of an unfamiliar female when that female is in the same estrous phase as their mate.

Table 4. Comparison on the duration of male's sniffing, flank scratching and self-grooming behaviors in experimental boxes with spouse female odor and unfamiliar female odor (mean±SE), n=10.

Experiment Odor Stimulus	Sniffing		Flank scratching		Self-grooming	
	Frequency (number)	Duration (s)	Frequency (number)	Duration (s)	Frequency (number)	Duration (s)
Estrous novel female	56.1±5.7	56.0±8.1	3.6±0.8	3.4±0.8	5.9±1.0	18.5±5.5
Estrous spouse female	28.8±2.2	17.5±2.3	1.6±0.4	0.8±0.2	3.4±0.7	7.7±1.4
	**	**	*	*	ns	*
Diestrous novel female	49.2±8.0	33.6±6.7	3.3±1.1	2.7±0.8	4.7±0.8	11.3±2.2
Estrous spouse female	68.7±4.3	53.7±3.9	5.7±1.4	6.0±2.1	9.1±1.3	26.2±4.5
	*	*	ns	ns	*	*
Estrous novel female	80.8±11.9	70.1±9.2	6.2±1.6	6.7±2.3	8.4±1.4	25.0±4.2
Diestrous spouse female	31.9±2.6	12.0±1.3	1.1±0.4	1.1±0.5	3.8±0.8	8.9±2.3
	**	**	*	**	*	**
Diestrous novel female	77.3±3.6	47.5±4.0	4.0±1.0	4.0±1.3	7.8±1.3	16.6±3.3
Diestrous spouse female	38.4±1.8	12.5±1.0	1.8±0.7	0.8±0.3	6.1±1.4	13.9±3.2
	**	**	ns	*	ns	ns

Wilcoxon test: * $p<0.05$, ** $p<0.01$, ns: no significance

PHOTOPERIOD INFLUENCES ODORS, ODOR PREFERENCES AND THE MALE'S PLASMA TESTOSTERONE LEVEL

Seasonal change of photoperiod is one of the main ecological factors that regulate seasonal breeding (Vandenbergh, 1988). The role of photoperiod on olfaction is an active area of research (Ferkin and Gorman, 1992; Ferkin et al., 1995). Our present experiments indicated that male voles living in a long-day photoperiod (LD) showed more social investigation of odors from LD males and females than of the odors of short-day photoperiod (SD) males and females (Table 5). However, SD males did not show any significant differences in their investigation of odors from LD and SD females.

Urine and feces from animals housed in LD and SD conditions were used as odor stimuli in 10-min discrimination tests of adult Brandt's voles. Both LD and SD voles spent more time near the urinary and fecal scents from LD donors. All respondents preferred the odors of the opposite-sex donors over those of the same sex. These results suggest that urinary and fecal chemosignals might convey information about sex and seasonal breeding conditions.

In a final experiment, conspecific novel male's substrate was given as individual odor stimulus to solitary adult male. The plasma testosterone of the male actors was determined by radioimmunoassay. We found that presenting the soiled bedding of an unfamiliar male caused an increase in the male's testosterone levels in the plasma, and that the increased levels varied with the length of the exposure. Within 30 min, LD males had larger increases in testosterone than did SD males (Table 6). One and two hours after the onset of exposure, the testosterone levels of LD males were higher, but not statistically significantly so, than those of SD males.

Table 5. The comparison on the odor preferences (mean±SE) of male actors to the substrate of LD and SD male and female donors, n=10.

Male actors	Odor	Sniffing F	Sniffing D	Digging F
LD male	LD male odor	21.6±3.1	380.8±101.6	27.4±2.8
	SD male odor	13.2±1.8	78.1±18.2	13.5±1.9
		*	**	**
	LD estrous female odor	26.6±4.2	244.3±45.0	11.1±1.8
	SD estrous female odor	11.3±2.3	69.0±21.8	5.4±1.8
		*	**	*
	LD diestrous female odor	16.5±3.0	78.6±19.3	14.6±3.3
	SD diestrous female odor	11.5±1.5	30.2±5.0	9.0±2.2
		ns	**	ns
SD male	LD male odor	13.9±2.7	128.1±48.4	12.0±2.3
	SD male odor	6.6±1.2	23.6±6.3	3.8±1.6
		*	*	*
	LD estrous female odor	18.8±2.8	88.2±18.7	6.7±1.6
	SD estrous female odor	14.9±2.8	73.4±23.9	7.0±1.7
		ns	ns	ns
	LD diestrous female odor	13.9±2.7	6.7±1.6	8.0±2.3
	SD diestrous female odor	6.6±1.2	7.0±1.7	7.7±2.2
		ns	ns	ns

Wilcoxon test: * $p<0.05$, ** $p<0.01$, ns: no significance; F= frequency (number), D= duration (s)

Table 6. The comparison on the changes of plasma testosterone level (mean±SE) between LD and SD adult male voles exposed in a novel adult male's substrate.

Photoperiod	Plasma testosterone level (ng/ml)			
	Control	30 minutes	1 hour	2 hours
LD male	1.620±0.950	3.754±2.297	5.900±2.631	13.005±4.196
SD male	0.077±0.026	0.102±0.039	3.579±2.543	7.426±3.247
Mann-Whitney test	ns	*	ns	ns

* $p<0.05$, ns: no significance

GENERAL DISCUSSION

All of these results suggest that both production and perception of sociosexual chemosignals are influenced by photoperiod. During LD, social odors presumably evoke greater levels of sexual interest than do odors produced during SD. In contrast, odors produced during SD suppress sexual attraction and might lack corresponding stimulatory effects on reproductive physiology. SD odors might be more compatible with group living, an important thermoregulatory strategy used by some mammals in temperate climates. Overall, our results reported here suggest that photoperiod's effects on chemosignal production, and perhaps perception, might be a consequence of its effects on underlying physiology and endocrine secretion.

REFERENCES

Brown, R. E., 1979, Mammalian social odours: a critical review, *Adv. Stud. Behav.* **10**:103-162.
Brown, R. E., 1991, Effects of rearing condition, gender, and sexual experience on odour preferences and urine marking in Long-Evans rats, *Anim. Learn. Behav.* **19**(1):18-28.
Ferkin, M. H., and Gorman, M. R., 1992, Photoperiod and gonadal hormones influence odor preferences of the male meadow vole, *Microtus pennsylvanicus*, *Physiol. Behav.* **51**:1087-1091.
Ferkin, M. H., and Johnston, R. E., 1995, Meadow voles, *Microtus pennsylvanicus*, use multiple sources of scent for sex recognition, *Anim. Behav.* **49**:37-44.
Ferkin, M. H., Sorokin, E. S., and Johnston, R. E., 1995, Seasonal changes in scents and responses to them in meadow voles: evidence for the co-evolution of signals and response mechanisms, *Ethology* **100**:89-98.
Halpin, Z. T., 1986, Individual odours among mammals: origins and functions, *Adv. Stud. Behav.* **16**:39-70.
Johnston, R. E., 1990, Chemical communication in golden hamsters: from behavior to molecules and neural mechanisms, in: *Contemporary Issues in Comparative Psychology*, (D. A. Dewsbury, ed.), Sinauer Press, New York.
Johnston, R. E., 1993, Memory for individual scent in hamsters (*Mesocricetus auratus*) as assessed by habituation methods. *J. Comp. Psychol.* **107**:201-207.
Sawrey, D. K., and Dewsbury, D. A., 1994, Conspecific odour preferences in montane voles (*Microtus montanus*): effects of sexual experience, *Physiol. Behav.* **56**:339-344.
Vandenbergh, J. G., 1988, Pheromones and mammalian reproduction, in: *The Physiology of Reproduction*, (E. Knobil, and J. Neill, eds.), Raven Press, New York.
Zhang, L., and Fang, J., 1996, Discrimination of male adult Brandt's vole (*Microtus brandti*) on group odour during nonbreeding period, *Acta Theriol. Sinica* **16**(4):285-290.
Zhang, L., and Fang, J., 1999, Odour preferences of female Brandt's vole (*Microtus brandti*): discrimination of male individual scents, *Acta Zool. Sinica* **45**(3):294-301.
Zhang, L., Fang, J., and Sun, R., 2000, Behavioural development on olfactory communication of Brandt's vole (*Microtus brandti*): the colony odor discrimination of pups, *Acta Theriol. Sinica* **20**(1):30-36.

EFFECTS OF SOCIAL DOMINANCE AND FEMALE ODOR ON SPERM ACTIVITY OF MALE MICE

Sachiko Koyama and Shinji Kamimura

Division of Biology, Department of Life Sciences
Graduate School of Arts and Sciences
University of Tokyo
Tokyo, 153-8902, Japan

1. INTRODUCTION

In the latter half of twentieth century, much has been learned about chemical communication in house mice (*Mus musculus*). It is known that urine of mice can signal information about strain, physiological state (estrus or diestrus, stressed or not), social status, and so on (Brown, 1979). Some of the volatile semiochemicals in urine have been identified (Novotny *et al.*, 1985; Harvey *et al.*, 1989; Novotny *et al.*, 1999b), and Major Urinary Protein (MUP) has also been found to function as a conveyance that delivers the semiochemicals to receptor (Novotny *et al.*, 1999a). Recent evidence suggests that MUP acts as a „storage and releasing system" that lengthens the persistence of such semiochemicals by binding with them and gradually releasing them (Hurst *et al.*, 1998; Novotny *et al.*, 1999a).

The odor of urine can release behaviors in recipient mice (releaser effect), and it can change physiological states of recipients (primer effect) (Brown, 1979, 1985a, 1985b; Cavaggioni *et al.*, 1999; Vandenbergh, 1994). At least ten different pheromonal effects have been identified, but there appear to be more demonstrations of such effects on females than on males (Cavaggioni *et al.*, 1999; Vandenbergh, 1994). Precisely, there have been few studies showing the influence of odor from female mice on the physiological state of male mice (Batty, 1978; Barnard *et al.*, 1997), and the results of those studies were contradictory. Testosterone concentration was higher in males that were kept with females in one study (Batty, 1978), and it was lower in another (Barnard *et al.*, 1997).

In the present study, we investigated the influences of the odor and physical proximity of female mice on sperm activity and weights of reproductive organs of paired male mice. Mice are frequently used in the study of spermatology (Si and Okuno, 1993). We applied the techniques of spermatology to evaluate the reproductive activity of male mice by comparing sperm motility and sperm density. Considering the results of our previous study (Koyama and Kamimura, 1999), we also investigated the dominant-subordinate relationships of the males and compared the influence of females according to the male's social status in the present study.

In our previous study, we observed that the sperm motility (percent of motile sperm) of subordinate mice was lower than that of dominant males cohabiting with the subordinate males (Figure 1). Furthermore, the sperm motility of isolated males was equivalent to that of dominant males and significantly greater than that of the subordinate males, suggesting that social subordination by another male leads to a decline in sperm motility.(Koyama and Kamimura, 1999).

Such phenomena of suppressed physiological states in socially subordinated mice have also been observed in some group living non-human primates (Epple and Katz, 1984; Snowdon, 1990; Wingfield *et al.*, 1994) and carnivores (Creel *et al.*, 1992). Female subordinates had suppressed ovarian cycles and decreased estrogen levels (Creel *et al.*, 1992; Snowdon, 1990). Male subordinates also showed low levels of testosterone concentration in some species (Abbott *et al.*, 1989; Arnold and Dittami, 1997). In talapoin monkeys, exposure of males to females caused an increase in serum testosterone in dominant, but not subordinate, males (Eberhart *et al.*, 1980; Wingfield *et al.*, 1994).

Feral house mice live in social groups of 2 to 10 individuals, as determined by food abundance in the environment (Berry and Bronson, 1992). As our previous study (Koyama and Kamimura, 1999) showed that sperm motility differed according to social status, and as there has been information on reproductive suppression according to social dominance in other group-living animals, we thought that it would be important to investigate the effect of female odor according to social status in the present study.

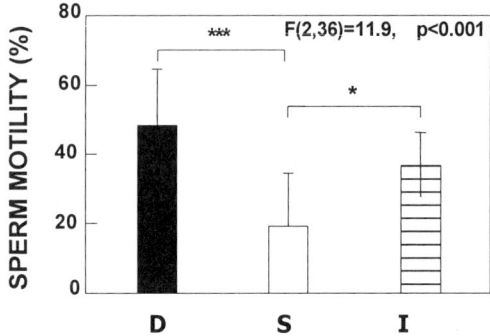

Figure 1. Sperm motility of dominant, subordinate, and isolated mice. D: dominant mice, S: subordinate mice, I: isolated mice. *: $p<0.05$, ***: $p<0.001$. (Used with permission from Koyama and Kamimura, 2000.)

2. METHODS

We used 3 groups of pair-housed male mice (a total of 118 males). Most, but not all, such pairs of males will readily establish social dominance relationships (sensu Uhrich, 1938). In order to get most simplified dominant-subordinate relationships in our study, we kept all the males in pairs in which only the „1 dominant and 1 subordinate" type of dominant-subordinate relationship could be possible to emerge. Treatments to all pairs of males started when the males were 5 weeks of age and continued until they were killed for physiological assessments at 15 weeks of age.

The 3 groups of male mice were as follows:

1) 36 males were pair housed and were supplied with soiled bedding from cages of isolated females every 3 or 4 days from 5 weeks of age to 15 weeks of age (Female

Bedding Group). At the time it was collected, the soiled bedding thus contained fully fresh urine and defecation to 3 to 4 days' old urine and defecation. Every female mouse was provided about 5 g of fresh bedding and 1 g of soiled bedding from male mice that were used for intruder tests (see section 2.1). The provision of male-soiled bedding to females helped ensure normal estrous cycles (the Whitten effect). The 6 g of fresh bedding turned to 12 to 16 g of soiled bedding when it was collected 3 or 4 days later. The soiled bedding was considered to contain high amount of volatile and nonvolatile, fresh and old, estrus and diestrus odorants from females. The female used as the source of the soiled bedding was randomized every time in order to avoid the possibility that a pair would receive soiled bedding from a female with abnormal estrous cycles.

2) 38 male mice were pair housed and were kept with cages of isolated female mice placed on top of their own cages (With Female Group). Cages were placed as one cage of female mouse per one cage of paired male mice. The males could see, smell, and reach to the female mouse placed on their cages by their forelegs or whiskers, but they could not copulate with the female mouse. Urine and defecation of the female mouse dropped down into the males' cages below.

3) The third group of paired male mice was the Control Group (number of pairs =16). These pairs did not have any physical contact with females or female-soiled bedding. As they were kept in the same room as the other 2 groups, volatile odor and vocalizations of females may have reached these males. So, this group is a group with fewer stimuli from females than others.

2.1. Determination of Dominant-Subordinate Relationships

Three times a week the paired males went through intruder tests. In mice and rats, it is known that the dominant male of a group attacks any intruders (Rowe and Redfern, 1969). So, placing an intruder animal into the home-cages of paired male mice and noting which male attacks to the intruder is a convenient method to check dominant-subordinate relationships of mice (Koyama and Kamimura, 1999). We used this intruder test method to our study and assigned points to the male that showed aggression: +1 point when he attacked to the intruder and +2 points when he attacked to his cohabitant (Koyama and Kamimura, 1999). Each intruder test terminated when one of the residents showed aggression, or when 10 min had passed without any aggression. Males used as intruders were kept in threes. They were used only for this purpose and were not used in the final investigations on sperm activity, etc.

Pairs with clear dominant-subordinate relationships were apparent when the difference of total points between the cohabitating 2 mice was large at the end of intruder test series. We used only the paired mice with clear dominant-subordinate relationships to investigate the influence of female mice. Even if the difference of total points between the cohabitants of a pair was large, the pair was excluded from the final investigation on sperm activity and weights of reproductive organ when the male, that usually did not show attack long through the intruder test series and thus was nearly determined to be subordinate, began to show attack at the end of intruder test series. In this case we considered the possibility that the dominant-subordinate relationships might have reversed.

2.2. Observation of Sperm and Measuring Weights of Reproductive Organs

At the end of 15 weeks of age, pairs with clear dominant-subordinate relationships went through final investigation of sperm activity and weight of reproductive organs. Testosterone and corticosterone concentrations were also investigated. Blood was immediately collected from the heart after the mice were euthanized by cervical dislocation. After the body was weighed, sperm was immediately collected from the right cauda epididymis, and was diluted with ca.100 volumes of modified B.W.W. medium (Si and Okuno, 1993; Koyama and Kamimura, 1999). The diluted suspension of sperm was observed under a phase-contrast microscope (Nikon Optiphoto) and was video-recorded using a Newvicon camera (Panasonic, WV-1850).

Sperm motility was determined by counting the number of motile spermatozoa and immotile spermatozoa in several frames of the video-records.

In order to determine sperm density, the left cauda epididymis was cut and sperm from 1 cm of the tubular duct was diluted in 0.5 ml of distilled water. 100 µl of the diluted sperm was assessed by counting the number of spermatozoa with a Thoma blood cell counter. Weights of testes and preputial glands were then measured.

3. RESULTS

The number of pairs that showed clear dominant-subordinate relationships had a tendency to be higher in With Female Group, although it was not significant (Chi-square=2.403, df=2, n.s.). Numbers of pairs with clear dominant-subordinate relationships were as follows: 5 (28%) for Female Bedding Group, 9 (47 %) for With Female Group, and 4 (25%) for Control Group.

Sperm motility was significantly higher in the dominants than the subordinates [$F(1,30)=18.753$, $p<0.001$] (Figure 2). Group differences were not significant [$F(2,30)=1.153$, $p=0.329$], but the interaction between the factors of social dominance and group differences was nearly significant [$F(2,30)=2.696$, $p=0.084$]. It was revealed from the posthoc tests (Fisher's LSD) that, although the results of subordinates of all the 3 groups did not significantly differ, the differences between the dominants and subordinates were significant only in the Control Group and Female Bedding Group ($p=0.001$ for Control Group and $p=0.023$ for Female Bedding Group). Sperm motility of the dominants of With Female Group was rather low, and did not have significant difference with the subordinates of the same group.

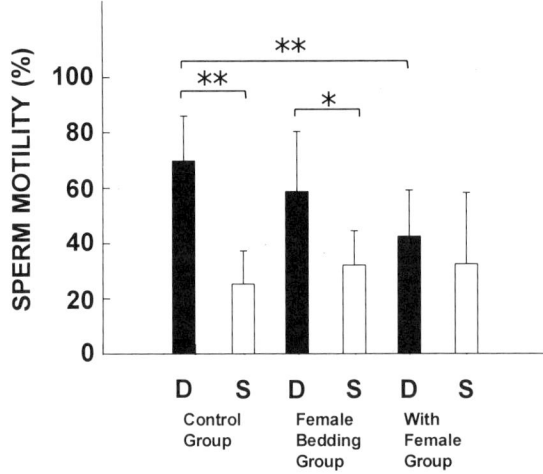

Figure 2. Sperm motility of dominant and subordinate mice according to housing condition. D: dominants, S: subordinates. *: $p<0.05$, **: $p<0.01$. (Used with permission from Koyama and Kamimura, 2000.)

Sperm density also showed significant difference between dominant-subordinate relationships [$F(1,26)=37.340$, $p<0.001$] (Figure 3). But there were remarkable contrasts to the results in sperm motility. Sperm density in Control Group was low both in dominants and subordinates. Posthoc tests (Fisher's LSD) showed that there were not a significant difference between the dominants and subordinates of Control Group. On the other hand, sperm density of the dominants of Female Bedding Group and With Female Group was significantly higher than the subordinates (Female Bedding Group, $p<0.001$, With Female Group, $p=0.001$).

Sperm density of the dominants of these groups was also significantly higher than the dominants of Control Group (Control Group vs. Female Bedding Group, $p<0.001$, Control Group vs. With Female Group, $p=0.002$). Sperm density of subordinates was low in all the 3 groups and there were not any significant differences among them.

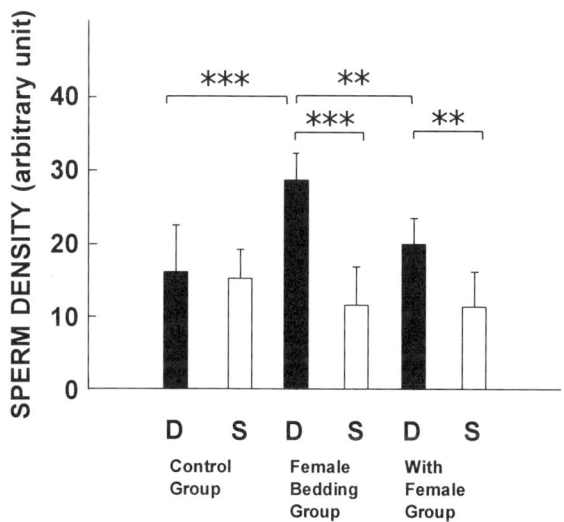

Figure 3. Sperm density of dominant and subordinate mice according to housing condition. D: dominants, S: subordinates. **: $p<0.01$, ***: $p<0.001$. (Used with permission from Koyama and Kamimura, 2000.)

The weights of preputial glands (data not shown) were significantly heavier in the dominants than in the subordinates [$F(1,30)=11.020$, $p=0.002$]. But posthoc tests revealed that such difference between the dominants and subordinates was clear only in the Female Bedding Group ($p=0.003$) and With Female Group ($p=0.007$). Although the analysis of variance on housing condition was not significant, the weights of preputial glands of the dominants and subordinates of the Control Group were light, and there was not a significant difference between them. Such tendency in the Control group to have lighter weights in the organs than the other 2 groups was also observed in the weights of testes [housing condition, $F(2,31)=2.964$, $p=0.066$]. But there were no significant differences in the weights of testes according to social status.

Testosterone concentration (data not shown) was significantly higher in the dominants than the subordinates [$F(1,22)=5.945$, $p=0.023$]. Corticosterone concentration did not show differences according to social status nor housing condition.

4. DISCUSSION

The results of the present study can be summarized into the following points: 1) Sperm density of the dominant mice are subject to being influenced by female odor and female existence. 2) Sperm motility and sperm density of the subordinate mice are suppressed, and female odor and existence have little influence on them. 3) As the results of dominant males of „Female Bedding Group" showed a clear increase in the sperm density, it would be able to say that the „odor" of females alone can influence reproductive activity of the dominant males.

The results of the dominant mice and subordinate mice were different as such. So, we would like to discuss on the dominants and subordinates respectively at first, and then make general discussion.

4.1. Effect of Female Odor on the Dominant Mice

Female odor was found to have an influence on the physiological state of dominant mice, especially on sperm density and preputial glands. The results on sperm density suggest that female odor enhances spermatogenesis of dominant male mice. On the other hand, the results on sperm motility suggest that female odor do not raise sperm motility.

Sperm motility is related to sperm maturation. And sperm maturation is a phenomenon different from spermatogenesis. Sperm maturation is completed in epididymis and spermatogenesis is conducted in testis (Sharpe, 1994; Yanagimachi, 1994). The results of the present study suggest that female odor and existence have influence on spermatogenesis rather than sperm maturation. In a study using rats, the number of spermatozoa was found to be larger in the males that were kept with females than in those that were kept without females (Taylor *et al.*, 1987). Such results on rats also suggest that female stimuli have influence on spermatogenesis.

The results of the weight of preputial gland in the present study resembled the results of sperm density (Koyama and Kamimura, submitted). These results suggested that 1) there would be some factor regulating spermatogenesis and the growth of preputial glands in common, 2) the factor would not be related to sperm maturation, and 3) the regulatory mechanism of the factor is susceptible to female odor.

4.2. The Subordinate Mice

Female odor did not have influence on subordinate mice. Their sperm motility and density were low as a whole.

In our previous study (Koyama and Kamimura, 1999), the sperm motility of the subordinates was found suppressed. This was also observed in the present study. On the other hand, the results of sperm density did not differ according to social status in our previous study (Koyama and Kamimura, 1999), but it did differ according to both of the social status and housing condition in the present study. As female stimuli were not provided in our previous study (Koyama and Kamimura, 1999), those results correspond to the results on males in the Control Group of the present study, which also lacked significant difference according to social status. The present study thus clarified that the suppressed state of reproduction in subordinate mice was not overcome by the provision of female stimuli.

Such low susceptibility in the subordinates to female odor is also observed in talapoin monkeys. The subordinate did not increase testosterone concentration when they were exposed to females, whereas females did enhance testosterone concentration in the dominants (Eberhart *et al.*, 1980; Wingfield *et al.*, 1994). So, why do the subordinates lack responsiveness to female odor? We consider two reasons for such low responsiveness in the subordinate male mice.

The first hypothesis is „the primer effect hypothesis." The odor of dominant male mice might have suppressed the sperm motility and density of subordinate mice. And such influence of the odor of dominant mice on subordinate mice may have been so strong as to suppress responsiveness to female odor as observed in the present study.

The second hypothesis is the „stress hypothesis." Although the subordinates did not show higher corticosterone concentration, the subordinates may have suffered from „subordinate stress" (Blanchard *et al.*, 1993; Miczek *et al.*, 1991; Wingfield *et al.*, 1994; Zayan, 1991), which could have worked to suppress their responsiveness to female odor.

4.3. General Discussion

It can be concluded that female odor have enhancing effect on male reproductive activity, but these effects are restricted to the dominant mice. As the previous studies on group-living non-human primates and carnivores had implicated (Epple and Katz, 1984; Snowdon, 1990; Wingfield *et al.*, 1994; Creel *et al.*, 1992; Abbott *et al.*, 1989; Arnold and Dittami, 1997; Eberhart *et al.*, 1980), the physiological state of mice should be considered differentially according to their social status.

The mechanism of how female odor enhances spermatogenesis in dominant males would be of great interest. What is the factor influenced by female odor, which is related with spermatogenesis, and not related with sperm maturation? What are the exact changes that female odor evokes in the vomeronasal organ, in the accessory olfactory bulb, and in the brain system, to reproductive system? Although there is not any study that would directly answer these questions, some of the recent studies are highly suggestive. A recent study revealed that volatile semiochemicals could evoke excitatory response in single vomeronasal neurons of mice, and the detection threshold was extremely low (Leinders-Zufall *et al.*, 2000). The exposure of females to urine or soiled bedding of male mice and male rats induced a lot of Fos-ir cells (Inamura *et al.*, 1999) and increase in c-fos mRNA level (Guo *et al.*, 1997) in the accessory olfactory bulb of the females. The removal of vomeronasal organ in female rats resulted in a large reduction of such Fos-ir cells in accessory olfactory bulbs (Inamura *et al.*, 1999). The odor of unfamiliar male odor has been found to increase secretion of dopamine from arcuate nucleus of hypothalamus, which functioned to reduce prolactin secretion from pituitary gland in female mice and induced Bruce effect (Kaba, 1994). How the results of these studies are related, or not related to our present study would be fascinating themes to be resolved in future.

REFERENCES

Abbott, D. H., Barrett, J., Faulkes, C. G., and George, L. M., 1989, Social contraception in naked mole-rats and marmoset monkeys, *J. Zool. Lond.* **219**:703-710.
Arnold, W., and Dittami, J., 1997, Reproductive suppression in male alpine marmots, *Anim. Behav.* **53**:53-66.
Barnard, C. J., Behnke, J. M., Gage, A. R., Brown, H., and Smithurst, P. R., 1997, Immunity costs and behavioural modulation in male laboratory mice (*Mus musculus*) exposed to the odours of females, *Physiol. Behav.* **62**:857-866.
Batty, J., 1978, Acute changes in plasma testosterone levels and their relation to measures of sexual behaviour in the male house mouse (*Mus musculus*), *Anim. Behav.* **26**:349-357.
Berry, R. J., and Bronson, F. H., 1992, Life history and bioeconomy of the house mouse, *Biol. Rev.* **67**:519-550.
Blanchard, D. C., Sakai, R. R., McEwen, B., Weiss, S. M., and Blanchard, R. J., 1993, Subordination stress: behavioral, brain, and neuroendocrine correlates, *Behav. Brain Res.* **58**:113-121.

Brown, R. E., 1979, Mammalian social odors: a critical review, in: *Advances in The Study of Behavior, Vol. 10*, (J. S. Rosenblatt, R. A. Hinde, C. Beer, and M.-C. Busnel, eds.) Academic Press, Inc., New York.

Brown, R. E., 1985a, The rodents I: effects of odours on reprocductive physiology (primer effects), in: *Social Odour in Mammals, Vol. 1*, (R. E. Brown and D. W. Macdonald, eds.) Clarendon Press, Oxford.

Brown, R. E., 1985b, The rodents II: suborder Myomorpha, in: *Social Odour in Mammals, Vol. 1*, (R. E. Brown and D. W. Macdonald, eds.) Clarendon Press, Oxford.

Cavaggioni, A., Mucignat, C., and Tirindelli, R., 1999, Pheromone signaling in the mouse: role of urinary proteins and vomeronasal organ, *Arch. Ital. Biol.* **137**:193-200.

Creel, S., Creel, N., Wildt, D. E., and Monfort, S. L., 1992, Behavioural and endocrine mechanisms of reproductive suppression in Serengeti dwarf mongooses, *Anim. Behav.* **43**:231-245.

Eberhart, J. A., Keverne, E. B., and Meller, R. E., 1980, Social influences on plasma testosterone levels in male talapoin monkeys, *Horm. Behav.* **14**:247-266.

Epple, G., and Katz, Y., 1984, Social influences on estrogen excretion and ovarian cyclicity in saddle-back tamarins (*Saguinus fuscicollis*). *Am. J. Primatol.* **6**:215-227.

Guo, J., Zhou, A. W., and Moss, R. L., 1997, Urine and urine-derived compounds induce c-fos mRNA expression in accessory olfactory bulb, *Neuroreport* **8**:1679-1683.

Harvey, S., Jemiolo, B., and Novotny, M., 1989, Pattern of volatile compounds in dominant and subordinate male mouse urine, *J. Chem. Ecol.* **15**:2061-2072.

Hurst, J., Robertson, D. H. L., Tolladay, U., and Beynon, R. J., 1998, Proteins in urine scent marks of male house mice extend the longevity of olfactory signals, *Anim. Behav.* **55**:1289-1297.

Inamura, K., Kashiwayanagi, M., and Kurihara, K., 1999, Regionalization of Fos immunostaining in rat accessory olfactory bulb when the vomeronasal organ was exposed to urine, *Eur. J. Neurosci.* **11**:2254-2260.

Kaba, H., 1994, Neural, synaptic and molecular mechanisms of an olfactory memory and its biological significance, *Jpn. J. Taste Smell Res.* **1**:23-33 (in Japanese).

Koyama, S., and Kamimura, S., 1999, Lowered sperm motility in subordinate social status of mice, *Physiol. Behav.* **65**:665-669.

Koyama, S., Kamimura, S., 2000, Influence of social dominance and female odor on the sperm activity of male mice, *Physiol. Behav.* **71**:415-422.

Leinders-Zufall, T., Lane, A. P., Puche, A. C., Ma, W., Novotny, M. V., Shipley, M. T., and Zufall, F., 2000, Ultrasensitive pheromone detection by mammalian vomeronasal neurons, *Nature* **405**:792-796.

Miczek, K. A., Thompson, M. L., and Tornatzky, W., 1991, Subordinate animals, in: *Neurobiology and Neuroendocrinology of Stress*, (M. R. Brown; C. River, and G. Koob, eds.) Marcel Dekker; New York.

Novotny, M. V., Harvey, S., Jemiolo, B., and Alberts, J., 1985, Synthetic pheromones that promote inter-male aggression in mice, *Proc. Natl. Acad. Sci. USA* **82**:2059-2061.

Novotny, M. V., Ma, W., Wiesler, D., and Zidek, L., 1999a, Positive identification of the puberty-accelerating pheromone of the house mouse: the volatile ligands associating with the major urinary protein, *Proc. R. Soc. Lond. B* **266**:2017-2022.

Novotny, M. V., Jemiolo, B., Wiesler, D., Ma, W., Harvey, S., Xu, F., Xie, T-M., and Carmack, M., 1999b, A unique urinary constituent, 6-hydroxy-6-methyl-3-heptanone, is a pheromone that accelerates puberty in female mice, *Chem. Biol.* **6**:377-383.

Rowe, R. P., and Redfern, R., 1969, Aggressive behaviour in related and unrelated wild house mice (*Mus musculus* L.), *Ann. Appl. Biol.* **64**:425-431.

Si, Y., and Okuno, M., 1993, Multiple activation of mouse sperm motility, *Mol. Reprod. Develop.* **36**:89-95.

Sharpe, R. M., 1994, Regulation of spermatogenesis, in: *The Physiology of Reproduction, Second Edition*, (E. Knobil and J. D. Neill, eds) Raven Press, Ltd., New York.

Snowdon, C. T., 1990, Mechanisms maintaining monogamy in monkeys, in: *Contemporary Issues in Comparative Psychology*, (D. A. Dewsbury, ed.) Sinauer Associates, Inc., Massachusetts.

Taylor, G. T., Weiss, J., and Rupich, R., 1987, Male rat behavior, endocrinology and reproductive physiology in a mixed-sex, socially stressful colony, *Physiol. Behav.* **39**:429-433.

Uhrich, J., 1938, The social hierarchy in albino mice, *J. Comp. Physiol. Psychol.* **25**:373-413.

Vandenbergh, J. G., 1994, Pheromones and mammalian reproduction, in: *The Physiology of Reproduction, Second Edition*, (E. Knobil and J. D.Neill, eds.) Raven Press, Ltd., New York.

Wingfield, J. C., Whaling, C. S., and Marler, P., 1994, Communication in vertebrate aggression and reproduction: the role of hormones, in: *The Physiology of Reproduction, Second Edition*, (E. Knobil and J. D. Neill, eds.) Raven Press, Ltd., New York.

Yanagimachi, R., 1994, Mammalian fertilization, in: *The Physiology of Reproduction, Second Edition*, (E. Knobil and J. D. Neill, eds.) Raven Press, Ltd., New York.

Zayan, R., 1991, The specificity of social stress, *Behav. Process* **25**:81-93.

EXPOSURE OF JUVENILE MALE CAMPBELLI'S HAMSTERS AND HOUSE MICE TO CAT URINE ELICITS SPECIES-SPECIFIC RESPONSES IN REPRODUCTIVE DEVELOPMENT

Natalija V. Sokolskaja,[1] Raimund Apfelbach,[2] Dietrich von Holst,[3] and Nina Y. Vasilieva[1]

[1]A.N.Severtzov Institute of Ecology and Evolution,
Russian Academy of Sciences,
Leninskiy pr., 33, Moscow 117071, Russia
[2]Department of Zoology, University of Tübingen,
Auf der Morgenstelle 28, 72076 Tübingen, Germany
[3]Department of Animal Physiology, University of Bayreuth,
Box 3008, D-8580, Bayreuth, Germany

INTRODUCTION

Predator-prey interactions have recently received increased attention from evolutionary biologists and ecologists (for review, Lima and Dill, 1990; Lima, 1998). It has been documented that the risk of predation elicits different behavioral and physiological responses in prey species. A limited number of studies have focused on the influence that predators or their odors exert on reproduction of the potential prey. For example, decreased reproductive organ masses were observed in male and female bank voles placed in conditions simulating high predation (Ylönen, 1989; Heikkila *et al.*, 1993). Exposure to cat urine caused a delay in sexual maturation of Mongolian gerbils (Vasilieva, 1995), and of golden (Vasilieva, 1997) and Campbelli's hamsters (Vasilieva *et al.*, 1999).

We decided to compare responses to predator odors between two species that vary in their reproductive strategies, and which have evolved in conditions with presumably different levels of predation risk. Our subjects were house mice, *Mus domesticus*, and Campbelli's hamsters, *Phodopus campbelli*. The population density of Campbelli's hamsters in the field is usually low and remains relatively stable over the years (Flint, 1966; Surov *et al.*, 1988). This observation suggests the existence of well-developed population regulatory mechanisms in this species, including ecological factors that influence the onset of puberty. The presence of the father, or exposure to male urine and midventral gland secretions, suppressed the reproductive development of juvenile males (Sokolov *et al.*, 1989, 1991, 1992; Kuznetzova and Vasilieva, 1997). In contrast, exposure to males or their odors accelerated sexual maturation in juvenile females (Levin and Johnston, 1986; Reasner and

Johnston, 1988; Sokolov et al., 1989, 1991, 1992; Gudermuth et al., 1992). Surprisingly, odor cues from males of a closely related species influenced the rate of sexual maturation to the same extent as those from conspecific males (Sokolov and Vasilieva, 1996; Kuznetzova and Vasilieva, 1997).

Our previous experiments showed that odor cues from adult males, including urine and midventral gland secretions, suppressed sexual maturation in young male Campbelli's hamsters, as evidenced by decreased mass of the testes and the epididymis (Sokolov et al., 1989, 1991, 1992; Kuznetzova and Vasilieva, 1997). Though exposure to cat urine also suppressed male sexual maturation, this was manifested as a reduction in the mass of the epididymis and increased mass of the adrenal glands, suggesting activation of the adrenocortical stress axis in the responses to cat urine (Vasilieva et al., 1999). For more than 40 years, house mice have been model species for the study of chemical cues involved in puberty regulation and estrous cycling (Whitten, 1956, 1959; Marsden and Bronson, 1964; Drickamer, 1986; Vandenbergh, 1983; Novotny et al., 1999). In this study, we compare the effects of cat urine on sexual maturation of male Campbelli's hamsters and house mice.

MATERIALS AND METHODS

The Campbelli's hamsters used in these experiments were laboratory stock derived from 5 pairs of breeders captured in northeastern Mongolia in 1985. The house mice were from an outbred laboratory stock kept in our laboratory for about 30 years. Breeding pairs of age 5 to 8 months were kept in 20 x 40 x 25 cm plastic cages. All animals were provided with rodent chow, grains, vegetables and water *ad libitum*. Shaved wood chips served as bedding. The colony was maintained on 14L:10D light cycle at 20±2 °C.

The test subjects were 30 male hamsters from 10 litters and 39 male mice from 9 litters. The 11-day-old hamsters and 12-day-old mice were randomly assigned to either the experimental (15 hamsters and 24 mice) or control (15 hamsters and 15 mice) group, within these constraints: 1) one half of the male pups from each litter served as experimental subjects; the second half served as controls; 2) the differences in body weight of pairs of siblings assigned to the experimental or control group did not exceed 0.5 g at the beginning of the experiment. Experimental and control animals were kept in subgroups of 4 - 5 hamsters or 7-8 mice in 20 x 40 x 25 cm plastic cages.

The group-housed juvenile males were treated every other day from the beginning of the experiment until 6 weeks of age by dropping 0.5 ml of cat urine (experimental group) or 0.5 ml of water (control group) on the bedding material. The cat urine had been frozen at −20°C after collection from 3 adult, unrelated male cats aged 5 to 7 years; we alternated which cat's urine was delivered at each delivery time.

After termination of the experiment at six weeks of age, urine from hamsters and blood from mice were collected for testosterone measurements, and then the body, testes and epididymis weights in unilaterally gonadectomized animals were obtained. Surgery was performed under Nembutal® (sodium pentobarbital) anesthesia (0.1 mg/kg).

The urine and serum testosterone concentrations were determined by radioimmunoassay specific for this steroid (Probst and Fenske, 1982). Hamster urine was collected in metabolic cages and mouse serum from about 1.5 ml of blood removed from the sublingual vein of ether-anesthetized mice. Urine and serum samples were kept frozen at -20 °C until radioimmunoassay. Statistical analyses used the Student t-test for unpaired comparisons (Plochinsky, 1970).

RESULTS

Male hamsters treated from postnatal day 11 until day 42 with cat urine had significantly lower testes (t=2.52, df=28, P<0.05) and epididymis weights (t=2.34, df=28, P<0.05) than control males (Table 1). Cat urine treatment also led a reduction in urinary testosterone levels (t=2.90, df=28, P<0.01; Table 1), but had no effect on body weight. These findings suggest that cat urine suppress sexual maturation in juvenile Campbelli's hamster males.

Table 1. Effects of treatment with cat urine on body weight, reproductive organ weights and urine testosterone level in 6 weeks old Campbelli's hamster males.

Parameter	Control group (n=15)	Experimental group (n=15)	P
Body weight (g)	28.4 ± 1.4	27.4 ±1.1	ns
Testis weight (mg)	760.0 ± 45.9	609.6 ± 38.3	< 0.05
Epididymis weight (mg)	50.8 ± 5.8	31.8 ± 5.5	< 0.05
Urine testosterone level (ng/ml)	0.2059 ± 0.002	0.1085 ± 0.027	< 0.005

Male house mice treated from postnatal day 12 until day 42 with cat urine had significantly greater epididymis weights than controls males (Table 2; t = 2.94, df=37, P<0.01). The increase in epididymis weight in house mice treated with cat urine indicates that, contrary to hamsters, it caused accelerated reproductive development. There were no significant differences among the other measures (Table 2).

Table 2. Effects of treatment with cat urine on body weight, reproductive organ weights and serum testosterone level in 6 weeks old male mice.

Parameter	Control group (n=15)	Experimental group (n=24)	P
Body weight (g)	24.7 ± 0.4	25.4 ± 0.8	Ns
Testis weight (mg)	84.9 ± 4.4	88.2 ± 2.5	Ns
Epididymis weight (mg)	7.5 ± 0.3	8.5 ± 0.2	< 0.01
Blood testosterone level (ng/ml)	5.0334 ± 0.8509	4.7213 ± 0.7404	Ns

DISCUSSION

The results of our experiments indicate that exposure to cat urine influences the developing reproductive system of male Campbelli's hamsters and house mice in an opposite manner. In agreement with our previous findings (Vasilieva et al., 1999), treatment of male Campbelli's hamsters with cat urine from age 11 to 42 days was associated with decreased epididymis and testes weights and a reduction in urinary testosterone concentration. In contrast, male house mice exposed to cat urine from age 12 to 42 days had heavier epididymis weights than control males.

The age of puberty onset is a major component of an animal's reproductive success and perhaps its life span. Precocial maturation may lead to different consequences in terms of individual fitness in males of different species that vary in the level of paternal investment into progeny. Campbelli's hamsters and house mice may represent extreme examples along the spectrum of male parental care. In laboratory observations of Campbelli's hamsters, male care was found to increase the survival of pups (Wynne-Edwards and Lisk, 1989). Despite a promiscuous mating system, Campbelli's hamster males maintain contact with particular females and their pups by regularly visiting their burrows, bringing food, scent marking pups and nearby areas (Vasilieva et al., 1988; Vasilieva, 1990; Sokolov and Vasilieva, 1993). Such paternal care might be costly enough in terms of metabolic energy that it impairs survival, e.g. due to insufficient accumulation of fat to survive hibernation. Delayed sexual maturation in male Campbelli's hamsters in condition of high predation risk, when the probability to successfully raise pups is low, might have an adaptive value by allowing energy conservation. Contrary to Campbelli's hamsters, house mice appear to be adapted to reproducing even when the density of predators is high, since it is clear that these mice can reach high reproductive success in a wide diversity of environmental conditions (Rowe, 1981). Since the house mouse breeding system involves communal nursing and low investment of paternal care (König, 1989), these characteristics might favor the early maturation of males. We, therefore, speculate that the high adaptivity of mice allow them to respond to conditions of high predation risk, as to any another environmental stress, by accelerating maturation which in turn leads to an increase in the reproduction.

ACKNOWLEDGMENTS

This work was supported by Russian Funds for Basic Research granted to Nina Yu. Vasilieva, no. 98-04-49435.

REFERENCES

Drickamer, L. C., 1986, Puberty-influencing chemosignals in house mice: ecological and evolutionary considerations, in: *Chemical Signals in Vertebrates 4* (D. Duvan, D. Müller-Schwarze, and R. M. Silverstein, eds.), Plenum Press, New York, London, pp. 441-455.
Flint, V.V., 1966, *Die Zwerghamster der Palaarktischen Fauna*, Ziemson Verlag, Wittenberg-Lutherstadt.
Gudermuth, D. F., Butler, W. R., and Johnston, R. E., 1992, Social influences on reproductive development and fertility in female Djungarian hamsters (*Phodopus campbelli*), *Horm. Behav.* **26**:308-329.
Heikkila, J., Kaarsalo, K., Mustonen, O., and Pekkarinen, P., 1993, Influence of predation risk on early development and maturation in tree species of *Clethrionomys* voles, *Ann. Zool. Fennici* **30**:153-161.
König, B., 1989, Behavioral ecology of kin recognition in house mice, *Ethol.Ecol.Evol.* **1**:99-110.
Kuznetzova, M. V., and Vasilieva, N. Y., 1997, Con- and heterospecific urinary chemosignals influence the rate of sexual maturation in Campbelli's hamster (*Phodopus campbelli*), in: *Chemical Signals in Vertebrates VIII*, Ithaca, NY, (abstract)
Levin, R. N., and Johnston, R. E., 1986b, Social mediation of puberty: An adaptive female strategy., *Behav. Neural. Biol.* **46**:308-324.
Lima, S. L., 1998, Nonlethal effects in the ecology of predator - prey interactions, *Bio-Science* **48**:25-34.
Lima, S. L., and Dill, L. M., 1990, Behavioral decisions made under the risk of predation: A review and prospectus, *Can. J. Zool.* **68**:619-640.
Marsden, H. M., and Bronson, F. N., 1964, Estrous synchrony in mice: alteration by exposure to male urine, *Science* **144**:1469.
Novotny, M. V, Weidong, M., Zidek, L., and Daev, E., 1999, Recent biochemical insights into puberty acceleration, estrus induction, and puberty delay in the house mouse, in: *Advances in Chemical Communication in Vertebrates,* (R. E. Johnston, D. Müller-Schwarze, P. W. Sorensen, eds.), Kluwer Academic/Plenum Publishers, New York, pp. 99-116.
Plochinsky, N. A., 1970, *Biometry*, University Press, Moscow.

Probst, B., and Fenske, M., 1982, Production of androgens by gerbil (*Meriones unguiculatus*) dispersed testicular interstitial cells: validation of a highly sensitive LH-bioassay, *Comp. Biochem. Physiol.[A]* **71**:113-117.

Reasner, D. S., and Johnston, R. E., 1988, Acceleration of reproductive development in female Djungarian hamsters by adult males, *Physiol. Behav.* **43**:57-64.

Rowe, F. P., 1981, Wild house mouse biology and control, in: *Biology of the House Mouse* (R. J. Berry, ed.), Academic Press, New York, pp. 575-589.

Sokolov, V. E., and Vasilieva, N. Yu., 1993, Djungarian hamsters (*Phodopus campbelli*) behavior in nature supports the theory of the "signaling biological field", *Dokl. Ross. Acad. Nauk* **332**:667-670 (in Russian).

Sokolov, V. E., and Vasilieva, N. Yu., 1996, Heterospecific social environment influences reproductive development in Campbelli's hamsters (*Phodopus campbelli* Thomas, 1905): The role of crossfostering and Djungarian hamster (*Phodopus sungorus* Pallas, 1773) males' midventral gland secretions, *Dokl. Ross. Acad. Nauk* (in press) (in Russian).

Sokolov, V. E., Vasilieva, N. Yu., and Zinkevich, E. P., 1989, Secretion of midventral glands of Djungarian hamsters (*Phodopus campbelli* Thomas, 1905) contains a factor for regulation of sexual maturation, *Dokl. Akad. Nauk SSSR.,* **308**:1274-1277 (in Russian).

Sokolov, V. E., Vasilieva, N. Y., and Zinkevich, E. P., 1991, The unknown function of specific skin glands: role of secretion of male midventral gland of Campbelli's hamsters (*Phodopus campbelli* Thomas, 1905) regulation of sexual maturation, in: *Problems of Animals Chemical Communication* (V.E. Sokolov, ed.), Nauka, Moscow, pp. 415-435 (in Russian).

Sokolov, V. E., Vasilieva, N. Yu., and Zinkevich, E. P., 1992, Influence of specific skin glands on the sexual maturation of male golden and Campbelli's hamsters, in: *Chemical Signals in Vertebrates VI.* (R. L. Doty, and D. Müller-Schwarze, eds.), Plenum Press, New York, pp. 259-262.

Surov, A. V., Vasilieva, N. Yu., and Telitzina, A. Yu., 1988, Population structure of Djungarian hamsters in the South of Tuva, in: *Rodents. The 6th All-Union Symposium,* (S. E. Ramensky, ed.), Acad. Nauk SSSR, Sverdlovsk **2**:61-62 (abstract) (in Russian).

Vandenbergh, J. G., 1983, Pheromonal regulation of puberty, in: *Pheromones and Reproduction in Mammals*, (J. G. Vandenbergh, ed.), Academic Press, New York, pp. 95-112.

Vasilieva, N. Yu, 1990, The Function of Specific Skin Glands. Functional and Evolutionary Aspects of Scent-Marking Behavior. Cricetinae Rodents as a Model, Unpubl. Dokt. Diss. Moscow, 381 pp.

Vasilieva, N. Yu, 1995, Factors influencing growth and reproductive development in cricetid rodents: the role of conspecific and heterospecific chemical cues, *7th European Ecological Congress*, Budapest, p. 201.

Vasilieva, N. Yu., 1997, Influence of predator odor on reproductive success in golden hamsters, *Chem. Senses* **22**:815.

Vasilieva, N. Yu., Surov, A. V., and Telitzina, A. Yu, 1988, Population structure of Djungarian hamsters in the south of Tuva, in: *Rodents. Proceedings of the VII All-Union Meeting* (S. E. Ramensky, ed.), Nauka, Sverdlovsk, pp. 61-62.

Vasilieva, N. Yu., Cherepanova, E. V., and Safronova, L. D.,1999, Influence of cat's urinary chemosignals on sexual maturation and meiosis in Campbelli's hamster males (*Phodopus campbelli*), in: *Advances in Chemical Communication in Vertebrates,* (R. E. Johnston, D. Müller-Schwarze, P. W. Sorensen, eds.), Kluwer Academic/Plenum Publishers, New York, pp.445-455.

Whitten, W. K., 1956, Modification of the estrous cycle of the mouse by external stimuli associated with the male, *J. Endocrinol.* **13**:399-404.

Whitten, W. K., 1959, Occurrence of anoestrus in mice caged in groups, *J. Endocrinol.* **18**:102-108.

Wynne-Edwards, K. E., and Lisk, R. D, 1989, Differential effects of parental presence on pup survival in two species of dwarf hamsters (*Phodopus sungorus* and *Phodopus campbelli*), *Physiol. Behav.* **45**:465-469.

Ylönen, H., 1989, Weasels suppress reproduction in cyclic bank voles *Clethrionomys glareolus*, *Oikos* **55**:138-140.

THE ROLE OF OLFACTION IN THE FEEDING BEHAVIOR OF HUMAN NEONATES

Richard H. Porter,[1] Heili Varendi,[2] and Jan Winberg[3]

[1] URA 1291 CNRS/INRA, 37380 Nouzilly, France
[2] Tartu University Children's Hospital, Estonia
[3] Karolinska Institute, Stockholm, Sweden

INTRODUCTION

Chemical cues have been implicated in the feeding behavior of species belonging to all five classes of vertebrates, including (perhaps most surprisingly) birds (Burghardt, 1990; Roper, 1999; Stoddart, 1980). In various mammals, olfaction plays a critical role in the mediation of sucking and milk ingestion by neonates. This is illustrated by the severe deficits in nipple localization and attachment displayed by rat and mouse pups following olfactory bulbectomy (Cooper and Cowley, 1976; Risser and Slotnick, 1987) or peripheral destruction of the nasal olfactory receptors by infusions of zinc sulfate solution (Singh *et al.*, 1976). Similar adverse effects were observed in young rabbits when their mother's nipples were covered with a thin rubber film (Distel and Hudson, 1985), and in rat pups after the dam's ventrum had been thoroughly washed (Teicher and Blass, 1976). The significance of the odor of mother's milk and breasts for successful human nursing has been suspected for centuries. For example, a survey of documents written by medical authorities between 1500-1800, found that 55% of the authors who discussed necessary attributes of wet nurses' breast milk included «good smell» as a required quality (Fildes, 1986). As seen below, such early beliefs are supported by recent empirical evidence indicating that components of the typical pattern of behavior involved in effective breastfeeding by newborn humans are activated and directed by maternal odors.

AROUSAL AND SUCKING ACTIVITY

Beginning shortly after birth, human infants are active participants in the breast-feeding process. Widstrom and her colleagues (Widstrom *et al.*, 1987) observed that healthy, full-term neonates that were placed on their mother's bare chest immediately after delivery displayed an organized sequence of prefeeding behavior, that included spontaneous sucking movements of the lips, tongue and mouth, and rooting activity (head turning accompanied by opening of the mouth). This pattern of behavior terminated when the baby grasped a nipple without maternal intervention and began to suck, which typically occurred before the end of the first hour of postnatal life.

Several studies indicate that odors emanating from the mother's breast region have a stimulating effect on newborn infants' prefeeding behavior. In an early experiment (Russell, 1976), infants were exposed to cotton pads (held 1-2 cm from the nose) that had been worn in contact with their mother's breast for 3-hrs. before the test session began. At 2 weeks postpartum, the maternal odor stimulus elicited a «general arousal» response. When the same babies were tested again at 6 weeks of age, sucking activity was frequently observed when their own mother's breast pad was presented, however, there was little response to the breast odor of a strange mother (see also Soussignan et al., 1997). More recently, a single–subject experiment examined a premature infant's pattern of sucking on an artificial nipple connected to a pressure transducer, during gavage feeding (Meza et al., 1998). Over repeated test trials starting at 32 weeks gestational age, the rate of high amplitude non-nutritive sucking increased in response to the odor of the mother's milk.

In an additional study, newborn infants were reported to increase their mouthing activity when exposed to the portion of their mother's hospital gown that had been worn over her breasts and axillae (Sullivan and Toubas, 1998). During this same series of 60-sec. tests, infants who were crying more quickly stopped vocalizing in response to the maternal breast odor as contrasted with control gown presentations. However, a more prolonged exposure to breast odor had the opposite effect; throughout a 60-min. period beginning 30 minutes after birth, babies spent more time crying when their mother's breast pad – rather than a clean control pad - was placed 2-cm from their nostrils (Varendi et al., 1998). Increased activation and crying were likewise observed by Macfarlane (1975) after babies maintained contact with their mother's breast pad «for an extended time». It therefore appears that infants may become disturbed or frustrated if they continually perceive breast odors in a context where nipple access and sucking are not permitted. Under more natural conditions, maternal breast odors would be perceived when the baby is in close physical proximity to the mother. Heightened crying when the baby is unable to grasp a nipple might then signal that he is distressed or hungry and thereby be a means of exerting some control over the mother's nursing behavior. Indeed, 2 and 4-month old infants who were frequently fed, cried less often than did babies with longer intervals between feeds (e.g. Barr and Elias, 1988).

An alternative method to investigate the influence of maternal breast- or milk-odors on newborns' sucking activity involves manipulation of those cues by the introduction of strong novel scents. In experiments conducted during the 19th century, neonates refused to suck after highly odorous substances, such as diluted petroleum or Asa foetida, were applied to their mother's breasts (Kroner, 1882; Preyer, 1885). Infants that were already sucking released the nipple or withdrew when similar olfactory stimuli were presented (Garbini, 1896). The odor of a mother's milk can be modified in a less intrusive manner by asking her to consume highly flavorful or odor producing substances. For example, 2-3 hrs. after a sample of nursing mothers had swallowed capsules of garlic extract, the characteristic odor of that plant could be detected in their breast milk (Mennella and Beauchamp, 1991). Furthermore, babies spent a longer period of time attached to the nipple, and sucked more frequently, after their mother had consumed a garlic capsule rather than a placebo. However, the amount of milk ingested did not differ reliably between the placebo and garlic trials.

NIPPLE ORIENTATION/LOCALIZATION

Based upon relevant data from non-human mammalian young, as discussed briefly above, we hypothesized that odors emanating from the mother's breast region may similarly contribute to the spontaneous physical movement and attachment to the nipple described by Widstrom *et al.* (1987) for newborn infants of our own species. To assess this hypothesis, we recorded the prefeeding behavior of newborn infants after one breast of each participating mother had been thoroughly washed with unscented soap (Varendi *et al.*, 1994). When placed in a prone position on their mother's chest and allowed to choose between the breast whose natural scent had been eliminated (washed) and the alternative (unwashed) breast, significantly more babies than expected by chance alone (22/30) selected the untreated breast for their first sucking bout. This experiment was subsequently replicated with a sample of 2-3 day-old infants, with similar results; i.e. once again, a statistically reliable preference for the unwashed breast was observed (Varendi *et al.*, 1997). Temperature readings taken at the surface of the nipple, areola and elsewhere on the breast at regular intervals over the first 55-min. following birth, indicated no effects of the washing procedure. It was therefore concluded that naturally occurring maternal odors help guide the neonate to the nipple.

In the normal breastfeeding situation, infants are exposed to an array of stimuli associated with the mother. That is, aside from producing discernible odors, the mother is the source of thermal/tactile, visual and auditory cues that could have a further initiating or directing effect on the overt behavior displayed by the infant when placed on her bare chest. Accordingly, to further investigate the influence of maternal breast odors per se (i.e. in the absence of additional sensory input arising from the mother) on the activation and oriented movement of neonates, 22 newborn infants were tested individually during 2 trials on a warming bed with a surface temperature = 29° C. Each 3-min. test session began when the infant was placed in a prone position with the nose touching the surface of the warming bed. For one trial, the mother's soiled breast pad was placed on the bed, 14-17 cm in front of the baby's nose; a clean control pad was used in the other trial. An observer who was unaware of the stimulus pad condition (control or breast pad) recorded the latencies for the following behavioral categories:
-Rooting
-Lifting the head
-Forward movement (> 7 cm in the direction of the stimulus pad).

Table 1. Responses of infants on a warming bed when exposed to their mother's Breast pad vs. a Clean control pad.

	Stimulus	
	Breast Pad	Control Pad
Median rooting latency (s)	8	13
Median latency to lift head (s)	15 *	63
Median latency to move forward (s)	60 **	180

*p<.05 ; **p<.001 (Wilcoxon test)

As seen in Table 1, the subject infants lifted their head and moved forward significantly sooner when exposed to the maternal breast pad than during the test trials with the odorless control pad. The scent of the mother's breasts was therefore sufficient to accelerate motor activity and physical movement towards the odor pad even when it was presented in isolation from other stimuli typically associated with the mother. Such oriented movement in the direction of olfactory cues emanating from the mother's breasts would facilitate nipple localization and milk ingestion when neonates are in bodily contact with that parent.

BIOLOGICAL SIGNIFICANCE?

There are several lines of evidence suggesting that the evolutionary adapted pattern of mother-infant interactions in humans involves constant bodily contact and frequent, on demand sucking throughout the day and night. For example, anthropological accounts of contemporary hunter-gatherers provide insights into the presumed natural history of human breastfeeding behavior. In a commonly cited study, !Kung infants were reported to remain «in immediate physical proximity with their mothers» during the first 2 years after birth and to breast feed intermittently, with a mean interval of only 13 min. between the brief nursing bouts (Konner and Worthman, 1980). It should be pointed out that high sucking frequencies have also been documented in other traditional populations, however, such temporal patterning is not always observed in such cultural groups (Woolridge, 1995). Rather, there is a wide variability in breastfeeding routines across non-technological societies that may reflect differing cultural norms and physiological factors.

More convincingly, comparative studies of numerous species of mammals have revealed a strong relationship between milk composition and feeding schedules (Ben Shaul, 1962; Blurton-Jones, 1972). The milk of mammals whose young feed very infrequently (e.g. rabbits and tree shrews) is characterized by a high content of protein and fat. In contrast, animals with offspring that are routinely kept in close contact with their mother and are able to feed frequently around the clock (e.g. rhesus macaques, chimpanzees) produce dilute milk that is relatively low in fat and protein, but has a higher carbohydrate content. Human breast milk falls into the latter category and is therefore similar to that of other species with young that feed continuously or according to their individual needs. Additional analyses indicate that the rate of sucking is also correlated with feeding frequency across different mammalian species (Wolff, 1968; Blurton-Jones, 1972). In general, young of species that feed infrequently tend to suck the most rapidly. The slow pace of sucking observed in human infants is once again consistent with short intervals between feeds and therefore the maintenance of close contact with the mother. Frequent sucking also stimulates the release of prolactin in the nursing mother, which in turn has a positive affect on milk production (Whitworth, 1988). Thus, women who nurse their infants infrequently may produce insufficient quantities of milk and are likely to experience problems with breast-feeding (Gussler and Briesemeister, 1980).

Infant-initiated nipple attachment and sucking (as discussed above) would greatly facilitate frequent day and night feeding bouts since active involvement of the mother, aside from her physical presence, would thereby not be required for effective nursing. Olfactory mediated nipple localization in this context has obvious adaptive value. Odors that help guide babies to the nipple region are continuously present and should be effective regardless of the mother's state and the ambient level of illumination. Thus, within a short distance from the source, infants could rely on salient maternal odors to orient to the breasts even when the mother is sleeping in a dark environment and she and her baby are covered with bedclothes. In support of this scenario, McKenna *et al.* (1997) recently reported that the duration of nocturnal breastfeeding by infants who slept in the same bed with their mother was three times greater than that of babies who slept separately. Moreover, the bedsharing infants spent much of the night «oriented toward their mother» while in close proximity to her. The authors suggested that the enhanced breastfeeding associated with mother-infant co-sleeping might reflect lowered hunger and arousal thresholds resulting from exposure of the babies to maternal odors.

REFERENCES

Barr, R. G., and Elias, M. F., 1988, Nursing interval and maternal responsivity: Effect on early infant crying, *Pediatrics* **81**:529-536.

Ben Shaul, D. M., 1962, The composition of the milk of wild animal, *Int. Zoo Year Book* **4**:333-342.

Blurton-Jones, N., 1972, Comparative aspects of mother-child contact, in: *Ethological Studies of Child Behaviour* (N. Blurton-Jones, ed.), Cambridge University Press, Cambridge, pp. 305-328.

Burghardt, G. M., 1990, Chemically mediated predation in vertebrates: Diversity, ontogeny, and information, in: *Chemical Signals in Vertebrates* 5 (D. W. Macdonald, D. Müller-Schwarze, and S. E. Natynczuk, eds.), Oxford University Press, Oxford, pp. 475-499.

Cooper, A. J., and Cowley, J. J., 1976, Mother-infant interaction in mice bulbectomized early in life, *Physiol. Behav.* **16**:453-459.

Distel, H., and Hudson, R., 1985, The contribution of the olfactory and tactile modalities to the nipple-search behaviour of newborn rabbits, *J. Comp. Physiol.* A **157**:599-605.

Fildes, V. A., 1986, *Breasts, Bottles and Babies*, Edinburgh University Press, Edinburgh.

Garbini, A., 1896, Evoluzione del senso olfattivo nella infanzia, *Arch. Antropol. Etnol. (Firenze)* **26**:239-286.

Gussler, J. D., and Briesemeister, L. H., 1980, The insufficient milk syndrome: A biocultural explanation, *Med. Anthro.* **4**:145-174.

Konner, M., and Worthman, C., 1980, Nursing frequency, gonadal function, and birth spacing among !Kung hunter-gatherers, *Science* **207**:788-791.

Kroner, T., 1882, Uber Sinnesempfindungen des Neugeborenen, *Breslauer arztl. Ztschr.* **4**:37-58.

Macfarlane, A., 1975, Olfaction in the development of social preferences in the human neonate, in: *Parent-Infant Interaction* (R. Porter, and M. O'Connor, eds.), Ciba Found. Symposium 33, American Elsevier, New York, pp. 103-113.

McKenna, J.J., Mosko, S. S., and Richard, C. A., 1997, Bedsharing promotes breastfeeding, *Pediatrics* **100**:214-219.

Mennella, J. A., and Beauchamp, G. K., 1991, Maternal diet alters the sensory qualities of human milk and the nursling's behavior, *Pediatrics* **88**:737-744.

Meza, C. V., Powell, N. J., and Covington, C., 1998, The influence of olfactory intervention on non-nutritive sucking skills in a premature infant, *Occup. Therapy J. Res.* **18**:71-83.

Preyer, W., 1885, *Die Seele des Kindes*, Alcan, Paris.

Risser, J. M., and Slotnick, B. M., 1987, Nipple attachment and survival in neonatal olfactory bulbectomized rats, *Physiol. Behav.* **40**:545-549.

Roper, T. J., 1999, Olfaction in birds, *Adv. Study Behav.* **20**:247-332.

Russell, M. J., 1976, Human olfactory communication, *Nature* **260**:520-522.

Singh, P. J., Tucker, A. M., and Hofer, M. A., 1976, Effects of nasal ZnSO4 irrigation and olfactory bulbectomy on rat pups, *Physiol. Behav.* **17**:373-382.

Soussignan, R., Schaal, B., Marlier, L., and Jiang, T., 1997, Facial and autonomic responses to biological and artificial olfactory stimuli in human neonates: Re-examining early hedonic discrimination of odors, *Physiol. Behav.* **62**:745-758.

Stoddart, D. M., 1980, *The Ecology of Vertebrate Olfaction*, Chapman and Hall, London.

Sullivan, R. M., and Toubas, P., 1998, Clinical usefulness of maternal odor in newborns: Soothing and feeding preparatory responses, *Biol. Neonate* **74**:402-408.

Teicher, M. H., and Blass, E. M., 1976, Suckling in newborn rats: Eliminated by nipple lavage, reinstated by pup saliva, *Science* **193**:422-425.

Varendi, H., Porter, R. H., and Winberg, J., 1994, Does the newborn baby find the nipple by smell? *The Lancet* **344**:989-990.

Varendi, H., Porter, R. H., and Winberg, J., 1997, Natural odour preferences of newborn infants change over time, *Acta Paediatr.* **86**:985-990.

Varendi, H., Christensson, K., Porter, R. H., and Winberg, J., 1998, Soothing effect of amniotic fluid smell in newborn infants, *Early Human Develop.* **51**:47-55.

Whitworth, N. S., 1988, Lactation in humans, *Psychoneuroendocrin.* **13**:171-188.

Widstrom, A.-M., Ransjo-Arvidson, A. B., Christensson, K., Matthiesen, A.-S., Winberg, J., and Uvnas-Moberg, K., 1987, Gastric suction in healthy newborn infants, *Acta Paediatr. Scand.* **76**:566-572.

Wolff, P. H., 1968, Sucking patterns of infants mammals, *Brain, Behav. Evol.* **1**:354-367.

Woolridge, M. W., 1995, Baby-controlled breastfeeding: Biocultural implications, in: *Breastfeeding: Biocultural Perspectives* (P. Stuart-Macadam, and K. A. Dettwyler, eds.), Aldine de Gruyter, New York, pp. 217-242.

PIG RESPONSES TO TASTE STIMULI

D. Glaser,[1] M. Wanner,[2] J. M. Tinti,[3] and C. Nofre[3]

[1] Anthropological Institute and Museum, University of Zürich,
Winterthurerstrasse 190, 8057 Zürich, Switzerland
[2] Institute of Animal Nutrition, University of Zürich, Switzerland
[3] Faculty of Medicine of Lyon Laennec, University of Lyon I, Lyon, France

INTRODUCTION

Since 1966, researchers have tested various primate species in experiments investigating their behavioral response to gustatory stimuli. We found a great diversity within the order of primates in their response to compounds sweet in man. Our studies focus on pigs, (*Sus scrofa domesticus*). Pigs are omnivores like humans, and their physiology shares many common points with humans. Expect for limited behavioral and electrophysiological studies in which pigs were exposed to familiar gustatory stimuli (sucrose, glucose, etc.), relatively few reports have described responses to wide diversity of gustatory stimuli (see Glaser *et al.*, 2000). The members of Suidae in order Artiodactyla, are known from the Lower Eocene ($\approx$ 55 MYA); although they probably emerged during the late cretaceous time. Several families of primitive pig-like forms are known in the Eocene, and therefore, they are more ancient than the catarrhine primates (of $\approx$ 37 MYA; Carroll, 1988). Importance of taste is indisputable in many aspects of behavior, particularly for the selection of food, which is essentially based on palatability. Therefore, in previous work, we determined the gustatory responses of pigs to various natural and artificial compounds known to be sweet in humans (Glaser, *et al.*, 2000). The purpose of the present work is to extend this knowledge of the pig taste by experimenting compounds representative of different taste qualities (salt, bitter and acid tastes) in addition to alpha-amino acids, many of which are known to be sweet in humans.

ANIMALS, METHOD AND MATERIALS

Between 1996 up to 1999, we tested a total of 123 pigs, 58 females and 65 males, aged 2 to 12 months, of body mass 15 to 120. We constructed special conical stainless steel drinking buckets, of the dimension seen in Figure 1.

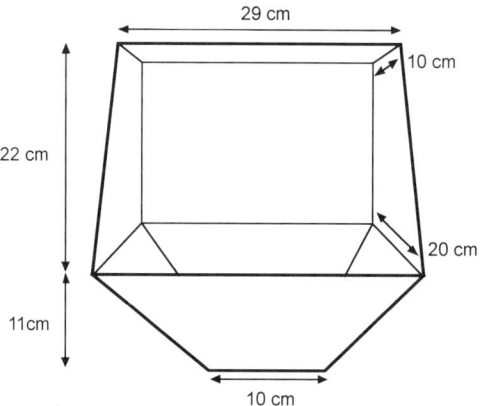

Figure 1. Conical stainless steel drinking bucket.

This conical shape was designed to reduce as much as possible, the volume of a taste solution presented to the animals, since pigs may ingest a considerable volume of an expensive taste solution if they find it to be palatable). The pigs were housed in cages (2 x 3 m), equipped with automatic waterers, and the subjects were never deprived of normal water intake. Pigs were tested individually, using a method based on a two-choice preference test as shown in Figure 2.

Figure 2. Two-choice preference test.

In these short-term preference tests, the pigs were videotaped when offered the choice between two solutions of 250 ml in each bucket, one containing water and the second containing the tested solution. A positive response (+) was concluded when 80 % or more of the tested solution was consumed while was not consumed more then the loss the loss of a small volume when it was sampled. Preferred compounds also elicitated eager drinking, quick swallowing, lip sucking/smacking, and orientation of the head toward the stimulus as videotaped (Glaser et al., 1997). Conversely, a negative response (-) was concluded when no more then a small sampling volume was consumed from the solution. Avoided compounds also elicitated withdrawal of the head and head shaking. The relative position of the water and the tested solution were randomized. Tests began at 9.00 pm, and following the conclusion of the experiments, the pigs received their daily feeding, composed of a commercial pelleted food (HOKOVIT-2150 NATURA). All the compounds tasted by the pigs were of commercial origin as noted in Table 3.

RESULTS

Compounds Known to be Sweet in Humans

In a previous paper it was shown that pigs have a gustatory preference for all carbohydrates tested over water (Glaser et al., 2000) and sucrose is the most preferred one. In humans and pigs, D-fructose is half as potent as sucrose. These animals also have a marked preference for all polyols examined versus water and xylitol is the most preferred polyol, being approximately as effective as sucrose on a molar basis.

The results obtained with artificial or natural compounds known to be sweet in humans show that only five compounds – i.e. acesulfame-K, alitame, dulcin, saccharin and sucralose – are able to elicit a preference in pigs. Aspartame, cyclamate, monellin, thaumatin, NHDHC, P-4000 and perillartine did not elicit any preference over water in these animals, even for solutions several tens of times more concentrated than needed to induce an explicit sweet perception in humans (except for P-4000 which is too poorly soluble to test concentrated solutions).

Comparative Gustatory Responses to Amino Acids in Pigs and Humans

From human studies, we already know that some amino acids taste sweet, while others bitter, sour, or sometimes salty bitter (Haefeli and Glaser, 1990). As mentioned above, we have found different taste reactions in the pigs too. A comparative overview of the taste of amino acids in humans (from different authors) in comparison to the pig responses is given in Table 1.

Compounds with a Salty and an Umami Taste

The responses to the salt taste were determined with five concentrations of sodium chloride solutions (25 to 400 mmol/l; succeeding increments doubled). A sodium chloride solution at the concentration of 400 mmol/l was rejected by all the pigs tested (12 animals) but was accepted at 200 mmol/l (11.7 g/l), while at 100 mmol/l (5.85 g/l) indifference dominates (attractiveness for only 3 out of 12 pigs). At the concentrations of 50 and 25 mmol/l, corresponding to the level at which monosodium-L-glutamate (MSG) induces a clear preference, no attraction for the salt solutions was noticed.

Monosodium-L-glutamate (monohydrate) (MSG), representative of the umami taste (Yamaguchi, 1987), was tested with two groups each of 8 animals. At the lowest concentration of 16 mmol/l this compound was found palatable by all the pigs tested.

Table 1. Tastes of amino acids in humans compared to the corresponding gustatory responses in pigs.

Amino acids	Taste in humans[a]				Pig gustatory responses[b]
	Schiffman et al. (1976, 1982)	Wieser et al. (1977)	Birch and Kemp (1989)	Haefeli and Glaser (1990)	
Compounds with a predominant sweet taste in humans					
Gly	S	S	S	S	+
L-Ala	S/B	S	S	S	+
D-Ala	S	S	S	S	+
D-Asn	S	S	----	S	+
D-Gln	S	S	----	S	+
D-His	S	S	S	S	-*
L-HPro	----	S	S	S	+
D-Ileu	flat/B	S	----	S	-
D-Leu	mod. S	S	S/B	S	-*
D-Lys	----	S	----	S	+
D-Phe	poss. S/B	S	B/S	S	+
L-Se	flat/S/compl.	S	S	S	+
D-Se	S	S	T	S	+
L-Thr	flat/S/poss. B	S	----	S	+
D-Thr	T/poss. S	S	----	S	+
D-Trp	S/poss. B	S	S/B	S	+
D-Tyr	----	S	----	S	-
D-Val	T/poss. S	S	S	S	-*
Compounds with a predominant acid taste in humans					
L-Asp	flat/sour/B	sour	sour	sour	+
D-Asp	sour/salty	sour	sour	sour	+
L-Glu	sour/compl.	sour	T	sour	+
D-Glu	sour/salty	----	----	sour	+
Compounds with a predominant bitter taste in humans					
L-Arg	flat/B/compl.	B	B	B	-*
L-His	flat/B	B	B/T	B	-
L-Ileu	flat/B	B	B	B	-
L-Leu	flat/B	B	B	B	-
L-Phe	B/compl.	B	B	B	-*
D-Pro	----	neutral	B	B	-
L-Trp	flat/B	B	B	B	-*
L-Tyr	flat/B	B	----	B	-
L-Val	flat/B/S	B	B/S	B	-
Compounds with a complex taste or no predominant taste in humans					
D-Arg	----	neutral	----	S/B	-*
L-Asn	flat/B	neutral	----	B/S	+
L-Cys	sulfur.	sulfur.	----	S/sulfur.	-
D-Cys	B	sulfur.	----	S/sulfur.	+
L-Gln	flat/S/meaty	neutral	----	B/S	+
D-HPro	----	neutral	T	----	-
L-Lys	flat/compl.	S/B	S/other	B/S	-
L-Met	compl./sulfur.	sulfur.	B	S/B	-
D-Met	flat/B/compl.	S/sulfur.	S	S/sulfur.	-
L-Pro	S/compl.	S/B	S/B	S	-

[a] Abbreviations: S = sweet; B = bitter; T = tasteless; Sulfur. = sulfurous; mod. = moderately; poss. = possibly; compl. = complex.
Taste conditions: Schiffman and Engelhard (1976) for L-amino acids: the composite taste is deduced from seven independent studies; Schiffman et al. (1982) for D-amino acids, Wieser et al. (1977) and Haefeli and Glaser (1990): determination at the threshold; Birch and Kemp (1989): predominant taste in solution at 6 g/l (2 g/l for L-aspartic acid).
[b] + indicates a preference; -, an indifference or a rejection; -*, a clear rejection.
More detailed results are published independently.

Compounds with a Bitter Taste

Six different concentrations between 0.50 to 0.000488 mmol/l were used to ascertain the lowest aversive concentration for quinine hydrochloride. The pigs (n = 12) displayed an aversion for quinine hydrochloride and to any concentration higher than 0.001902 mmol/l. They strongly rejected the solutions and they seemed indifferent to solutions from 0.000976 to 0.000488 mmol/l.

Table 2. Gustatory responses of pigs to a 5 g/l sucrose solution in mixture with some non attractive compounds.

Solutions	Concentrations		Number of pigs	Gustatory responses[a]
	mmol/l	g/l		
Sucrose 5 g/l (14.6 mmol/l)			9	9+
Sucrose (5 g/l) in mixture with:				
L-Arginine	250.0	21.75	9	3+, 6-
D-Arginine	250.0	43.55	9	2+, 7-
D-Histidine	240.0	37.23	9	3+, 6-
D-Leucine	250.0	32.79	9	1+, 8-
L-Phenylalanine	30.2	5.0	9	3+, 6-
L-Tryptophan	14.6	2.98	9	3+, 6-
D-Valine	250.0	29.29	9	2+, 7-
Quinine HCl, 2H$_2$O	0.50	0.20	9	9-
	0.125	0.05	9	3+, 6-
	0.0075	0.0031	9	9+

[a] '+' means that 80 % of the tested solution is consumed and not the water control simultaneously presented; '-' means that none of the tested solution and water solution were consumed.

All amino acids which are clearly bitter for humans (L-Arg, L-His, L-Ileu, L-Leu, L-Phe, D-Pro, L-Trp, L-Tyr, L-Val) are not preferred by pigs (Table 1). Their mixture with a sucrose solution (Table 2) is even rejected. This is probably due to their bitter taste as pigs detect and respond aversively to bitter compounds (Nelson and Sanregret, 1997).

Compounds with a Sour Taste

In contrast to results from studies on other mammals, pigs consumed solutions of various acids presented at the concentration of 50 mmol/l. Table 3 gives a clear picture of a preference behavior. The approximate threshold value in our experimental conditions for acetic acid in pigs was found between 12.5 and 6.2 mmol/l.

Table 3. Gustatory preference of pigs to dicarboxylic amino acids and various acids.

Compounds[a]	Concentrations		pH	Number of pigs	Gustatory responses[b]
	mmol/l	g/l			
L-Aspartic acid	50.0	6.65	3.08	4	4+
	25.0	3.12	3.22	9	9+
	12.5	1.56	3.44	9	9+
D-Aspartic acid	50.0	6.65	3.04	4	4+
L-Glutamic acid	50.0	7.35	3.40	4	4+
D-Glutamic acid	50.0	7.35	3.40	4	4+
Acetic acid	50.0	3.14	3.39	11	11+
	25.0	1.57	3.65	4	3+, 1-
	12.5	0.78	4.03	4	3+, 1-
	6.2	0.39	4.46	8	1+, 7-
Citric acid	50.0	10.50	2.27	12	12+
D-Malic acid	50.0	6.70	2.53	5	5+
L-Malic acid	50.0	6.70	2.53	5	5+

[a] All compounds are from Fluka, except for L-aspartic acid from Merck and D-aspartic acid from Sigma.
[b] '+' means that 80 % of the tested solution is consumed and not the water control simultaneously presented; '-' means that none of the tested solution and water were consumed.

DISCUSSION

If we consider the main nutritive carbohydrates, their molar order of effectiveness in pigs roughly mirrors the relative molar sweetness potency order in humans. For humans, the order is: sucrose > D-fructose > maltose = lactose > D-glucose > D-galactose, while for pigs the order is: sucrose > D-fructose > maltose = lactose > D-glucose = D-galactose. However, in pigs the relative response intensities (by comparison with sucrose, our standard reference) are notably lower except for D-fructose and sucrose. For example, lactose and maltose are approximately six times less preferred then sucrose by pigs, while, to humans both of these compounds are of equivalent sweetness, about one third that of sucrose.

Pigs also have a marked preference for all the polyols examined. Xylitol is the most potent of these compounds in pigs and in humans; but, while xylitol is in humans about one third less sweet than sucrose on a molar basis, it is roughly as preferred as sucrose on a molar basis in pigs. After xylitol, sorbitol was the second most favored polyol in pigs; it is approximately four times less preferred in pigs than sucrose or xylitol, but is still about twice as preferred as D-glucose, while in humans sorbitol is isosweet with D-glucose. Note that sorbitol is common in many fruits, often at a concentration of about 10-30 g/l of fresh fruit juice. Although slowly absorbed by the intestine, this polyol may be considered as an effective energetic sweetener, being metabolically converted into D-fructose at the hepatic level. The comparison of the taste sensitivity for the polyols between pigs and the data of humans (Haefeli, 1983) shows that the pigs are a slightly more sensitive for xylitol and sorbitol than humans, but for the other polyols, humans are more sensitive than pigs.

The results obtained with twelve artificial or natural compounds known to be sweet in humans are more disparate. Only acesulfame-K, alitame, dulcin, saccharin and sucralose – were able to elicit a preference in pigs.

The indifference of pigs towards aspartame is not surprising. All mammals tested so far - with the unique exception of the Catarrhini (the Old World primates, including humans) - do not give any explicit 'sweet' gustatory responses to aspartame, as observed in hamsters, gerbils, rats, or in dogs, cows or horses (unpublished results of our group), in Prosimii (prosimians) and Platyrrhini (the New World monkeys). Similarly to aspartame, sodium cyclamate, which is sweet to all the catarrhine primates tested until now, is 'unsweet' to pigs, and to all the mammals studied so far, such as hamsters, gerbils, rats, cats, tree shrews, and non-catarrhine primates.

Our pigs displayed a preference for sodium chloride (NaCl) only at the concentration of 200 mmol/l (11.69 g/l), they strongly avoided all the other concentrations tested. The animals consumed MSG at relatively low level, an observation that corroborates an early mention that MSG "can be used as a substitute for sucrose in pig rations" (Klay, 1964). This glutamate effect cannot be confused with a salt effect that requires higher concentrations to be attractive. These findings also agree with neural activity in whole chorda tympani and glossopharyngeal nerves in pigs, with large responses recorded for MSG, and small ones for sodium chloride (Danilova et al., 1999). Interestingly, a significant stimulation of single fibres best responding to sucrose is also reported for MSG in pigs (Danilova et al., 1999), suggesting an additional gustatory dimension in the palatability of this compound.

The attractiveness of acids to pigs is striking. Aspartic and glutamic acids are known, under their acid form, to have an acid taste in humans (Birch and Kemp, 1989; Haefeli and Glaser, 1990; Schiffman et al., 1982; Schiffman et al., 1981; Wieser et al., 1977). Solutions of these acids induced a strong preference in pigs (Table 3). Additional experiments with solutions of acetic, citric, D- and L-malic acid confirmed previous observations (Falkowski and Aherne, 1984) that pigs are sensitive to acid solutions (Table 3). These results also agree with the large responses to citric and ascorbic acids recently noticed in recordings of pig whole nerve chorda tympani activity (Danilova et al., 1999).

Gustatory discrimination or electrophysiological responses to sourness in diverse animals have been reported (Contreras et al., 1985; Gilbertson and Gilbertson, 1994; Hellekant et al., 1997; Goatcher and Church, 1970; Harriman, 1976; Harriman and Nevitt, 1978; Pritchard and Norgren, 1991), but seldom in relation with a gustatory preference. For instance, in the house musk shrew, L-asp and L-Glu were strongly rejected at 20 mmol/l, similarly to hydrochloric acid solutions (Iwasaki and Sato, 1982). Most of the primate species tested are indifferent to, or will avoid, 5-10 mmol/l of acetic acid solutions (Glaser, 1977). Only one species, *Aotus trivirgatus*, a South American primate, was found to prefer acetic and citric acid solutions over water (Glaser, 1977). Cattle are moderately attracted by a 0.2 ml/l solution of acetic acid (~3.5 mmol/l) (pH 4.4), but avoid the more concentrated solutions (Goatcher and Church, 1970), those, which were avidly consumed by our pigs.

No apparent correlation seems to exist between the attractiveness of amino acids and the fact they must be regularly provided in the diet (free, or under peptide or protein forms). The contrary is even found, essential amino acids (L-Arg, L-Ileu, L-Leu, L-Phe, L-Trp or L-Val) are not attractive, and many of those are not essential (L-Ala, L-Asn, L-Gln, Gly, L-Ser) elicited a gustatory preference.

Our data in different species enlarge our knowledge of the mechanism in taste reception, but will be also helpful for a better nutrition and care of animals. The strong preference of pigs to acid solutions was for instance found very useful for masking the bitter taste frequently associated with medications. It was possible to render very palatable for pigs a mixture previously rejected of sucrose and of a specific antibiotic by lowering the pH of the solution with citric acid. At least all our data in different species will be a help in confirming the multipoint attachment (MPA) model, the theory proposed to explain the sweet taste of many substances in humans (Nofre and Tinti; 1996).

ACKNOWLEDGMENTS

The authors are grateful to Prof. Dr. Hans Ulrich Bertschinger and Dr. med. vet. Esther Bürgi for making animals available and for kind assistance in carrying out the experiments to Mr. Roland Liechti and Mrs. Helena Ciolarro. Special thanks are extended for the illustrations to Mrs. Jeanne Peter. The present research is a contribution to the EU-project: 'The Optimisation of Sweet Taste Quality' (TOSTQ) (PL98-4040) and was financially supported by the 'Bundesamt für Bildung und Wissenschaft' (BBW Nr. 98.0259) Bern, Switzerland and by The NutraSweet Company, Chicago,IL., (U.S.A.).

REFERENCES

Birch, G. G., and Kemp, S. E., 1989, Apparent specific volumes and tastes of amino acids, *Chem. Senses* **14**:249-258.

Carroll, R. L., 1988, *Vertebrate Paleontology and Evolution*, Freeman, New York.

Contreras, R. J., Bird, E., and Weisz, D. J., 1985, Behavioral and neural gustatory responses in rabbit, *Physiol. Behav.* **34**:761-768.

Danilova, V., Roberts, T., and Hellekant, G., 1999, Responses of single taste fibers and whole chorda tympani and glossopharyngeal nerve in the domestic pig, *Sus scrofa*, *Chem. Senses* **24**:301-316.

Falkowski, J. F., and Aherne, F. X., 1984, Fumaric and citric acid as feed additives in starter pig nutrition, *J. Anim. Sci.* **58**:935-938.

Gilbertson, D. M., and Gilbertson, T. A., 1994, Amiloride reduces the aversiveness of acids in preference tests, *Physiol. Behav.* **56**:649-654.

Glaser, D., 1977, Geschmacksleistungen bei nachtaktiven Primaten, *Ztscht. Morphol.Anthropol.* **68**:241-246.

Glaser, D., Tinti, J. M., Nofre, C., and Wanner, M., 1997, Gustatory responses in pigs to compounds sweet in man, *Video tape*, SVHS, 5.5 min.

Glaser, D., Wanner, M., Tinti, J. M., and Nofre, C., 2000, Gustatory responses of pigs to various natural and artificial compounds known to be sweet in man, *Food Chem.* **68**:375-385.

Goatcher, W. D., and Church, D. C., 1970, Taste responses in ruminants. IV. Reactions of pigmy goats, normal goats, sheep and cattle to acetic acid and quinine hydrochloride, *J. Anim. Sci.* **31**:373-382.

Haefeli, R. J., 1983, Vergleichende Betrachtungen über die Geschmacksschwellen von Zuckern und Zuckeralkoholen beim Menschen, *Lebensmittel-Wissenschaft und -Technologie*, **16**:48-50.

Haefeli, R. J., and Glaser, D., 1990, Taste responses and thresholds obtained with the primary amino acids in humans, *Lebensmittel-Wissenschaft und -Technologie* **23**:523-527.

Halpern, B. P., Bernard, R. A., and Kare, M. R., 1962, Amino acids as gustatory stimuli in the rat, *J. Gen. Physiol.* **45**:681-701.

Harriman, A. E., 1976, Preferences by northern grasshopper mice for solutions of sugars, acids, and salts in Richter-type drinking tests, *J. Gen. Psychol.* **95**:85-92.

Harriman, A. E., and Nevitt, J. R., 1978, Preferences by deer mice for solutions of sugars, salts, and acids in Richter-type drinking tests, *J. Gen. Psychol.* **98**:207-214.

Hellekant, G., Danilova, V., and Ninomiya, Y., 1997, Primate sense of taste: behavioral and single chorda tympani and glossopharyngeal nerve fiber recordings in the rhesus monkey, *Macaca mulatta*, *J. Neurophysiol.* **77**:978-993.

Klay, R. F., 1964, Monosodium glutamate in pig creep rations, *J. Anim. Sci.* **23**:598.

Nelson, S. L., and Sanregret, J. D., 1997, Response of pigs to bitter-tasting compounds, *Chem. Senses* **22**:129-132.

Nofre, C., and Tinti, J. M., 1996, Sweetness reception in man: the multipoint attachment theory, *Food Chem.* **56**:263-274.

Pritchard, T. C., and Norgren, R., 1991, Preference of Old World monkeys for amino acids and other gustatory stimuli: the influence of monosodium glutamate, *Physiol. Behav.* **49**:1003-1007.

Schiffman, S. S., and Engelhard, H. H., 1976, Taste of dipeptides, *Physiol. Behav.* **17**:523-535.

Schiffman, S. S., Sennewald, K., and Gagnon, J., 1981, Comparison of taste qualities and thresholds of D- and L-amino acids, *Physiol. Behav.* **27**:51-59.

Schiffman, S. S., Clark, T. B., and Gagnon, J., 1982, Influence of chirality of amino acids on the growth of perceived taste intensity with concentration, *Physiol. Behav.* **28**:457-465.

Wieser, H., Jugel, H., and Belitz, H.-D., 1977, Zusammenhänge zwischen Struktur und Süssgeschmack bei Aminosäuren, *Zeitschrift für Lebensmittel-Untersuchung und-Forschung*, **164**:277-282.

Yamaguchi, S., 1987, Fundamental properties of umami in human taste sensation, in: *Umami: a basic Taste* (Y. Kawamura, and M. R. Kare, eds.), New York and Basel: Dekker.

IMPRINTING ON NATIVE POND ODOUR IN THE POOL FROG, *RANA LESSONAE* CAM.

Sergei V. Ogurtsov and Vladimir A. Bastakov

Institute for Information Transmission Problems
Russian Academy of Sciences
Bolshoy Karetney per., 19, Moscow 101447, Russia

INTRODUCTION

Imprinting on environmental stimuli in early development is a common ecological feature of vertebrate ontogeny. Among fishes and reptiles, most of which grow without maternal care, imprinting to the chemical stimuli present at the place of birth is not unusual (Hasler and Scholz, 1978; Grassman, 1993). For example, froglets of the pool frog *Rana lessonae* normally remain near their native pond even after they have undergone metamorphosis. This pattern of habitat selection is presumably based on the froglets' preference for the odor of their native pond, which they learn during larval development (Bastakov, 1986, 1992). The goal of the present study was to determine whether or not natural and artificial chemical markers in pond water at different stages of larval development would be able to modify the discriminations of adult animals.

MATERIALS AND METHODS

The study was performed with the pool frog, *Rana lessonae* Cam. All frogs and eggs used in the experiment were collected from ponds in the Moscow region.

To measure the unmanipulated reactions of wild pool frogs to native pond water, frogs in Group 1 were caught near their native ponds at the end of metamorphosis, i.e., in developmental stages 43-46 (Gosner, 1960). Discriminatory preferences were monitored in a white plastic test chamber of size 76 cm long, 12 cm wide and 15 cm high, covered with transparent glass. Twenty ml each of a pair of familiar and unfamiliar odorants were added to Petri dishes placed at opposite sides of the test chamber (see Figure 1). A group of 4-11 frogs was placed in the center of the test chamber and were left to move about freely in the test chamber. To describe the spatial distribution of frogs between two odorants, the test-chamber was divided into 5 sections, as shown in Figure 1. The number of frogs in each section was counted at 5 min intervals during a 40 min test. Thus, 8 blocks of 5 numbers were generated, in which each frog's position was counted 5 times.

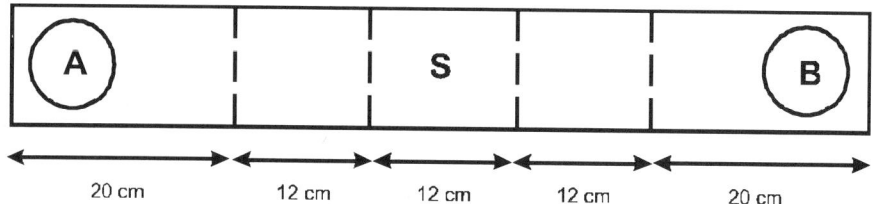

Figure 1. Scheme of a test-chamber, view from above. A and B – Petri dishes with chemical stimuli, S – starting position of frogs.

To guard against a frog bias toward one end of the test chamber, the locations of the odorants were reversed in successive tests, and the results of each group of test were combined for analysis according to the position of the familiar odorant. The chamber was washed with tap water after each test. Tests were conducted at night with illumination from a 40 W incandescent lamp placed 40 cm from the middle of the long axis of the test chamber. A reaction to the odorants was evidenced by asymmetry in the distribution of frogs in the test chamber. Using nonparametric Wilcoxon matched pairs test, we compared the number of frogs in the two outermost sections and then in two sections adjacent to the start section. In the first round of tests of frogs in Group 1, the choices were native pond water versus tap water that had been aged for 2 or 3 days. Wild frogs in test 2 of Group 1 had to discriminate between water from their native pond and from an unfamiliar pond.

We then carried out a developmental study to determine the age at which the frogs might learn what constitutes the native odor of their ponds. As noted in Table 1, 11 separate groups (Groups 2-12) of frogs underwent exposures to chemical stimuli at variuos developmental stages. For rearing tadpoles oviposited eggs were collected in ponds before hatching and the larvae were grown in aquaria in the laboratory; all groups were fed leaves of a common boiled nettle beginning at stage 21, the first stage of active feeding. The water used to rear the tadpoles during specific developmental stages was conditioned with particular odorants, as noted in Table 1. The odorants included morpholine at 10^{-8} mole/l concentration, β-phenylethanol at 10^{-7} or 10^{-8} mole/l, natural pond water, or boiled nettle (one boiled leaf left in 200 ml of water for 30 min). These odorants were used later to test frogs 1-4 months after metamorphosis, after which the frogs were kept in terrariums and had no contact with the odorants used for exposure. Additional tests were conducted 7-9 and 19 months after metamorphosis to determine how long preferences persisted.

RESULTS

Wild frogs captured near their native ponds shortly after the end of metamorphosis exhibited a significant preference for water from those ponds when it was paired with tap water as the alternative choice (test 1 of Group 1 in Table 1). The froglets also preferred native pond water over water from an unfamiliar pond (test 2 of Group 1 in Table 1).

Recently metamorphosed frogs reared in laboratory conditions and exposed to native pond water before stage 21 or exposed to boiled nettle after stage 21 also revealed preferences for these odorants (Group 2 and 3 in Table 1). Thus, exposure to chemical markers during these periods of larval development caused the frogs to develop a preference for these stimuli.

Frogs in Group 4, which had been exposed to pond water only before hatching during stages 1-18, for 4-6 days, subsequently failed to discriminate between native and unfamiliar pond water. But incubation of the tadpoles in Groups 5 and 6 in water solutions of morpholine or β-phenylethanol for about 4-7 days between hatching and the beginning of active feeding, stages 18-21, resulted in the formation of preferences for these odorants when the alternate choice was tap water (Table 1). Frogs in Group 7, exposed to β-phenylethanol for 24 days during stages 25-31, were indifferent to this stimulus after metamorphosis. On the contrary, all the frogs reared in contact with β-phenylethanol or morpholine or with mixture of these odorants after stage 32 or 40 subsequently exhibited a preference for the the stimuli they had been reared in, i.e., see data for Groups 8-10 of Table 1; the minimum time of preference-inducing exposure was 12-15 days in Group 9.

Table 1. Treatment conditions during larval development and results of frog preferences in the test chamber when preferences were tested shortly after metamorphosis.

Group	Treatment[1] Stages	Odorants[2]	Test number	Odorants tested	Score for[3] Odorant	Water	p[4]	No. of frogs
1	1-46	PW	1	PW	92	7	<0,00001	21
			2	PW	80	15	<0,0001	25
2	1-21	PW	1	PW	31	10	<0,05	18
3	21-43	Nt	1	Nt	93	61	<0,05	28
4	1-18	PW	1	PW	21	30	n.s.	17
5	18-21	M	1	M	34	9	<0,01	6
6	18-21	P	1	P	24	3	<0,001	6
7	25-31	P	1	P	14	15	n.s.	8
8	32-43	M+Nt	1	M	47	23	<0,05	12
			2	M+Nt	24	7	<0,05	13
9	40-43	P	1	P	55	37	<0,05	17
10	32-43	M+P	1	M+P	49	19	<0,05	7
			2	M	18	20	n.s.	7
			3	P	20	23	n.s.	7
11	Contr.	PW	1	Nt	83	94	n.s.	21
12	Contr.	Nt	1	M	10	12	n.s.	7
			2	P	8	6	n.s.	12
			3	M+P	12	12	n.s.	11

[1] Contr. – frogs which had no contact during larval development with the odorants they were tested with
[2] PW - native pond water (tested in pair with strange pond water, except test 1 in group 1), M – morpholine, P - β-phenylethanol, Nt - boiled nettle, «+» – two odorants were present simultaneously
[3] All observations during a test are summarized, thus each frog was counted several times in a test
[4] Normal type – strong reaction, comparison of the number of frogs in outermost sections of the chamber, boldface type – weak reaction, significant when the number of frogs are compared in middle sections only

Exposure of tadpoles to chemical markers on stages 18-21 and 32(40)-43, but not during stages 1-18 or 25-31, resulted in the development of froglets' preferences for those markers. Exposure of tadpoles to one stimulus during stages 18-21 and to another during stages 32-43 resulted in froglets which had developed preferences for both stimuli.

Frogs which had contacted morpholine or β-phenylethanol during stages 18-21 or 32-43 preferred these odorants in concentrations of 10^{-7}-10^{-8} mole/l, equal to or lower than that used for incubation, since nearly 70% of frogs were in the 3 sections close to the native odorant. High concentrations of β-phenylethanol, such as 10^{-4}-10^{-5} mole/l, were rejected by the frogs, i.e., 70% were in the 3 sections close to the tap water. Frogs were also indifferent to the intermediate concentration of β-phenylethanol, with 70% found in the 3 central sections. Indifferent reaction was observed in all groups of frogs to the lowest concentration used in tests - 10^{-9} mole/l, which might be below their olfactory threshold for this odorant.

To determine whether frogs discriminate single odorants in a scent mixture, Group 10 was exposed to a mixture of morpholine and β-phenylethanol. When tested, these frogs showed preference for the mixture but not for morpholine or β-phenylethanol alone (see Table 1). Frogs in Group 8, which had been exposed to morpholine in the presence of boiled nettle developed a weak preference for morpholine, significant only when the number of frogs in the middle sections were compared. That means most frogs moved only one section towards the native stimulus and during the second half of the test they distributed themselves randomly in the chamber (test 1 of Group 8 in Table 1, boldface type). When the same group was tested on morpholine together with native foodstuff, there was a strong preference for the native scent mixture, significant when the number of frogs were compared in the outermost sections. That means the majority of frogs moved to the section containing the native odorant and stayed there during the whole test (test 2 of Group 8 in Table 1, normal type).

Frogs were indifferent to chemical stimuli which were not present in the water during their larval development. Thus, wild frogs in Group 11 were indifferent to an unfamiliar laboratory foodstuff, boiled nettle. Similarly, frogs of Group 12 were indifferent to morpholine, β-phenylethanol, and their mixture. When tested on a series of concentrations of artificial odorants, frogs of this group showed indifference to all, even when high concentrations (10^{-4} mole/l for β-phenylethanol) were presented. It is also worth mentioning that frogs which preferred native odorants visited the outermost sections of the test-chamber more frequently than did frogs from the control groups (40% versus 20% of all visits during a test, $p<0,01$), as if a familiar odour increased their locomotor activity.

The reactions of the frogs revealed that preference for native odorants changed with time (data not shown). One to four months after metamorphosis, the majority of groups preferred native stimuli, though Group 1 of wild frogs became indifferent to native pond water on the third month of terrestrial life. Seven months after metamorphosis, the two groups of frogs reared in laboratory conditions still showed preferences for the native stimuli, but 8-9 months after metamorphosis the majority of groups were indifferent to native odorants. At the same time, Group 1 of wild frogs and Group 3 of frogs reared in laboratory conditions revealed weak negative reaction to native odorants. Frogs of control groups and of experimental groups 4 and 7, which were indifferent to the stimuli when tested after metamorphosis, were still indifferent to them on the ninth month of the terrestrial life. The only group, Group 9, tested after the second wintering period, 19 months after metamorphosis, showed preference for the native odorant (β-phenylethanol). When tested on a series of concentrations of the odorant these frogs revealed preference for the very concentration the group has been exposed to.

DISCUSSION

Froglets of the pool frog, *Rana lessonae* Cam., keep close to their native ponds after metamorphosis. When caught near the pond and tested in laboratory, these young frogs reveal strong preferences for their native pond water, and were able to discriminate between water from the native pond and water from an unfamiliar pond. The latter observation is interpreted to imply that the preference for the native pond odorants is specific.

The present data were obtained while simultaneously testing groups of frogs, rather than individuals. Preferences of individual wild froglets for native odorants were shown earlier (Bastakov, 1986), so similarities between results of tests of individuals and of groups indicates that our testing method is adequate.

A series of experiments with frogs reared in laboratory conditions showed that they could learn various chemical stimuli dissolved in water during larval development. Frogs learn not only natural stimuli such as pond water or boiled nettle, but also artificial ones such as morpholine and β-phenylethanol. Learning took place during stages 18-21 and 32(40)-43, and led to preferences for familiar odorants. The minimum time of exposure for chemical stimuli was 4 days, a very small part of the nearly 2.5 months duration of the entire larval period. Frogs were able to learn two different odorants, even when one was present only during the former stages and the other was present only during the latter stages. Frogs exposed to stimuli on stages 1-18 and 25-31 did not learn the odorants in our study and reacted indifferently to them. The same was true for the frogs which had never contacted any of the odorants used in our experiments. Thus learning is possible during two separate periods of larval development. It is worth mentioning that morphological development of frog's olfactory system also has two stages (Spaeti, 1978). The developmental stages of learning in our study correspond well with the two stages of active neurogenesis.

The fact that frogs reacted to very low concentrations of odorants (10^{-8} mole/l) suggest that olfactory reception mediated these responses. Preferring low, and rejecting high, concentrations of a familiar stimulus, frogs could choose an optimal position in a concentration gradient of an odorant. Frogs that learned chemical stimuli through exposure during larval development reacted differentially to various concentrations of the familiar odorants, but frogs which never had contact with the same stimuli were indifferent to all concentrations tested. The situation resembles the phenomenon of experience-induced sensitivity in young mammals (Hudson, 1999).

We showed that if frogs are reared in a mixture of chemical stimuli during larval development, they react more strongly to the scent mixture than to each of its components by itself. During stages 18-21, larvae do not feed and so are exposed only to odorants dissolved in the water, but during stages 32-43, odorants mixed with foods provide the experience of a scent mixture. One can also propose that frogs could discriminate between water from native and unfamiliar ponds by the complex mixtures of odorants, rather than by a single component.

It is known that frogs and toads use olfaction for spatial orientation (Sinsch, 1992). Whether the frogs prefer or reject the native pond odour largely depends on the biology of a species. Common toads (*Bufo bufo* L.) leave their native ponds soon after metamorphosis, and might be expected to move away from, or discriminate against, their native pond odorants. On the contrary, froglets of the pool frog, *R. lessonae*, are strongly attached to the native pond and did indeed reveal their preferences for native pond odorants (Bastakov, 1992).

The present study shows that reactions of pool frogs to native pond odour changes gradually with age and these changes correspond with the strategy of behaviour of different age groups of this species. According to the results of our censuses (unpublished), young frogs, 8-9 months after metamorphosis, after wintering, usually occupy unfamiliar ponds and constantly move from one pond to another. At these ages, we observed indifference or even negative reactions to native pond odours. However, it appears that pool frogs return for first breeding, at the age of 2-3 years, to their native ponds (Breden, 1987). Adult pool frogs are also strongly attached to their breeding ponds (Sjögren, 1994). In our experiment we were able to register the preference for native pond odour in frogs nearly 2 years old. Thus it is very likely that pool frogs could use native pond odour for spatial orientation in returning to breed when they become mature.

Our data confirm the possibility of learning and use of odorants in pool frogs. Frogs learn different odorants during larval development. No visible reward is necessary for such learning. The periods of learning are rather short, but the reaction to this odour is maintained for a long time after metamorphosis. All of these observations indicate that pool frogs undergo olfactory imprinting, as discovered earlier in fish (Hasler and Scholz, 1978) and reptiles (Grassman, 1993).

ACKNOWLEDGMENTS

This study was partly supported by Russian Fund of Basic Research, grant number 00-04-58754. We wish to thank Dr. Ivan N. Pigarev for critical comments on the manuscript.

REFERENCES

Bastakov, V. A., 1986, Preference by youngs-of-the-year of the edible frog (Rana esculenta complex) for their own reservoir ground smell, *J. Zoology* **65**(12):1864-1868 (in Russian).

Bastakov, V. A., 1992, Experimental study of the memorizing of pond odour during larval development of two anuran species, *J. Zoology* **71**(10):123-127 (in Russian).

Breden, F., 1987, The effect of post-metamorphic dispersal on the population genetic structure of Fowler's toad, *Bufo woodhousei fowleri*, *Copeia* **1987**:386-394.

Gosner, K. L., 1960, A simplified table for staging anuran embryos and larvae with notes on identification, *Herpetologica* **16**:183-190.

Grassman, M., 1993, Chemosensory orientation behaviour in juvenile sea turtles, *Brain, Behav. Evol.* **41**:224-228.

Hasler, A. D., and Scholz, A. T., 1978, Olfactory imprinting in Coho Salmon (Oncorhynchus kisutch), in: *Animal Migration, Navigation and Homing* (K. Schmidt-Koenig, and W. T. Keeton, eds), Springer-Verlag, Berlin, pp. 356-369.

Hudson, R., 1999, From molecule to mind: the role of experience in shaping olfactory function, *J. Comp. Physiol. A* **185**:297-304.

Sinsch, U., 1992, Amphibians, in: *Animal Homing* (F. Papi, ed.), Chapman and Hall, London, pp. 213-233.

Sjögren, P., 1994, Distribution and extinction patterns within a northern metapopulation of the pool frog, *Rana lessonae*, *Ecology* **75**:1357-1367.

Spaeti, U., 1978, Development of the sensory systems in the larval and metamorphosing European grass frog (Rana temporia L.), *Hirnforschung* **19**(6):543-575.

CHEMICAL DISCRIMINATION IN *LIOLAEMUS* LIZARDS: COMPARISON OF BEHAVIORAL AND CHEMICAL DATA

Antonieta Labra[1,2], Carlos A. Escobar[1], and Hermann M. Niemeyer[1]

[1]Facultad de Ciencias, Laboratorio de Química Ecológica,
Universidad de Chile, Casilla 653, Santiago, Chile
[2]Departamento de Ecología, Facultad de Ciencias Biológicas,
Universidad Católica de Chile, Santiago, C.P. 651-3677, Chile

INTRODUCTION

Chemical signals produced by lizards are involved in different aspects of their interactions at intra and interspecific levels, including sex recognition, territory marking, and individual self-recognition (Mason, 1992; Cooper, 1994; Font, 1996), and have also been proposed as pre-reproductive barriers for sympatric congeneric species (Cooper and Vitt, 1986). The range of functions of chemical recognition would suggest that lizards should be able to discriminate many different patterns of chemical signals. We postulate that it would be „more economical" for lizards to discriminate only one chemical signal pattern, the one used by individuals in their self-recognition, and that signals not matching with the basic chemical pattern will not be discriminated.

To support this proposition we studied self-recognition at individual and specific levels in different lizard species of the genus *Liolaemus* of Chile. If self-recognition predominates at individual level, lizards would show lower number of tongue flicks in their own enclosures than in other enclosures (*i.e.* Alberts, 1992; Cooper *et al.*, 1999). At specific level, it is expected that lizards behave differently in the presence of chemical signals of conspecifics than of congeneric species. Finally, we analyzed the relationship between behavior and chemical composition of secretions from precloacal pores, since these are a source of chemical signals (Labra *et al.*, submitted).

MATERIAL AND METHODS

Lizards were carried to the laboratory, maintained individually in enclosures with sand as substrate, a bowl of water and a rock, and were tested after one week of acclimation (Labra *et al.*, 2001). Prior to the experiments, lizards were removed from their enclosures and placed randomly in different enclosures (treatments, see below). In order to avoid the potential use of visual signals in the discrimination, the enclosures were partially emptied before the experiments: the lizard, the rock and the bowl of water were withdrawn, and the sand remained. The number of tongue-flicks (TF) that lizards made during 10 min was recorded. Records began after lizards made their first TF. Tongue-flicks were recorded since traditionally they have been used as the way to determine if lizards are able to discriminate among different chemical signals (Cooper, 1998).

Behavior

Individual self-recognition. To test self-recognition at the individual level, we studied five *Liolaemus* species: *L. bellii*, *L. constanzae*, *L. fitzgeraldi*, *L. eleodori* and *L. tenuis*. The site of capture and the type of habitat of the species used in behavioral assays are shown in Table 1. Treatments consisted in placing each lizard in the enclosure of the tested lizard (own), that of a conspecific, and an unused enclosure (control). Unless otherwise stated, conspecific treatments referred to the use of an enclosure whose owner was of the same sex of the tested individual (*i.e.* male in a male's enclosure).

Specific self-recognition. To test self-recognition at the species level, two focal species were used, *L. jamesi* and *L. bellii*. They were tested in enclosures of conspecifics and congeneric species, both sympatric and allopatric. For *L. jamesi* its sympatric species was *L. alticolor*, and the allopatric one was *L. ornatus*. In the case of *L. bellii* the sympatric species was *L. nigroviridis* and the allopatric one was *L. fitzgeraldi* (Labra and Niemeyer, in prep.). The localities where different species were collected are indicated in Table 1.

Table 1. Distribution and characteristic habitat of the *Liolaemus* species included in this study.

Liolaemus	Altitude (m.a.s.l.)	Latitude/Longitude (°S; °W)	Characteristic habitat
alticolor	4350	18°10'; 69°25'	Andean Plateau
bellii	2300	33°20'; 70°19'	High mountain
constanzae	2250	23°46'; 68°14'	Atacama Desert
eleodori	3670	27°04'; 69°10'	Andean Plateau
fitzgeraldi	3500	23°50'; 70°09'	High mountain
jamesi	4350	18°10'; 69°25'	Andean Plateau
lemniscatus	950	33°35'; 70°28'	Chaparral
nigroviridis	2300	33°20'; 70°19'	High mountain
ornatus	3710	19°15'; 68°43'	Andean Plateau
tenuis	1850	33°46'; 70°15'	Chaparral

Chemical Data

The chemical analysis of the precloacal secretions of 20 *Liolaemus* species from different localities of Chile, was performed using GC-MS. The secretions obtained were weighed and dissolved in *n*-hexane. The secretions of three individuals of each species were pooled prior to the analysis. In addition, nine males of *L. bellii* were analyzed individually to determine the intraspecific variation in the chemical composition of the secretions. The presence or absence of a given compound in the chromatographic profile of each *Liolaemus* species and of the different individuals of *L. bellii*, were determined by comparing retention times and mass spectra with a mass spectra library and commercial standards (Escobar et al., submitted).

Statistics

Behavioral data were analyzed using ANOVA for repeated measurements and Tukey a posteriori tests. The similitude in the chemical composition of the secretions, among species and among individuals of L. bellii, was analyzed with unweighted pair-group average (Zar, 1984).

RESULTS

Behavior

Individual self-recognition. For *L. tenuis* (Labra and Niemeyer, 1999) and *L. bellii* (Labra et al., 2001) the analysis of the effects of season and sex of the enclosure owner, showed a lower number of TF in the own enclosures in both seasons in *L. tenuis*, and only in the autumn in *L. bellii*. In both species, female enclosures elicited a higher number of TF during the reproductive season. However, the reanalysis of results for the reproductive season using only data of conspecifics of the same sex, showed no differences in TF between conspecific and control enclosures, and a lower number of TF in the own enclosures.

The other three species showed the same trends, i.e. lower number of tongue-flicks in their own enclosures. For *L. eleodori* differences were found among treatments, but TF was similar in conspecific and control enclosures (Figure 1A). *L. lemniscatus* showed a lower number of TF in the own enclosure in comparison with the control condition, and *L. constanzae* also showed lower number of TF in the own enclosure, in comparison with enclosures of conspecifics (Labra and Niemeyer, in preparation).

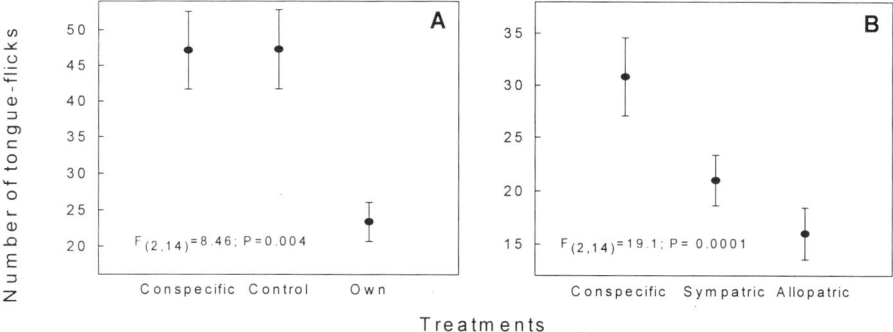

Figure 1. Self-recognition: A. At individual level for *L. eleodori*, B: At specific level in *L. jamesi*. Treatments are explained in the text. Points represent mean values; bars represent one standard error. The values from the statistical tests are given in the figures.

Specific self-recognition. *Liolaemus jamesi* and *L. bellii* (Labra and Niemeyer, in preparation) showed similar patterns of chemical exploratory behavior. The individuals of the focal species showed a higher number of TF in enclosures of conspecifics than in those of any of the congeneric species. There were no differences in the behaviors recorded in the enclosures of sympatric and allopatric species. Figure 1B shows the results for *L. jamesi*.

CHEMICAL DATA

The interspecific chemical analysis of male precloacal secretions of 20 *Liolaemus* species showed the presence of 50 different compounds distributed among all the species. The compounds belonged to three main categories: *n*-alkanes, long chain carboxylic acids, and steroids. Only six of these 50 compounds were present in all the species analyzed: cholesterol and five carboxylic acids (Escobar *et al.*, submitted). Using the information of presence or absence of the different compounds, preliminary analyses of similitude were performed among the species, and also among the individuals of *L. bellii*. The individuals of *L. bellii* and the pooled sample of this species were grouped together, the 20 species were well differentiated among them, and the distance indices among species were higher than those obtained between individuals of *L. bellii* (Figure 2).

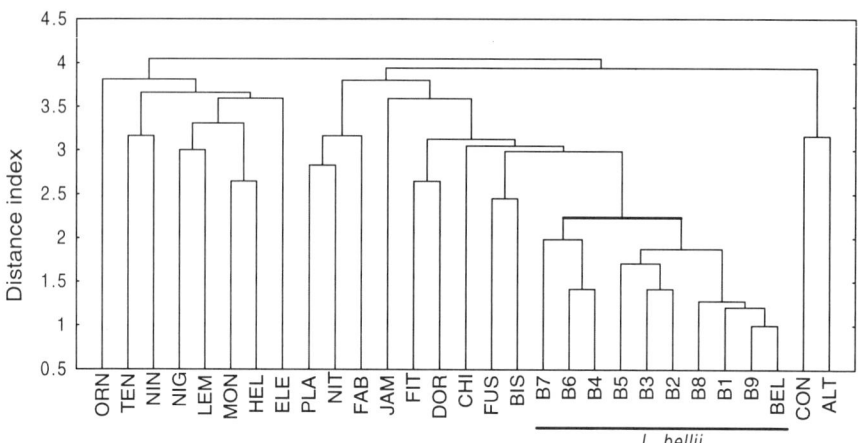

Figure 2. Phenogram showing the similitude among 20 species of *Liolaemus* and nine individuals of *L.bellii* (B1- B9). *Liolaemus* species codes are: ALT = *alticolor*; BEL = *bellii*; BIS = *bisignatus*; CHI = *chiliensis*; CON = *constanzae*; DOR = *dorbigni*; ELE = *eleodori*; FAB = *fabiani*; FIT = *fitzgeraldi*; FUS = *fuscus*; HEL = *hellmichi*; JAM = *jamesi*; LEM = *lemniscatus*; MON = *monticola*; NIG = *nigroviridis*; NIN = *nigroviridis nigroroseus*; NIT = *nitidus*; ORN = *ornatus*; PLA = *platei* and TEN = *tenuis*.

DISCUSSION

The five species of *Liolaemus* studied showed self-recognition at the individual level, as manifested by a lower number of TF in their own enclosures. The similar values of TF in control and conspecific treatments may be due to TF not being equally distributed along the duration of the experiment; this hypothesis was not tested. In addition, in *L. tenuis* and *L. bellii* female enclosures triggered higher number of TF during the reproductive season

(Labra and Niemeyer, 1999; Labra et al., 2001). Self-recognition at the species level occurred, as manifested by a higher chemical exploratory behavior in enclosures of conspecific, rather than in those of congeneric species. No differences were observed between the behaviors recorded in enclosures of sympatric or allopatric species. Therefore, if chemical signals do constitute reproductive barriers, they are useful in the distinction of an individual from its conspecifics but not from individuals of any other species.

Self-recognition in *Liolaemus* at individual and specific levels seems to be quite a generalized phenomenon since the species studied came from widely different environments and belong to different clades (Etheridge, 1995). The phenogram of Figure 2 shows that chemical composition of precloacal secretions from all species are different, and that there are clear differences between intraspecific and interspecific variability. This suggests that: i) the six chemical compounds common to all species may be the means to chemically identify the genus *Liolaemus*; ii) that each species has its own characteristic pattern of chemical signals, as a result of qualitative and/or quantitative differences in the chemical constituents of precloacal secretions; and iii) that each species has certain variability in the basic pheromonal pattern that would enable individuals to exhibit self-recognition. In addition, lizards would also be able to discriminate (innately or through learning) other variants of the basic pattern, as it occurs in the cases of recognition of familiar individuals (Cooper, 1996) and in mother-offspring recognition (*i.e.* Bull *et al.*, 1994), and also between certain variants of pheromones from conspecifics, under particular circumstances. Thus, female enclosures elicited a higher number of TF during the breeding season (Labra and Niemeyer, 1999; Labra *et al*, 2001). In this case, physiological changes due to the reproductive condition would allow animals to detect (in the case of males) and/or to produce (in the case of females) the basic signal pattern with some modifications, and thus trigger an increase in chemical exploratory behavior. If the signal comes from a congeneric, so it does not match the basic chemical pattern, there is no discrimination and no particular behavior is triggered.

From the behavioral data presented we conclude that self-recognition is the basic „type of recognition" in these lizard species, at individual and specific levels. The next step will be to test self-recognition at populational level. Based on the present results, it is highly probable that self-recognition also occurs at populational level, thus constituting another mechanism that would favor the high rate of speciation found in this genus (Lamborot, 1993).

ACKNOWLEDGMENTS

We are grateful to M. P. Aguilar, D. Benítez, S. Cortéz, R. Irisarri, B. López, D. Macari and L. Ovalle, for field and laboratory assistance. This study was funded by IFS grants 2933-1 to AL and 2934-1 to CAE, FONDECYT grant 3990021 to AL, and Cátedra Presidencial en ciencias to HMN. The Corporación Nacional Forestal allowed us to collect some species within the National Park system; Servicio Agrícola y Ganadero gave us all the authorizations to collect the lizards, and Carabineros de Chile allowed the use of their infrastructure during some collection campaigns.

REFERENCES

Alberts, A. C., 1992, Pheromonal self-recognition in desert iguanas, *Copeia* **1992**:229-232.
Bull, C. M., Doberty, M., Schulze, L. R., and Pamula, Y., 1994, Recognition of offspring by females of the Australian skink, *Tiliqua rugosa, J. Herpetol.* **28**:117-120.

Cooper, Jr., W. E., 1994, Chemical discrimination by tongue-flicking in lizards: a review with hypothesis on its origin and its ecological and phylogenetic relationships, *J. Chem. Ecol.* **20**:439-487.

Cooper, Jr., W. E., 1996, Chemosensory recognition of familiar and unfamiliar conspecifics by the scincid lizard *Eumeces laticeps, Ethology* **102**:454 – 464.

Cooper, Jr., W. E., 1998, Evaluation of the swap and related tests as a bioassay for assessing responses by squamata reptiles to chemical stimuli, *J. Chem. Ecol.* **24**:841- 866.

Cooper, Jr., W. E., and Vitt, L. J., 1986, Interspecific odour discrimination among syntopic congeners in scincid lizards (Genus *Eumeces*), *Behaviour* **97**:1 – 9.

Cooper, Jr., W. E., Van Wyk, J. H., and Mouton, P. Le F. N., 1999, Discrimination between self-produced pheromones and those produced by individuals of the same sex in the lizard *Cordylus cordylus, J. Chem. Ecol.* **25**:197 – 208.

Etheridge, R., 1995, Redescription of *Ctenoblepharys adspersa* Tschudi, 1845, and the taxonomy of Liolaeminae (Reptilia: Squamata: Tropiduridae), *Am. Mus. Novitates* **3142**:1 – 34.

Font, E., 1996, Los sentidos químicos de los reptiles. Un enfoque etológico, in: *Psicología Comparada y Comportamiento Animal* (F. Colmenares, ed.), Editorial Síntesis, Madrid. S. A., España, pp. 197-259.

Labra, A., and Niemeyer, H. M., 1999, Intraspecific chemical recognition in the lizard *Liolaemus tenuis* (Tropiduridae), *J. Chem. Ecol.* **25**:1799-1812.

Labra, A., Beltrán, S., and Niemeyer, H. M., 2001, Chemical exploratory behavior in the lizard *Liolaemus bellii. J. Herpetol.* **35**:51-55.

Lamborot, M., 1993, Chromosomal evolution and speciation in some Chilean lizards. *Evol. Biol.* **7**:133 – 151.

Mason, R. T., 1992, Reptilian pheromones, in: *Biology of Reptilia. Hormones, Brain and Behavior*. Volume 18 (E. C. Gans, and D. Crews, eds.), The University of Chicago Press, Chicago, US, pp. 114-228.

Zar, J., 1984, *Biostatistical Analysis*, 2nd ed., Prentice-Hall, Inc. New Jersey, 718 pp.

SIMPLIFIED TESTS OF AGGRESSION CHEMOSIGNALS IN MALE HOUSE MICE SUGGEST THAT A MELANOCORTIN-DEPENDENT PRODUCT OF THE PREPUTIAL GLAND REDUCES ATTACKS

Heather K. Caldwell,[1] Liyong Wang,[2] and John J. Lepri[3]

[1]Biology Department, Georgia State University,
Atlanta, GA 30302
[2]Department of Foods and Nutrition, Purdue University,
West Lafayette, IN 47907-1264
[3]Department of Biology, The University of North Carolina at Greensboro,
Greensboro, NC 27402

INTRODUCTION

Investigations of aggression chemosignals in mammals have mostly involved lesioning a chemosensory system, e.g., chemically-induced anosmia, or surgically eliminating the glandular source of putative pheromones, e.g., preputial-glandectomy (Lee and Ingersoll, 1983). Typically, interactive behavioral tests with the altered animals or their modified chemosignals have been used to assess the possible chemosensory coordination of aggression. The recent development of genetically-engineered house mice having targeted mutations or disruptions of single genes presents an innovative and highly focused level of analysis in chemosignal studies of sociosexual behaviors, not only in ascertaining the appropriate chemosensory receptors (Mombaerts, 1999), but also in identifying the cellular sources and chemical identities of pheromones (Kingsley, 1998). This report is a description of efforts to develop a behavioral test for aggression chemosignals in male house mice.

Chemosignals are released from a variety of sources, including sebaceous glands in the skin and in the Harderian, lacrimal, and preputial glands. The preputial glands of male house mice are paired, symmetrical, glands located under the pubic skin near the base of the penis. The preputial glands contain sebatocytes that synthesize a mixture of waxy, lipid-rich, products, including sebum, that enter excreted urine and can influence intermale aggression.

Androgens promote the growth and activity of the preputial glands, which enlarge with the onset of puberty, and atrophy following castration (Ebling *et al.*, 1970). Secretion from the preputial glands occurs in response to hormones from the pituitary gland (Thody and Shuster, 1973). In particular, the melanocortin α-melanocyte stimulating hormone (hereafter α-MSH), initiates the lysis of sebatocytes in the preputial glands, causing sebum and cellular debris from the lysed cell to exit the gland via a duct (Albone, 1984). Correspondingly, the sebatocytes in the preputial glands and elsewhere contain the melanocortin receptor identified as MC5-R (Chen et al., 1997; Tatro and Reichlin, 1987).

Melanocortin secretion from the hypothalamic-pituitary-adrenal „stress" axis during male-male interactions can presumably alter the excretion of chemosignals that impact aggressive behaviors (Jones and Nowell, 1973a; McKinney and Christian, 1970; Thody and Shuster, 1973; Thody et al., 1976). Dominant male mice rendered anosmic by intranasal treatment with zinc sulfate no longer demonstrated differential aggressiveness responses to subordinates treated with or without α-MSH (Nowell et al., 1980). Similarly, male mice whose preputial glands had been surgically removed did not evoke as much aggression as males with intact glands (Hayashi, 1987; Ingersoll et al., 1986). Even though these and other studies suggest that chemosignals from the preputial glands cause an increase in aggression, conflicting evidence suggests that preputial products actually inhibit aggression, and thus serve as „aversion" rather than „aggression" chemosignals (Jones and Nowell, 1973b; Novotny et al., 1990). This conflict is perhaps partially the result of uncertainties about which participant has the larger effect on the interactive behaviors observed when two animals are placed in physical contact with each other.

The production of mice with a genetic disruption of the fifth melanocortin receptor (Chen et al., 1997), hereafter called MC5-RKO mice, allowed the studies described here. Phenotypically, MC5-RKO mice have a fully functional hypothalamic-pituitary-adrenocortical axis, as neither basal- nor stress-induced levels of corticosteroids differ from those of non-altered, wild-type (hereafter WT) mice. When the pelage is wet, however, the knock-outs have impaired thermoregulation, presumably because they do not secrete enough sebaceous gland products into the pelage to effectively repulse water (Chen et al., 1997).

The experiments described here address the hypothesis that functional melanocortin receptors are required to undergo melanocortin- and stress-induced changes in chemosignal excretion. Male mice in the first experiment were anesthetized sufficiently to make them immobile, injected with melanocortins, and then delivered as „intruders" to the cages of subject „resident" males. A second procedure was then developed in which non-anesthetized, castrated, „surrogate" males were placed inside a small, screen-enclosed protective cage, upon which were applied chemosignals that had been collected from either WT or MC5-RKO mice. The scented, protective cage, hereafter called the „corral," was then placed into the home cages of resident males whose behavioral responses were scored. This simplified resident-intruder test allowed the analysis of aggression-related responses to chemosignals, which had been presented, in a biologically relevant context that prevented injury to the animals.

METHODS AND RESULTS

Strains of *Mus musculus domesticus* were used. The WT and MC5-RKO mice were kindly provided by Roger Cone and Wen-biao Chen at the Vollum Institute of the Oregon Health Sciences University (OHSU). For experiments 2 and 3, male 'retired-breeders' (sexually-proven sires of the ICR strain, more than 6 months of age, Taconic Laboratories) were used as resident males, whereas castrated males of the S129/SvEv strains (Taconic) were used as intruders. All of the observed mice were individually housed at the University of North Carolina at Greensboro (UNCG) Animal Facility in polypropylene cages. Food and water were provided ad libitum. All behavioral testing was completed during the dark part of a 14L:10D reverse light cycle in recognition of the nocturnal behavior patterns of house mice. A small electric desk lamp provided dim illumination during testing. Proposals for this work were institutionally approved.

Experiment 1

Adult resident WT (n=12) and MC5-RKO (n=13) male mice were tested for their behavioral responses to an anesthetized WT or MC5-RKO conspecific intruder that had been injected with either α-MSH (10μg, in 100μl saline) or saline. The intruders were anesthetized (50mg pentobarbital/kg body mass) to prevent melanocortin-induced changes in their behavior from influencing the behavior of the residents, and then placed into the home cage of a resident. The following behaviors were scored for 15 minutes: activity (resident's rostrum crosses a midline across the short axis of the cage), and latency and duration of rostral and caudal proximity (resident's rostrum within 1 cm of the intruder's rostral or caudal half, respectively). Tests were terminated if the intruder males were attacked and at risk of injury.

Data were analyzed by a 2 x 2 ANOVA for the main effects of genotype (WT vs. MC5-RKO) and treatment (α-MSH vs. saline) and the interaction between these. The main effects of genotype and hormone treatment on all behaviors failed to reach statistical significance (data not shown). There was one significant interaction: mean latency for rostral proximity for both groups of residents was lower in response to WT intruders treated with α-MSH than it was for the other three types of intruders (means: 5.4 sec for WT intruder+α-MSH; versus 13.1, 11.9, and 10.1 sec, respectively, for WT+saline, MC5-RKO+α-MSH, and MC5-RKO+saline; $F_{1,23}=4.9$, $p<0.04$).

Experiment 2

Adult WT male residents (n=13) were tested for their behavioral responses to urine collected from WT or MC5-RKO mice that had either received an injection of histamine (stressed) or had been injected with vehicle only (unstressed); the injections and urine collections were completed at OHSU, and „blind" samples were sent to UNCG for tests. WT males were selected for use as residents in this experiment based on their having initiated aggressive reactions within one minute during at least one of four daily, consecutive trials in which an intact male was dangled in front of them; aggression screening preceded testing by one week. Two days before testing, resident males were habituated to the testing apparatus by having an empty wire corral placed into their cage for 5 minutes while being exposed to the sounds heard during actual testing, i.e., beeps from stop watches and clicks from tally counters.

The 'resident-intruder' model used here involved placing a castrated 'intruder' male inside a cylindrical, double-walled, corral made of wire screen (see Figure 1). The corral had external dimensions of approximately 14 cm length and 6 cm diameter. For each test, the exterior of the corral was fitted with new paper-and-metal twist-ties at opposite ends, onto which 8 μl of urine had been applied by pipette. The corral was then placed in the resident's cage for 15 minutes while the following behavioral measurements were made: latencies and frequencies of bites and attacks, locomotor activity, duration on top of the corral, latencies and durations of 1-cm proximity of the resident's rostrum to the corral and to the stimuli. A latency score of 900 seconds was assigned for any behavior that did not occur. In a randomized order, each resident male was exposed to one stimulus per day over several days. The corral was washed with 70% ethanol and dried between uses.

Data were analyzed by a 2 x 2 ANOVA for the main effects of genotype (WT vs. MC5-RKO) and treatment (stressed vs. unstressed) and the interactions (Table 1). The statistically significant effects of genotype are in general agreement with the conclusion that residents were slower to approach and attack when the stimulus urine was from WT males. There were no statistically significant interactions between genotype and treatment.

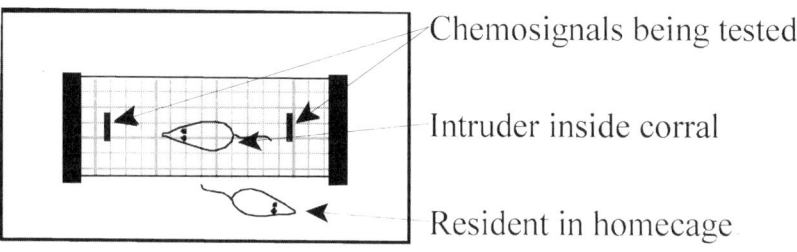

Figure 1. In the modified resident-intruder test, chemosignals to be tested are applied to paper-and-wire tabs attached to a double-walled, wire corral housing a castrated male. This provides both protection from injury and a biologically relevant setting for tests of chemosignal modulation of inter-male aggression.

Table 1. Mean (± SEM) and p values for behaviors in Experiment 2 (n=13 subjects).

	Type of Urine Tested				Stress main effect	Genotype main effect
	WT		MC5-RKO			
	unstressed	stressed	unstressed	stressed		
Latency to bite (sec)	147.5 (± 42.4)	314.4 (± 82.0)	82.2 (± 13.5)	144.5 (± 42.4)	$p<0.02$	$p<0.02$
Number of bites (freq)	30.9 (± 3.6)	24.8 (± 6.1)	34.9 (± 6.2)	40.6 (± 6.5)	$p<0.96$	$p<0.03$
Latency to attack (sec)	608.2 (± 67.2)	707.2 (± 71.4)	541.4 (± 95.5)	435.5 (± 86.2)	$p<0.99$	$p<0.03$
Number of attacks (freq)	6.7 (± 3.0)	4.8 (± 2.1)	10.6 (± 3.7)	14.2 (± 3.5)	$p<0.73$	$p<0.01$
Time spent on corral (sec)	311.8 (± 21.3)	257.1 (± 23.6)	297.1 (± 30.8)	273.9 (± 31.6)	$p<0.02$	$p<0.02$

Experiment 3

Using the corral system, adult WT male residents (n=16) were tested for responses to preputial-gland homogenates in saline. Preputial glands were from WT or MC5-RKO mice that had been restrained for 60 minutes in a centrifuge tube (stressed) or gland collections were done at OHSU; „blind" samples were sent to UNCG for tests. The homogenate (20 μl) was applied to each twist-tie in the resident-intruder behavioral model.

A 2 x 2 ANOVA revealed that the homogenate from the WT mice was associated with a decreased latency to bite ($F_{1,36}=5.11$; $p=0.03$; means are 428.7 vs. 291.5 sec, MC5-RKO vs. WT respectively) compared to that from the MC5-RKO mice; no others reached significance.

DISCUSSION

Communication is broadly defined as anything a sender does that influences the behavior or physiology of a recipient. The careful analysis of communication requires a keen awareness that interacting animals frequently exchange sender and recipient roles, but it is widely acknowledged that few experimental designs in chemical communication studies successfully resolve this role reversal. The problem is particularly acute when the scenario is as complex and sequentially rapid as the aggressive interactions between male house mice.

The experimental approach used in Experiment 1 allowed only the non-anesthetized resident mouse to generate behavioral responses, because the intruder was anesthetized. There were no statistically significant differences due to the main effects of genotype or hormone treatment; the only significant interaction showed a more rapid approach to the rostral areas of WT males treated with α-MSH than to the other three groups, which were statistically equivalent to each other. There are shortcomings of this experiment that discount the validity of these results. First, the anesthesia prevented urination by the intruder mouse. Hence, even if melanocortin injections had elicited sebatocyte lysis and chemosignal excretion from preputial glands of WT males, they failed to urinate, so the residents did not have an opportunity to sample changes in urinary chemosignal excretion. A second problem is that MC5-RKO mice, presumably lacking melanocortin-induced pheromone excretion from their own bodies, might lack prerequisite behavioral experience to respond to such chemosignals when they emanate from conspecifics (*sensu* Todrank *et al.*, 1999).

Experiment 2 described a novel approach in which a castrated male accompanied the presentations of chemosignals, thus including a biological context for the investigation of, and response to, chemosignals related to the coordination of aggression. Moreover, the protective corral surrounding the castrated male assured that all of the observation sessions went uninterrupted by violent attacks, because there was no risk of injury. A potential shortcoming of this procedure is that castrated males in the corral might have altered their behavior or other characteristics such as vocalizations in response to the chemosignals placed on the outside of the corral. If so, differences in residents' responses still could have been interactively driven by chemosignal-driven changes in the castrated intruder. In a preliminary experiment using an unscented corral enclosing a castrated male, the level of interest shown by residents was low (data not shown).

The results of Experiment 2 (Table 1) strongly support the hypothesis that males with functional MC5-R excrete urinary chemosignals that delay or inhibit aggressive responses from other males. Notably, the statistically significant main effects of genotype on latency to bite, number of bites, latency to attack, number of attacks, and time spent on top of the corral, are all in accordance with a conclusion that aggressive responses were more rapid and vigorous when urine from MC5-RKO males was placed onto the corral.

The results of experiment 3 suggest that isolated secretions from the preputial glands do not have the same stimulus strength as urine in causing males to avoid the testing corral. It is possible that preputial products gain their pheromonal potency as a consequence of their interactions with urinary proteins (Beynon *et al.*, 1999). Anecdotally, it was noted that resident males in Experiment 3 appeared less interested 'over-all' than in the first experiment. Since the same resident males were used in this experiment as in the previous experiment, they may have habituated to the 'resident-intruder' model, thereby reducing responsiveness.

The experiments presented in this report demonstrate the development of a simplified test for observations of intermale aggression in a setting that is biologically relevant without risk of injury. In addition, the data represent some of the first studies on the endocrine modulation of chemosignal excretion in genetically engineered house mice. The primary conclusion of these studies is that a melanocortin-dependent chemosignal, presumably from the melanocortin-5-receptor rich preputial glands, reduces aggression among male mice.

ACKNOWLEDGMENTS

Thanks to Roger Cone (OHSU, Vollum Institute) and his research associate, Wen-Biao Chen, for the use of the knockout mice and the samples they provided. Marjut Mäkelä and Dave Ludwig provided technical assistance, and Cheryl Logan commented extensively on the work. This research was supported by the UNCG Department of Biology, and portions of it constitute the master's thesis work of the senior author (Kingsley, 1998).

REFERENCES

Albone, E. S., 1984, Semiochemistry: The Investigation of Chemical Signals Between Mammals. John Wiley and Sons Limited, New York.
Beynon, R. J., Roberston, D. H. L., Hubbard, S. J., Gaskell, S. J., and Hurst, J. L., 1999, The role of protein binding in chemical communication: major urinary proteins in the house mouse, in: *Advances in Chemical Signals in Vertebrates,* (R. E. Johnston, D. Müller-Schwarze, and P. W. Sorensen, eds.), Kluwer Academic/ Plenum Publishers, New York, pp. 137-148.
Chen, W., Kelly, M. A., Opitz-Araya, X., Thomas, R. E., Low, M .J., and Cone, R. D., 1997, Exocrine gland dysfunction in MC5-R-deficient mice: evidence for coordinated regulation of exocrine gland function by melanocortin peptides, *Cell* **91**:789-98.
Ebling, F. J., Ebling, E., Skinner, J., and White, A., 1970, The response of the sebaceous glands of hypophysectomized-castrated male rats to andrenocorticotrophic hormone and to testosterone, *J.Endocr.* **48**:73-81.
Hayashi, S., 1987, The effects of preputialectomy on aggression in male mice, *Zool. Sci.* **4**:551-555.
Ingersoll, D. W., Morley, K. T., Benvenga, M., and Hands, C., 1986, An accessory sex gland aggression promoting chemosignal in male mice, *Behav. Neurosci.* **100**:777-782.
Jones, R. B., and Nowell, N. W., 1973, Effects of preputial and coagulating gland secretions upon aggressive behaviour in male mice: a confirmation, *J.Endocr.* **59**:203-204.
Kingsley, H. K., 1998, Disruption of the fifth melanocortin receptor alters intermale aggression pheromones in house mice. unpublished M.S. thesis, University of North Carolina at Greensboro, Department of Biology, 69 pages.
Lee, C., and Ingersoll, D. W., 1983, Pheromonal influence on aggressive behavior, in: *Hormones and Aggressive Behavior*, (B. B. Svare, ed.), Plenum Press, New York, pp. 373-392.
McKinney, T. D., and Christian, J. J., 1970, Effect of preputialectomy on fighting behavior in mice, *Proc. Soc. Exp. Bio. Med.* **134**:291-293.
Mombaerts, P., 1999, Seven-transmembrane proteins as odorant and chemosensory receptors. *Science* **286**:707-711.
Novotny, M., Harvey, S., and Jemiolo, B., 1990, Chemistry of male dominance in the house mouse, *Mus domesticus, Experientia* **46**:109-113.
Nowell, N. W., Thody, A. J., and Woodley, R., 1980. α-Melanocyte stimulating hormone and aggressive behavior in the male mouse. *Physiol.Behav.* **24**:5-9.
Tatro, J. B., and Reichlin, S., 1987, Specific receptors for α-melanocyte-stimulating hormone are widely distributed in tissues of rodents. *Endocrinology* **121**:1900-1907.
Thody, A. J., and Shuster, S., 1973, Possible role of MSH in the mammal, *Nature* **245**:207-209.
Thody, A. J., Cooper, M. F., Bowden, P. E., Meddis, D., and Shuster, S., 1976, Effect of α-melanocyte-stimulating hormone and testosterone on cutaneous and modified sebaceous glands in the rat, *J.Endocr.* **71**:279-288.
Todrank, J., Heth, G., and Johnston, R. E., 1999, Kin and Individual Recognition: odor signals, social experience, and mechanisms of recognition, in: *Advances in Chemical Signals in Vertebrates,* (R. E. Johnston, D. Müller-Schwarze, and P. W. Sorensen, eds.), Kluwer Academic/ Plenum Publishers, New York, pp. 289-298.

REMOTE MONITORING OF BADGERS (*MELES MELES*) FOR TESTING DISCRIMINATION BETWEEN URINE SAMPLES FROM DONORS OF DIFFERENT AGE AND SEX CATEGORIES

Katrina M. Service[1] and Stephen Harris[2]

[1]Department of Zoology
University of Oxford
Oxford, OX1 3PS, England, UK
[2]Department of Biological Sciences
University of Bristol
Bristol, BS8 1UG, England, UK

INTRODUCTION

Responses exhibited by animals as they investigate chemical signals deposited by conspecifics can provide insight into the information content of scent marks. Depending on the species and type of scent mark under consideration, the information content can include identification of species, sex, age, reproductive state, individual identity, group membership and an individual's dominance status (Ralls, 1971; Brown, 1979). Here we describe an automated method for field measurements of badgers' responses to urinary chemosignals of conspecifics. We found that the badgers' responses to scent marks were determined in part by the type of scent donor, supporting the hypothesis that information on age, sex, and seasonality are included in badger urine scent marks. In the British Isles badgers live predominantly in social groups, the members of which defend a shared territory (Neal and Cheeseman, 1995). Badgers scent mark using a variety of glandular secretions and excretory products, and although many aspects of olfactory communication in badgers have already been studied, the role of scent marking with urine in badger social behaviour remains unclear.

During this study, the responses shown by badgers towards a range of conspecific urine samples were monitored, using a novel experimental set up which allowed the frequency and duration of investigations directed towards the samples to be recorded automatically. The pattern of investigations badgers directed towards urine samples from different age and sex categories of conspecifics were compared in order to determine whether differential patterns of investigation occurred; if they did this would provide evidence that badgers could discriminate between the categories.

METHODS

The Remote-Monitoring Apparatus

The design of the apparatus was such that when a urine sample was investigated, the passage of a monitored infra-red beam between a bulb and detector was impeded by the investigating animal's head. Circuit diagrams for the electronics used are given in Service (1998). Six weatherproof units, each incorporating a scent sample dish, were constructed from Perspex (Figure 1) and set up so that each unit was held 30 cm off the ground, approximately at the height of a badger's head.

All six units were connected to a single Squirrel 12000 series datalogger (Grant Instruments Ltd., Barrington, Cambridge, CB2 5QZ, U.K.) recording on six separate channels, each corresponding to an individual monitoring unit. The time for which infra-red detection was impeded was recorded as the time for which a sample had been investigated. The datalogger recorded investigations in three-second units, so the minimum duration of any logged investigation was three seconds ("short" duration investigations). Investigations for longer than three seconds duration were "long" investigations. Recordings were downloaded from the datalogger each week onto Microsoft Excel, using a Spreadsheet Transfer Programme (Grant Instruments Ltd., Cambridge). The power source for the infra-red system was provided by two 6-volt motorcycle batteries.

Urine samples were positioned in the apparatus using disposable scent dishes made from 11 cm petri dishes. Holes were drilled in each dish to allow attachment of a circular cotton pad, onto which samples were soaked. Sample dishes were used once only.

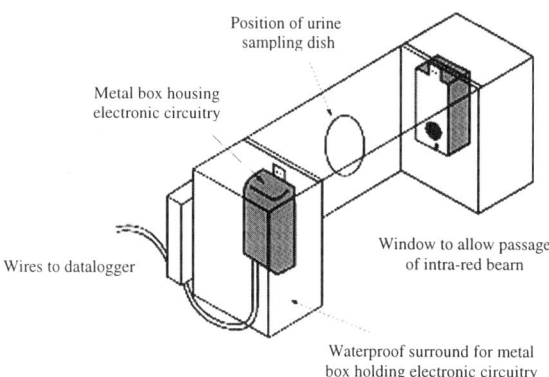

Figure 1. Diagram of one infra-red monitoring unit.

Badger Urine Samples and Controls

Badger urine was obtained from the following sources:
1) Badgers routinely live-trapped at the Central Science Laboratories (CSL) study site at Woodchester Park in Gloucestershire.
2) Badgers killed in traffic accidents and during the Ministry of Agriculture, Fisheries and Food (MAFF) badger removal operations, collected at the MAFF Animal Health Unit in Stroud, Gloucestershire.
3) Freshly road-killed badgers from the Bristol area.

Most samples used were from source (1), but as trapping is largely suspended at Woodchester Park during spring, additional spring urine samples were obtained from the other two sources.

The urine samples were categorised based on the age and sex of the urine donor animal into the six categories shown in Table 1. Information on age of animals was provided from Woodchester Park records or, in the case of road kills, was assessed by toothwear (Harris et al., 1992). Where possible, urine samples were also categorised by season, i.e. spring (March-May); summer (June-August); autumn (September-November); winter (December-February), giving 24 separate classes in total. Due to trapping restrictions, availability of urine samples from certain age/sex categories was limited: winter samples from cubs and spring samples from cubs and yearlings. The number of samples from each category presented during the autumn and spring trials are shown in Tables 1 and 2. Novel badger urine samples were of limited availability, so on certain days fresh aliquots of urine samples, which had been presented during previous trials, were used.

Table 1. Urine samples presented during autumn 1996 trials.

	Spring	Summer	Autumn	Winter	Total
Adult Male	13	6	2	2	23
Yearling Male	1	1	3	1	6
Cub Male	1	2	0	1	4
Adult Female	10	12	6	3	31
Yearling Female	1	1	0	2	4
Cub Female	4	1	1	0	6

Table 2. Urine samples presented during spring 1997 trials.

	Spring	Summer	Autumn	Winter	Total
Adult Male	21	9	4	4	38
Yearling Male	1	3	6	5	15
Cub Male	1	5	3	1	10
Adult Female	12	4	11	3	30
Yearling Female	1	6	5	4	16
Cub Female	4	2	6	1	13

Study Site, Animals and Experimental Procedure

The experiment was set up in the badger enclosure at New Forest Nature Quest, a wildlife park in Hampshire, England. A social group of four adult animals, two boars and two sows, were housed in an artificial sett sited in an outdoor enclosure covering half a hectare. The animals were viewed only by video-link, so aside from daily feeding and occasional maintenance work in the enclosure, they were undisturbed.

The experiment ran during October and November in autumn 1996 and during spring (March – May) 1997. Trials were resumed in spring because seasonal differences in response to odours have been documented in other animals (Ferkin and Johnston, 1995). The infra-red units were individually numbered from one to six and were sited around the badger enclosure. The urine samples were delivered to the wildlife park weekly, as 4 ml aliquots, and were stored frozen at -18°C before use. Each day's samples were defrosted as required. Urine samples were replaced between 5:00 and 6:30 p.m. daily by staff at the wildlife park. Each day's samples consisted of three badger urine samples, marked „A", „B" or „C", and three distilled water controls, all marked „0". The three urine samples used on any one day were selected so that a random adult male and adult female's urine comprised two of the samples. The third sample urine was randomly selected from another category. The samples were placed in the numbered infra-red units according to a set weekly schedule, which had been drawn up to randomly rotate urine samples and controls between infra-red units day by day.

RESULTS

Pattern of Investigation During Trials

The total number of investigations, towards urine and controls, declined sharply after the first two weeks of the experiment. The same trend was evident during both sets of trials.

To determine the extent of the trends, auto-correlation coefficients were analysed (Chatfield, 1989). The resulting correlograms showed that, though the number of investigations was higher in the first two to three weeks after this time there was no trend, i.e., the level of investigations stayed relatively constant after initial high interest.

Autumn Trials

562 investigations were logged over eight weeks, 416 of these were directed towards badger urine samples and 146 towards distilled-water controls. Of the badger urine investigations, 222 (53%) were investigations of the first presentation of a sample. First-day investigations of urine logged during autumn trials are shown in Table 3.

In total, badger urine samples were investigated significantly more often than the controls (Wilcoxon Matched Pairs Test, $z = -5.64$, $p<0.0001$). 170 long duration investigations were logged; of these 131 were directed towards badger urine samples and 39 towards controls.

Spring Trials

856 investigations were logged over 12 weeks. 656 of these were directed towards badger urine, and 200 were directed towards controls. Of the 656 investigations of urine, 450 (69%) were investigations of the first presentation of a urine sample (Table 4).

As in the autumn trials, urine samples were investigated significantly more often than the control samples (Wilcoxon Matched Pairs Test, $z = -6.86$, $p<0.0001$). During the spring 1997 trials a total of 145 long duration investigations were recorded. 126 of these were directed towards badger urine samples, and 19 towards controls.

Table 3. Autumn trials: number of investigations recorded for each urine category (first urine presentations only are listed).

	Spring	Summer	Autumn	Winter	Total
Adult Male	69	18	3	9	99
Yearling Male	3	2	7	3	15
Cub Male	2	1	4	n/a	7
Adult Female	31	29	8	12	80
Yearling Female	1	2	n/a	4	7
Cub Female	3	11	n/a	n/a	14

Table 4. Spring trials: number of investigations recorded for each urine category (first urine presentations only are listed).

	Spring	Summer	Autumn	Winter	Total
Adult Male	89	32	14	16	151
Yearling Male	7	19	18	21	65
Cub Male	3	8	13	7	31
Adult Female	61	6	32	11	110
Yearling Female	8	18	9	7	42
Cub Female	7	22	17	5	51

Investigation Patterns Towards Urine from Different Age and Sex Categories of Badgers

Only data from the first presentation of any urine sample were used in these analyses.

For the autumn trials, a Mann-Whitney U-test showed that significantly more investigations were directed towards urine samples from adult males than towards samples from adult female badgers (Mann-Whitney U-test, z= -2.3, p<0.05). Comparisons were not made between males and females in yearling and cub age categories due to the small sample sizes. However, in the spring trials there were no significant differences between the categories (Kruskal-Wallis test, χ^2 = 1.3, n.s.).

Long investigations occurred relatively infrequently. Data from autumn and spring trials were combined to increase the sample size. A Kruskal-Wallis test revealed a significant different between the six categories (χ^2 = 15.7, p<0.01). This difference was caused by male cubs having a significantly lower proportion of long duration investigations directed towards them than any other category (male cubs vs. all other categories: Mann-Whitney U-test, z= -3.71, p<0.001).

The only age category of badger for which sufficient urine samples had been presented to allow comparison of urine between season of collection was adults. Of urine samples from adults, approximately half were collected in spring, therefore the main seasonal comparison made was between spring and the rest of the year.

There was no significant difference between the proportion of total investigations directed towards adult male spring urine samples between the spring and autumn trials (z= -0.92, n.s.) or, again comparing spring and autumn trials, between investigations towards adult male summer urine samples (z= -1.39, n.s.). Therefore, data from the autumn and spring trials were pooled to increase the sample size. Using combined data from the spring and autumn trials, based on proportion of total investigations, adult male spring urine was investigated significantly more than adult male summer urine (z= -1.97, p<0.05).

Differences in Investigation of Urine Samples Collected During Different Seasons: Urine from Females

No significant differences were recorded between either proportion of total urine investigations or proportion of long investigations for any age category of female badger urine, in either the autumn (total investigations; Kruskal-Wallis test, χ^2 = 0.72, n.s.; long investigations; χ^2 = 0.71, n.s.); or spring trials (total investigations; χ^2 = 0.6, n.s.; long investigations; χ^2 = 0.65, n.s.). Considering investigations of adult female urine samples collected during spring and summer, there were no significant differences. There were no differences in proportion of total investigations between spring and summer urine samples, either during the autumn trial (z= -0.35, n.s.) or the spring trial (z= -0.47, n.s.).

DISCUSSION

Monitoring the behaviour of carnivores in the field can be difficult due to their secretive and often nocturnal habits. The equipment used in this experiment was specifically developed to overcome these problems and provide an alternative to direct observation. Use of this system had four distinct advantages:

1) Badger investigations of samples could be monitored at up to six different recording posts simultaneously.
2) There was no limit on the duration of recording / monitoring sessions – which effectively lasted all through the night.
3) No observer was present so during the trials the badgers were left undisturbed.
4) Badgers did not have to be in view during the trials, allowing the experiment to be carried out at night, the time at which badgers are usually most active, in semi-natural conditions similar to a field environment.

The infra-red monitoring and remote recording equipment successfully provided a full record of the number, frequency and duration of investigations badgers directed towards the experimental samples. Differential patterns of investigation were logged between certain categories of badger urine and this was taken to be evidence that badgers could potentially discriminate between these categories. The results obtained from the remote monitoring equipment demonstrate that badgers differentiated between badger urine from both adults and cubs of different sexes, between urine samples from badgers from different age categories and between urine samples collected during different seasons of the year. This is evidence that badger urine potentially contains information on the sex of the urine-donor, information on its age, and also cues that vary with season. Seasonally varying cues may be related to reproductive condition as well as to diet; spring is the main season for both badger mating and territorial marking (Neal and Cheeseman, 1995) and male urine samples from this season received more investigations than samples from the rest of the year. Though it has not been demonstrated that information on sex, age and seasonality is actually perceived by badgers, these types of information are commonly hypothesised to be present in scent marks, and a potential for badger urine to contain information of this type has already been demonstrated by GCMS analysis (Service, 1998).

Scent discrimination experiments have been successfully carried out on free-ranging animals but only in situations where the animals are clearly and consistently visible to observers (e.g. Mertl, 1977; Harris and Murie, 1982; Passanisi and Macdonald, 1990). Use of remote-monitoring equipment would be of great practical benefit during field studies. Though, as shown in the current experiment, scent discrimination can be demonstrated when total investigations carried out by whole groups of animals are recorded (also see Brown and Johnston, 1987), it would be useful to record additional information, e.g. on sex-based differences in investigation patterns. Incorporation of an active transponder system would allow identity of individual animals during their investigations of scent to be logged: a system of this type has been used with badgers (Hutchings and Harris, 1996). In addition, the infra-red units themselves could be modified to any specifications of robustness or scale, and adapted for used on a number of target species. The supports for the units also could be modified so that the whole apparatus was portable, allowing monitoring of responses to scent samples to be carried out in different parts of an animal's home range or territory. As well as its usefulness in scent discrimination experiments, this equipment could also be used to monitor the effectiveness of attractant and / or deterrent chemicals being used to monitor populations or in control of free-ranging wildlife.

ACKNOWLEDGMENTS

We wish to thank the following people who helped with this study: the researchers at Woodchester Park, especially Dr P. J. Mallinson, for collecting samples; Mrs L. Teagle and Mr J. M. Service, who built the equipment; Mr D. Gow and Mr T. Cullen for access and assistance at New Forest Nature Quest; Mr P. Gillett at First Creative Ltd., Preston, for drawing the figure; Dr G. W. McLaren for his practical help and comments on the paper; the anonymous reviewer, whose comments improved this manuscript.

REFERENCES

Brown, R. E., 1979, Mammalian social odours: a critical review, *Adv. Stud. Behav.* **10**: 103-162.
Brown, D. S., and Johnston, R. E., 1987, Individual discrimination on the basis of urine in dogs and wolves, in: *Proceedings of the Third International Symposium on Chemical Signals in Vertebrates*, (D. Müller-Schwarze and R. M. Silverstein, eds.), Plenum Press, London, pp. 343-356.
Chatfield, C., 1989, *The Analysis of Time Series*: an introduction, Chapman and Hall, London.
Ferkin, M. H., and Johnston, R. E., 1995, Effects of pregnancy, lactation and postpartum oestrus on odour signals and the attraction to odours in female meadow voles, *Microtus pennsylvannicus*, *Anim. Behav.* **49**: 1211-1217.
Harris, M. A., and Murie, J. O., 1982, Responses to oral gland scents from different males in Columbian ground squirrels, *Anim. Behav.* **30**: 140-148.
Harris, S., Cresswell, W. J., and Cheeseman, C. L., 1992, Age-determination of badgers (*Meles meles*) from toothwear – the need for a pragmatic approach, *J. Zool., Lond.* **228**: 679-684.
Hutchings, M. R., and Harris, S., 1996, An active transponder system for remotely monitoring animal activity at specific sites, *J. Zool., Lond.* **240**: 798-804.
Mertl, A. S., 1977, Habituation to territorial scent marks in the field by *Lemur catta*, *Behav. Biol.* **21**: 500-507.
Neal, E., and Cheeseman, C. L., 1995, *Badgers,* T. and A. D. Poyser, London.
Passanisi, W. C., and Macdonald, D. W., 1990, Group discrimination on the basis of urine in a farm cat colony, in: *Chemical Signals in Vertebrates V*, (D.W. Macdonald, D. Müller-Schwarze and S. E. Natynczuk, eds.), Oxford University Press, Oxford, pp. 336 – 345.
Ralls, K., 1971, Mammalian scent marking, *Science, N. Y.* **171**: 443-449.
Service, K. M., 1998, *Properties of badger (*Meles meles*) urine as a substance used in scent marking*, unpublished PhD thesis, University of Bristol.

THE DEVELOPMENT OF A SIMPLE ASSOCIATIVE TEST OF OLFACTORY LEARNING AND MEMORY

Heather M. Schellinck, Catherine A. Forestell, Vincent M. LoLordo, Patti Guidry, and Richard E. Brown

Department of Psychology, Dalhousie University
Halifax, Nova Scotia, B3H 4J1, Canada

INTRODUCTION

The information provided by olfactory signals has traditionally been investigated to gain an understanding of social interactions among mammals (Brown and Macdonald, 1985; Schellinck and Brown, 1999). Olfactory paradigms are also important for the study of basic memory processes in rodents (Otto and Eichenbaum, 1992). When tested with olfactory stimuli, rats appear to process information in a way similar to humans. For example, they can develop a learning set (Slotnick and Katz, 1974; Slotnick et al. 2000), appear to understand the relational organization among a series of odor items, i.e., transitive inference (Bunsey and Eichenbaum, 1996) and can perform sophisticated matching:non-matching-to-sample tasks (Winters et al., 2000).

Because mice have become the animal of choice in behaviour genetics, it has become important to develop new behavioural tests which take into consideration the specific characteristics of mice (Brown et al., 2000; Crawley, 2000; Malakoff, 2000). For example, compared with rats, mice have very high activity levels, are more vulnerable to ill effects from food or water deprivation and may be more stressed by social isolation (Brain, 1975). Few olfactory learning tests have been specifically designed to assess memory in mice (but see Bodyak and Slotnick, 1999). This paper discusses the methodological issues which must be addressed when developing a new behavioral test, describes a new procedure for testing olfactory learning and memory in mice and summarizes the results of our experiments which demonstrate that both male and female mice can learn the task and that mice can learn the task when trained for one trial a day for four days.

DEVELOPING A METHODOLOGY

The goal of behavioural testing with knockout and mutant mice is to determine what genes control the differences between strains. Unless careful attention is paid to detail, it is possible that differences attributed to genetic manipulations may be an experimental artifact of the conditions under which the animals were housed or differences in procedural variables between labs (Crabbe *et al.*, 1999; Brown *et al.*, 2000). Changes in behaviour may be related to background variables such as the size of the housing cage (Poon *et al.*, 1997) or the number of animals housed in a cage (Koolhaas *et al.*, 1997). During the experiment such procedural variables as time of the light:dark cycle that the mice are tested (Kreigsfeld *et al.*, 1999), the size of the apparatus, how the animals are handled and whether both males and females are tested may also influence the results of the test.

In developing our test, there were several specific points to consider. We were particularly concerned with how the mice were housed. Social isolation appears to increase corticosteroid levels and decrease serotonin and norephinephrine in mice (Brain, 1975) whereas housing male mice together leads to the formation of dominance hierarchies and subsequent stress in the subordinate animal (Bronson and Eleftheriou, 1965; Miczek *et al.*, 1982). Because of the negative effects of social isolation on mice, we initially housed our mice in pairs. Over the course of developing the test, however, we determined that it may not be appropriate to combine social housing with food restriction. When the mice were housed in pairs, male mice appeared to compete for food; as a result, the more restricted mouse would become less active. If this level of activity was suppressed in the subordinate mouse, it would be impossible to know whether lack of digging behaviour in the test phase of the experiment was related to lack of learning or to lack of motivation. The negative effects of weight loss were obvious in terms of reduced activity levels, however, it is possible that the accompanying social defeat could also result in inappropriate stressors (Koolhaas *et al.*, 1997). This could also influence the outcome of the test even if changes in activity levels are not demonstrated. If, for example, the mice were anxious, then, they might show less investigatory behaviour.

The size of the apparatus can also influence test results (Crawley and Paylor, 1997). For our experiments, we trained mice in small cages designed for housing mice and tested them in a scaled down preference apparatus modelled on one generally used for rats. To avoid any influence of handling on the behaviour of the mice (Barry, 1957), the mice were not handled prior to the experiment and were picked up by the tail to transport them in and out of the training and test chambers. We choose to test the mice during the dark phase of their light-dark cycle when they were most active because our measure of learning and memory i.e., digging behaviour, was correlated with activity. We also tested males and females to determine if there were sex differences associated with learning.

PROCEDURAL DETAILS

We have developed a test which takes advantage of the mouse's tendency to use olfactory cues to forage for food. The task involves a simple odor discrimination and is based upon the principles of associative learning. During training, mice are presented with one odor (CS+) paired with a sugar reward and a second odor (CS-) with no reinforcement. During the test phase, the odors are presented simultaneously but in the absence of sugar. The animals demonstrate that they have learned the discrimination by digging in the bedding odor previously paired with small pieces of sugar.

Four days prior to training, the mice were placed on a food restriction schedule to maintain them at 80 - 85 percent of their free feeding weight. During training two odor stimuli were used. Phenyl acetate, i.e., rose, (Aldrich Chemicals) and linalool, i.e., lemon, (Aldrich Chemicals) were diluted to a concentration of 15 percent in propylene glycol (Caledon Chemical Co.) and presented (0.03 ml) on filter paper (Whatman, #1, 55 mm in diameter) on the bottom of a 1.5 cm. plastic beverage cup. A petri dish cover with predrilled holes to allow odor diffusion kept the filter paper in place and out of reach of the subject. Pro chip bedding (PWI Industries, St. Hyacinthe, Quebec) was placed on top of the cover to fill the odor pot. When an odor stimulus was paired with a reinforcement, sugar was buried in the Pro chips.

The mice were trained several hours after lights off in rooms illuminated with dim red light. During training mice were exposed to the odor stimuli in polycarbonate opaque cages with stainless steel tops identical to their home cages. The odor pots were placed in the center of the back third of the cage. Subjects received four ten-minute trials per day for four days, two trials of rose and two trials of lemon, i.e., 16 trials in total. The order of odor presentation was randomized across days. The odors were presented in different rooms but the same odor pots and cages were kept in a particular room. Four mice were trained in each room simultaneously. For those mice receiving sugar reinforcement, the sugar was placed on the top of the Pro chip in the odor pot for the first odor-sugar pairing; for the remaining CS+ trials, the sugar was buried in the Pro chip. At the end of each trial, the mice were returned briefly to the colony room and the material in the odor pots and cages discarded. The odor pots and cages were not washed between trials.

On the test day, preference tests were conducted in a 68.0 x 20.0 x 20 cm. acrylic box divided into three equal compartments by clear acrylic walls. Holes were placed in each dividing wall to provide an entrance way from the center of the box to either end. Pro chip bedding was distributed over the floor of the chamber. New odor pots identical to those used in training were used in the test phase. The odor pots used in training were never reused to avoid the possibility that even minute amounts of sugar could be present during the test phase of the experiment. Two stopwatches were used to record digging behaviour; the test phase was also videotaped to safeguard against potential loss of data.

Preference tests were performed in a clean vacant animal colony room on the day following the last training trials. Prior to testing each animal was habituated to the apparatus for two minutes by being placed individually in the center of the apparatus and given the opportunity to explore the compartments and an empty odor pot placed in the middle of each end compartment.

Following habituation, the subject was briefly removed and a rose odor pot placed in one end of the chamber and a lemon odor pot in the other. Sugar was never present during preference testing. The time the mouse spent digging in each odor pot during a two minute period was recorded by an observer blind to the conditions of the experiment. Digging was defined as digging with forepaws or "nosing" in the bedding but only if the bedding moved during the investigation. At the end of each trial, the test chamber was washed with hot water. The preference chamber was rotated 180 degrees between subjects to counteract the influence of any room cues.

EXPERIMENT 1: ARE THERE SEX DIFFERENCES IN LEARNING THE TASK?

In the first use of this task, we showe that female BALB/c mice learned the task following 4 trials of the CS+ odor and 4 trials of the CS- odor per day for four days (Brennan et al., 1998). As males and females may differ in performance on learning tasks (Mishima et al., 1986; Douhet et al., 1997), the purpose of the first experiment was to determine whether there were sex differences in learning the task. In addition, we trained the mice for fewer trials per day to determine if the test could be made more efficient.

We used 16 male and 16 female CD-1 mice (Charles River, Quebec), weighing 20-25 grams. The animals were maintained on a 12:12 reversed light/dark schedule, weighed daily and fed approximately 4 grams Laboratory Rodent Chow #5001 (Agribrand Purina) to maintain 80 - 85 per cent of their free feeding weight. Water was available ad lib. They were housed individually in polyethylene cages (30.5 x 6.2 cm x 6.2 cm) with stainless steel tops and pine chip bedding. The mice were fed at the same time each day such that during training, they received their food several hours following training.

Mice were assigned to groups as follows: Group R+/L+: 8 animals received roseodor paired with sugar but lemon odor alone; 8 animals received lemon odor paired with sugar but rose odor alone; Group R-/L-: 16 animals received both odors alone, i.e. neither were paired with sugar.

The mice learned to discriminate between the odor which had been reinforced with sugar and the odor presented alone. They dug exclusively in the presence of the CS+ odor when it was presented simultaneously with the CS- odor (Figure 1). Wilcoxon matched-pairs signed-ranks indicated that the experimental animals (R+/L- and L+/R-) spent more time digging in the CS+ odor than in the CS- odor [T(16)=0, p<.01]. The control subjects which were exposed to both odors but no sugar reward during training never dug in either odor during testing. There were no differences between the male R+ animals and the female R+ animals in time spent digging in the CS+ odor [Mann Whitney U (4, 4) = 7, p>.1]. There were no differences between the male L+ animals and the female L+ animals in time spent digging in the CS+ odor (Mann Whitney U (4, 4) = 5, p>.1).). Moreover, there were no overall differences between the L+ animals and the R+ animals in time spent digging in the CS+ odor [Mann Whitney U (8, 8) = 20, p>.1]. These results indicate that males and females both learn the task and do not differ in their digging behaviour toward the S+ and S-. The task can be learned in 2 trials of each odor per day rather than 4 trials of each odor type per day used by Brennan *et al.*, (1998). Therefore, this paradigm still provides extremely consistent results even when the mice are given few exposures to the odors.

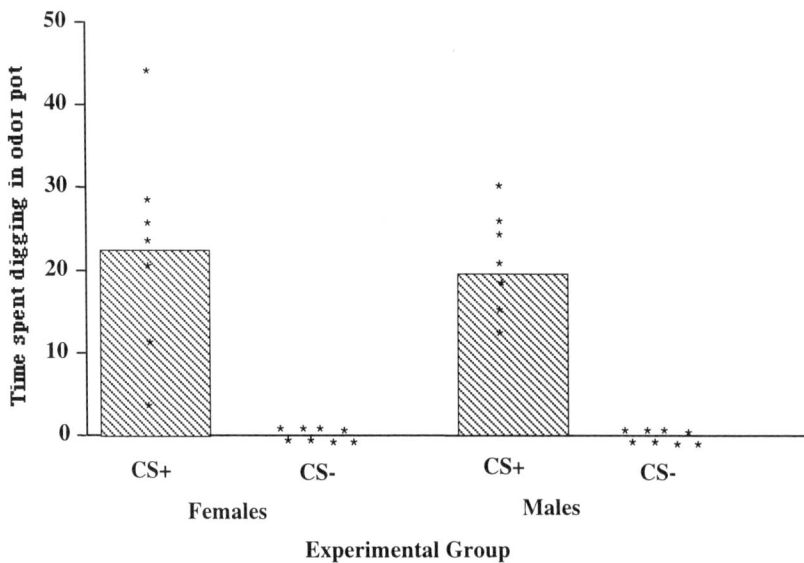

Figure 1: The amount of time spent digging in the CS+ and CS- odors by individual male and female mice during a two minute preference test. The bar graphs represent the median score.

EXPERIMENT 2: CAN MICE LEARN THE TASK WITH ONE TRIAL PER DAY?

The goal of developing this olfactory learning and memory paradigm is to provide a rapid test that will give reliable results. If the mice could learn and remember the task with one training trial per day, the test would be more useful to those investigators administering a battery of tests. To determine if one trial of training per day is sufficient with each odor type, we repeated the experiment and assessed the ability of mice to learn the task using the preference task. We obtained eight CD1 adult male and eight female mice, weighing approximately 20 gms from Charles River Canada (St. Constant, Que.) and housed them in the same conditions as described in Experiment 1. The training apparatus, odor stimuli and testing arena used in Experiment 1 were used in Experiment 2.

Four days prior to training, the mice were placed on a food restriction schedule to maintain them at 80 - 85 percent of their free feeding weight. They were then assigned to two groups: Group 1 (R+/L- and L+/R-) consisted of two males and two females that received rose odor with sugar and lemon odor with no sugar and another two males and two females received lemon odor paired with sugar and rose odor alone. Group 2 (LR) contained 8 animals (4 males and 4 females) that received lemon and rose For this group, neither odor was paired with reinforcement. All animals in both groups were given one training trial of each odor type per day for four days. The training and testing procedures were identical to those described in Experiment 1.

Both the female and male mice learned the odor discrimination. All of the experimental subjects dug in the bedding in the presence of the CS+ odor and none dug in the presence of the CS- odor (Figure 2). Wilcoxon matched-pairs signed-ranks indicated that the rewarded animals (R+/L- and L+/R-) spent more time digging in the CS+ odor than in the CS- odor [T(8)=0, p. <.01]. Thus, we have demonstrated that this task will produce extremely reliable results when mice receive one presentation of the CS+ and CS- odor per day for four days.

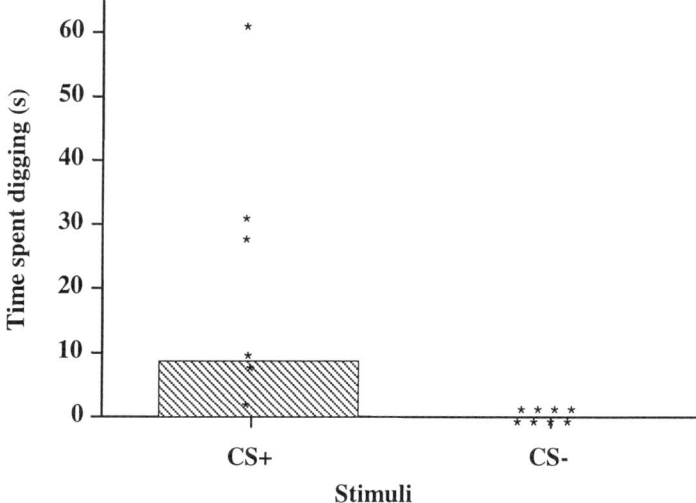

Figure 2: The amount of time spent digging in the odor pots during a two minute preference test, following one exposure to the CS+ and CS- odors per day for four days. The bar graphs represent the median score.

GENERAL DISCUSSION AND CONCLUSIONS

In this paper we have demonstrated that mice can rapidly and reliably learn an olfactory discrimination task when given either one or two exposures per day to the CS+ and CS- odors over four days. Also, males and females learn the task equally well. More recent experiments from our laboratory suggest that food deprivation is necessary prior to the test day but not the training days (Forestell *et al.*, in press). Moreover, mice can remember the task for at least 90 days. This is a simple test which may be easily and accurately administered by scientists who are inexperienced with the subtleties of behavioural testing. We expect that it will be useful for understanding the mechanisms of olfactory association learning and olfactory memory. In particular, it will be of interest to determine whether the same parameters governing conditioned odor aversion learning also govern this form of learning.

REFERENCES

Barry III, H., 1957, Habituation to handling as a factor in retention of maze performance in rats, *J. Comp. Physiol. Psychol.* **50**:366-367.

Bodyak, N., and Slotnick, B., 1999, Performance of mice in an automated olfactometer: odor detection, discrimination and odor memory, *Chem. Senses* **24**(6):637-645.

Brain, P., 1975, What does individual housing mean to a mouse? *Life Sci.* **16**:187-200.

Brennan, P. A., Schellinck, H. M., De La Riva, C., Kendrick, K. M., and Keverne, E. B., 1998, Changes in neurotransmitter release in the main olfactory bulb following an olfactory conditioning procedure in mice, *Neuroscience* **87**:583-590.

Bronson, F. H., and Eleftheriou, B. E., 1965, Adrenal response to fighting in mice: separation of physical and psychological causes, *Science* **147**:627-628.

Brown, R. E., and Macdonald, D. W. (eds)., 1985, *Social odours in mammals*, Volume 1-2. Clarendon Press, Oxford.

Brown, R. E., Stanford, L. and Schellinck, H., 2000, Developing standardized behavioral tests for knockout and mutant mice, *ILAR Journal* **41**(3): 163-174.

Bunsey, M., and Eichenbaum, H., 1996, Conservation of hippocampal memory function in rats and humans, *Nature* **379**:255-257.

Crabbe, J. C., Walhsten, D., and Dudek, B. C., 1999, Genetics of mouse behavior: Interactions with laboratory environment, *Science* **284**:1670-1672.

Crawley, J. N., 2000, Behavioral phenotyping of transgenic and knockout mice: experimental design and evaluation of general health, sensory functions, motor abilities, and specific behavioral tests, *ILAR Journal* **41**(3):136-143.

Crawley, J. N., and Paylor, P. R., 1997, A proposed test battery and constellations of specific behavioral paradigms to investigated the behavioral phenotypes of transgenic and knockout mice, *Horm. Behav.* **31**:197-211.

Douhet, P., Bertaina, V., Durkin, T., Calas, A., and Destrade, C., 1997, Sex-linked behavioral differences in mice expressing a human insulin transgene in the medial habenula, *Behav. Brain Res.* **89**:259-266.

Forestell, C. A., Schellinck, H. M., Boudreau, S., and LoLordo, V. M., Effect of deprivation on the acquisition and expression of a conditioned odor discrimination task, *Physiol. Behav.* (in press).

Koolhaas, J. M., DeBoer, S.F., DeRuiter, A.J.H., Meerlo, P., and Sgoifo, A., 1997, Social stress in rats and mice, *Acta Physiol Sc. Suppl.* **640**:69-72.

Kriegsfield, L .J., Eliasson, M. J. L., Demas, G. E, Blackshaw, S., Dawson, T. M., Nelson, R. J., and Snyder, S. H., 1999, Nocturnal motor coordination deficits in neuronal nitric oxide synthase knock-out mice, *Neuroscience* **89**:311-315.

Malakoff, D., 2000, The rise of the mouse, biomedicine's model mammal, *Science* **288**:248-253.

Miczek, K. A.,1982, Opioid-like analgesia in defeated mice, *Science* **215**:1520-1522.

Mishima, N., Higashitani, F., Teraoka, K., and Yoshioka, R., 1986, Sex differences in appetitive learning of mice, *Physiol. Behav.* **37**:263-268.

Otto, T. and Eichenbaum., H. E., 1992, Olfactory learning and memory in the rat: a "model system" for studies of the neurobiology of memory, (K. L. Chober, and Serby, M., eds.), *The Science of Olfaction*, Springer Verlag, New York, pp. 213-244.

Poon, A. M. S., Wu, B. M., Poon, W. F., Cheung, E. P. W., Chan, F. H. Y., and Lam, F. K., 1997, Effect of cage size on ultradian locomotor rhythms of laboratory mice, *Physiol. Behav.* **62**(6):1253-1258.

Schellinck, H. M., and Brown, R. E., (1999) Searching for the source of urinary odours of individuality in rodents, (R. E. Johnston, D. Müller-Schwarze, P. W. Sorenson, eds.), *Advances in Chemical Signals in Vertebrates,* Kluwer Academic/Plenum Publishers Corporation, New York, pp. 267-280.

Slotnick, B., Hanford, S., and Hodos, W., 2000, Can rats acquire an olfactory learning set?, *J. Exp. Psych: Anim. Behav. Proces.* **26**(4):399-415.

Slotnick, B., and Katz, H., 1974, Can rats acquire an olfactory learning set? *Science* **185**:796-798.

Winters, B. D., Matheson, W., McGregor, I. S., and Brown, R. E., 2000, An automated two choice test of olfactory working memory: The effect of scopolamine, *Psychobiology* **28**:21-31.

OLFACTORY COMMUNICATION BETWEEN MAN AND OTHER ANIMALS

Barbara A. Sommerville and Donald M. Broom

Animal Welfare and Human-Animal Interactions Group
Department of Clin. Vet. Med.
University of Cambridge
Cambridge CB5 8AF, UK

When an animal communicates with another, it transmits a signal in such a way that the sender usually benefits from the response of the receiver. In most situations involving humans and companion animals, the word „communication" is used to describe a dialogue, the human assuming that each participant has some awareness of the other. By „awareness" we mean a state in which complex brain analysis is used to process sensory stimuli or constructs based on memory. Most people would contend that their companion animal is capable of a degree of „assessment awareness", the ability to assess and deduce the significance of a situation in relation to itself over a short time span, if not „executive awareness", the ability to assess, deduce and plan in relation to long term intention. In this paper we examine evidence of olfactory communication in dogs and cats. We show that dogs can use human odour to gain relevant social information and consider evidence relating to their level of awareness when they interact with humans.

The definition of awareness used here is: a state in which complex brain analysis is used to process sensory stimuli or constructs based on memory (Broom, 1998). The term „complex brain analysis" implies that there is some degree of interpretative thought over and above perceptual processing. It is impossible to be certain of the degree of awareness in any organism other than oneself but we consider man to be capable of the highest level of awareness „executive awareness". This means that the individual is able to assess, deduce and plan in relation to long term intention. It may involve deductions about the feelings of others, imagination, and the mental construction of elaborate sequences of events. Examples include complex communication with conspecifics in humans and possibly also in other social mammals and the making and using of tools in an original way. Evidence suggests that chimpanzees are probably aware at an executive level. Wolves and dogs modify their behaviour as if extrapolating their feelings and thoughts to others and so they may have executive awareness.

It maybe debatable whether there is any qualitative distinction between the executive awareness of man and that of apes. There is certainly a quantitative distinction. Human language provides an unique medium of communication and it is used to convey abstract thought. An individual can be aware of an idea or a mental image triggered by memory rather than current sensory input. For example, animals which are aware in this way may feel fear or anxiety in an unfamiliar situation or when they remember a similar scenario which resulted in a bad experience. However, man can also function at lower levels of awareness which potentially can involve olfaction. Although no releaser pheromones have been demonstrated in humans, smells and tastes often have strong emotive associations related to a memory of some sort but one not easily recalled unless triggered by that sensation. An example is the madeleine-triggered childhood memory recounted by Proust (1932) in "Remembrance of Things Past". Smells are very evocative but can only be described indirectly by similes. Both these phenomena are probably related to the limbic rather than cortical nature of primary olfactory processing and contrast markedly with human visual experiences which can be described directly and precisely.

In several species of birds smell is used to detect and monitor food such as chickens, gulls and vultures. Domestic chicks can be trained to discriminate odours and pigeons, starlings and swifts use local environmental scents for navigation (Papi and Wallraff, 1992). Although it can hardly be described as interspecific communication, man has effectively employed the olfactory guided homing of carrier pigeons for human communication in past times.

Alarm pheromones signalling danger are used extensively in shoaling fish and herding mammals. Experiments with mice and rats have shown that some volatile signal emitted by a frightened animal will induce fear in conspecifics in adjacent cages or when they walk across the place where an animal has recently received an electric shock (Mackay-Sim and Laing, 1981; Fanselow, 1985). The human adrenergic response to fear affects the quantity of ecrine sweat and could affect the quality and quantity of apocrine sweat through the stimulation of mucosa and smooth muscle associated with the glands and their duct system. Such signals could well be interspecific among mammals

Most mammals have several anatomical sources of semiochemicals. Pheromones may be metabolites in faeces or urine and saliva or secretions from glands in the urogenital system, anus and skin. The domestic cat, for instance, has at least three anatomically distinct glands on its head used in scent marking objects, conspecifics and humans (Bradshaw and Cameron-Beaumont, 2000). The chemical identity of most mammalian pheromones is still unknown although a number of compounds are known from carnivore anal glands (Albone and Gronneberg, 1977; Banks *et al.*, 1992) and fluids or secretions used for marking such as from the circumgenital glands of tamarins (Belcher *et al.*, 1988). The presence of a substance in an active secretion is not proof of its activity and behavioural bioassays have often proved negative despite compelling analytical evidence. The boar pheromone, androstenone, is an exception, possibly because activity is restricted to a few related compounds rather than depending upon the synergistic influence of many components. Androstenone was the first mammalian pheromone to be chemically identified (Patterson, 1968; Perry *et al.*, 1972). Pigs are aggressive within hours of birth and there are at least three pheromones, in addition to androstenone, exerting a subtle control over social encounters by balancing individual aggression and submission (McGlone *et al.*, 1987; McGlone, 1988). The chemical identity of these other volatiles is not yet known but they could be described as releaser pheromones. In the wild pig, aggressive encounters between boars are common and the victor is the largest dominant boar which produces the most copious flow of saliva with the highest androstenone content. Slavering is provoked by any aggressive interaction with another wild boar or with men (Booth and Baldwin,1980; Pearce *et al.*, 1988). This last observation is interesting because human male armpit sweat contains androstenone and the level is higher than in women's sweat (Gower and Booth, 1986) so this may be an example of interspecific pheromonal communication. More work on the interactions between pigs and people might reveal other examples. An interesting exercise would be to identify and synthesise the submissive pheromone with a view to artificially reducing aggression levels in group housed pigs and humans. Should it prove effective in humans, one would have a valuable instrument for crowd control.

The domestic dog has lived with man for many millennia and shows interesting modifications of the social behaviour seen in its wild relative, the wolf. Fox (1975, 1978) has studied the social behaviour of a range of canids and concludes that the pattern of peer interaction is similar in the wolf and dog each showing a wider repertoire than the less social canids. Individual breeds of dog show the effect of selection for behavioural aptitudes, for example, the sight dominated herding seen in sheep and cattle dogs contrasts with the olfactory dominated searching seen in the retrievers. There is a very large range of physical characteristics in domestic dog breeds, for example, the contrast in size between a Chihuahua and a Great Dane. But despite domestic selection, the social behavioural repertoire of the dog remains essentially similar to the wolf and coyote perhaps because it is directed not only to conspecifics but also to humans.

The social behaviour of humans and wolves has many features in common. They both form self-supportive and caring communities where some resources are shared. This necessitates subtle communication to recognise individuals, their sexual and social status, and their intentions. The social similarity goes a long way towards explaining the close relationships which can exist between dog and man. In the closest human-dog attachments, the dog is treated as a surrogate child or companion. Dogs may respond socially to humans because they spend their socially formative period (3-14 weeks) in the presence of humans whom they come to regard as members of their pack. There is some evidence to support this idea such as the finding that male dogs use visual and olfactory cues to establish the sex of an unfamiliar human and then approach the man, but not the woman, with the caution that they use approaching an unfamiliar male dog. Millot et al., (1988) showed that dogs interacted with familiar and unfamiliar children using olfactory investigative behaviour similar to that used with conspecifics. There is as yet no hard experimental evidence that the dog ranks its human cohabitants in a canine pack dominance hierarchy but this assumption does explain the interactive behaviour remarkably well and is accepted by most consultants dealing with dog behavioural problems (Appleby, 1990; O'Farrell, 1992). Typically the dog will show less respect for those of lower rank, probably the children, and may not obey their commands or may even be aggressive towards them. In an extreme but not uncommon situation, a domineering dog may subjugate the entire family. Human visitors elicit varying responses depending upon their interaction with the human host and the relative dominance of the resident dog.

There is clearly a great deal of subtle communication taking place between humans and dogs and the relative importance of different categories of cues is not known (Levinson, 1980; Van Leeuwen, 1981). It is likely that visual, auditory and tactile cues given by the human may not be as immediately informative to the dog as an interspecific olfactory cue. It seems possible that they also may be able to detect subtle changes in the spectrum of volatiles emitted by the body as a result of emotional disturbance. If dogs can detect such changes, it could influence their response to disturbed or frightened people and be a relevant factor in dog attack. Both dogs and cats emit anal gland odour in acute fear.

Dogs can accurately distinguish and identify a large number of people by general body scent alone (Sommerville et al., 1993; Schoon and de Bruin, 1994; Settle et al., 1994; Sommerville et al., 1995). In standard competition trials, bloodhounds can follow the trail of a person who walked over rough moorland 24 hours before and identify the person they have tracked. If a dog approaches a recently (3-20 minutes old) laid human track at right-angles it spends 3-5 seconds sniffing intensively at 2-5 footprints and then begins to track rapidly in the direction in which the trail layer moved (Thesen et al., 1993). This necessitates detecting a difference in scent between footprints made 1-4 seconds apart, 3-20 minutes previously.

Unlike dogs, cats use scent to mark home ranges, particularly the paths they use and also a smaller area of territory within the home range which they may defend against intruders. In various felid species urine or marking fluid is sprayed at objects used as territory markers by both sexes with the tail held in a typical vertical position (Bramachary and Dutta, 1981). Interestingly, this tail signal is also used by the domestic cat when greeting a human as if the message has become modified from an olfactory to a visual signal during the course of domestication (Cameron-Beaumont, 1997). Inanimate objects in the home, humans and cohabiting conspecifics are scent-marked with secretion from the various facial glands, probably to establish a reassuring home and group odour (Leyhausen, 1979). A cat transplanted to a strange environment, particularly if it smells of a strange cat, will be uneasy until it has thoroughly scent-marked the place. A synthetic spray of marking-pheromones is now available and these seem to reassure the cat and reduce its need to urine mark (Pageat and Tessier 1998). Analytical chemistry and behavioural work to establish the significance of each component of this effect have not yet been carried out.

Dogs are very scent-oriented and easily transfer the ability to identify each member of their pack by smell to their human companions (Kalmus, 1955; Sommerville *et al.*, 1990, Settle et al, 1993;). Such information is important for them to respond appropriately in the high density, complex human society in which they live. The way in which dogs and even cats appear to anticipate the way their owners will behave or their emotional state, encourages humans to attribute high levels of awareness to their pets. It could be that a relatively anosmic species like man misses a lot of canine information sources. The care devoted to sniffing scent marks and the enormous olfactory acuity of dogs suggests that much information is potentially available but its precise nature is still uncertain. Perhaps the appropriate changes in behaviour noted in the dog are based entirely upon evidence conveyed by the body language of their human and so they do not need executive awareness. Considering the close relationship between dog, cat and man, the lack of firm information on such abilities is surprising. It emphasises the way our poor sense of smell limits our intellectual curiosity about olfactory matters.

REFERENCES

Albone, E. S., and Gronneberg, T. O., 1977, Lipids of the anal sac secretion of the red fox, *Vulpes vulpes*, and of the lion, *Panthera leo, J. Lipid Res.* **18**:474-476.

Appleby, D., 1990, The Good Behaviour Guide, Publ. Dog Help,Worcester, UK.

Banks, G. R., Buglas, A. J., and Waterhouse, J. S., 1992, Amines in the marking fluid and anal sac secretion of the Tiger, *Panthera tigris, Z. Naturforsch.* **47c**:618-622.

Belcher, A., Epple, G., Kuderling, I., and Smith, A. B., 1988, Volatile components of scent material from cotton-top tamarin *(Saguinus o. oedipus)*: a chemical and behavioral study, *J. Chem Ecol.* **14**:1367-1372.

Booth, W. D., and Baldwin, B. A., 1980, Lack of effect on sexual behaviour or the development of testicular function after removal of olfactory bulbs in prepubertal boars, *J. Reprod. Fertil.* **58**:173-178.

Bradshaw, J. W. S., and Cameron-Beaumont, C. L., 2000, The domestic cat: the biology of its behaviour, (D. C.Turner and P. Bateson eds.), Cambridge University Press.

Broom, D. M., 1998, Welfare, stress and the evolution of feelings. *Adv. Study Behav.* **27**:371-403.

Bramachary, R. L., and Dutta, J., 1981, On the pheromones of tigers - experiments and theory, *Am. Nat.* **118**:561-569.

Cameron-Beaumont, C. L., 1997, Visual and tactile communication in the domestic cat *(Felis sylvestris catus)* and undomesticated small felids. Ph.D. Thesis, University of Southampton, UK.

Fanselow, M., 1985, Stressed rat odour effect in unstressed rats, *Behav. Neurosci.* **99**:589-592.

Fox, M. W., 1975, *The Wild Canids; their systematics, behavioural ecology and evolution.* London: Van Nostrand Reinhold, UK.

Fox, M. W., 1978, *The dog: its domestication and behaviour,* New York: Garland STPM Press.

Gower, D. B., and Booth, W. D., 1986, Salivary pheromones in the pig and human in relation to sexual status and age, in *Ontogeny of Olfaction,* (W. Breipohl ed.), Springer-Verlag, Berlin, pp. 255-258.

Kalmus, H., 1955, The discrimination by the nose of the dog of individual human odours and in particular of the odour of twins, *Br. J. Anim. Behav.* **3**:25-29.

Levinson, B. M., 1980, The child and his pet: a world of non-verbal communication, in *Ethology and Nonverbal Communication in Mental Health,* (S. A. Corson and E. O'Leary Corson eds.), Pergamom Press, Oxford, UK

Leyhausen, P., 1975, *Verhaltensstudien an Katzen,* 4th ed., Paul Parey, Berlin.

Mackay-Sim, A., and Laing, D. G., 1981, Rats' responses to blood and body odours of stressed and non-stressed conspecifics, *Physiol. Behav.* **27**:503-510.

McGlone, J. J., Stansbury, W. F., and Tribble, L. F., 1987, Effects of heat and social stressors and within-pen weight variation on young pig performance and agonistic behaviour. *J. Anim. Sci.,* **63**: 456.

McGlone, J. J., 1988, Olfactory signals that modulate pig aggression and submissive behavior. in *Social Stress in Domestic Animals,* (R. Zayan and R. Dantzer, eds.), Dordrecht: Kluwer.

Millot, J. L., Filiatre, J. C., Gagnon, A. C., Eckerlin, A., and Montagner, H., 1988, Children and their pet dogs: how they communicate, *Behav. Process.* **17**: 1.

O'Farrell, V., 1992, The effect of owner attitudes and personality on dog behaviour. In:The Domestic Dog (J.Serpell, ed.), Cambridge University Press, UK.

Pageat, O. P., and Tessier, Y., 1998, The use of a female facial pheromone analogue to prevent intraspecific aggression in domestic cats, in *8th International Conference on Human-Animal Interactions,* Prague (poster, abstract).

Papi, F., and Wallraff, H. G., 1992, Birds. In Animal Homing (F. Papi. ed.), London: Chapman and Hall, UK.

Patterson, R. L. S., 1968, Identification of 3-hydroxy-5-androst-16-ene as the musk odour component of boar submaxillary gland and its relationship to the sex odour taint in pork meat, *J. Sci. Food Agric.* **19**:434-436.

Pearce, G. P., Hughes, P. E., and Booth, W. D., 1988, The involvement of boar submaxillary salivary gland secretion in boar induced precocial puberty attainment in the gilt, *Anim. Reprod. Sci.* **6**:125-129.

Perry, G. C., Patterson, R. L. S., and Stinson, C. G., 1972, Submaxillary salivary gland involvement in porcine mating behaviour, in: *7th International Congress of Animal Reproduction and Artificial Insemination,* Munich, pp.395-398.

Schoon, G. A. A., and de Bruin, J. C.,1994, The ability of dogs to recognise and cross-match human odours, *Forensic Sci. Int.* **69**:111.

Settle, R. H., Sommerville, B. A., McCormick, J. P., and Broom, D. M., 1994, Human scent matching using specially trained dogs, *Anim. Behav.* **48**:1443-1448.

Sommerville, B. A., Settle, R. H., Darling, F. M. C., and Broom, D. M., 1993, The use of trained dogs to discriminate human scent, *Anim. Behav.* **46**:189-190.

Sommerville, B. A., Green, M. A., and Gee, D. J., 1990, Using chromatography and a dog to identify some of the compounds in human sweat which are under genetic influence, in: *Chemical Signals in Vertebrates 5* (D. W. Macdonald, D. Müller-Schwarze and S. E. Natynczuk eds.), Oxford University Press, Oxford, UK, pp. 634-639.

Sommerville, B. A., Wobst, B., McCormick, J. P., Eggert, F., Zavazava, N., and Broom, D. M., 1995, Volatile identity signals in human axillary sweat: the possible influence of MHC class I genes, in: *Chemical Signals in Vertebrates 7,* (R. Apfelbach, D. Müller-Schwarze, K. Reutter, and E. Weiler eds), Oxford: Pergamon, pp. 535-539.

Thesen, A., Steen, J. B., and Doving, K. B., 1993, Behaviour of dogs during olfactory tracking, *J. Exp. Biol.* **180**:247-250.

Van Leeuwen, J., 1981, A child psychiatrist's perspective on children and their companion animals, in: *Interrelationships Between People and Pets* (B. Fogle, ed.), Thomas: Springfield.

AUTHOR INDEX

Adrian J. C. Jr., 305-312
Apfelbach R., 347-352; 411-415

Barnard C. J., 225-231
Bastakov V. A., 433-438
Beckmann M., 313-319
Berger E., 377-384
Beynon R. J., 43-52; 149-156; 157-163; 169-176; 217-224; 225-231; 233-240; 353-360
Bhatnagar K. P., 93-99, 101-106
Borowski Z., 289-293
Bradbrook C., 233-240
Bradshaw A., 313-319
Breer H., 177-181
Broom D. M., 467-471
Brouette-Lahlou I., 269-275
Brown G. E., 305-312
Brown R. E., 459-465
Buesching C. D., 321-327
Burrows A. M., 93-99; 101-106

Caldwell H. K., 445-450
Caretta A., 165-167
Cavaggioni A., 165-167
Chanel J., 269-275
Chastrette F., 269-275
Chivers D. P., 19-26; 277-284
Claus R., 133-140
Coppola D. M., 189-196
Coureaud G., 197-209

Dehnhard M., 125-132; 133-140; 141-148
Dluzen D. E., 107-115
Drickamer L. C., 35-41; 211-216

East M., 141-148
Erickson J. L., 305-312
Escobar C. A., 439-444

Fang J., 397-401
Ferkin M. H., 343-346
Ferrara L., 177-181
Fini C., 177-181

Forestell C. A., 459-465

Galizia C. G., 77-84
Gaskell S. J., 149-156; 157-163
Gaugler C., 205-209
Gershaneck D., 305-312
Glaser D., 423-431
Göritz F., 125-132
Gosling L. M., 11-17
Götz U., 133-140
Greene M. J., 361-368
Grzegorzewski W. 117-123
Guidry P., 459-465
Gutzke W. H. N., 285-288

Haber H., 125-132
Harris S., 451-457
Hayes R. A., 335-341
Heistermann M., 125-132
Hermes R., 125-132
Hildebrandt T., 125-132
Hofer H., 141-148
Hotchkiss A. K., 183-187
Hubbard S. J., 149-156
Humphries R. E., 43-52; 149-156; 233-240; 353-360
Hurst J. L., 43-52; 149-156; 157-163; 169-176; 217-224; 225-231; 233-240; 353-360

Ianovschi I., 329-333

Johnson B. A., 85-91
Johnston R. E., 61-68

Kamimura S., 403-410
Kapusta J., 241-247
Kaufman I. H., 305-312
Kaulfuß K.-H., 377-384
Kolosova I., 249-253
Koyama S., 403-410
Krzymowski T., 117-123

Labra A., 439-444
Lacorn M., 133-140
Landgraf R., 107-115

LeMaster M. P., 369-376
Leon M., 85-91
Lepri J. J., 445-450
Litvinova K., 249-253
Loebel D., 177-181
LoLordo V. M., 459-465

Macdonald D. W., 321-327
Madison D. M., 295-304
Maico L. M., 101-106
Mak V., 249-253
Malone N., 43-52; 149-156; 217-224; 233-240
Marchese S., 177-181
Marchlewska-Koj A., 391-396
Marie A. D., 149-156; 169-176
Marlier L., 197-204; 205-209
Martinsen L., 149-156
Mason R. T., 361-368; 369-376
Messer J., 205-209
Minaev A. N., 255-262
Mirza R. S., 19-26; 277-284
Mooney M. P., 93-99; 101-106
Moshkin M., 249-253
Mucignat-Caretta C., 165-167
Müller-Schwarze D., 1-10
Mustaparta H., 77-84

Nevison C. M., 43-52; 149-156; 225-231; 353-360
Nevitt G. A., 27-33
Niemeyer H. M., 439-444
Nofre C., 423-431

Ogurtsov S. V., 433-438
Owadowska E., 289-293

Paolini S., 177-181
Parzefall J., 263-268
Payne C. E., 43-52; 149-156; 157-163; 217-224; 233-240
Pelosi P., 177-181
Porter R. H., 417-422

Richardson B. J., 335-341
Roberts S. C., 11-17
Robertson D. H. L., 43-52; 149-156; 157-163; 169-176; 353-360
Rohr J. R., 295-304
Roslinski D. L., 101-106

Sachse S., 77-84
Sämmang I., 141-148
Scaloni A., 177-181
Schaal B., 197-204; 205-209
Schellinck H. M., 459-465
Schenk P., 377-384
Schmidt U., 347-352
Service K. M., 451-457
Shang Y., 107-115
Siegel M. I.,93-99; 101-106
Skipor J., 117-123

Slater F., 313-319
Smith T. D., 93-99, 101-106
Sokolskaja N. V., 411-415
Solovieva A. V., 255-262
Sommerville B. A., 467-471
Soussignan R., 197-204
Stefańczyk-Krzymowska S., 117-123
Stevens R., 313-319
Strauss G., 125-132
Strotmann J., 69-75
Sun R., 397-401
Surov A. V., 255-262
Süß R., 377-384
Szentgyörgyi H., 241-247

Tinti J. M., 423-431

Vandenbergh J. G., 183-187
Varendi H., 417-422
Vasilieva N. Y., 347- 352; 411-415
Vasilieva V. S., 53-60
Veggerby C., 43-52; 149-156; 157-163; 169-176; 233-240
Vernet-Maury E. 269-275
von Borell E., 377-384
von Holst D. 411-415

Wang L., 445-450
Wang Z.-W., 385-390
Wanner M., 423-431
Warbeck A., 263-268
Wąsowska B., 117-123
Weisgerber C., 125-132
Winberg J., 417-422
Wyllie S. G., 335-341

Zhang J.-X., 385-390
Zhang L., 397-401
Zhang Z.-B., 385-390
Zinkevich E. P., 53-60

SUBJECT INDEX

Accessory olfactory system (vomeronasal system), 62-63, 165, 189, 192, 195
  see also Vomeronasal organ
Adaptation, 27, 32, 79, 118, 197
Adrenal glands, 185, 186, 412
Aeshna umbrosa, see: dragonfly
Agkistrodon piscivorus, see: Snakes
Aggregative behavior, 28, 363
Aggression, 147, 230, 394, 468, 469;
  see also Behavior of mice 165-167, 211, 218-219, 225, 226, 229-230, 250, 445-450
Alarm pheromones, 9, 295, 305, 306, 311, 468, 473; see also Alarm signals
Alarm signals, 19-24, 295, 305, 311
  alarm substance, 305-311, 322, 409
  club cells, 305, 306, 310, 311
Alburnus alburnus: see Fishes
Alces alces: see Moose
Allo-marking, 322, 324, 325
Ambystoma macrodactylum: Salamanders
American mink: *see* Mustelids
Amniotic fluid, 183, 190, 196, 198, 199, 200, 202, 206
Amniotic odors, 198, 200-202
Amniotic sacs, 191, 192
Amygdala, 120, 165
  medial amygdala, 65-66
Anal glands, 8, 55, 141-147, 317, 318, 321-323, 329, 330, 468, 469
Androgen, 63, 185, 211, 219-222, 133;
  see also Testosterone
Androstenol (5α-androst-2-en-17β-ol), 117, 119-122,
Androstenone (5 α-androst-2-en-17-one), 58, 117-119, 126-131, 134, 468
  boar pheromone, 120, 126, 130, 131, 473, 468
Anogenital distance (AGD), 183, 184, 211-215
Anosmia, 445
Antarctic krill (*Euphausia superba*), 28
Antelopes, 2, 3
  African antelopes, 15
  klipspringer antelopes, 12
  oribi (*Ourebia ourebi*), 3
Antennae, 78

Antennal lobe (AL), 77-80, 82
Antipredatory behavior (anti-predator behaviour), 6, 19-21, 23, 24, 293, 295, 296, 303, 305, 306
Aphrodisin, 178
Apis mellifera: see Honeybee
Artiodactyla, 118, 423
Axillae, 418

Badger (*Meles meles*), 6, 7, 321-325, 451, 452, 455, 456
  feces of, 5, 6, 321-323
  urine of, 321-322, 451-456
Baiomys taylori: see Mouse
Bark beetle (*Ips paraconfusus*), 4
Beavers
  Castor canadensis (North American beaver), 1, 3, 8, 297
  Castor fiber (Eurasian beaver), 8, 9
Behavior (behaviour), 1, 9, 20, 21, 43-50, 53, 58, 62, 147, 157, 169, 189, 201, 218, 225, 226, 264, 318, 369, 390, 418, 419, 443, 447, 449, 437, 451, 456, 459-462, 467, 469
  aggregative, 28, 363
  aggressive, 28, 229, 395
  antipredator, 6, 19-21, 23, 24, 293, 295, 296, 303, 305, 306
  combat, 365, 366
  copulatory, 63
  counter marking: *see* Counter marking
  courtship, 266, 363-365, 369, 370, 374
  feeding, 5, 417
  fetal, 193
  marking, 2, 213, 215, 322, 324, 325, 355
  maternal, 269-274
  mating, 263, 364, 366
  neonatal, 197
  parental, 105, 321, 398, 414
  reproductive, 101, 102, 105, 219, 363
  scent marking: *see* Scent marking behavior
  sexual: *see* Sexual behavior
  territorial, 313
  trailing, 366-367
Benzaldehyde
  in otter, 315-317

Birds, 27-32
    Daption capense (cape petrel), 32
    Diomedea chrysostoma (gray headed albatros), 28
    Diomedea exulans (wondering albatros), 28
    Diomedea melanophris (black-browed albatrosses), 28, 32
    Pachyptila sp. (prion), 28
    Procellariiform seabirds, 27-30
Boar pheromone: *see* Adrostenone
Boiga irregularis: *see* Snakes
Brevicomin (3,4-dehydro-*exo*-brevicomin), 45, 46, 50, 150, 152, 153, 157, 165, 169-173, 219, 234, 353, 354, 358,
Bruce effect, 4, 393, 409
Bufo americanus: *see* Toads
Bufo boreas: *see* Toads
Bufo bufo: *see* Toads
Bufo terrestris: *see* Toads

c-fos mRNA, 409
Caesarian sections, 183
Camel (*Camelus dromedarius*), 126
Cape petrel: see Birds
Carassius auratus: *see* Fishes
Carassius carassius: *see* Fishes
Carboxylic acids, 322, 442
Carotid rete, 118
Carp *see*: Fishes
Castor canadensis: *see* Beavers
Castor fiber: *see* Beavers
Cat (*Felis catus*), 330
    urine of, 412, 413
Catecholamine, 109
Catfish: *see* Fishes
Cattle, 4, 58, 70, 429, 469
Cheek pouches, 347, 350, 351
Chelydra serpentina: *see* Turtles
Chemical signals, 1, 58, 61, 62, 77, 117, 183, 246, 255, 266, 267, 329, 385, 391, 395, 397, 439, 440, 443, 451; *see also* Pheromones
    alarm, 19-24, 295, 310, 311
    attractiveness and, 220, 222, 249, 250-252, 260, 363, 365, 369, 370-374, 385, 388
    endocrine response to, 118, 165, 249
    individuality and, 154, 155, 226, 239
    neuroendocrine reaction and, 125, 134, 184, 374, 382, 391, 401
    social status and, 37, 38, 43-45, 49, 62, 143, 147, 217-222, 225-227, 229, 251, 335-340, 394, 404, 407-409
    species recognition, 54
    in urine, 4, 7, 8, 36, 38, 45-48, 55, 56, 62, 64, 75, 119, 125-131, 134-138, 149, 153, 157-162, 165, 166, 169-174, 177, 212-215, 218-222, 226-228, 233-236, 238, 256, 257, 269, 274, 321, 329-331, 353-359, 387-389, 391- 393, 403,405, 409, 412, 413, 447-449, 451, 456, 468

Chemosensory epithelium, 69, 94; *see also* Epithelium
Chin glands, 335-337, 340
Chinook salmon: *see* Fishes
Chromosomes 4, 162
Chrosomus neogaeus: *see* Fishes
Chrysemys picta; *see* Turtles
Clethrionomys glareolus: *see* Voles
Clethrionomys rufocanus: *see* Voles
Coolidge effect, 64
Common toads: *see* Toads
Congenital adrenal hyperplasia (CAH), 186
Copulatory behavior 63; *see also* Behavior
Corticosterone, 405, 407, 409
Cottonmouth (water) moccasin: *see* Snakes
Cotton-top tamarin, 125
Cottus cognatus; *see* Fishes
Counter-marking, 45-47, 221, 226, 227, 229, 353-358; *see also* Behavior
Courtship behavior, 266, 363-365, 369, 370, 374; *see also* Behavior
Cricetulus triton: *see* Hamsters
Crocuta crocuta: *see* Hyenas
Cross-fostering, 201, 255, 258, 259,
Culaea inconstans: *see* Fishes
Cutthroat trout: *see* Fishes
Cyprinus carpio: *see* Fishes

Daption capense: *see* Birds
Deer
    black-tailed deer, 12,
Delayed puberty of females, 7, 292; *see also* Puberty
2-Deoxyglucose (2-DG), 85, 87-90, 190, 192, 194, 195
Dihydrothiazole, 45, 149, 153, 157, 165, 169, 219, 233, 353; *see also* Thiazole
Dimethyl disulfide, 201
Dimethyl sulfide, (DMS), 30, 31
Dimethylsulfionopropionate (DMSP), 31
Diomedea melanophris: *see* Birds
Di (2-ethyl-hexyl) adipate, 269, 274
Dodecyl propionate (DP), 269
Dominant male, 45-47, 49, 50, 56, 218-220, 222, 225-227, 353, 408, 409, 446
DNA 71, 152
    cDNA, 69, 150-152, 160, 179
Dog, 57, 469, 470
Dominance/subordination, 35-38, 44, 49, 218, 226, 251, 318, 335, 337, 339, 340, 366, 403-406, 408, 451, 460, 469; *see also* Social status
Dragonfly (*Aeshna umbrosa*), 21

Elephants, 30, 130, 131
    Elaphas maximus (Asian elephant), 125-131
    Elephas maximus: *see* Elephants
    Loxodonta africana (African elephant), 130
    oestrus cycle 126
Endopeptidase LysC, 151, 152
Entorhinal cortex, 65, 66, 69, 73

Epidermal club cells, 305, 306, 310
Epithelium, 72, 73, 90, 120, 191, 192, 194;
  see also Chemosensory epithelium
Esox lucius: see Fishes
Esox niger: see Fishes
EST clones, 150
EST (expressed sequence tag), 150, 151
Estradiol (oestradiol), 126, 386, 388, 389
Estrogen (oestrogen), 3, 125, 274, 374, 385, 388, 404
Estrus (oestrus), 58, 219, 234, 250, 251, 257-260, 377, 378, 382, 383, 388
  delay, 35, 414, 449
Estrous cycle, 35, 53, 62, 118, 119, 126, 314, 386, 405, 412
Etheostoma exile: see Fishes
2-ethylfuran
  in otter, 315, 317
Euphausia superba: see Antarctic krill
Ewe: see Sheep

Fetal olfaction, 119, 190, 193
Farnesenes, 46, 50, 219
  E,E-alpha-farnesene, 45
  E-beta farnesene, 45
Feeding behavior, 5, 417
Felis catus: see Cat
Fetus, 94, 95, 183, 184, 191-195, 197, 199, 200, 202, 211, 241
  behavior, 193
  olfaction, 119, 190, 193
  perception, 197
Fishes
  Alburnus alburnus (bleak), 279
  Carassius auratus (goldfish), 279, 281
  Carassius carassius (crucian carp), 278-279, 281
  Chrosomus neogaeus (finescale dace), 279, 305, 306, 310
  Cottus cognatus (slimy sculpin), 279, 282
  Culaea inconstans (brook stickleback), 21, 24, 279, 281
  Cyprinus carpio (carp), 4
  Etheostoma exile (Iowa darters), 20
  Esox lucius (pike), 20, 21, 24, 278- 282
    feces, 292
  Esox niger, 21
  Hemigrammus erthrozonus (glowlight tetra), 279, 306-310
  Lepomis macrochilus (bluegill sunfish), 280, 300
  Notemigonus chrysoleucas (golden shiner), 279
  Oncorhynchus clarkii (cutthrout trout), 5, 22
  Oncorhynchus mykiss (rainbow trout), 21
  Oncorhynchus tshawytscha (Chinook salmon), 5, 21, 22
  Phoxinus neogaeus (Finescale dace), 279, 305, 306, 310
  Paracheirodon innesi (neon tetras), 306, 308, 309

  Phoxinus phoxinus (European minnow), 21, 24
  Pimephales promelas (fathead minnow), 20, 21, 24, 279-281, 305, 306, 310, 311
  Salmo salar (Atlantic salmon), 279, 281
  salmonids, 20
  sticklebacks, 20, 21, 24, 25, 279, 281
  Salvelinus fontinalis (brook trout) 20-22, 279, 281, 282, 284
  Xiphophorus helleri (swordtails), 21, 22, 281, 306-308
Follicle-stimulating hormone (FSH), 379, 380, 382, 383
Follicular phase, 127, 128, 130
Food-odor-coding patterns, 80
Frogs, 280, 433-438
  Pacific tree frogs, 6
  Rana aurora (red-legged frog), 20, 278, 280, 436
  Rana cascade (cascade frog), 23, 295
  Rana esculenta, 295
  Rana lessonae (pool frog), 295, 433, 436-438
  Rana luteiventris (Columbia spotted frog), 280
  Rana sylvanica *(wood frog), 280*
  Rana temporaria (common frog), 280

Garlic, 200, 418,
Gazella thomsoni: see Thomson gazelles
Genes
  Major Urinary Proteins, 150, 151, 155, 162, 237, 238
  OR genes *(OR 37)*, 69, 71-74
Gerbil, Mongolian, 70, 184, 185, 212, 411, 429
Glands; see also Scent glands
  adrenal, 185, 186, 412
  anal, 8, 55, 141-147, 317, 318, 321-323, 329, 330, 468, 469
  axillae, 418
  chin, 335-337, 340
  Harderian, 63
  interdigital, 55
  lacrimal, 445
  preputial, 7, 269, 270, 274, 406-408,445, 446, 448-450
  subcaudal, 55, 321-325
  submaxilary, 126, 178
  tarsal, 55
Glowlight tetras: see Fishes
Goats, 118, 133, 134, 139
  urine of, 135-138
Gonadotropin releasing hormone (GnRH), 120, 134
Guanine, 306-308
Gustatory responses, 423-428
Gustatory stimuli, 423
Gymnophiona, 263

Hamsters, 44, 47, 63-66, 125, 429
  Cricetulus triton (rat-like hamsters), 386-388
    estrus, 382-383, 389
      feces of, 385-400
      urine of, 385, 387, 389
  Mesocricetus auratus (golden hamster), 55, 56, 61, 62, 64, 66, 178, 260, 343, 255-261
  Phodopus campbelli (Campbelli's hamster), 347-351, 411-414
    urine of, 412, 413
  Phodopus sungorus (dwarf hamster), 347
Heliothis virescens: see Moth
Helogale pervula, (dwarf mongoose) 330, 332
Hemigrammus erythrozonus: see Fishes
Heptanone (6-hydroxy-6-methyl-3-heptanone), 157
Hexanoic acid, 146
Home ranges, 38, 213, 289-293, 298, 456, 470
Honeybee (*Apis mellifera*), 78, 80-82
Honest signals, 12, 16, 43-45, 49, 50; see also Chemosignals
Hormones, 237, 445, 447, 449
  responses to odors, 63, 118, 133, 134, 377-383, 392
  and scent marking, 229
Humans, 126, 185, 186, 459, 467-470
  behavior response to odor, 198-202, 417-420
  odor, 57, 58, 90, 205
  taste stimuli, 423, 425-429
  VNO of, 93-98
6-Hydroxy-6-methyl-3-heptanone: see Heptanone
6-hydroxydopamine, 109, 111
Hypoxanthine-3-*N*-oxide, 305, 307, 308, 310, 311
Hypoxanthine, 306-308, 310, 311
Hyena brunnea: see Hyenas
Hyena hyena; see Hyenas
Hyenas (*Hyaenidae*)
  Crocuta crocuta (spotted hyena), 3, 141-144, 146, 147
  Hyaena brunnea (brown hyena), 3, 141
  Hyaena hyaena (striped hyena), 3, 141, 142
  Proteles cristatus (aardwolf), 141, 142

Immune system: see Major Histocompatibility Complex
Immunocompetence handicap hypothesis, 221
Imprinting, 1, 433, 438
Individual odors, 6, 64-66, 81, 88, 349, 351, 400; see also Chemosignals
Infanticide, 166, 318
Inter-fetal communication, 183
Intrauterine environment, 183
Intrauterine position, 7, 184, 185, 211, 212
Insects, 77-82, 278, 362, 363, 374
Intermale aggression, 445, 449; see also Aggression
Ips paraconfusus: see Bark beetle
Iso-propylmyristate (traces), 269

Kafue lechwe (*K. leche kafuensis*), 3
Ketones, 55, 139, 335, 365, 369, 371-374
Kobus kob thomasi: see Uganda kob

Lacrimal glands, 445
Lactation, 107, 143, 200, 270-274
Lampropeltis getula: see Snakes
Lateral olfactory tract, 65
Learning, 437, 438, 459-464
Learning mechanism, 202
Learning olfactory, 229, 260, 261, 437
Lemurs
  Lemur catta (ring-tailed lemur), 12, 94-97
  Microcebus murinus (mouse lemur), 93-95
  Microcebus myoxinus, 94-97
Lemur catta (ring-tailed lemur): see Lemurs
Lepomis macrochilus: see Fishes
Ligand, 63, 70, 88, 120, 177-180
Ligand binding, 70, 118, 179
Ligand release rate, 157, 162, 170-172, 174, 234, 353, 355, 358
Liolaemus alticolor: see Lizards
Liolaemus bellii: see Lizards
Liolaemus constanzae: see Lizards
Liolaemus eleodori: see Lizards
Liolaemus fitzgeraldi: see Lizards
Liolaemus jamesi: see Lizards
Liolaemus nigroviridis: see Lizards
Liolaemus ornatus: see Lizards
Liolaemus tenuis: see Lizards
Lipocalins, 48, 149, 157, 165, 169, 177, 178, 180, 219, 233
Lizards, 439, 440, 442, 443
  Liolaemus alticolor, 440
  Liolaemus bellii, 440, 441, 442
  Liolaemus constanzae, 440, 441
  Liolaemus eleodori, 440, 441
  Liolaemus fitzgeraldi, 440
  Liolaemus jamesi, 440
  Liolaemus nigroviridis, 440
  Liolaemus ornatus, 440
  Liolaemus tenuis, 441, 442
Locomotor activity, 289, 290, 293
Loxodonta africana: see Elephants
Luteal phase, 126-131
Luteinizing hormone (LH), 118, 133, 134, 379, 380, 382, 383
Lutra lutra: see Otters, European

Macroglomerular complex (MGC), 78- 80
Major Histocompatibility Complex (MHC), 6, 7, 37, 45, 48, 49, 155, 221, 222, 233, 358
Major Urinary Proteins (MUPs), 45-50, 149-155, 157-162, 165, 166, 169-174, 177, 178, 180, 219, 221, 233, 234, 237-239, 353-355, 358
  heterogeneity of, 49, 150, 154, 157- 159, 161, 162, 170, 174, 221, 233, 234, 236, 239
Main olfactory bulb, 65, 66, 69, 85, 115, 190; *see also* Olfactory bulb

Mate choice, 37, 39, 165, 212, 217, 218, 222, 264, 265, 267, 397
Marking behavior, 2, 213, 215, 322, 324, 325, 355
Mate selection, 35, 37, 38, 43, 49, 64, 221, 263-265
Maternal behavior, 264-274
Mating behavior, 263, 364, 366
Mating system, 101, 325, 344, 345, 362-365, 394, 395, 456
Medial preoptic area, 108
Medial basal hypothalamus, 134
Meles meles: see Badgers
Memory, 64, 65, 107, 110, 111, 205, 340-346, 459, 460, 463, 464, 467, 468
Menadione, 150, 152, 153, 355, 356, 358
Merino sheep: *see* Sheep
2-methylbutanal
    in otter, 315, 317
3-methylbutanal
    in otter, 315, 317
2-methylpropanal
    in otter, 315, 317
Mesocricetus auratus: *see* Hamsters
MHC: *see* Major Histocompatibility Complex
Mice
    Baiomys taylori (pigmy mice), 255, 259
    Mus musculus: *see* Mouse
    Onyhomys torridus (grasshopper mice), 259
    Peromyscus leucopus, 259, 260
    Peromyscus maniculatus (deer mouse), 36
Microcebus murinus: *see* Lemurs
Microcebus myoxinus: *see* Lemurs
Microtus agrestis: *see* Voles
Microtus brandti: *see* Voles
Microtus canicaudus: *see* Voles
Microtus oeconomus: *see* Voles
Microtus ochrogaster: *see* Voles
Microtus pennsylvanicus: *see* Voles
Microtus townsendi: *see* Voles
Minnow, fathead: *see* Fishes
Mitral cells, 82
Monosodium-L-glutamate (monohydrate) (MSG), 425, 429
Moth (*Heliothis virescens*), 78, 79
Moose (*Alces alces*), 126
Mouse (*Mus musculus*), 7, 8, 35-39, 44-50, 55, 56, 62, 64, 70-74, 93-95, 97, 112, 125, 126, 149-151, 153-155, 157-159, 161, 162, 165, 166, 169-172, 174, 177-180, 183-185, 189-195, 211-215, 217-222, 225-229, 233-239, 241-246, 249-252, 255, 259, 353-355, 357, 358, 393, 403-409, 411-414, 417, 445-450, 459-464
    strains of, 250, 255
    BALB/C, 150, 157-162, 170-174, 226, 228, 229, 354, 461
    CBA, 241-246
    C_{57}BL, 241-246
    C57BL/6, 150, 157, 160-162, 170, 171, 174
    MC5-RKO, 446, 447-449

Mungos mungo: *see* Striped mongooses
Mus musculus: *see* Mouse
Mus domesticus: *see* Mouse
Mustelids, 289, 292, 313, 318, 321
    American mink, 55
    Lutra lutra (European otter) 313-318
    Mustela erminea (stoats), 318
    Mustela nivalis (weasel), 289-293
MUPs: *see* Major Urinary Proteins
    MUP-Phe$_{56,}$ 152, 153
    MUP-Val$_{56,}$ 152, 153

N-phenyl-naphthylamine (NPN), 152, 170-172
Nasal cavity, 192
Nasal septum, 70, 192
Neon tetras: *see* Fishes
Neonates, 189, 197-201, 241, 242, 246, 269, 286-288, 417-420
Neonatal behavior, 197
Neonatal perception, 201, 208
Newts, 278, 295-303
    Notophthalmus viridescens (red-spotted newt), 6, 21, 295, 296
    Taricha granulosa (rough skin newt), 280, 302
    Taricha torosa (California newt), 302
Nipple, 125, 189, 197, 198, 201, 202, 270, 417-421
Nitrogen-oxides, 305-307, 310, 311
Norepinephrine (NE), 107, 109, 111-114
Notemigonus chrysoleucas: *see* Fishes
Notophthalmus viridescens: *see* Newts

Odor (odour)
    index of odor attractiveness, 250
    baited trap technique, 7, 212
    coding, 57, 71, 77-82, 85, 87-90, 260
    of plumes, 29, 30, 78, 79
    preferences of, 35, 37-39, 65, 66, 212, 214, 397, 398, 400
Odorant-binding proteins (OBPs), 177, 178, 180; *see also* MUPs; Lipocaline
Old World primates, 98, 428
Olfactory bulb, 63-66, 69, 73, 74, 77, 81, 85, 87-91, 107, 117, 120, 189, 190, 195, 394, 417
    tufted cells in, 74, 77
Olfactory receptor (OR), 69-71, 77
Olfactory system (main olfactory system), 53, 54, 56, 62, 63-66, 69, 74, 80, 241, 394, 395, 437
    coding of information, 57, 71, 77-82, 85, 87, 88, 90, 260
    epithelium, 69-73, 90; *see also* Epithelium
    glomeruli, 69, 73, 74, 77-82, 87-90
    mucosa, 64, 468
    pattern (of input), 69-74, 77, 79-82, 85-87, 90, 91
Oncorhynchus clarkii: *see* Fishes
Oncorhynchus mykiss: *see* Fishes
Onyhomys torridus: *see* Mice

Oryctolagus cuniculus: see Rabbit
Ostariophysi, 20, 305, 306, 310
Ostariophysan alarm pheromones, 305, 306, 310, 311; *see also* Alarm signals
Otter, European (*Lutra lutra*), 313-318
Ourebia ourebi: see Antylopes
Over-marking: *see* Counter-marking
Ovulation, 380-383, 393

Paraventricular nucleus, 65, 66, 409
Parental behavior, 105, 321, 398, 414
Parental care, 255, 414
2-pentylfuran
  in otter, 315
Peromyscus leucopus: see Mice
Peromyscus maniculatus (deer mouse): *see* Mice
Pest control, 361-363, 367
Phenol, 349, 350, 351
2-phenoxy ethanol, 339, 340
Phoxinus neogaeus: see Fishes
Phoxinus phoxinus: see Fishes
Pheromones, 6, 7, 53-55, 56, 58, 61-64, 66, 77-80, 101, 105, 117-121, 125-131, 133-139, 149-154, 157, 165, 166, 169, 177-180, 183, 184, 186, 195, 199, 219, 239, 260, 266, 269, 281, 282, 305, 306, 310, 311, 314, 362, 363, 365, 367, 369-374, 377-382, 392, 393, 443, 468-470; *see also* Chemical signals
  endocrine dependence, 118, 125, 134, 184, 249, 391, 401, 450
  peptides, 45, 48, 49, 120, 150, 152, 154, 155, 158, 159, 161, 162, 178, 180, 195, 221, 233, 234, 267
  primers: *see* priming
  priming, 6, 7, 9, 117, 120-122, 157, 169, 183, 184, 186, 314, 353, 403, 409
  proteins and, 45-50, 55, 63, 69, 70, 81, 88, 118, 120, 149-153, 157-162, 177-180, 220, 222, 225, 233-239, 267, 314, 322, 335-358, 429, 449, 403; *see also* Histocompatibility Complex
  receptor cells of, 95, 97
  releaser, 62, 105, 403, 468
  sexual behavior and, 63, 369; *see also* Behavior
  sexual receptivity and, 58, 385, 386
  signaling, 121
  steroids, 118, 119, 125-128, 178, 179, 412; *see also* Androgen
  vomeronasal organ and, 48, 62-64, 66, 93, 96, 97, 101-104, 117, 118, 165, 178, 180, 190, 213, 263, 286, 409
Phodopus campbelli: see Hamsters
Phodopus sungorus: see Hamsters
Photoperiod, 133, 397, 400, 401
Pig (*Sus scrofa domesticus*), 58, 70, 117-122, 126, 130, 131, 177-180, 423-429, 468
  piglet, 198
Plant odors, 80, 82
Population dynamics, 185

Predator avoidance, 286, 295, 299
Predator detection, 295, 301-303
Predator diet, 277-283
Predator recognition, 20-23, 278, 280-282, 286, 288, 290-293
Pregnancy, 35, 102-104, 186, 199, 201, 263, 394, 395
Pregnancy-block effect, 64, 112, 219, 234, 393
Preferences odor, 35, 37-39, 65, 66, 212, 214, 397, 398, 400
  endocrine effects on, 387, 388
  for familiar vs. unfamiliar, 39
Pimephales promelas: see Fishes
Piscivorous, 278, 281; *see also* Fishes
Primary spermatocytes, 102, 103,
Procellariiform seabirds, 27-30; *see also* Birds
Progesterone, 118, 125, 126, 131, 380-383
Prolactin, 409, 420
Proteins, 55, 63, 69, 70, 81, 88, 118, 120, 159, 235, 236, 314, 322, 335, 403, 420, 429, 449; *see also* Major Urinary Proteins
  G-proteins, 70, 118, 180
  glycoproteins, 233
  lipocalins, 48, 149, 157, 165, 169, 177, 178, 180, 219, 233
Proteles cristatus: see Hyenas
Puberty, 7, 35, 38, 39, 62, 183, 185, 215, 234, 392, 395, 411-413, 445; *see also* Sexual maturation
  puberty-delaying chemosignal, 7, 214, 335
  puberty-accelerating chemosignal, 35, 38, 121, 122, 125, 154, 214, 234
Pyridine, 305, 307, 308, 310
Pyridine-*N*-oxide, 305, 307, 308, 310
4(3H)-Pyrimidone, 307, 308

Rabbit *(Oryctolagus cuniculus)*, 82, 118, 125, 198, 330, 335-340, 392, 417, 420
  pups, 199, 201, 202
Rana aurora: *see* Frogs
Rana cascade: *see* Frogs
Rana esculenta: *see* Frogs
Rana lessonae: *see* Frogs
Rana luteiventris: *see* Frogs
Rana sylvanica: see *Frogs*
Rana temporaria: *see* Frogs
"Ram effect", 377, 382, 383; *see also* Sheep
Ram's sebaceous gland, 377
Ram-pheromone, 378; see also "Ram effect"
Rats, 57, 69, 70, 85, 86, 107-109, 154, 155, 189, 200, 201, 255-257, 260, 261, 269
  fetal VNO, 189-195
  maternal behavior, 269-274
  pups, 189-195
Rat-like hamster: *see* Hamsters
Rattus norvegicus: see Rat
Reproductive behavior, 101, 102, 105, 219, 363
Receptivity signals, 260, 261, 385, 386
Rhamphotyphlops braminu: *see* Snakes
Rostral modules, 86, 87

Salamanders, 6, 20
　　Ambystoma macrodactylum (long-toed salamanders), 280
Salmo salar: see Fishes
Salvelinus fontinalis: see Fishes
Scent glands, 142, 385; see also Glands
　　anal, 141-144, 147
　　chin, 335-337, 340
　　sebaceous, 133, 134, 377, 445, 446
　　sex differences, 64, 233, 234, 237, 460, 461
　　subcaudal, 141, 321-325
　　vaginal, 56, 63, 387, 388
Scent marking, 1, 2, 3, 219, 220, 414, 468
　　badger, 321, 322, 451
　　beaver, 8, 9
　　black tailed deer, 12
　　discrimination, 65, 66
　　hyenas, 141-144
　　mice, 44-46, 49, 50, 219-222, 225, 353-359
　　striped mongoose, 329, 331
Sea urchins, 295
Sebaceous glands, 133, 134, 377, 445, 446
Sex differences, 64- 66, 234, 237, 460, 461
　　odor preferences, 35, 37-39, 214, 397, 398, 400
Sexual behavior (and odor stimuli), 4, 54, 56, 63, 118, 131, 255, 274, 394, 445
　　courtship, 263, 363-365, 367, 374
Sexual maturation, 5, 53, 58, 180, 219, 392, 411-414
Sheep, 55, 109, 118, 125, 133, 134, 337-382, 464; see also "Ram effect"
Shrews, 420, 429
Snakes, 5, 279-282, 285-288, 301, 361-367, 369-374
　　Agkistrodon piscivorus (cottonmouth (water) moccasins), 6, 285, 286
　　Boiga irregularis (brown tree snake), 5, 361-363, 365, 367
　　Lampropeltis getula (kingsnake), 285-288
　　Rhamphotyphlops braminu (flowerpot snake), 361, 363
　　Thamnophis radix (plains garter snake), 374
　　Thamnophis sirtalis (eastern garter snake), 280, 296, 302
　　Thamnophis sirtalis parietalis (red-sided garter snake), 365, 369, 370
Social recognition, 61, 65, 107-109, 111, 112, 114
Social status, 43, 143, 147, 394, 469,
　　of mice, 44, 45, 49, 217-222, 225-227, 403, 404, 407-409
　　of rabbits, 335-340
Spatial distribution, 44, 50, 433
Spatial patterns, 69, 82, 85, 90, 91
Sperm density, 403, 406-408
Sperm motility, 403, 404, 406- 409
Spermatids, 102, 103
Spermatogenic index (A/B), 102, 103
Splenic plague-forming cells (PFC), 250, 251
Stress, 199, 201, 412, 414

Striped mongooses (*Mungos mungo*), 239, 330-332
　　urine, 329
Subcaudal glands, 141, 321-325; see also Scent glands
Submaxillary glands, 126, 178; see also Glands
Subordinate males, 35-38, 49, 50, 218, 220, 226-228, 337, 250-252, 394, 395, 404, 409
Supplementary sacs (sacculi), 347-351
Sus scrofa domesticus: see Pig

Taricha granulosa: see Newts
Taricha torosa: see Newts
Territorial behavior, 313
Territorial marking, 1, 2, 3, 456; see also Scent marking
Territoriality, 12, 16
Testosterone, 134, 183-186, 212-214, 234, 237, 238, 274, 337, 377, 386, 397, 400, 401, 403-405, 407, 408, 412, 413
Thamnophis radix: see Snakes
Thamnophis sirtalis: see Snakes
Thamnophis sirtalis parietalis: see Snakes
Thiazole (2-*sec* butyl-4,5 dihydro-thiazole), 45, 46, 50, 149, 153, 157, 165, 169, 170, 172-174, 219, 233, 234, 353, 354, 358
Thomson gazelles (*Gazella thomsoni*), 2
Toads
　　Bufo americanus (American toad) 4, 280
　　Bufo boreas (western toad), 6, 21, 280
　　Bufo bufo (common toads), 437
　　Bufo terrestris (southern toad), 280
Tongue-flicks (TF), 363-365, 369, 440-443
Trailling behavior, 366, 367
Trigeminal system, 193, 194, 208
Tufted cells in OB, 74, 77
Turtles, 301, 302
　　Chelydra serpentina (snapping turtles), 296, 302
　　Chrysemys picta (painted turtles), 296, 302
Typhlonectes natans, 263
　　feces of, 265
　　kin recognition, 263-267

Uganda kob (*Kobus kob thomasi*), 3
Ultrasonic calls, 241-244, 246, 381
Ultrasonic vocalization, 269, 274, 394
Umami taste, 425
Urine and
　　marking, 3, 4, 8, 36, 45-50, 226, 227, 321-325, 330-332, 353-359, 468
　　reproductive activity, 119, 134-138, 233, 234, 411-413
　　sex signals, 38, 55, 125-131, 212-215, 233, 249, 257-260, 269, 388-400, 452-456
　　social status, 212-214, 218-220, 226-228
Urinary proteins
　　lipocalins, 48, 149, 165, 169, 177, 178, 180, 219

Major Urinary Proteins, 45-50, 149-155, 157-162, 165, 166, 169-174, 177, 178, 180, 219, 221, 233, 234, 237-239, 353-355, 358
Uterine weight, 184, 392
Vaginal secretions, 62-64, 119, 178, 385-388
Vesicular sacs 313; *see also* Otter, European
Visual cues, 13, 19, 29-31, 44, 125, 295, 419, 440, 469, 470
Voles, 4, 5, 36, 101-105, 184, 289-293, 343-346, 391-396, 397-401
   Clethrionomys glareolus (bank vole), 241, 246, 289, 391-395
     urine of, 391-393
   Clethrionomys rufocanus, 391
   Microtus agrestis (field vole), 289-293, 393
   Microtus brandti (Brandt's vole), 397 - 400
   Microtus canicaudus (gray-tailed vole), 4, 5, 393
   Microtus oeconomus (root vole), 289, 292, 293
   Microtus ochrogaster (prairie voles), 101, 102, 105, 343-346
   Microtus pennsylvanicus (meadow vole), 44, 47, 101-103, 105, 343-346
   Microtus townsendi (Townsend's vole), 5
Vomeronasal organ (VNO), 62-66;
   see also Accessory olfactory system
   epithelium, 93, 98, 101
   in caecilians, 263
   in hamsters, 62-66
   in human, 93, 94, 97
   in mice, 48, 62, 64, 65, 180, 189-191, 409
   in *Microtus sp.*, 101-105, 459, 461
   in pigs, 117, 118
   in primates, 93-98; *see also Microcebus murinus*
   in rats, 178, 189, 192-195
   in snakes, 286, 363
   in ungulates, 12, 198
   neuroepithelium (VNNE), 94-98, 102-105
   receptor proteins, 48
   sexual dimorphism, 105
Vomeronasal system: *see* Accessory olfactory system; Vomeronasal organ

Weasel (*Mustela nivalis*) *see*: Mustelids
Wolf, 2, 5, 8, 55, 469
   urine of, 7
Womb, 189, 191, 195

Xiphophorus helleri: see *Fishes*

(Z)-7-dodecenyl acetate
   in elephant, 125,
Z11-16:AL (*cis*-11-hexadecenal)
   in moth, 79
   cis-11-hexadecenyl acetate (Z11-16:AC), 79
   cis-11-hexadecenol (Z11-16:OH), 79
Z9-14:AL (*cis*-9-tetradecenal)
   in moth, 79